TENTH EDITION

Laboratory Manual in PHYSICAL GEOLOGY

PRODUCED UNDER THE AUSPICES OF THE

American Geosciences Institute
www.agiweb.org

AND THE

National Association of Geoscience Teachers
www.nagt.org

Richard M. Busch, Editor
West Chester University of Pennsylvania

ILLUSTRATED BY
Dennis Tasa
Tasa Graphic Arts, Inc.

PEARSON

Boston Columbus Indianapolis New York San Francisco
Upper Saddle River Amsterdam Cape Town Dubai London
Madrid Munich Paris Montréal Toronto Delhi Mexico City
São Paulo Sydney Hong Kong Seoul Singapore Taipei Tokyo

Acquisitions Editor: *Andrew Dunaway*
Senior Marketing Manager: *Maureen McLaughlin*
Senior Project Editor: *Crissy Dudonis*
Director of Development: *Jennifer Hart*
Executive Development Editor: *Jonathan Cheney*
Editorial Assistant: *Sarah Shefveland*
Senior Marketing Assistant: *Nicola Houston*
Content Producer: *Timothy Hainley*
Team Lead, Geosciences and Chemistry: *Gina M. Cheselka*
Production Project Manager: *Connie M. Long*
Full Service/Composition: *GEX Publishing Services*
Full Service Project Manager: *GEX Publishing Services*
Illustrations: *Dennis Tasa*
Image Lead: *Maya Melenchuk*
Photo Researcher: *Lauren McFalls*
Text Permissions Manager: *Timothy Nicholls*
Text Permissions Researcher: *William Opaluch*
Design Manager: *Marilyn Perry*
Interior and Cover Designer: *Elise Lansdon*
Operations Specialist: *Christy Hall*
Cover Photo Credit: *Hikers on Matanuska Glacier, Alaska, USA. Getty Images/Noppawat Tom Charoensinphon*

Credits and acknowledgments borrowed from other sources and reproduced, with permission, in this textbook appear on the appropriate page within the text.

Library of Congress Cataloging-in-Publication Data
Laboratory manual in physical geology / produced under the auspices of the American Geological Institute, and the National Association of Geoscience Teachers; Richard M. Busch, editor, West Chester University of Pennsylvania; illustrated by Dennis Tasa, Tasa Graphic Arts, Inc.–Tenth edition.
pages cm.
1. Physical geology–Laboratory manuals. I. Busch, Richard M. II. American Geological Institute. III. National Association of Geology Teachers.
QE44.L33 2015
551.078--dc23

2014045736

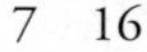

7 16

ISBN-10: 0-321-94451-8; ISBN-13: 978-0-321-94451-1
(Student Edition)

ISBN-10: 0-321-95821-7; ISBN-13: 978-0-321-95821-1
(Instructor's Review Copy)

www.pearsonhighered.com

CONTRIBUTING AUTHORS

Thomas H. Anderson
University of Pittsburgh

Harold E. Andrews
Wellesley College

James R. Besancon
Wellesley College

Jane L. Boger
SUNY–College at Geneseo

Phillip D. Boger
SUNY–College at Geneseo

Claude Bolze
Tulsa Community College

Jonathan Bushee
Northern Kentucky University

Roseann J. Carlson
Tidewater Community College

Cynthia Fisher
West Chester University of Pennsylvania

Charles I. Frye
Northwest Missouri State University

Pamela J.W. Gore
Georgia Perimeter College

Anne M. Hall
Emory University

Edward A. Hay
De Anza College

Charles G. Higgins
University of California, Davis

Michael F. Hochella, Jr.
Virginia Polytechnic Institute and State University

Michael J. Hozik
Richard Stockton College of New Jersey

Sharon Laska
Acadia University

David Lumsden
University of Memphis

Richard W. Macomber
Long Island University, Brooklyn

Garry D. Mckenzie
Ohio State University

Cherukupalli E. Nehru
Brooklyn College (CUNY)

John K. Osmond
Florida State University

Charles G. Oviatt
Kansas State University

William R. Parrott, Jr.
Richard Stockton College of New Jersey

Raman J. Singh
Northern Kentucky University

Kenton E. Strickland
Wright State University

Richard N. Strom
University of South Florida, Tampa

James Swinehart
University of Nebraska

Raymond W. Talkington
Richard Stockton College of New Jersey

Margaret D. Thompson
Wellesley College

James Titus*
U.S. Environmental Protection Agency

Nancy A. Van Wagoner
Acadia University

John R. Wagner
Clemson University

Donald W. Watson
Slippery Rock University

James R. Wilson
Weber State University

Monte D. Wilson
Boise State University

C. Gil Wiswall
West Chester University of Pennsylvania

*The opinions contributed by this person do not officially represent opinions of the U.S. Environmental Protection Agency.

Contents

Preface vii
Measurement Units xi
Mathematical Conversions xii
Laboratory Equipment xiii
World Map xiv

LABORATORY 1
Thinking Like a Geologist 1

ACTIVITY 1.1 Geologic Inquiry 3
ACTIVITY 1.2 Spheres of Matter, Energy, and Change 9
ACTIVITY 1.3 Modeling Earth Materials and Processes 14
ACTIVITY 1.4 Measuring and Determining Relationships 14
ACTIVITY 1.5 Density, Gravity, and Isostasy 20
ACTIVITY 1.6 Isostasy and Earth's Global Topography 22

LABORATORY 2
Plate Tectonics and the Origin of Magma 39

ACTIVITY 2.1 Plate Motion Inquiry Using GPS Time-Series 43
ACTIVITY 2.2 Is Plate Tectonics Caused by a Change in Earth's Size? 45
ACTIVITY 2.3 Lava Lamp Model of Earth 46
ACTIVITY 2.4 Paleomagnetic Stripes and Seafloor Spreading 47
ACTIVITY 2.5 Atlantic Seafloor Spreading 49
ACTIVITY 2.6 Using Earthquakes to Identify Plate Boundaries 50
ACTIVITY 2.7 San Andreas Transform-Boundary Plate Motions 50
ACTIVITY 2.8 Hot Spots and Plate Motions 50
ACTIVITY 2.9 The Origin of Magma 51

LABORATORY 3
Mineral Properties, Identification, and Uses 73

ACTIVITY 3.1 Mineral and Rock Inquiry 74
ACTIVITY 3.2 Mineral Properties 77
ACTIVITY 3.3 Determining Specific Gravity (SG) 86
ACTIVITY 3.4 Mineral Identification and Uses 88
ACTIVITY 3.5 The Mineral Dependency Crisis 89
ACTIVITY 3.6 Urban Ore 99

LABORATORY 4
Rock-Forming Processes and the Rock Cycle 111

ACTIVITY 4.1 Rock Inquiry 112
ACTIVITY 4.2 What Are Rocks Made Of? 113
ACTIVITY 4.3 Rock-forming Minerals 117
ACTIVITY 4.4 What Is Rock Texture? 117
ACTIVITY 4.5 Rocks and the Rock Cycle Model 119

LABORATORY 5
Igneous Rocks and Processes 129

ACTIVITY 5.1 Igneous Rock Inquiry 130
ACTIVITY 5.2 Minerals That Form Igneous Rocks 130
ACTIVITY 5.3 Estimate Rock Composition 131
ACTIVITY 5.4 Glassy and Vesicular Textures of Igneous Rocks 133
ACTIVITY 5.5 Crystalline Textures of Igneous Rocks 134
ACTIVITY 5.6 Rock Analysis, Classification, and Origin 135
ACTIVITY 5.7 Thin Section Analysis and Bowen's Reaction Series 135
ACTIVITY 5.8 Analysis and Interpretation of Igneous Rocks 141
ACTIVITY 5.9 Geologic History of Southeastern Pennsylvania 142

LABORATORY 6
Sedimentary Processes, Rocks, and Environments 153

ACTIVITY 6.1 Sedimentary Rock Inquiry 154
ACTIVITY 6.2 Mount Rainier Sediment Analysis 154
ACTIVITY 6.3 Clastic and Detrital Sediment 154
ACTIVITY 6.4 Biochemical and Chemical Sediment and Rock 155
ACTIVITY 6.5 Sediment Analysis, Classification, and Interpretation 155
ACTIVITY 6.6 Hand Sample Analysis and Interpretation 160
ACTIVITY 6.7 Grand Canyon Outcrop Analysis and Interpretation 163
ACTIVITY 6.8 Using the Present to Imagine the Past—Dogs to Dinosaurs 163
ACTIVITY 6.9 Using the Present to Imagine the Past—Cape Cod to Kansas 166
ACTIVITY 6.10 "Reading" Earth History from a Sequence of Strata 167

LABORATORY 7
Metamorphic Rocks, Processes, and Resources 187

ACTIVITY 7.1 Metamorphic Rock Inquiry 188
ACTIVITY 7.2 Metamorphic Rock Analysis and Interpretation 189
ACTIVITY 7.3 Hand Sample Analysis, Classification, and Origin 196
ACTIVITY 7.4 Metamorphic Grades and Facies 198

LABORATORY 8
Dating of Rocks, Fossils, and Geologic Events 207

ACTIVITY 8.1 Geologic Inquiry for Relative Age Dating 208
ACTIVITY 8.2 Determining Sequence of Events in Geologic Cross Sections 208
ACTIVITY 8.3 Using Index Fossils to Date Rocks and Events 212
ACTIVITY 8.4 Absolute Dating of Rocks and Fossils 214
ACTIVITY 8.5 Infer Geologic History from a New Mexico Outcrop 216
ACTIVITY 8.6 CSI (Canyon Scene Investigation) Arizona 216

LABORATORY 9
Topographic Maps and Orthoimages 227

ACTIVITY 9.1 Map and Google Earth™ Inquiry 228
ACTIVITY 9.2 Map Locations, Distances, Directions, and Symbols 228
ACTIVITY 9.3 Topographic Map Construction 239
ACTIVITY 9.4 Topographic Map and Orthoimage Interpretation 239
ACTIVITY 9.5 Relief and Gradient (Slope) Analysis 246
ACTIVITY 9.6 Topographic Profile Construction 246

LABORATORY 10
Geologic Structures, Maps, and Block Diagrams 259

ACTIVITY 10.1 Geologic Structures Inquiry 260
ACTIVITY 10.2 Visualizing How Stresses Deform Rocks 260
ACTIVITY 10.3 Map Contacts and Formations 261
ACTIVITY 10.4 Determine Attitude of Rock Layers and a Formation Contact 262
ACTIVITY 10.5 Cardboard Model Analysis and Interpretation 263
ACTIVITY 10.6 Block Diagram Analysis and Interpretation 263
ACTIVITY 10.7 Nevada Fault Analysis Using Orthoimages 263
ACTIVITY 10.8 Appalachian Mountains Geologic Map 263

LABORATORY 11
Stream Processes, Landscapes, Mass Wastage, and Flood Hazards 283

ACTIVITY 11.1 Streamer Inquiry 284
ACTIVITY 11.2 Introduction to Stream Processes and Landscapes 284
ACTIVITY 11.3 Escarpments and Stream Terraces 284
ACTIVITY 11.4 Meander Evolution on the Rio Grande 284
ACTIVITY 11.5 Mass Wastage at Niagara Falls 292
ACTIVITY 11.6 Flood Hazard Mapping, Assessment, and Risk 295

LABORATORY 12
Groundwater Processes, Resources, and Risks 311

ACTIVITY 12.1 Groundwater Inquiry 312
ACTIVITY 12.2 Karst Processes and Topography 312
ACTIVITY 12.3 Floridan Limestone Aquifer 314
ACTIVITY 12.4 Land Subsidence from Groundwater Withdrawal 317

LABORATORY 13
Glaciers and the Dynamic Cryosphere 329

ACTIVITY 13.1 Cryosphere Inquiry 330
ACTIVITY 13.2 Mountain Glaciers and Glacial Landforms 330
ACTIVITY 13.3 Continental Glaciation of North America 330
ACTIVITY 13.4 Glacier National Park Investigation 334
ACTIVITY 13.5 Nisqually Glacier Response to Climate Change 334
ACTIVITY 13.6 The Changing Extent of Sea Ice 335

LABORATORY 14
Dryland Landforms, Hazards, and Risks 357

ACTIVITY 14.1 Dryland Inquiry 358
ACTIVITY 14.2 Mojave Desert, Death Valley, California 358
ACTIVITY 14.3 Sand Seas of Nebraska and the Arabian Peninsula 363
ACTIVITY 14.4 Dryland Lakes of Utah 365

LABORATORY 15
Coastal Processes, Landforms, Hazards, and Risks 375

ACTIVITY 15.1 Coastline Inquiry 376
ACTIVITY 15.2 Introduction to Shorelines 376
ACTIVITY 15.3 Shoreline Modification at Ocean City, Maryland 381
ACTIVITY 15.4 The Threat of Rising Seas 381

LABORATORY 16
Earthquake Hazards and Human Risks 391

ACTIVITY 16.1 Earthquake Hazards Inquiry 392
ACTIVITY 16.2 How Seismic Waves Travel through Earth 392
ACTIVITY 16.3 Locate the Epicenter of an Earthquake 393
ACTIVITY 16.4 San Andreas Fault Analysis at Wallace Creek 394
ACTIVITY 16.5 New Madrid Blind Fault Zone 394

Graph Paper

Cardboard Models

GeoTools

About our Sustainability Initiatives

Pearson recognizes the environmental challenges facing this planet, and acknowledges our responsibility in making a difference. This book has been carefully crafted to minimize environmental impact. The binding, cover, and paper come from facilities that minimize waste, energy consumption, and the use of harmful chemicals. Pearson closes the loop by recycling every out-of-date text returned to our warehouse.

Along with developing and exploring digital solutions to our market's needs, Pearson has a strong commitment to achieving carbon neutrality. As of 2009, Pearson became the first carbon- and climate-neutral publishing company. Since then, Pearson remains strongly committed to measuring, reducing, and offsetting our carbon footprint.

The future holds great promise for reducing our impact on Earth's environment, and Pearson is proud to be leading the way. We strive to publish the best books with the most up-to-date and accurate content, and to do so in ways that minimize our impact on Earth. To learn more about our initiatives, please visit **www.pearson.com/responsibility**.

PEARSON

Preface

Laboratory Manual in Physical Geology is produced under the auspices of the American Geosciences Institute (AGI) and the National Association of Geoscience Teachers (NAGT). For decades it has been the most widely adopted manual available for teaching laboratories in introductory geology and geoscience. This new edition is more user-friendly than ever, with a new pedagogical format and many more teaching and learning options. It is now backed by MasteringGeology, the most effective and widely used online homework, tutorial, and assessment platform in the Geosciences, GeoTools (ruler, protractor, UTM grids, sediment grain size scale, etc.), an Instructor Resource Guide, and resources on the Instructor's Resource DVD.

The idea for this jointly sponsored laboratory manual was proffered by Robert W. Ridky (past president of NAGT and a member of the AGI Education Advisory Committee), who envisioned a manual made up of the "best laboratory investigations written by geology teachers." To that end, this product is the 28-year evolution of the cumulative ideas of more than 225 contributing authors, faculty peer reviewers, and students and faculty who have used past editions.

New to This Edition

In the tenth edition there are numerous new activities and figures, dynamic pedagogical changes, and practical revisions that have been made at the request of faculty and students who have used past editions. The new features in this edition are listed below:

Pre-Lab Videos

Pre-Lab videos are found on the chapter-opening spreads of each lab, and are accessed via QR code or direct web-link. These videos will ensure students come to lab better prepared and ready to immediately jump into the lab exercise. No longer do instructors have to spend the first portion of hands-on lab time lecturing. The videos will review key concepts relevant to the lab exercise during the students' own preparatory time. The videos, created by Callan Bentley, are personable and friendly, and assure students that they will be able to successfully complete the lab activities by following a clear series of steps.

New Format and Pedagogical Framework

- **Big Ideas and Engaging Chapter Openers.** Every laboratory opens with an engaging image and Big Ideas, which are the overall conceptual themes upon which the laboratory is based. Big Ideas are concise statements that help students understand and focus on how all parts of the laboratory are related.
- **Think About It—Key Questions.** Every activity is based on a key question that is linked to the Big Ideas and can be used for individualized or think-pair-share learning before or after the activity. Think About It questions function as the conceptual "lenses" that frame student inquiry and promote critical thinking and discourse. ***Schematic questions*** target the cognitive domain of psychology, are meant to help students assimilate to the topic (apply their existing schemata), and lay a constructivist conceptual foundation for scaffolding to new concepts and skills of the topic. ***Engaging questions*** target the affective domain of psychology, are open ended questions with more than one possible answer, and are meant to foster interest in the topic. When used in a brief think-pair-share or class brainstorming activity at the start of an activity, these questions often foster curiosity and cognitive disequilibrium that leads to ***authentic questions*** by students.
- **Guided and Structured Inquiry Activities.** Every laboratory begins with a guided inquiry activity. It is designed to be both engaging and schematic and could be used for individualized or cooperative learning. The guided inquiry activity is followed by activities that are more structured, as in past editions. All of the activities have objectives framed in relation to Bloom et al. levels of critical thinking.
- **Reflect and Discuss Questions.** Every activity concludes with a Reflect and Discuss question designed to foster greater accommodation of knowledge by having students apply what they learned to a new situation or to state broader conceptual understanding.
- **Manipulatives.** Manipulatives are integrated with most activities. They provide opportunities for assimilation and accommodation based on real objects or models, are designed to target the psychomotor domain of psychology, and can be used to foster curiosity and cognitive disequilibrium that leads to *authentic questions* by students.

- **Continuous Assessment Options.** The new pedagogical framework and organization provides many options for continuous assessment such as Think About It questions and guided inquiry activities that provide options for pre-assessment, activity worksheets, and the Reflect and Discuss questions. When students tear out and hand in an activity for grading, their manual will still contain the significant text and reference figures that they need for future study. Grading of students' work is easier because all students submit their own work in a similar format. Instructors save time, resources, and money because they no longer need to photocopy and hand out worksheets to supplement the manual.

New and Revised Activities and Text

- **20% More Activities.** Abundant new activities provide more options for students to learn content that was ranked "essential" or "most important" by faculty peer reviewers and past faculty and student users of the manual. There are now 98 activities that can be mixed or matched at the instructor's discretion acccording to course content and level of difficulty. And because many activities do not require sophisticated equipment, they can also be assigned for students to complete as pre-laboratory assignments, lecture supplements, or recitation topics.
- **Every Past Activity Has Been Revised.** Every past activity has been revised with new directives, questions, clarity, or format on the basis of user input. All activities now follow the new pedagogical framework and have at least one summative Reflect and Discuss question.
- **35% of Written Materials Are New or Rewritten.** These changes have been made on the basis of reviews by faculty and students, current trends in the geosciences, and the new pedagogical framework.

Revised Art Program and Enhanced Learning Options

- **Greater Visual Clarity and Appeal.** This edition contains almost twice as many phtographs and images as the ninth edition. One-fourth of the figures are new, and one-fourth have been revised. Many maps have been revised or replaced. Dennis Tasa's brilliant artwork reinforces the visual aspect of geology and enhance student learning.
- **Transferable Skill Development and Real-World Connections.** Many activities have been newly designed or revised for students to develop transferrable skills and make real-world connections to their lives and the world in which they live. For example, they learn how to obtain and use data and maps that will enable them to make wiser choices about where they live and work. They evaluate their use of Earth resources in relation to questions about resource management and sustainability. They make real-world connections using U.G. Geological Survey, JPL-NASA, NOAA, Google Earth™, and other online sources for analysis and evaluation of Earth and its resources, hazards, changes, and management.
- **The Math You Need (TMYN) Options.** Throughout the laboratories, students are refered to online options for them to review or learn mathematical skills using *The Math You Need, When You Need it* (TMYN). TMYN consists of modular math tutorials that have been designed for students in any introductory geoscience course by Jennifer Wenner (University of Washington–Oshkosh) and Eric Baer (Highline Community College).
- **QR Codes.** Quick Response (QR) codes have been added to give students with smartphones or other mobile devices rapid access to supporting content and websites.
- **Enhanced Instructor Support.** Free instructor materials are available online in the Instructor Resource Center (IRC) at www.pearsonhighered.com/irc, and Instructor Resource DVD. Resources include the enhanced Instructor Resource Guide (answer key and teaching tips), files of all figures in the manual, PowerPoint™ presentations for each laboratory (including video clips), the Pearson Geoscience Animation Library (over 100 animations illuminating the most difficult-to-visualize geological concepts and phenomena in Flash files and PowerPoint™ slides), and Mastering Geology™ options.

NEW! MasteringGeology

The MasteringGeology™ platform delivers engaging, dynamic learning opportunities—focused on course objectives and responsive to each student's progress—that are proven to help students absorb course material and understand difficult concepts. Robust diagnostics and unrivalled gradebook reporting allow instructors to pinpoint the weaknesses and misconceptions of a student or class to provide timely intervention.

- **NEW! Pre-lab video quizzes** can be assigned. These will ensure students come to lab better prepared and ready to immediately jump into the lab exercise.
- **NEW! Post-lab quizzes** assess students' understanding and analysis of the lab content.

Learn More at www.MasteringGeology.com

Learning Catalytics

Learning Catalytics™ is a "bring your own device" student engagement, assessment, and classroom intelligence system. With Learning Catalytics you can

- assess students in real time, using open-ended tasks to probe student understanding.
- understand immediately where students are and adjust your lecture accordingly.
- improve your students' critical-thinking skills.
- access rich analytics to understand student performance.

- add your own questions to make Learning Catalytics fit your course exactly.
- manage student interactions with intelligent grouping and timing.

Learning Catalytics is a technology that has grown out of twenty years of cutting edge research, innovation, and implementation of interactive teaching and peer instruction. Available integrated with MasteringGeology. www learningcatalytics.com

Outstanding Features

This edition contains the tried-and-tested strengths of nine past editions of this lab manual published over nearly three decades. The outstanding features listed below remain a core part of this title.

Pedagogy for Diverse Styles/Preferences of Learning

Hands-on multisensory-oriented activities with samples, cardboard models, and GeoTools appeal to *concrete /kinesthetic learners*. High quality images, maps, charts, diagrams, PowerPoints™, cardboard models, and visualizations appeal to *visual/spatial learners*. Activity sheets, charts, lists, supporting text, and opportunities for discourse appeal to *linguistic/verbal/read-write learners*. PowerPoints™ and video clips appeal to *auditory/aural learners*. Numerical data, mathematics, models, graphs, systems, and opportunities for discourse appeal to *logical/abstract learners*.

Terminology of the American Geosciences Institute (AGI)

All terms are consistent with AGI's latest *Glossary of Geology*, which was developed by the AGI federation of 48 geoscientific and professional associations. The glossary is available in print, online for a 30-day free trial period, or as an app for the iPhone, iPod, and iPad from the App Store. See http://www.agiweb.org/pubs/glossary.

Materials

Laboratories are based on samples and equipment normally housed in existing geoscience teaching laboratories (page xiii).

GeoTools, GPS, and UTM

There are rulers, protractors, a sediment grain size scale, UTM grids, and other laboratory tools to cut from transparent sheets at the back of the manual. No other manual provides such abundant supporting tools! Students are introduced to GPS and UTM and their application in mapping. UTM grids are provided for most scales of U.S. and Canadian maps.

Support for Geoscience!

Royalties from sales of this product support programs of the American Geosciences Institute and the National Association of Geoscience Teachers.

Acknowledgments

We acknowledge and sincerely appreciate the assistance of many people and organizations who have helped make possible this tenth edition of *Laboratory Manual in Physical Geology*. Revisions in this new edition are based on suggestions from faculty who used the last editon of the manual, feedback from students using the manual, and market research by Pearson. New activities were field tested in Introductory Geology laboratories at West Chester University.

Development and production of this highly-revised 10th edition of *Laboratory Manual in Physical Geology* required the expertise, dedication, and cooperation of many people. The very talented publishing team at Pearson Education led the effort. Andy Dunaway's knowledge of market trends, eternal quest to meet the needs of faculty and students, and dedication to excellence guided the vision for this extraordinary 10th edition. Jonathan Cheney's pre-revision memos and developmental editing framed the revision goals for each topic and ensured that all writing was practical and purposeful. Crissy Dudonis set revision schedules, tracked revision progress, and efficiently coordinated the needs and collaborative efforts of team members. Sarah Shefveland managed accuracy reviews of revision drafts. Connie Long managed the production process. Her expertise and dedication to excellence enabled her to locate, manage, and merge disparate elements of lab manual production. Page design and proofing was expertly managed by Jacki Russell, GEX Publishing Services. The team at GEX Publishing Services, lead by Alison Smith and Erin Hernandez, composited pages for publication. This process was especially difficult when it came to the activity worksheets, and we thank Alison and Erin for addressing every challenge and achieving our product goals.

We thank the following individuals for their constructive criticisms and suggestions that led to improvements for this edition of the manual:

Mark Boryta–*Mount San Antonio College*
Cinzia Cervato–*Iowa State University*
James Constantopoulos–*Eastern New Mexico University*
Raymond Coveney–*University of Missouri–Kansas City*
John Dassinger–*Chandler-Gilbert Community College*
Meredith Denton-Hendrick–*Austin Community College*
Kelli Dilliard–*Wayne State College*
Richard Dunning–*Normandale Community College*
Carol Edson–*Las Positas College*
Eleanor Gardner–*University of Tennessee–Martin*
Alessandro Grippo–*Santa Monica University*
Ruth Hanna–*Las Positas College*

Bruce Harrison–*New Mexico Technical University*
Michael Heaney–*Texas A&M University*
Martin Helmke–*West Chester University of Pennsylvania*
Beth Johnson–*University of Wisconsin–Fox Valley*
Amanda Julson–*Blinn College*
Dan Kelley–*Bowling Green State University*
Brendan McNulty–*California State University–Dominquez Hills*
James Puckette–*Oklahoma State University*
Randye Rutberg–*Hunter College*
Michael Rygel–*SUNY Potsdam*
Jennifer Sembach–*Indiana University–Purdue University Indianapolis*
Amy Smith–*Central Michigan University*
Amanda Stahl–*Washington State University*
Nicole Vermillion–*Georgia Perimeter College*
Julia Wellner–*University of Houston*
Kurt Wilkie–*Washington State University*

We thank Carrick Eggleston (University of Wyoming), Randall Marrett (University of Texas at Austin), Sergei Lebedev (Dublin Institute for Advanced Studies), Rob D. van der Hilst (MIT/IRIS Consortium), LeeAnn Srogi (West Chester University of Pennsylvania), and Tomas McGuire for the use of their personal photographs. Photographs and data related to St. Catherines Island, Georgia, were made possible by research grants to the editor from the St. Catherines Island Research Program, administered by the American Museum of Natural History and supported by the Edward J. Noble Foundation.

Maps, map data, aerial photographs, and satellite imagery have been used courtesy of the U.S. Geological Survey; Canadian Department of Energy, Mines, and Resources; Surveys and Resource Mapping Branch; U.S. Bureau of Land Management; JPL-NASA; NOAA; and the U.S./Japan ASTER Science Team.

We also thank Jennifer Wenner (University of Washington–Oshkosh) and Eric Baer (Highline Community College) for making possible the online options for students to review or learn mathematical skills using *The Math You Need, When You Need It* (TMYN) modules.

The continued success of this laboratory manual depends on criticisms, suggestions, and new contributions from persons who use it. We sincerely thank everyone who contributed to this project by voicing criticisms, suggesting changes, and conducting field tests.

Unsolicited reactions to the manual are especially welcomed as a barometer for quality control and the basis for many changes and new initiatives that keep the manual current. Please continue to submit your frank criticisms and input directly to the editor: Rich Busch, Department of Geology and Astronomy, Merion Hall, West Chester University, West Chester, PA 19383 (rbusch@wcupa.edu).

P. Patrick Leahy, *Executive Director, AGI*
Ann Benbow, *Director of Education, Outreach, and Development, AGI*
Richard M. Busch, *Editor*

Measurement Units

People in different parts of the world have historically used different systems of measurement. For example, people in the United States have historically used the English system of measurement based on units such as inches, feet, miles, acres, pounds, gallons, and degrees Fahrenheit. However, for more than a century, most other nations of the world have used the metric system of measurement. In 1975, the U.S. Congress recognized that global communication, science, technology, and commerce were aided by use of a common system of measurement, and they made the metric system the official measurement system of the United States. This conversion is not yet complete, so most Americans currently use both English and metric systems of measurement.

The International System (SI)

The International System of Units (SI) is a modern version of the metric system adopted by most nations of the world, including the United States. Each kind of metric unit can be divided or multiplied by 10 and its powers to form the smaller or larger units of the metric system. Therefore, the metric system is also known as a "base-10" or "decimal" system. The International System of Units (SI) is the official system of symbols, numbers, base-10 numerals, powers of 10, and prefixes in the modern metric system.

SYMBOL	NUMBER	NUMERAL	POWER OF 10	PREFIX
T	one trillion	1,000,000,000,000	10^{12}	tera-
G	one billion	1,000,000,000	10^{9}	giga-
M	one million	1,000,000	10^{6}	mega-
k	one thousand	1000	10^{3}	kilo-
h	one hundred	100	10^{2}	hecto-
da	ten	10	10^{1}	deka-
	one	1	10^{0}	
d	one-tenth	0.1	10^{-1}	deci-
c	one-hundredth	0.01	10^{-2}	centi-
m	one-thousandth	0.001	10^{-3}	milli-
μ	one-millionth	0.000001	10^{-6}	micro-
n	one-billionth	0.000000001	10^{-9}	nano-
p	one-trillionth	0.000000000001	10^{-12}	pico-

Examples

1 meter (1 m) = 0.001 kilometers (0.001 km), 10 decimeters (10 dm), 100 centimeters (100 cm), or 1000 millimeters (1000 mm)
1 kilometer (1 km) = 1000 meters (1000 m)
1 micrometer (1μm) = 0.000,001 meter (.000001 m) or 0.001 millimeters (0.001 mm)
1 kilogram (kg) = 1000 grams (1000 g)
1 gram (1 g) = 0.001 kilograms (0.001 kg)
1 metric ton (1 t) = 1000 kilograms (1000 kg)
1 liter (1 L) = 1000 milliliters (1000 mL)
1 milliliter (1 mL or 1 ml) = 0.001 liter (0.001 L)

Abbreviations for Measures of Time

A number of abbreviations are used in the geological literature to refer to time. Abbreviations for "years old" or "years ago" generally use SI prefixes and "a" (Latin for *year*). Abbreviations for intervals or durations of time generally combine SI symbols with "y" or "yr" (*years*). For example, the boundaries of the Paleozoic Erathem are 542 Ma and 251 Ma, so the Paleozoic Era lasted 291 m.y.

ka = kiloannum—thousand years old, ago, or before present
Ma = megannum—million years old, ago, or before present
Ga = gigannum—billion years old, ago, or before present
yr (or y) = year or years
Kyr or k.a. = thousand years
Myr or m.y. = million years
Gyr (or Byr or b.y) = gigayear—billion years

Mathematical Conversions

To convert:	To:	Multiply by:	
kilometers (km)	meters (m)	1000 m/km	LENGTHS AND DISTANCES
	centimeters (cm)	100,000 cm/km	
	miles (mi)	0.6214 mi/km	
	feet (ft)	3280.83 ft/km	
meters (m)	centimeters (cm)	100 cm/m	
	millimeters (mm)	1000 mm/m	
	feet (ft)	3.2808 ft/m	
	yards (yd)	1.0936 yd/m	
	inches (in.)	39.37 in./m	
	kilometers (km)	0.001 km/m	
	miles (mi)	0.0006214 mi/m	
centimeters (cm)	meters (m)	0.01 m/cm	
	millimeters (mm)	10 mm/cm	
	feet (ft)	0.0328 ft/cm	
	inches (in.)	0.3937 in./cm	
	micrometers (μm)*	10,000 μm/cm	
millimeters (mm)	meters (m)	0.001 m/mm	
	centimeters (cm)	0.1 cm/mm	
	inches (in.)	0.03937 in./mm	
	micrometers (μm)*	1000 μm/mm	
	nanometers (nm)	1,000,000 nm/mm	
micrometers (μm)*	millimeters (mm)	0.001 mm/μm	
nanometers (nm)	millimeters (mm)	0.000001 mm/nm	
miles (mi)	kilometers (km)	1.609 km/mi	
	feet (ft)	5280 ft/mi	
	meters (m)	1609.34 m/mi	
feet (ft)	centimeters (cm)	30.48 cm/ft	
	meters (m)	0.3048 m/ft	
	inches (in.)	12 in./ft	
	miles (mi)	0.000189 mi/ft	
inches (in.)	centimeters (cm)	2.54 cm/in.	
	millimeters (mm)	25.4 mm/in.	
	micrometers (μm)*	25,400 μm/in.	
square miles (mi2)	acres (a)	640 acres/mi^2	AREAS
	square km (km^2)	2.589988 km^2/m	
square km (km2)	square miles (mi^2)	0.3861 mi^2/km^2	
acres	square miles (mi^2)	0.001563 mi^2/acr	
	square km (km^2)	0.00405 km^2/acr	
gallons (gal)	liters (L)	3.78 L/gal	VOLUMES
fluid ounces (oz)	milliliters (mL)	30 mL/fluid oz	
milliliters (mL)	liters (L)	0.001 L/mL	
	cubic centimeters (cm^3)	1.000 cm^3/mL	
liters (L)	milliliters (mL)	1000 mL/L	
	cubic centimeters (cm^3)	1000 cm^3/mL	
	gallons (gal)	0.2646 gal/L	
	quarts (qt)	1.0582 qt/L	
	pints (pt)	2.1164 pt/L	
grams (g)	kilograms (kg)	0.001 kg/g	WEIGHTS AND MASSES
	pounds avdp. (lb)	0.002205 lb/g	
ounces avdp (oz)	grams (g)	28.35 g/oz	
ounces troy (ozt)	grams (g)	31.10 g/ozt	
pounds avdp. (lb)	kilograms (kg)	0.4536 kg/lb	
kilograms (kg)	pounds avdp. (lb)	2.2046 lb/kg	

To convert from degrees Fahrenheit (°F) to degrees Celsius (°C), subtract 32 degrees and then divide by 1.8 To convert from degrees Celsius (°C) to degrees Fahrenheit (°F), multiply by 1.8 and then add 32 degrees.

*Formerly called microns (μ)

LABORATORY EQUIPMENT

Also refer to the GeoTools provided at the back of this laboratory manual.

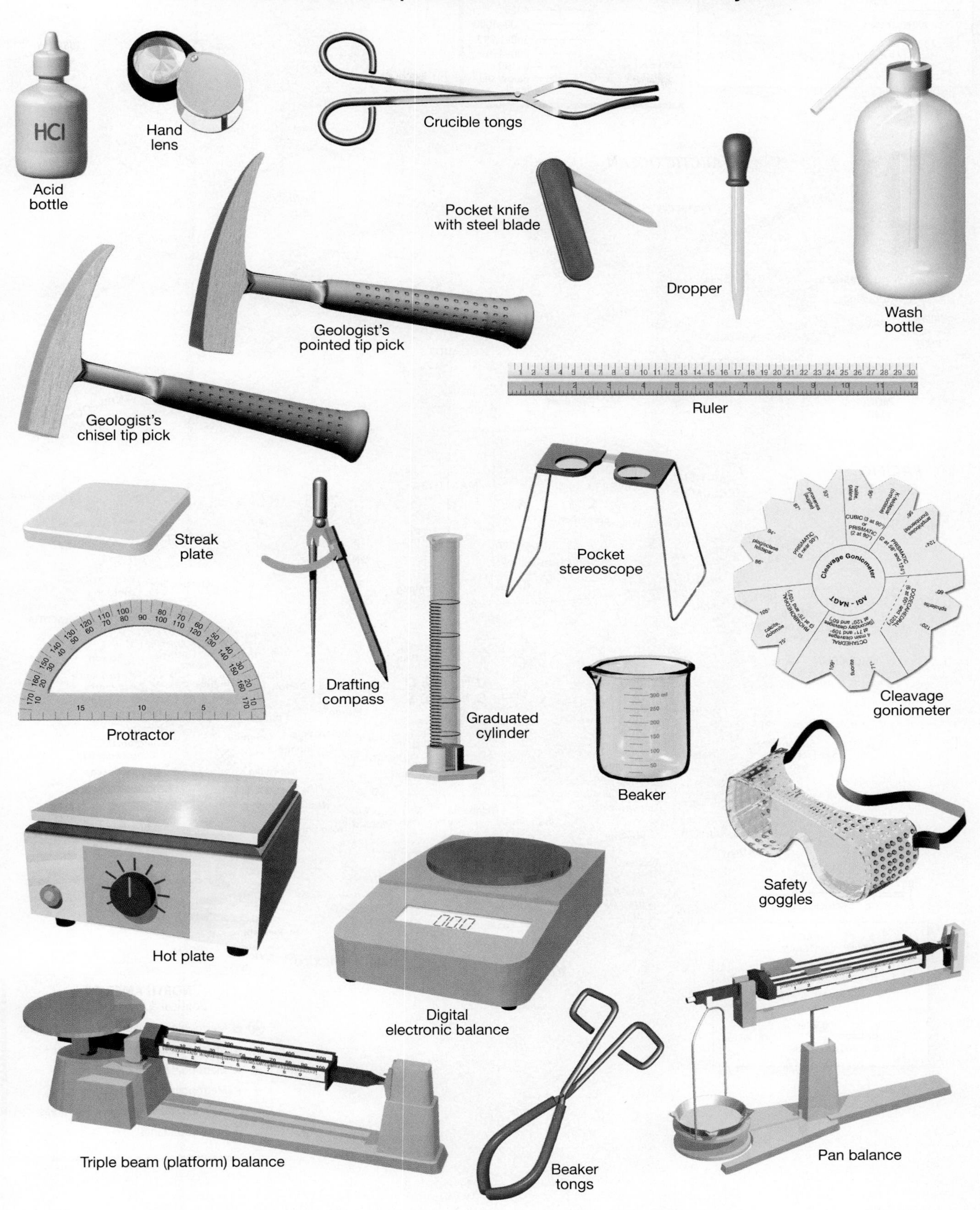

ELEVATION IN METERS
4000+
2000–4000
500–1,999
200–499
0–199
Below sea level
Sea Level
Anchorage
Bering Sea
Aleutian Islands
300 Miles
300 Kilometers
ARCTIC OCEAN
Greenland
EURASIAN PLATE
Baffin Bay
Baffin Island
Bering Strait
Brooks Range
Prudhoe Bay
Alaska Range
NORTH AMERICAN PLATE
Gulf of Alaska
YUKON TERRITORY
Whitehorse
NORTHWEST TERRITORIES
NUNAVUT
Great Slave Lake
Hudson Bay
NEWFOUNDLAND AND LABRADOR
Newfoundland
PACIFIC PLATE
PACIFIC OCEAN
Coast Mountains
BRITISH COLUMBIA
ALBERTA
MANITOBA
C A N A D A
Edmonton
Lake Winnipeg
Saskatoon
Calgary
SASKATCHEWAN
Regina
Winnipeg
ONTARIO
QUÉBEC
Prince Edward Island
P.E.I.
NEW BRUNSWICK
Saint John
Halifax
NOVA SCOTIA
Quebec City
Montréal
Ottawa
Vancouver Island
Vancouver
Puget Sound
Seattle
JUAN DE FUCA PLATE
Portland
ROCKY MOUNTAINS
GREAT PLAINS
L. Superior
Minneapolis-St. Paul
Adirondack Mtns.
Boston
Providence
Toronto
L. Ontario
Hamilton
Buffalo
Bridgeport
New York City
Detroit
Milwaukee
L. Michigan
L. Huron
L. Erie
Cleveland
UNITED STATES
Chicago
Pittsburgh
Philadelphia
Baltimore
Washington, D.C.
Chesapeake Bay
Columbus
Indianapolis
Cincinnati
Louisville
Virginia Beach
APPALACHIAN HIGHLANDS
Piedmont
Boise
Teton Range
Cascade Range
Coast Ranges
Central Valley
Sierra Nevada
Great Basin
Great Salt Lake
Salt Lake City
Sacramento
San Francisco
San Jose
Denver
Kansas City
St. Louis
Ozark Mtns.
Nashville
Charlotte
Las Vegas
Grand Canyon
Colorado Plateau
Los Angeles
Riverside
Salton Sea
San Diego
Albuquerque
Oklahoma City
Ouachita Plateau
Memphis
Atlanta
ATLANTIC OCEAN
Phoenix
Tucson
El Paso
Dallas-Fort Worth
Jacksonville
Austin
Houston
New Orleans
San Antonio
Orlando
Tampa-St. Petersburg
Miami
Florida Keys
Gulf of Mexico
MEXICO
Missouri
Platte R.
Arkansas R.
Red R.
Ohio R.
Mississippi
Rio Grande
Colorado R.
Mackenzie R.
St. Lawrence R.
500 Miles
500 Kilometers
PACIFIC OCEAN
Kauai
Oahu
Honolulu
Maui
HAWAII
Hawaii
100 Miles
100 Kilometers
NORTH AMERICA
Political & Physical Map
Metropolitan areas more than 20 million
Metropolitan areas 10–20 million
Metropolitan areas 5–9.9 million
Metropolitan areas 1–4.9 million
Selected smaller metropolitan areas
Plate boundaries

PRE-LAB VIDEO

LABORATORY 1

Thinking Like a Geologist

CONTRIBUTING AUTHORS

Cynthia Fisher • West *Chester University of Pennsylvania*

C. Gil Wiswall • *West Chester University of Pennsylvania*

Global Positioning System (GPS) devices, like this one on Mount St. Helens, are used to detect changes in the elevation of the volcano as magma moves beneath it. A rise in elevation of the volcano often precedes an eruption.

BIG IDEAS

Geology is the science of Earth, so geologists are Earth scientists or "geoscientists." Geologists observe, describe, and model the materials, energies, and processes of change that occur within and among Earth's spheres over time. They apply their knowledge to understand the present state of Earth, locate and manage resources, identify and mitigate hazards, predict change, and seek ways to sustain the human population.

FOCUS YOUR INQUIRY

THINK About It How and why do geologists observe Earth materials at different scales (orders of magnitude)?

ACTIVITY 1.1 Geologic Inquiry *(p. 3)*

THINK About It What materials, energies, and processes of change do geologists study?

ACTIVITY 1.2 Spheres of Matter, Energy, and Change *(p. 9)*

THINK About It How and why do geologists make models of Earth?

ACTIVITY 1.3 Modeling Earth Materials and Processes *(p. 14)*

THINK About It How and why do geologists measure Earth materials and graph relationships among Earth materials and processes of change?

ACTIVITY 1.4 Measuring and Determining Relationships *(p. 14)*

THINK About It How is the distribution of Earth materials related to their density?

ACTIVITY 1.5 Density, Gravity, and Isostasy *(p. 20)*

ACTIVITY 1.6 Isostasy and Earth's Global Topography *(p. 22)*

Introduction

Regardless of your educational background or interests, you probably have already done some thinking like a geologist. This lab will help you think and act even more like a geologist.

What Does It Mean to Start Thinking Like a Geologist?

You start thinking like a geologist when you focus on questions about planet Earth and try to answer them. You were thinking like a geologist if you ever observed an interesting landform, rock, or fossil, and wondered about how it formed. You were also thinking like a geologist if you ever wondered where your drinking water comes from, the possibility of earthquakes or floods where you live, where to find gold, how to vote on environmental issues, or what environmental risks may be associated with buying or building a home.

Wondering or inquiring about such things leads one to fundamental questions about Earth and how it operates. **Science** is a way of answering these questions by *gathering data* (information, evidence) based on investigations and careful observations, *thinking critically* (applying, analyzing, interpreting, and evaluating the data), *engaging in discourse* (verbal or written exchange, organization, and evaluation of information and ideas), and *communicating inferences* (conclusions justified with data and an explanation of one's critical thinking process). **Geology** is the science of Earth. Its name comes from two Greek words, *geo* = Earth and *logos* = discourse. So geologists are also Earth scientists or geoscientists.

Why Think Like a Geologist?

The products of geologic science are all around you—in the places where you live, the products you enjoy, the energy you use, and the government's environmental codes and safety policies that you must follow. For example, your home contains bricks, concrete, plaster wall boards (sheetrock), glass, metals, and asphalt roof shingles made with raw materials that were located by geologists. The safe location of your home was likely determined with the help of geologists. The wooden materials and foods in your home were processed with tools and machines containing metals that were extracted from ore minerals found by geologists. The electricity you use comes from generating plants that are fueled with coal, gas, oil, or uranium that was found by geologists. The safe location of the generating plants was evaluated by geologists, and the electricity is transported via copper wires made from copper ore minerals located by geologists. Even your trash and sewage are processed and recycled or disposed of in accordance with government policies developed with geologists and related to surface and groundwater. So geologists are Earth detectives who try to locate and manage resources, identify and mitigate hazards, predict change, and help communities plan for the future. These things lay the foundation upon which all industrial societies are based. Yet the growing societies of the world are now testing the ability of geoscientists to provide enough materials, energy, and wisdom to sustain people's wants and needs. Now, more than ever, geologists are addressing fundamental questions about natural resources, the environment, and public policies in ways that strive to ensure the ability of Earth to sustain the human population.

How to Start Thinking and Acting Like a Geologist. As you complete exercises in this laboratory manual, think and act like a geologist or Earth detective. **Focus on questions** about things like Earth materials and history, natural resources, processes and rates of environmental change, where and how people live in relation to the environment, and how geology contributes to sustaining the human population. **Conduct investigations and use your senses and tools to make observations** (determine and characterize the qualities and quantities of materials, energies, and changes). As you make observations, **record data** (factual information or evidence used as a basis for reasoning). **Engage in critical thinking**—apply, analyze, interpret, and evaluate the evidence to form tentative ideas or conclusions. **Engage in discourse** or collaborative inquiry with others (exchange, organization, evaluation, and debate of data and ideas). **Communicate inferences**—write down or otherwise share your conclusions and justify them with your data and critical thinking process.

These components of doing geology are often not a linear "scientific method" to be followed in steps. You may find yourself doing them all simultaneously or in odd order. For example, when you observe an object or event, you may form an initial interpretation of it or a hypothesis (tentative conclusion) about it. However, a good geologist (scientist) would also question these tentative conclusions and investigate further to see if they are valid or not. Your tentative ideas and conclusions may change as you make new observations, locate new information, or apply a different method of thinking.

How to Record Your Work. When making observations, you should observe and record **qualitative data** by describing how things look, feel, smell, sound, taste, or behave. You should also collect and record **quantitative data** by counting, measuring or otherwise expressing in numbers what you observe. Carefully and precisely record your data in a way that others could use it.

Your instructor will not accept simple yes or no answers to questions. He or she will expect your answers to be complete inferences justified with data and an explanation of your critical thinking (in your own words). Show your work whenever you use mathematics to solve a problem so your method of thinking is obvious.

ACTIVITY 1.1 Geologic Inquiry

THINK About It **How and why do geologists observe Earth materials at different scales (orders of magnitude)?**

OBJECTIVE Analyze and describe Earth materials at different scales of observation, then infer how they are related to you and thinking about geology.

PROCEDURES

1. **Before you begin**, do not look up definitions and information. Just focus on FIGURE 1.1. Use your current knowledge to start thinking like a geologist, and complete the worksheet with your current level of ability. Also, this is **what you will need** to do the activity:

   ____ Activity 1.1 Worksheets (pp. 25–26) and pencil with eraser
2. **Answer every question on the worksheets in a way that makes sense to you** and be prepared to compare your ideas with others.
3. **After you complete the worksheets**, read below about scales of observation and direct and remote observation of geology. Be prepared to discuss your observations and inferences with others.

Scales of Earth Observation

The most widely known geologic feature in the United States is undoubtedly the Grand Canyon. This canyon cuts a mile deep, through millions of rock layers that are like pages of an immense stone book of geologic history called the **geologic record.** The layers vary in thickness from millimeters to meters. Each one has distinguishing features—some as tiny as microscopic fossils or grains of sand and some as large as fossil trees, dinosaur skeletons, or ancient stream channels. Yet when one measures and describes the layers, it can be done at the scale of a single page or at the scale of many pages, much the same as one might describe a single tree or the entire forest in which the tree is found. Each successive layer also represents a specific event (formation of the layer), which occurred at a specific time in Earth's long geologic history. Therefore, geologists are concerned with scales of observation and measurement in both space and time.

Spatial Scales of Observation and Measurement

Geologists study all of Earth's materials, from the spatial scale of atoms (atomic scale) to the scale of our entire planet (global scale). At each spatial scale of observation, they identify materials and characterize relationships. Each scale is also related to the others. You should familiarize yourself with these **spatial scales of observation** as they are summarized in FIGURE 1.2 and the tables of quantitative units of measurement, symbols, abbreviations, and conversions on pages xi and xii at the front of this manual. Terms such as regional, local, hand sample, and microscopic are hierarchical *levels of scale*, not measurements. When making measurements, geologists use these kinds of scales:

- **Bar scale**—A bar scale is a small ruler printed on an image or map. You use it to measure distances on the image or map. For example, all of the images in FIGURE 1.1 are accompanied by bar scales so you can make exact measurements of features within them. If a bar scale is given in one unit of measurement, like miles, and you want to know distances in kilometers, then you must convert the measurement using the table on page xii at the front of the manual.
- **Magnification scale**—This scale tells you how many times larger or smaller an object is in a picture compared to its actual size in real life. Magnification scale can be expressed as a percentage or a multiplication factor. For example, if you take a picture of a rock and enlarge it to twice its actual (normal) size, then you should note a scale of 200%, 2x, or x2 on the picture. If you reduce the picture of the rock so it appears only half of its actual size, then you should note a scale of 50%, 0.5x, or x1/2 on the picture. It does not matter which units of measurement you magnify (multiply). For example, if you measure a distance of 6 millimeters on the image that has a scale of 200% or x2, then the distance is actually 12 millimeters in real life.
- **Fractional scale**—A fractional scale is used to indicate how much smaller something is than its actual size. It is like the magnification scale, but expressed as a fraction. Therefore, if a picture shows a rock at only half of its actual size, then you can use a fractional scale of *½ scale* to indicate it. It does not matter which units of measure you use, the actual size would still be half of what you measure in any units.
- **Ratio scale**—A ratio scale is commonly used when making models. The scale represents the proportional ratio of a linear dimension of the model to the same feature in real life. If a toy car is 20 centimeters long and the actual car was 800 centimeters long, then the ratio scale of model to actual car is 20:800, which reduces to 1:40. (Note: this is the same as a fractional scale of 1/40.) Ratio scales are commonly provided on maps, as well as three-dimensional models.

A. ASTER satellite images of Escondida open-pit copper mines region, Chile

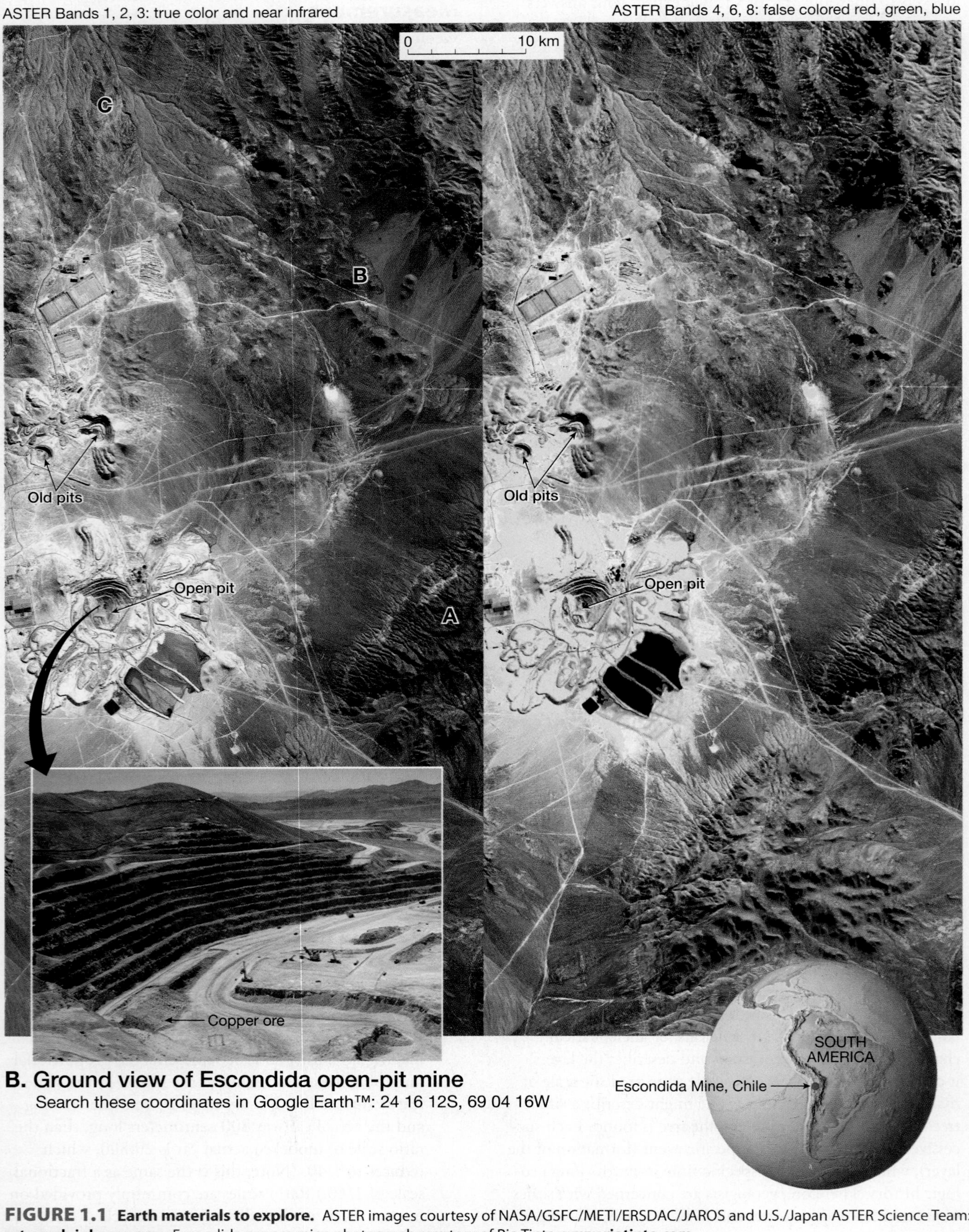

ASTER Bands 1, 2, 3: true color and near infrared

ASTER Bands 4, 6, 8: false colored red, green, blue

0 10 km

C

B

Old pits

Open pit

A

Old pits

Open pit

Copper ore

SOUTH AMERICA

Escondida Mine, Chile

B. Ground view of Escondida open-pit mine

Search these coordinates in Google Earth™: 24 16 12S, 69 04 16W

FIGURE 1.1 Earth materials to explore. ASTER images courtesy of NASA/GSFC/METI/ERSDAC/JAROS and U.S./Japan ASTER Science Team: **asterweb.jpl.nasa.gov.**; Escondida copper mine photograph courtesy of Rio Tinto: **www.riotinto.com**

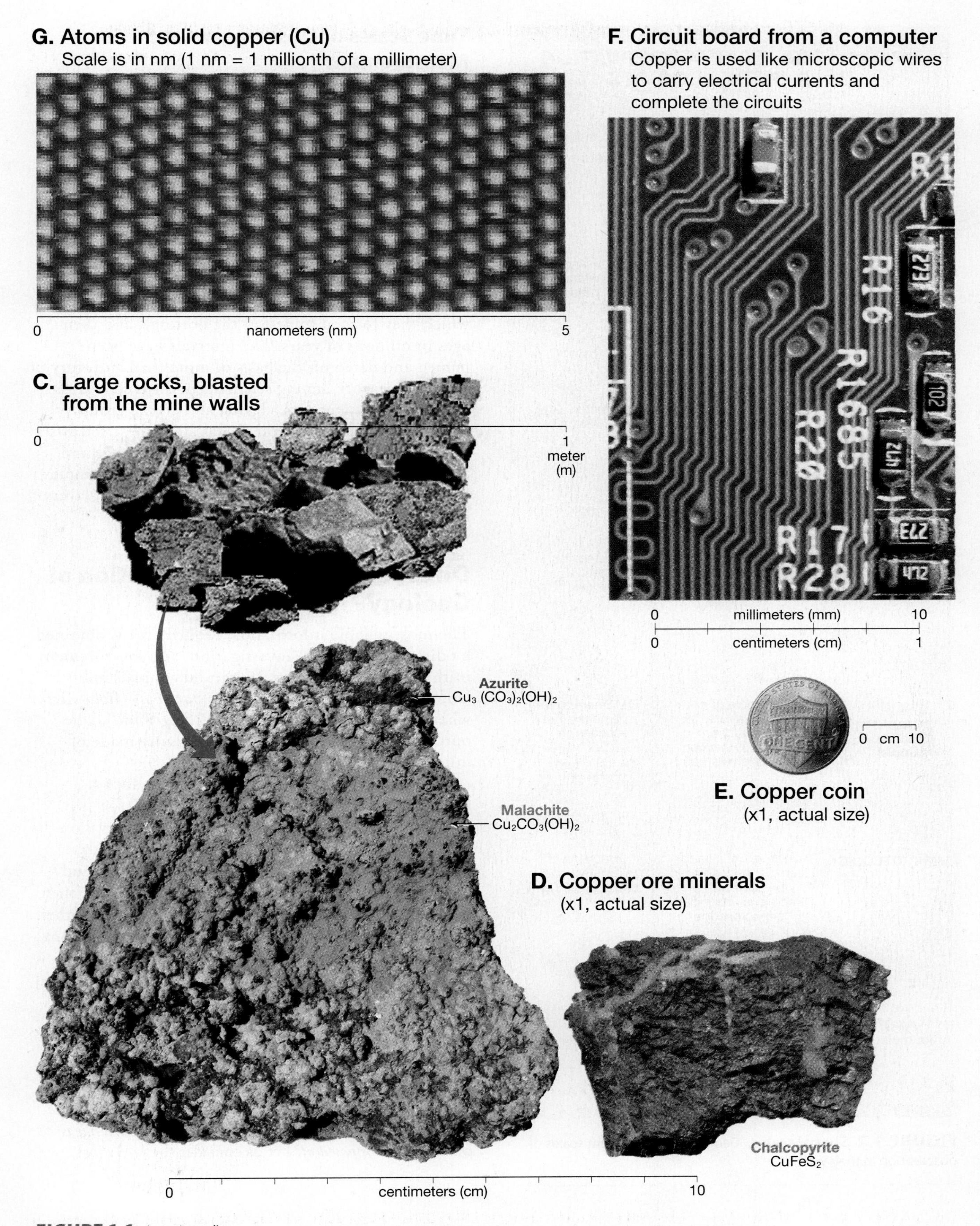

FIGURE 1.1 (continued)

SPATIAL SCALES OF OBSERVATION USED BY GEOLOGISTS

Scale of observation		Used to study things like...	Measured in...
Global	Macroscopic: visible with the naked eye	Entire planet and its interactive "spheres"	Thousands of kilometers (km) or miles (mi)
Regional	Macroscopic: visible with the naked eye	Portions of oceans, continents, countries, provinces, states, islands	Kilometers (km), miles (mi)
Local (outcrop or field site)	Macroscopic: visible with the naked eye	Specific locations that can be "pin-pointed" on a map	Meters (m), feet (ft)
Hand sample (field/lab. sample)	Macroscopic: visible with the naked eye	Sample of a mineral, a rock, air, water, or an organism that can be held in your hand	Centimeters (cm), millimeters (mm), inches (in.) 0 1 cm / 0 10 mm
Microscopic		Features of a hand sample that can only be seen with a hand lens (magnifier) or microscope	Fractions of millimeters (mm), micrometers (μm)
Atomic (or molecular)		Arrangements of the atoms or molecules in a substance	Nanometers (nm), angstroms (Å)

FIGURE 1.2 Spatial scales. Geologists use different scales of observation in their work.

Time Scales of Observation and Measurement

Geologists also think about **temporal scales of observation.** As geologic detectives, they analyze rock layers as stone pages of the geologic record for evidence of events and relationships. As geologic historians, they group the events and relationships into paragraphs, chapters, sections, and parts of geologic history that occurred over epochs, periods, eras, and eons of time. The index to this book of geologic history is called the **geologic time scale** (FIGURE 1.3). Notice that the geologic time scale is a chart showing named intervals of the geologic record (rock units), the sequence in which they formed (oldest at the bottom), and their ages in millions of years. The intervals have been named and dated on the basis of more than a century of cooperative work among scientists of different nations, races, religions, genders, classes, and ethnic groups from throughout the world. What all of these scientists have had in common is the ability to do science and an intense desire to decipher Earth's long and complex history based on evidence contained in the rock layers that are the natural record of geologic history.

Direct and Remote Investigation of Geology

The most reliable information about Earth is obtained by direct observation, investigation, and measurement in the field (out of doors, in natural context) and laboratory. Most geologists study *outcrops*—field sites where rocks *crop out* (stick out of the ground). The outcrops are made of rocks, and rocks are made of minerals.

Samples obtained in the field (from outcrops at field sites) are often removed to the laboratory for further analysis using basic science. Careful observation (use of your senses, tactile abilities, and tools to gather information) and critical thought lead to questions and hypotheses (tentative ideas to test). Investigations are then designed and carried out to test the hypotheses and gather data (information, evidence). Results of the investigations are analyzed to answer questions and justify logical conclusions.

Refer to the example of field and laboratory analysis in FIGURE 1.4. Observation 1 (in the field) reveals that Earth's rocky geosphere crops out at the surface of the land. Observation 2 reveals that outcrops are made of rocks. Observation 3 reveals that rocks are made of mineral crystals such as the mineral *chalcopyrite*. This line of reasoning leads to the next **logical question:** *What is chalcopyrite composed of?* Let us consider the two most

THE GEOLOGIC TIME SCALE

A chart showing the sequence, names, and ages of Earth's rock layers (oldest at the bottom)

Eon of time Eonothem of rock	Era of time Erathem of rock	Period of time System of rock**		Epoch of time Series of rock	Millions of years ago (Ma)	Some notable fossils in named rock layers
Phanerozoic	Cenozoic: (new life) Age of Mammals	Quaternary (Q)		Holocene	.0117	First *Homo* fossils, 70–100% extant mollusks⁺
				Pleistocene	2.6	
		Tertiary	Neogene (N)	Pliocene	5.3	First humans (Hominidae), 15–70% extant mollusks⁺
				Miocene	23	
			Paleogene (P_G)	Oligocene	34	More mammals than reptiles, <15% extant mollusks⁺
				Eocene	56	
				Paleocene	66	
	Mesozoic: (middle life) Age of Reptiles	Cretaceous (K)			145	Last dinosaur fossils: including *Tyrannosaurus rex*
		Jurassic (J)			201	First bird fossil: *Archaeopteryx*
		Triassic (TR)			252	First dinosaur, mammal, turtle, and crocodile fossils
	Paleozoic: (old life) Age of Trilobites	Permian (P)			299	Last (youngest) trilobite fossils
		Carboniferous (C)*	Pennsylvanian (IP)		323	First reptile fossils
			Mississippian (M)		359	First fossil conifer trees
		Devonian (D)			419	First amphibian, insect, tree, and shark fossils
		Silurian (S)			443	First true land plant fossils
		Ordovician (O)			485	First fossils of coral and fish
		Cambrian (Є)			541	First trilobite fossils First abundant visible fossils
Proterozoic	Precambrian: An informal name for all of this time and rock.				2500	Oldest fossils: mostly microscopic life, visible fossils rare
Archean		Oldest fossils of visible life (stromatolites)			3500	
					4000	
Hadean		Acasta Gneiss, northwestern Canada			4030	
		Nuvvuagittuq greenstone belt, Quebec, Canada			4280	
		Zircon mineral crystals in the Jack Hills Metaconglomerate, Western Australia			4400	
		Oldest meteorites			4550	

*European name

**Symbols in parentheses are abbreviations commonly used to designate the age of rock units on geologic maps.

⁺Extant mollusks are mollusks (clams, snails, squid, etc.) found as fossils and still living today.

FIGURE 1.3 The geologic time scale. Absolute ages in millions of years ago (Ma) follow the International Commission on Stratigraphy, 2013. See their website for more detailed versions and recent updates of the international geological time scale (**http://www.stratigraphy.org/index.php/ics-chart-timescale**).

Example of Geologic Field and Laboratory Investigation

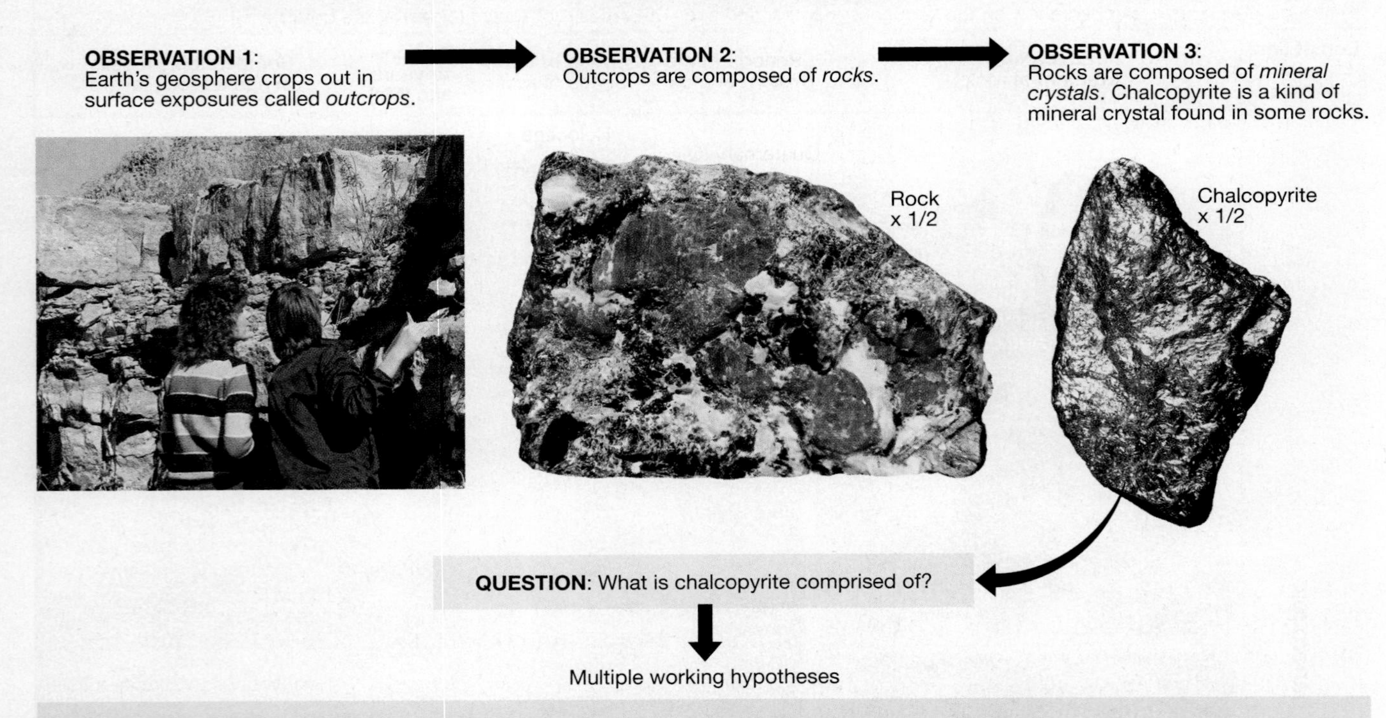

QUESTION: What is chalcopyrite comprised of?

Multiple working hypotheses

HYPOTHESIS 1: Chalcopyrite could be a native element—a pure, natural occurrence of an element. An *element* is a chemical substance that cannot be separated into simpler chemical substances by processes such as heating, leaching (dissolving) with acid, or electrolysis. There are 92 naturally occurring elements, which scientists refer to by name or symbol: e.g., hydrogen (H), oxygen (O), carbon (C), copper (Cu), iron (Fe), sulfur (S), gold (Au).

HYPOTHESIS 2: Chalcopyrite could be a *compound*—a chemical substance that can be separated (decomposed or dissociated, into its constituent elements by processes such as heating, leaching (dissolving) with acid, or electrolysis. Scientists represent compounds by their chemical formulas, which denote the elements and how they are chemically combined (bonded) in fixed proportions. For example: H_2O (water), CO_2 (carbon dioxide).

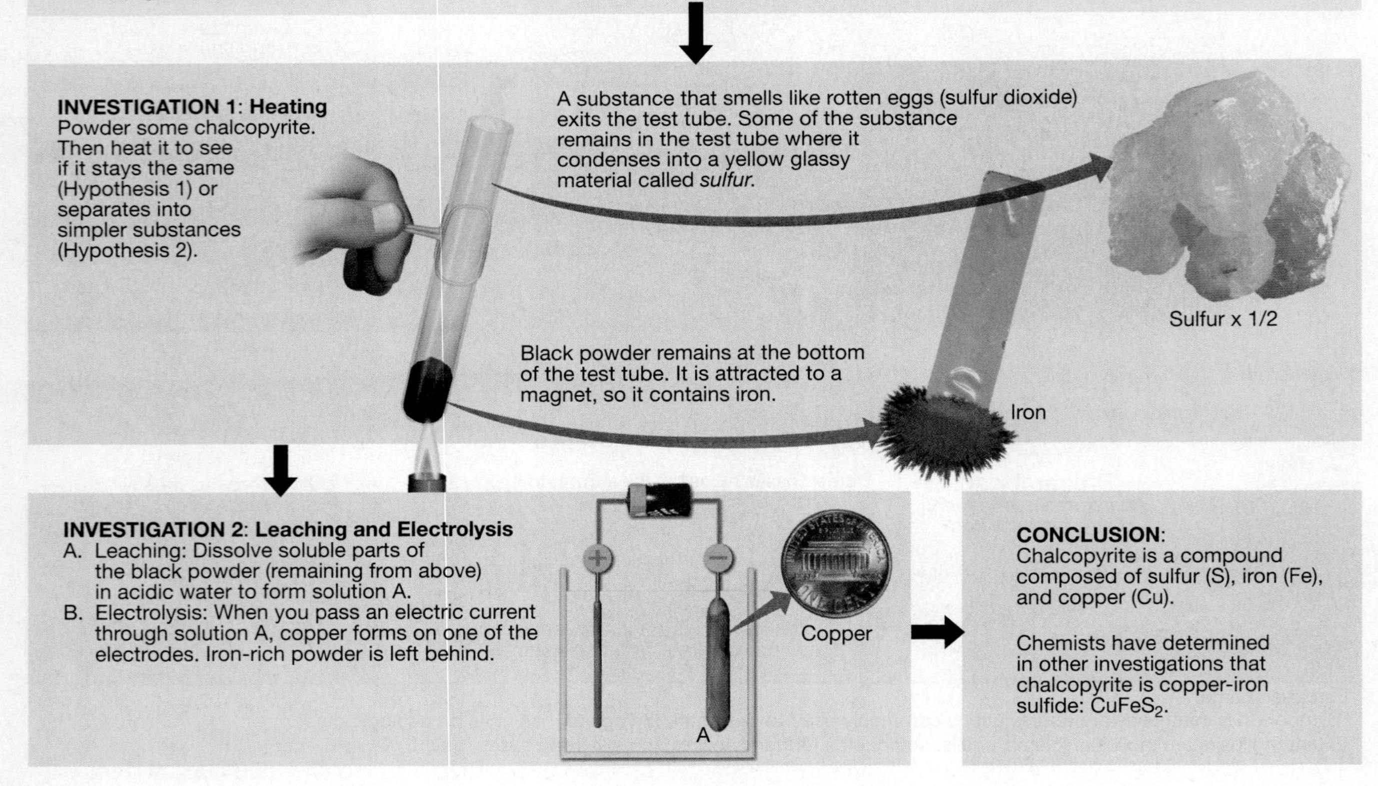

CONCLUSION:
Chalcopyrite is a compound composed of sulfur (S), iron (Fe), and copper (Cu).

Chemists have determined in other investigations that chalcopyrite is copper-iron sulfide: $CuFeS_2$.

FIGURE 1.4 **Example of geologic field and laboratory investigation.**

logical possibilities, or **working hypotheses** (tentative ideas to investigate, test). It is always best to have more than one working hypothesis.

1. Chalcopyrite may be a pure substance, or chemical *element.* What investigating and gathering of evidence could we do to reasonably determine if this is true or false?
2. Chalcopyrite may be a *compound* composed of two or more elements. What investigating and gathering of evidence could we do to reasonably determine if this is true or false? If true, then how could we find out which elements make up chalcopyrite?

Let us conduct two **investigations** (activities planned and conducted to test hypotheses, gather and record data, make measurements, or control and explore variables). In Investigation 1, the chalcopyrite is ground to a powder and heated. This investigation reveals the presence of sulfur and at least one other substance. The remaining substance is attracted to a magnet, so it may be iron or a compound containing iron. When the powder is leached (dissolved in acidic water) and subjected to electrolysis (Investigation 2), copper separates from the powder. The remaining powder is attracted to a magnet, indicating the presence of iron. **Analysis of the results** of these two investigations leads us to the **logical conclusion** that Hypothesis 2 was correct (chalcopyrite is a compound). The results are also evidence that chalcopyrite is composed of three different elements: sulfur (S), iron (Fe), and copper (Cu). Chemists call chalcopyrite copper-iron sulfide ($CuFeS_2$). Since chalcopyrite contains a significant proportion of copper, it is also a *copper ore* (natural material from which copper can be extracted at a reasonable profit).

This same laboratory procedure is applied on a massive scale at copper mines. Because most copper-bearing rock contains only a small percentage of chalcopyrite or another copper-bearing mineral, the rock is mined, crushed, and powdered. It is then mixed with water, detergents, and air bubbles that float the chalcopyrite grains to the surface of the water. When these grains are removed, they are smelted (roasted and then melted) to separate impure copper from the other parts of the melted chalcopyrite (that cool to form *slag*). The impure copper is then leached in sulfuric acid and subjected to electrolysis, whereupon the copper is deposited as a mass of pure copper on the positive electrode (cathode).

Satellite Remote Sensing of Geology

There are times when geologists cannot make direct observations of Earth and must rely on a technology to acquire and record information remotely (from a distance, without direct contact). This is called *remote sensing.* One of the most common kinds of remote sensing used by geologists is satellite remote sensing.

Electromagnetic (EM) Radiation. The electromagnetic (EM) spectrum of radiation is a spectrum of electric and magnetic waves that travel at the speed of light (300,000,000 meters/second, or 3×10^8 m/s). The spectrum is subdivided into **bands**—parts of the EM spectrum that are defined and named according to their wavelength (distance between two adjacent wave crests or troughs).

Instruments aboard satellites scan information from not only the visible bands of the electromagnetic spectrum, but also parts of the spectrum that are not visible to humans (e.g., infrared).The ASTER instrument scans 14 bands of electromagnetic radiation. Bands 1, 2, and 3 are visible (blue-green, red) bands (left side, **FIGURE 1.1**). Bands 4–9 are short wave infrared bands (SWIR) that are invisible to humans (right side, **FIGURE 1.1**). Bands 10–14 are thermal infrared (TIR) bands that are also invisible to humans.

True Color and False Color Images. Data from environmental satellite instruments must be rendered into an image that humans can see, either by giving objects in the image their true color or a false color. **True color** photographs and satellite images show objects in the colors that they would appear to be if viewed by the human eye However, since many bands of radiation detected by satellites are not visible to humans, the bands are given a **false color** in satellite images (right side of **FIGURE 1.1**).

ACTIVITY

1.2 Spheres of Matter, Energy, and Change

THINK About It **What materials, energies, and processes of change do geologists study?**

OBJECTIVE Analyze and describe the materials, energies, and processes of change within and among Earth's spheres.

PROCEDURES

1. **Before you begin**, read the following background information on matter and spheres, energy sources and sinks, and processes and cycles of change. This is **what you will need:**

   ___ Activity 1.2 Worksheets (pp. 27–29) and pencil with eraser

2. Then, **follow your instructor's directions** for completing the worksheets.

Matter and Spheres

Everything on Earth is made of matter and energy. Matter is anything that takes up space and has a mass that can be weighed. It is tangible materials and substances. At the global scale of observation, geologists conceptualize Earth as a dynamic planetary system composed of interacting *spheres* (subsystems) of living and nonliving materials.

Geosphere

The **geosphere** is Earth's rocky body (FIGURE 1.5). The inner core has a radius of 1196 km and is composed mostly of iron (Fe) in a solid state. The outer core is 2250 km thick and is composed mostly of iron (Fe) and nickel (Ni) in a liquid state. The mantle is 2900 km thick and is composed mostly of oxygen (O), silicon (Si), magnesium (Mg), and iron (Fe) in a solid state. The crust has an average thickness of about 25 km and is composed mostly of oxygen (O), silicon (Si), aluminum (Al), and iron (Fe) in a solid state. Some people consider the cryosphere as a sub-sphere of the geosphere. The **cryosphere** is composed of snow crystals and ice that form from freezing parts of the hydrosphere or atmosphere. Ice is a rock made of mineral crystals (like snowflakes), so the cryosphere is actually a sub-sphere of the geosphere. Most of it exists in the polar ice sheets (continental glaciers), permafrost (permanently frozen moisture in the ground), and sea ice (ice on the oceans).

Hydrosphere

The **hydrosphere** is all of the liquid water on Earth's surface and in the ground (groundwater). Most of the hydrosphere is salt water in the world ocean, which has an average depth (thickness) of 3.7 km. However, the hydrosphere also includes liquid water in lakes, streams, and the ground (called *groundwater*).

Atmosphere

The **atmosphere** is the gaseous envelope that surrounds Earth. It consists of about 78% nitrogen (Ni), 21% oxygen (O), 0.9% argon (Ar) and trace amounts of other gases like carbon dioxide, water vapor, and methane. About 80% of these gases (including nearly all of the water vapor) occur in the lowest layer of the atmosphere (troposphere), which has an average thickness of about 16 km (10 miles). From there, the atmosphere thins and eventually ends (no air) at about 1000 km above sea level.

Biosphere

The **biosphere** is the living part of Earth, the part that is organic and self-replicating. It includes all bacteria, plants, and animals, so you are a member of the biosphere.

Magnetosphere

Earth's **magnetosphere** is its magnetic force field; not a material. It is generated from the core of the planet, and it is important because it shields Earth from the solar wind (a radiation of energy and particles from the Sun) that would otherwise make our planet lifeless.

Energy Sources and Sinks

Energy is the capacity to be active or do work, so matter does not move unless it has energy.

Earth's spheres would never change without their energy.

Forms of Energy

Here are some of the forms of energy that power you and the Earth system around you.

- *Thermal (heat) energy* is the energy of moving or vibrating atoms in matter related to its temperature. The higher the temperature, the greater the vibration or motion of its molecules. A hot cup of tea has a lot of thermal energy, but a cup of iced tea has less thermal energy. Cups of tea at the same temperature have equal thermal energy. One of Earth's two main sources of energy is the heat energy of its core (called **geothermal energy**).
- *Electromagnetic energy* is light, an oscillating (wave) form of energy perpetuated by coupled electronic and magnetic fields emitted from and reflected by objects. The distance between two crests in the waves of electromagnetic energy is called *wavelength*. Humans can only see a small part of the spectrum of electromagnetic wavelengths that exist in nature. Electromagnetic energy from our Sun is called **solar energy** and is the other primary source of Earth's energy (other than geothermal energy).
- **Nuclear energy** is energy stored in the nuclei (plural of nucleus) of atoms. Inside the Sun, hydrogen atoms are heated and energized so much that collisions among hydrogen atoms can fuse their nuclei together (*nuclear fusion*). This thermonuclear reaction creates one larger helium atom from every four hydrogen atoms, but it also converts some of the nuclear energy into electromagnetic energy (sunlight). Thus, the sunlight warming Earth's surface was transformed from nuclear energy in atoms of the Sun. In Earth's core, nuclei of abundant unstable atoms eventually decay (split apart into smaller nuclei, a process called nuclear fission). This transforms energy from the atomic nuclei into thermal energy. So the two main sources of energy that power Earth (solar energy, geothermal energy) have actually been transformed from nuclear energy.
- **Potential energy** is energy stored in an object because of its position in a force field. A force is a push or a pull, and the most dominant force field affecting Earth materials is Earth's gravity. Think of a small rock perched on the edge of a cliff. The rock has energy stored within it as a result of the fact that it is being pulled by Earth's gravitational force field. Gravity will cause the rock to fall if it happens to drop off the edge of the cliff (whereupon the potential energy is converted to kinetic energy).

Sometimes, objects change shape (i.e., they experience elastic strain) as potential energy builds up within them, and their potential energy can also be called elastic energy (instead of just potential energy). For example, if you bend a ruler, the ruler has energy stored within it because of the

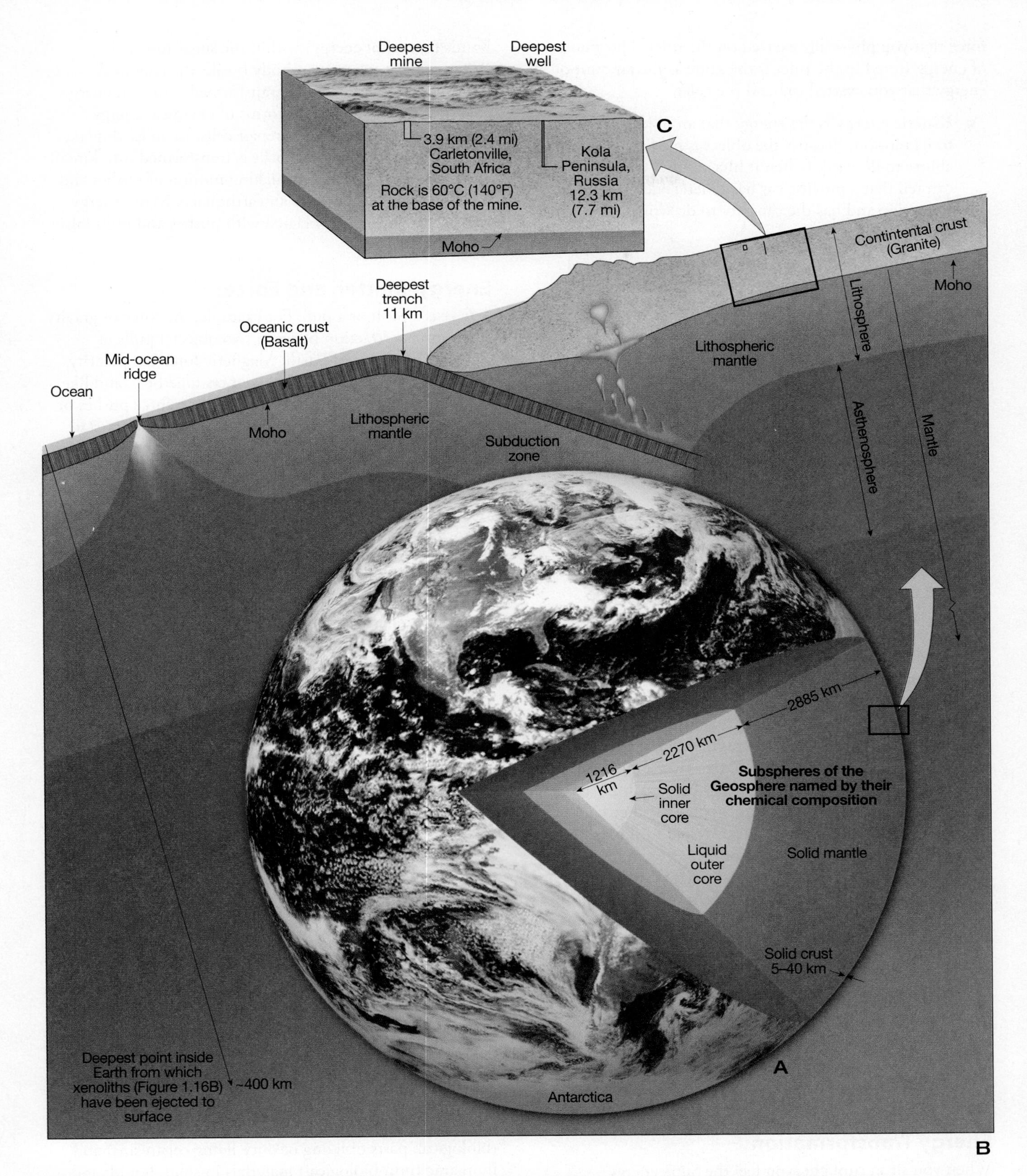

FIGURE 1.5 Global perspective of Earth. A. Earth photographed by Apollo 17 astronauts from about 37,000 km (23,000 mi) away in 1972. **(Courtesy of NASA)** Note compositional subspheres of the geosphere (rocky body of Earth) in the cutaway view: solid inner core, liquid outer core, solid mantle, and crust. B. Hypothetical cross section of the edge of the geosphere. Note the locations of thick continental crust, thin oceanic crust, Moho (base of the crust), lithosphere (crust + lithospheric mantle), and asthenosphere. C. Depths of the deepest mine and well ever drilled into Earth.

force that you physically exerted on the ruler. The amount of energy stored in the ruler is the same as the amount of energy that you exerted to bend the ruler.

- **Kinetic energy** is the energy that an object has due to its motion. Because the object is moving, it has the ability to do work (when it hits another object). You can tell that a moving car has kinetic energy because it is moving and has the capacity to do work when it hits something.
- **Mechanical energy** is the sum of energy associated with the motion (kinetic energy) and position (potential energy) of an object.
- **Chemical energy** is the energy stored in bonds of molecules and chemical compounds. The glucose sugar in your body stores energy that is released when its bonds are broken during respiration.
- **Electrical energy** is the energy carried by a flow of electrons, as in lightning and electrical currents flowing through wires and circuit boards in your electrical devices.

Energy Transfer and Equilibrium

Energy cannot be destroyed or used up, but it can be transferred from a *source* (place with more energy) to a *sink* (place with less energy). For example, energy from the Sun (a source) is transferred to Earth (the sink) by *radiation* (transfer of energy through space; not via materials), just as you can feel heat energy radiating from a hot stove. Energy from Earth's core (source) is transferred to the mantle above it (the sink) by *conduction* (energy transferred by direct contact between molecules of two stationary materials), just as you feel heat energy being conducted into your hand when you pick up a pan by its hot handle. Energy from Earth's core is also transferred to the mantle by *convection* (energy transferred or conveyed in moving molecules of flowing materials), just as you can see energy being transferred by convection in the motion of boiling water or in the hot globs of lava moving in a lava lamp. If energy of the source equals energy of the sink, then energy transfer stops, and the relationship is said to be in *equilibrium*. For example, if you turn off a boiling pot of water, its thermal energy will be dispersed into the materials around it until the pot of water and everything else in the room reach the same temperature (room temperature). When they reach room temperature, they are in thermal equilibrium.

Energy Transformation

When you sit in sunlight, you feel the Sun's energy being transformed into heat. Plants transform the Sun's electromagnetic energy into chemical energy to make their food. The plants apply the chemical energy when they use it to do work (cause an action; use it in photosynthesis to transform water and carbon dioxide into their food: $C_6H_{12}O_6$, glucose sugar). Energy is stored as chemical bonds (chemical energy) within the sugar molecules. When you eat sugar, your body breaks the chemical bonds so the energy can be transformed into heat energy, electrical energy, and other forms of chemical energy used to energize, grow, and repair cells. In an earthquake, potential energy stored in rocks is transformed into kinetic energy, which leads to the quaking motion of pushes and pulls that you feel and call an earthquake. Many energy transformations are associated with pushes and pulls (also called forces).

Energy, Matter, and Force

A *force* is a push or a pull. For example, the force of gravity (the mutual attraction between two objects) pulls us towards the center of Earth. Magnetic force has polarity. Unlike poles attract (pull the magnets together), and like poles repel (push the magnets apart). As a force pushes or pulls, it causes objects to build up potential energy, start moving, change direction of movement, or stop moving. Unlike matter, which you can see and feel, you cannot see a force. But you can feel the push or pull of a force, and you can see how it affects the motion or change in shape (deformation) of objects. When a force acts on matter, the matter has potential energy (energy stored in an object because of its position in a force field) or kinetic energy (the capacity to work as a function of its motion). So when we say that energy is the capacity to do work, it means that energy is the potential to exert a force. The force is what does the moving. The force also transfers energy and transforms energy.

When you push a heavy object, some of your chemical energy is converted to mechanical energy, which powers the force. During the force, the mechanical energy is transformed into potential energy until the object moves (whereupon it is transformed into kinetic energy). When you stop pushing, there is no more force (the force is destroyed), but the energy was transformed and conserved (not created or destroyed). This is a basic *law* (fundamental principle) of nature. Of course, matter cannot be created or destroyed either, so the two concepts are combined into one law. The Law of Conservation of Matter and Energy is that matter and energy can be transferred and transformed but cannot be created or destroyed.

Processes and Cycles of Change

Earth is characterized by the transfer (flow) of matter (materials) and energy during processes of change (**FIGURE 1.6**) at every spatial and temporal scale of observation. Most of these processes involve organic (biological; parts of living or once living organisms) and inorganic (non-biological) materials in solid, liquid, and gaseous states, or *phases*. Note that many of the processes have opposites depending on the flow of energy to or from a material: melting and freezing, evaporation and condensation, sublimation and deposition, dissolution and chemical precipitation, photosynthesis (food energy storage) and respiration (food energy release or "burning"

COMMON PROCESSES OF CHANGE		
Process	**Kind of Change**	**Example**
Melting	Solid phase changes to liquid phase.	Water ice turns to water.
Freezing	Liquid phase changes to solid phase.	Water turns to water ice.
Evaporation	Liquid phase changes to gas (vapor) phase.	Water turns to water vapor or steam (hot water vapor).
Condensation	Gas (vapor) phase changes to liquid phase.	Water vapor turns to water droplets.
Sublimation	Solid phase changes directly to a gas (vapor) phase.	Dry ice (carbon dioxide ice) turns to carbon dioxide gas.
Deposition	The laying down of solid material as when a gas phase changes into a solid phase or solid particles settle out of a fluid.	Frost is the deposition of ice (solid phase) from water vapor (gas). There is deposition of sand and gravel on beaches.
Dissolution	A substance becomes evenly dipersed into a liquid (or gas). The dispersed substance is called a solute, and the liquid (or gas) that causes the dissolution is called a solvent.	Table salt (solute) dissolves in water (solvent).
Vaporization	Solid or liquid changes into a gas (vapor), due to evaporation or sublimation.	Water turns to water vapor or water ice turns directly to water vapor.
Reaction	Any change that results in formation of a new chemical substance (by combining two or more different substances).	Sulfur dioxide (gas) combines with water vapor in the atmosphere to form sulfuric acid, one of the acids in rain.
Decomposition reaction	An irreversible reaction. The different elements in a chemical compound are irreversibly split apart from one another to form new compounds.	Feldspar mineral crystals decompose to clay minerals and metal oxides (rust).
Dissociation	A reversible reaction in which some of the elements in a chemical compound are temporarily split up. They can combine again under the right conditions to form back into the starting compound.	The mineral gypsum dissociates into water and calcium sulfate, which can recombine to form gypsum again.
Chemical precipitation	A solid that forms when a liquid solution evaporates or reacts with another substance.	Salt forms as ocean water evaporates. Table salt forms when hydrochloric acid and sodium hydroxide solutions are mixed.
Photosynthesis	Sugar (glucose) and oxygen are produced from the reaction of carbon dioxide and water in the presence of sunlight (solar energy).	Plants produce glucose sugar and oxygen.
Respiration	Sugar (glucose) and oxygen undergo combustion (burning) without flames and change to carbon dioxide, water, and heat energy.	Plants and animals obtain their energy from respiration.
Transpiration	Water vapor is produced by the biological processes of animals and plants (respiration, photosynthesis).	Plants release water vapor to the atmosphere through their pores.
Evolution	Change over time (gradually or in stages).	Biological evolution, change in the shape of Earth's landforms over time.
Crystallization	Atoms, ions, or molecules arrange themselves into a regular repeating 3-dimensional pattern. The formation of a crystal.	Water vapor freezes into snowflakes. Liquid magma cools into a solid mass of crystals.
Weathering	Materials are fragmented, worn, or chemically decomposed.	Rocks break apart, get worn into pebbles or sand, dissolve, rust, or decompose to mud.
Transportation	Materials are pushed, bounced, or carried by water, wind, ice, or organisms.	Sand and soil are blown away. Streams push, bounce, and carry materials downstream.
Radiation	Transfer of energy through space; not via materials.	Sunlight radiates from the Sun to Earth.
Conduction	Transfer of energy by direct contact between molecules of two stationary materials.	A pan conducts heat from the hot stove top that it sits on.
Convection	Transfer of energy in moving molecules of flowing materials.	Thermal energy in lava is transferred as the lava flows from a volcano.
Convection cycling	Cyclic current motion (and heat transfer) within a flowing body of matter due to unequal heating and cooling. As part of the material is heated and rises, a cooler part of the material descends to replace it (whereupon it is reheated and rises again to form a convection cell.	Warm air in the atmosphere rises and cooler air descends to replace it; water boiling in a pot.

FIGURE 1.6 Some common processes of change on Earth.

ACTIVITY 1.3 Modeling Earth Materials and Processes

THINK About It **How and why do geologists make models of Earth?**

OBJECTIVE Make models and use them to understand Earth processes and the relative proportions of Earth's physical spheres.

Geologists make models of things that are too large or small to visualize and study. A scale model is a physical representation of something that is actually much larger or smaller and has the same proportions as the actual object. For example, a toy car is a small model of an actual car. The *ratio scale* of the model is the ratio by which the actual object was enlarged or reduced to make the scale model. If a toy car is 20 centimeters long and the actual car was 800 centimeters long, then the ratio scale of model to actual car is 20:800, which reduces to 1:40. A 1:40 scale model has a *fractional scale* of 1/40, meaning that the actual car is 40 times (40x) larger than the model.

Geologists also make models to study how things work. They design laboratory experiments in which they can control variables and test ideas before they implement them in real life. For example, it is much cheaper and safer to build and test many models of earthquake-proof designs before constructing an actual earthquake-proof building.

PROCEDURES

1. **Before you begin**, this is **what you will need**:
   - ___ blue pencil or pen
   - ___ calculator, ruler, drafting compass
   - ___ several coins
   - ___ Activity 1.3 Worksheets (pp. 30–31) and pencil with eraser
2. **Then follow your instructor's directions** for completing the worksheets.

ACTIVITY 1.4 Measuring and Determining Relationships

THINK About It **How and why do geologists measure Earth materials and graph relationships among Earth materials and processes of change?**

OBJECTIVE Measure Earth materials using basic scientific equipment and techniques, and determine relationships using rates and graphs.

PROCEDURES

1. **Before you begin**, read the following background information on measuring Earth materials, determining rates, and graphing relationships. This is **what you will need**:
   - ___ Activity 1.4 Worksheets (pp. 32–34) and pencil with eraser
   - ___ calculator, ruler
   - ___ other materials provided in the lab: 10 mL and 500 mL or 1000 mL graduated cylinders, small piece of grease-based modeling clay, gram balance or scale, and basin of water
2. **Then follow your instructor's directions** for completing the worksheets.

without flames). And while some chemical reactions are irreversible (decomposition reactions, like tooth decay), many are reversible (under changing conditions, the chemicals react again and recombine back into the starting compounds). So these processes of change, powered by energy transfer and transformation, cause chemical materials to be endlessly cycled and recycled. One of these cycles is the *hydrologic cycle*, or "water cycle."

The hydrologic cycle (**FIGURE 1.7**) involves several processes and changes in relation to all three phases of water and all of Earth's spheres (global subsystems). It is one of the most important cycles that geologists routinely consider in their work. The hydrologic cycle is generally thought to operate like this: liquid water (hydrosphere) evaporating from Earth's surface produces water vapor (atmospheric gas). The water vapor eventually condenses in the atmosphere to form aerosol water droplets (clouds). The droplets combine to form raindrops or snowflakes (atmospheric precipitation). Snowflakes can accumulate to form ice (cryosphere) that sublimates back into the atmosphere or melts back into water. Both rainwater and meltwater soak into the ground (to form groundwater), evaporate back into the atmosphere, drain back into the ocean, or are consumed by plants and animals (which release the water back to the atmosphere via the process of transpiration).

In addition to water that is moving about the Earth system, there is also water that is stored and not circulating at any given time. For example, a very small portion of Earth's water (about 2% of the water volume in oceans) is currently stored in snow and glacial ice at the poles and on high mountaintops. Additional water (perhaps as much as 80% of the water now in oceans) is also stored in "*hydrous*" (water-bearing) minerals inside Earth. When glaciers melt, or rocks melt, the water can return to active circulation.

The endless exchange of energy and recycling of water undoubtedly has occurred since the first water bodies formed on Earth billions of years ago. Your next drink may include water molecules that once were part of a hydrous (water-bearing) mineral inside Earth or that once were consumed by a thirsty dinosaur!

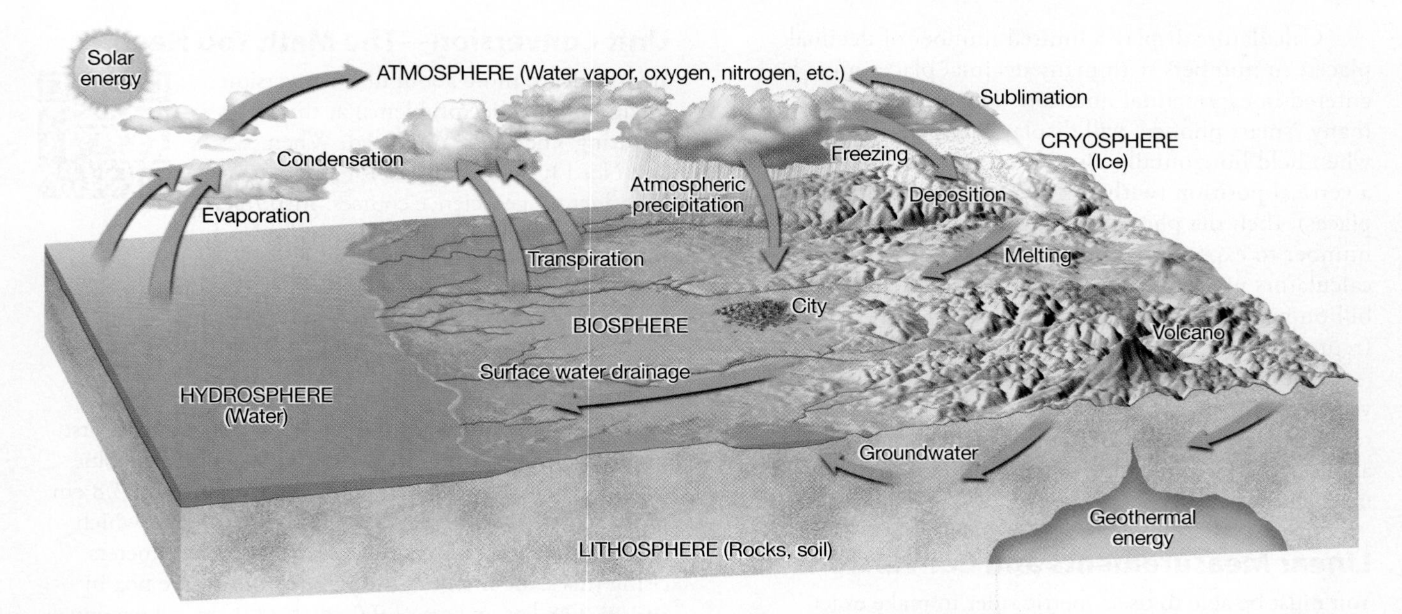

FIGURE 1.7 Hydrologic cycle (water cycle). Note the relationship of processes of change, and states of water, to Earth's spheres. Also note that the hydrologic cycle is driven (forced to operate) by energy from the Sun (solar energy), energy from Earth's interior (geothermal energy), and gravity.

Measuring Earth Materials

Every material has a *mass* that can be weighed and a *volume* of space that it occupies. An object's mass can be measured by determining its weight under the pull of Earth's gravity (using a balance). An object's volume can be calculated by determining the multiple of its linear dimensions (measured using a ruler) or directly measured by determining the volume of water that it displaces (using a graduated cylinder). In this laboratory, you will use metric balances, rulers, and graduated cylinders to analyze and evaluate the dimensions and density of Earth materials. Refer to page xiii at the front of this manual for illustrations of this basic laboratory equipment.

Metric System of Measurement

People in different parts of the world have historically used different systems of measurement. For example, people in the United States have historically used the English system of measurement based on units such as inches, feet, miles, pounds, gallons, and degrees Fahrenheit. However, for more than a century, most nations of the world have used the metric system of measurement based on units such as meters, liters, and degrees Celsius. In 1975, the U.S. Congress recognized the value of a global system of measurement and adopted the metric system as the official measurement system of the United States. This conversion is not yet complete, so Americans currently use both English and metric systems of measurement. In this laboratory we will only use the metric system.

Each kind of metric unit can be divided or multiplied by 10 and its powers to form the smaller or larger units of the metric system. Therefore, the metric system is also known as a base-10 or decimal system. The International System of Units (SI) is the modern version of metric system symbols, numbers, base-10 numerals, powers of ten, and prefixes (see page xi).

Orders of Magnitude and Scientific Notation

Differences of scale are sometimes expressed by powers (multipliers) of ten as **orders of magnitude**. For example, if object "A" is 10 times larger than object "B," then it is one order of magnitude larger (one power of ten, or 10 times larger). If object "A" is 100 times larger than object "B," then it is two orders of magnitude larger (two powers of ten, or 100 times larger).

Scientific (exponential) notation is a compact way of expressing very large or small numbers using base-ten orders of magnitude. For example, in scientific notation, three orders of magnitude larger would be "10 raised to a power of three" and written as 10^3. (This is also called "ten to the three" or "ten to the third.") The superscript "3" is called the exponent. So 3800 can be expressed in scientific notation as 3.8×10^3. For very small numbers, the exponent is negative, so 0.0038 is written 3.8×10^{-3} ("3.8 times ten to the negative three").

Scientific notation simplifies very large or small numbers by getting rid of zeros. For example, one billion is written as 1×10^9 instead of 1,000,000,000. Notice that the exponent signifies how many places to move the decimal place to the right (larger). One-millionth is written as 1×10^{-6} instead of 0.000001. In this case the exponent signifies how many places to move the decimal point to the left (smaller).

Calculators display a limited number of decimal places, so numbers with many decimal places must be entered in exponential notation. The calculators on many "smart phones" will display many decimal places when held horizontally. But if you turn the phone to a vertical position (with less space to display decimal places), then the phone will automatically change the number to exponential notion. Smart phones and most calculators use an "E" to signify the exponent, so one billion would be displayed as something like "1e+9" (representing 1×10^9).

The International System of Units (SI) is the modern version of the metric system and is based on powers of ten. See page xi at the front of this manual to learn more about it and how scientific notation is used to express large metric units.

Linear Measurements and Conversions

You must be able to use a metric ruler to make exact measurements of **length** (how long something is). This is called *linear measurement.* Most rulers in the United States are graduated in English units of length (inches) on one side and metric units of length (centimeters) on the other. For example, notice that one side of the ruler in **FIGURE 1.8A** is graduated in numbered inches, and each inch is subdivided into eighths and sixteenths. The other side of the ruler is graduated in numbered centimeters (hundredths of a meter), and each centimeter is subdivided into ten millimeters. The ruler provided for you in GeoTools Sheets 1 and 2 at the back of this manual are graduated in exactly the same way.

Review the examples of linear metric measurement in **FIGURE 1.8A** to be sure that you understand how to make *exact* metric measurements. Note that the length of an object may not coincide with a specific centimeter or millimeter mark on the ruler, so you may have to estimate the fraction of a unit as exactly as you can. The length of the red rectangle in **FIGURE 1.8A** is between graduation marks for 106 and 107 millimeters (mm), so the most exact measurement of this length is 106.5 mm. Also be sure that you measure lengths starting from the zero point on the ruler and *not from the end of the ruler.*

There will be times when you will need to convert a measurement from one unit of measure to another. This can be done with the aid of the mathematical conversions chart on page xii at the front of the manual. For example, to convert millimeters (mm) to meters (m), divide the measurement in mm by 1000 (because there are 1000 millimeters per meter):

$$\frac{106.5 \text{ mm}}{1000 \text{ mm/m}} = 0.1065 \text{ m}$$

Thus, 106.5 millimeters is the same as 0.1065 meters.

Unit Conversion—The Math You Need

You can learn more about unit conversion (including practice problems) at this site featuring The Math You Need, When You Need It math tutorials for students in introductory geoscience courses: **http://serc.carleton.edu/mathyouneed/units/index.html**

Area and Volume

An **area** is a two-dimensional space, such as the surface of a table. The long dimension is the *length*, and the short dimension is the *width*. If the area is square or rectangular, then the size of the area is the product of its length multiplied times its width. For example, the blue rectangular area in **FIGURE 1.8A** is 7.3 cm long and 3.8 cm wide. So the size of the area is 7.3 cm $\times$ 3.8 cm, which equals 27.7 cm^2. This is called 27.7 square centimeters. Using this same method, the yellow front of the box in **FIGURE 1.8B** has an area of 9.0 cm $\times$ 4.0 cm, which equals 36.0 cm^2. The green side of this same box has an area of 4.0 cm $\times$ 4.0 cm, which equals 16.0 cm^2.

Three-dimensional objects are said to occupy a **volume** of space. Box shaped objects have *linear volume* because they take up three linear dimensions of space: their length (longest dimension), width (or depth), and height (or thickness). So the volume of a box shaped object is the product of its length, width, and height. For example, the box in **FIGURE 1.8B** has a length of 9.0 cm, a width of 4.0 cm, and a height of 4.0 cm. Its volume is 9.0 cm $\times$ 4.0 cm $\times$ 4.0 cm, which equals 144 cm^3. This is read as "144 cubic centimeters."

Most natural materials such as rocks do not have linear dimensions, so their volumes cannot be calculated from linear measurements. However, the volumes of these odd-shaped materials can be determined by measuring the volume of water they displace. This is often done in the laboratory with a *graduated cylinder* (**FIGURE 1.8C**), an instrument used to measure volumes of fluid (fluid volume). Most graduated cylinders are graduated in metric units called milliliters (mL or ml), which are thousandths of a liter. *You should also note that 1 mL of fluid volume is exactly the same as 1 cm^3 of linear volume.*

When you pour water into a graduated cylinder, the surface of the liquid is usually a curved *meniscus*, and the volume is read at the bottom of the curve (**FIGURE 1.8C**: middle and left-hand examples). In some plastic graduated cylinders, however, there is no meniscus. The water level is flat (**FIGURE 1.8C**: right-hand example).

If you drop a rock into a graduated cylinder full of water, then it takes up space previously occupied by water at the bottom of the graduated cylinder. This displaced water has nowhere to go except higher into the graduated cylinder. Therefore, the volume of an object such as a rock is exactly the same as the volume of fluid (water) that it displaces.

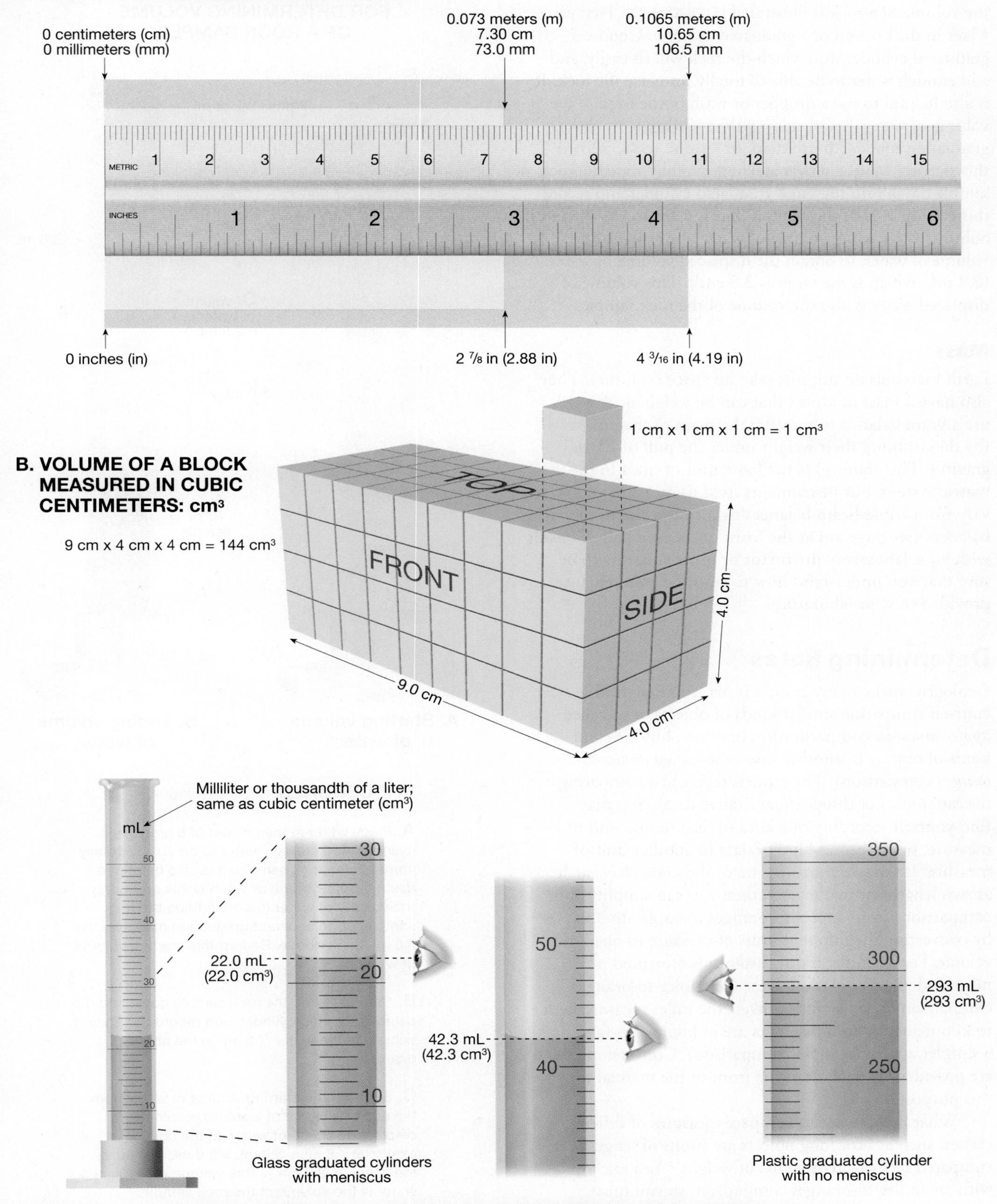

FIGURE 1.8 Tools and scales of measurement. A. Linear measurement using a ruler. B. Linear volume measured in cubic centimeters. C. Fluid volume measured with graduated cylinder (at base of meniscus). A milliliter (mL or ml) is the same as a cubic centimeter (cm^3).

The water displacement procedure for determining the volume of a rock is illustrated in **FIGURE 1.9**. First place water in the bottom of a graduated cylinder. Choose a graduated cylinder into which the rock will fit easily, and add enough water to be able to totally immerse the rock. It is also helpful to use a dropper or wash bottle to raise the volume of water (before adding the rock) up to an exact graduation mark (5.0 mL mark in **FIGURE 1.9A**). Record this starting volume of water. Then carefully slide the rock sample down into the same graduated cylinder and record this ending level of the water (7.8 mL mark in **FIGURE 1.9B**). Subtract the starting volume of water from the ending volume of water, to obtain the displaced volume of water (2.8 mL, which is the same as 2.8 cm^3). This volume of displaced water is also the volume of the rock sample.

Mass

Earth materials do not just take up space (volume). They also have a mass of atoms that can be weighed. You will use a gram balance to measure the **mass** of materials (by determining their weight under the pull of Earth's gravity). The gram (g) is the basic unit of mass in the metric system, but instruments used to measure grams vary from triple-beam balances to spring scales to digital balances (see page xiii at the front of the manual). Consult with your laboratory instructor or other students to be sure that you understand how to read the gram balance provided in your laboratory.

Determining Rates

Geologists make many comparisons. You may find yourself comparing similar kinds of objects (a so-called *apples-to-apples* comparison) in one case, but different kinds of objects in another case (a so-called *apples-to-oranges* comparison). The same is true when comparing measurements of things (quantitative data). You may find yourself recording one kind of data in one unit of measure, but a second kind of data in another unit of measure. If the two measures are of the same class, such as two lengths or two masses, then you can simplify your comparison (from apples-to-oranges to apples-to-apples) by converting the different units of measure to one kind of unit. For example, if one distance is measured in miles and another in kilometers (an apples-to-oranges comparison), then simply convert the miles measurement to kilometers so both distances are in kilometers (and make a simpler apples-to-apples comparison). Conversion tables are provided on page xii at the front of the manual for this purpose.

What if you want to compare measures of different classes, such as how long objects are (units of length) compared to their mass (units of weight)? You are "stuck" with an apples-to-oranges comparison, so you must determine a rate. **Rate** is a mathematical expression of how much an amount determined in one unit of measure varies "per" (divided by) an amount determined in a different unit of measure.

WATER DISPLACEMENT METHOD FOR DETERMINING VOLUME OF A ROCK SAMPLE

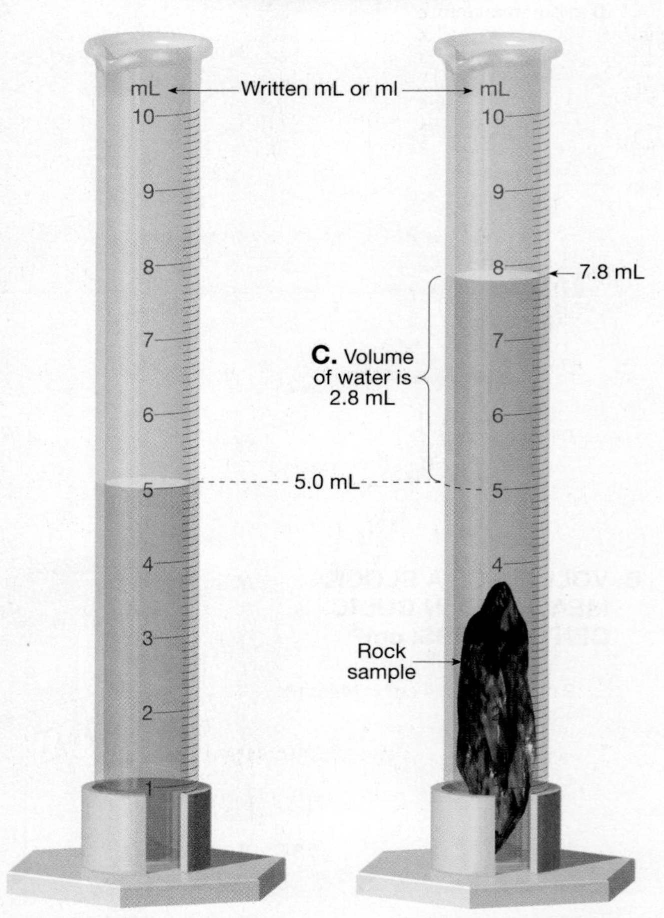

PROCEDURES

A. Place water in the bottom of a graduated cylinder. Add enough water to be able to totally immerse the rock sample. It is also helpful to use a dropper bottle or wash bottle and bring the volume of water (before adding the rock sample) up to an exact graduation mark like the 5.0 mL mark above. Record this starting volume of water.

B. Carefully slide the rock sample down into the same graduated cylinder, and record the ending volume of the water (7.8 mL in the above example).

C. Subtract the starting volume of water from the ending volume of water to obtain the displaced volume of water. In the above example: 7.8 mL – 5.0 mL = 2.8 mL (2.8 mL is the same as 2.8 cm^3). This volume of displaced water is the volume of the rock sample.

FIGURE 1.9 Procedure for determining volume of a rock sample by water displacement.

Rates Involving Time

The most common kind of rate is an expression of how something changes per unit of time, such as miles per hour. For example, if you drive 100 miles in 2 hours, then your rate of travel is 100 miles ÷ 2 hours, or 50 miles per hour (50 mi/h or 50 MPH).

- Knowing the rate of change per unit of time enables you to predict (calculate) how much change will occur by a future time. For example, if your rate of travel is 50 mi/h, then how far will you travel in four hours?

$$\frac{50 \text{ mi}}{1 \text{ h}} \times \frac{4 \text{ h}}{1} = 200 \text{ mi}$$

- Knowing the rate of change per unit of time also enables you to predict (calculate) how long it will take for a given amount of change to occur. If your rate of travel is 50 mi/h, then how far will it take you to travel 400 miles?

$$\frac{400 \text{ mi}}{50 \text{ mi/h}} = 8 \text{ h}$$

Calculating Rates—The Math You Need

You can learn more about calculating rates (including practice problems) at this site featuring The Math You Need, When You Need It math tutorials for students in introductory geoscience courses: **http://serc.carleton.edu/mathyouneed/rates/index.html**

Density: Mass per Volume

Every material has a *mass* that can be weighed and a *volume* of space that it occupies. However, the relationship between a material's mass and volume tends to vary from one kind of material to another. For example, a bucket of rocks has much greater mass than an equal-sized bucket of air. Therefore a useful way to describe an object is to determine its mass per unit of volume, which is a rate called **density**. *Per* refers to division (as in miles *per* hour). So, density is a measure (rate) of an object's mass divided by its volume (density = mass ÷ volume). Scientists and mathematicians use the Greek character rho (ρ) to represent density. Also, the gram (g) is the basic metric unit of mass, and the cubic centimeter is the basic unit of metric volume (cm^3), so density (ρ) is usually expressed in grams per cubic centimeter (g/cm^3).

Calculating Density—The Math You Need

You can learn more about calculating density (including practice problems) at this site featuring The Math You Need, When You Need It math tutorials for students in introductory geoscience courses: **http://serc.carleton.edu/mathyouneed/density/index.html**

Graphing Relationships

Graphs are useful for visualizing relationships in your data. So before you can make a graph, you need a set of data. Data makes more sense if you organize it into a chart. If you use Excel™, then your chart of data is called a spreadsheet. When organizing data, scientists normally have a column of data representing an independent variable. The independent variable is what you control, to see how it affects the dependent variable. The dependent variable is the variable being tested for, so you do not know what the values will be until you do an experiment. As you change the independent variable, you do a test or experiment to observe and record the dependent variable data. For example, if you want to characterize how fast a plant grows, then you may decide to measure its height every week (time here being the independent variable). When you measure and record the plant's height every week, then you are recording the dependent variable data. Geologists often want to characterize how something changes over time, so time is usually the independent variable. If you are placing your data in an Excel™ spreadsheet, then the independent variable data are entered in the first column (left-hand column, column A). The dependent variable data are entered in column B. Once you have an organized chart of data, then you can use it to construct a graph.

X-Y graphs are graphs with two axes, a horizontal X-axis and a vertical Y-axis. Scientists universally plot the independent variable along the X-axis and the dependent variable on the Y-axis. In Excel™, X-Y graphs are called scatter graphs or line graphs.

- **Scatter graphs** are X-Y graphs on which points are plotted (paced on the graph) but not joined into a line. Picture holes in a dart board. They are like the points on a scatter plot. By analyzing relationships among the points you can determine if they are widely scattered (a weak relationship to one another or no relationship at all) or closely concentrated (a strong relationship to one another). You can also look for patterns in the graph such as whether or not the points form one concentration or two or three concentrations of related points.

- **Line graphs** are X-Y graphs on which points are plotted and joined to form a line. If the points form a line that runs from lower left to upper right, then there is what is called a *positive* or *direct relationship*. This means that as the values of X (independent variable) increase, so do the values of Y. If the points form a line that runs from upper left to lower right, then there is what is called a *negative* or *inverse relationship*. This means that as the values of X (independent variable) increase, so do the values of Y. In both cases just described, the closer the points are to the line, the stronger is the relationship. Also, some line graphs compare two kinds of dependent variables to the independent variable. For example, you may want to know how plant height and number of leaves vary over the same time intervals. By plotting the data (two sets of dependent variable data) against the time axis (independent variable), you get two lines. So this kind of graph is called a *two-line graph*.

- **Bar graphs or histograms** use the length/height of a bar or column to show how frequently a measurement occurs. The bars are labeled according to the *class interval* of measurements that you choose (independent variable). The length/height of the bars is their *frequency*, how many times (how frequently) data values occur in each class interval. In Excel™, histograms are called bar graphs if the bars are horizontal and column graphs if the bars are vertical.

Graphing—The Math You Need

You can learn more about graphing and how to use graphs in the geosciences at this site featuring The Math You Need, When You Need It math tutorials for students in introductory geoscience courses: **http://serc.carleton.edu/mathyouneed/graphing/index.html**

ACTIVITY 1.5 Density, Gravity, and Isostasy

THINK About It | **How is the distribution of Earth materials related to their density?**

OBJECTIVE Develop and test models of isostasy, measure rock densities, and calculate the isostasy of oceanic and continental crust.

PROCEDURES

1. **Before you begin**, read the following background information on density, gravity, and isostasy, and equations.This is **what you will need**:
   - ____ Activity 1.5 Worksheet (p. 35) and pencil with eraser
   - ____ calculator, ruler
   - ____ other materials provided in the lab: wood blocks (oak, pine), basin of water, gram balance or scale
2. **Then follow your instructor's directions** for completing the worksheets.

Density, Gravity, and Isostasy

Scientists have wondered for centuries about how the distribution of Earth materials is related to their density and gravity. Curious about buoyancy, the Greek scientist and mathematician, Archimedes, experimented with floating objects around 225 B.C. When he placed a block of wood in a bucket of water, he noticed that the block floated and the water level rose (**FIGURE 1.10A**). When he pushed down on the wood block, the water level rose even more. And when he removed his fingers from the wood block, the water pushed it back up to its original level of floating. Archimedes eventually realized that every floating object is pulled down (toward Earth's center) by gravity, so the object displaces fluid and causes the fluid level to rise. However, Archimedes also realized that every floating object is also pushed upward by a buoyant force that is equal to the weight of the displaced fluid. This is now called Archimedes' Principle.

Buoyant force (buoyancy) is caused as gravity pulls on the mass of a fluid, causing it to exert a *fluid pressure* on submerged objects that increases steadily with increasing depth in the fluid. The deeper (greater amount of) a fluid, the more it weighs, so deep water exerts greater fluid pressure than shallow fluid. Therefore, the lowest surfaces of a submerged object are squeezed more (by the fluid pressure) than the upper surfaces. This creates the wedge of buoyant force that pushes the object upward and opposes the downward pull of gravity (white arrows in **FIGURE 1.10B**). An object will sink if it is heavier than the fluid it displaces (is denser than the fluid it displaces). An object will rise if it is lighter than the fluid it displaces (is less dense than the fluid it displaces). But a floating object is balanced between sinking and rising. The object sinks until it displaces a volume of fluid that has the same mass as the entire floating object. When the object achieves a motionless floating condition, it is balanced between the downward pull of gravity and the upward push of the buoyant force.

Isostasy

In the 1880s, geologists began to realize the abundant evidence that levels of shoreline along lakes and oceans had changed often throughout geologic time in all parts of the world. Geologists like Edward Suess hypothesized that changes in sea level can occur if *the volume of ocean water changes* in response to climate. Global atmospheric warming leads to sea level rise caused by melting of glaciers (cryosphere), and global atmospheric cooling leads to a drop in sea level as more of Earth's hydrosphere gets stored in thicker glaciers. However, an American geologist named Clarence Dutton suggested that shorelines can also change *if the level of the land changes* (and the volume of water remains the same).

Dutton reasoned that if blocks of Earth's crust are supported by the mantle beneath them (solid rock capable of a slow flow) then they must float in the mantle according to Archimedes' Principle (like wood blocks, icebergs, and boats floating in water). Therefore, he proposed that Earth's crust consists of buoyant blocks of rock that float in gravitational balance in the top of the mantle. He called this floating condition **isostasy** (Greek for "equal standing"). Loading a crustal block (by adding lava flows, sediments, glaciers, water, etc.) will decrease its buoyancy, and the block will sink (like pushing down on a

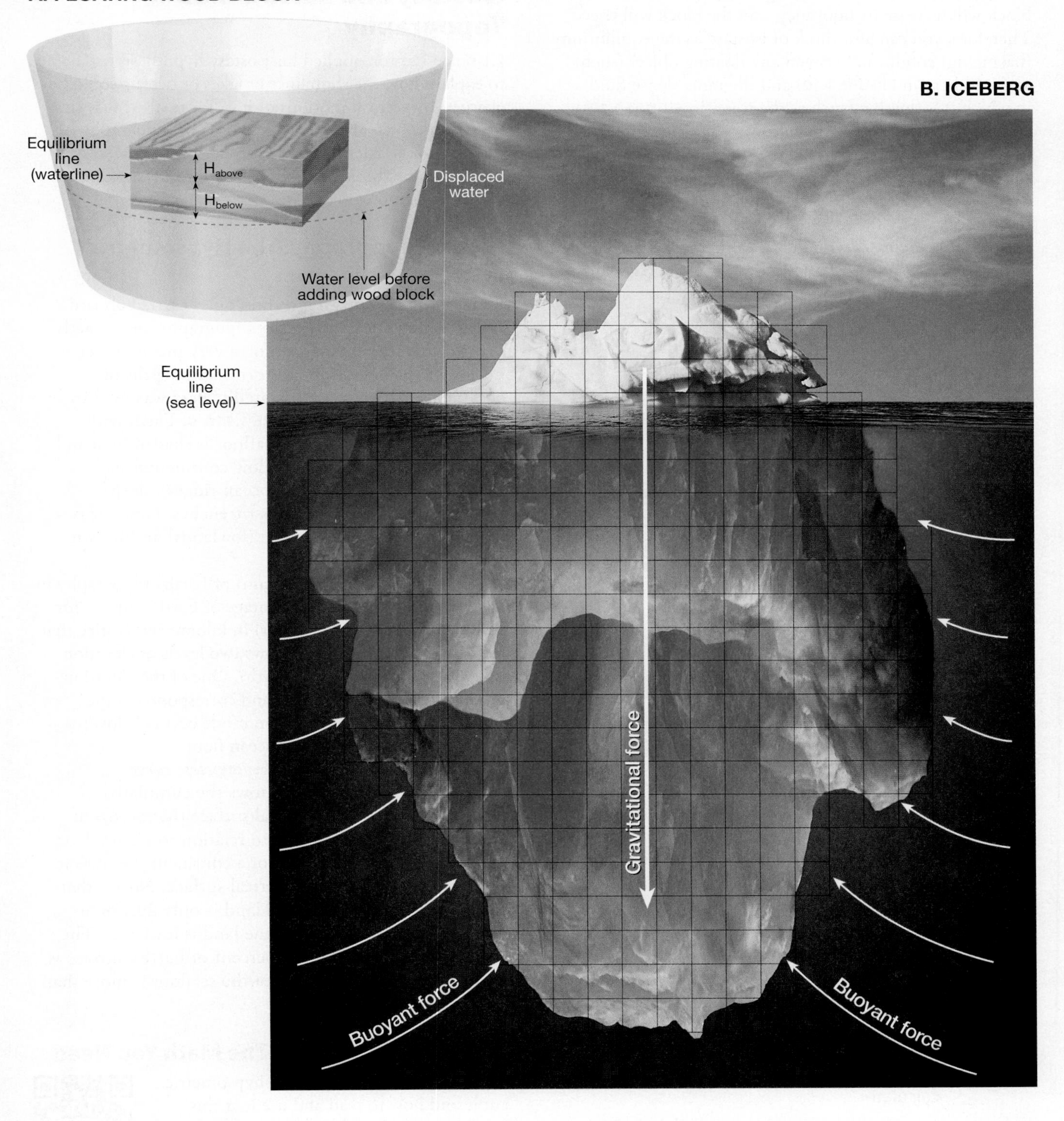

FIGURE 1.10 Isostasy relationships of a floating wood block (A) and iceberg (B). Refer to text for discussion.
(Iceberg image © Ralph A. Clavenger/CORBIS. All rights reserved)

floating wood block). Unloading materials from a crustal block will increase its buoyancy, and the block will rise. Therefore, you can also think of isostasy as the equilibrium (balancing) condition between any floating object (such as the iceberg in **FIGURE 1.10**) and the more dense fluid in which it is floating (such as the water in **FIGURE 1.10**). Gravity pulls the iceberg down toward Earth's center (this is called *gravitational force*), so the submerged root of the iceberg displaces water. At the same time, gravity also tries to pull the displaced water back into its original place (now occupied by the iceberg's root). This creates fluid pressure that increases with depth along the iceberg's root, so the iceberg is squeezed and wedged (pushed) upward. This squeezing and upward-pushing force is called *buoyant force*. **Isostatic equilibrium** (balanced floating) occurs when the buoyant force equals (is in equilibrium with) the gravitational force that opposes it. An **equilibrium line** (like the waterline on a boat) separates the iceberg's submerged root from its exposed top.

Equations—The Math You Need

Activity 1.6 involves writing and rearranging equations. You can learn more about equations (including practice isostasy problems) at this site featuring The Math You Need, When You Need It math tutorials for students in introductory geoscience courses: **http://serc.carleton.edu/mathyouneed/equations/ManEqSP.html**

ACTIVITY 1.6 Isostasy and Earth's Global Topography

THINK About It How is the distribution of Earth materials related to their density?

OBJECTIVE Analyze Earth's global topography and infer how the presence of continents and oceans it may be related to isostasy.

PROCEDURES

1. **Before you begin**, read the following background information on isostasy and Earth's global topography. This is **what you will need**:
   - ____ Activity 1.6 Worksheets (pp. 36–38) and pencil with eraser
   - ____ calculator
   - ____ other materials provided in the lab: 500 mL or 1000 mL graduated cylinder, small samples (about 30–50 g) of basalt and granite that fit into the graduated cylinder, a gram balance or scale, and water
2. **Then follow your instructor's directions** for completing the worksheets.

Isostasy and Earth's Global Topography

Clarence Dutton applied his isostasy hypothesis in 1889 to explain how the shorelines of lakes or oceans could be elevated by vertical motions of Earth's crust. At that time, little was known about Earth's mantle or topography of the seafloor. Modern data show that Dutton's isostasy hypothesis has broader application for understanding global topography.

Global Topography: The Hypsometric Curve

Radar and laser imaging technologies carried aboard satellites now measure Earth's topography very exactly, and the data can be used to form very precise relief images of the height of landforms and depths of ocean basins. For example, satellite data was used to construct the image in **FIGURE 1.11A** of Earth with ocean water removed. The seafloor is shaded blue and includes features such as shallow continental shelves, submarine mountains (mid-ocean ridges), deep abyssal plains, and even deeper trenches. Land areas (continents) are shaded green (lowlands) and brown (mountains).

The histogram (bar diagram) of Earth's topography in **FIGURE 1.11B** shows the percentage of Earth's surface for each depth or height class (bar) in kilometers. Notice that the histogram is bimodal (shows two levels of elevation that are most common on Earth). One of the elevation modes occurs above sea level and corresponds to the continents. The other elevation mode occurs below sea level and corresponds to the ocean floor.

FIGURE 1.11C is called a *hypsometric curve* (or *hypsographic curve*) and shows the cumulative percentage of Earth's spherical surface that occurs at specific elevations or depths in relation to sea level. This curve is not the profile of a continent, because it represents Earth's entire spherical surface. Notice that the cumulative percentage of land is only 29.2% of Earth's surface, and most of the land is lowlands. The remaining 70.8 cumulative percent of Earth's surface is covered by ocean, and most of the seafloor is more than 3 km deep.

Hypsometric Curve—The Math You Need

You can learn more about the hypsometric curve and how to read and use it at this site featuring The Math You Need, When You Need It math tutorials for students in introductory geoscience courses: **http://serc.carleton.edu/mathyouneed/hypsometric/index.html**

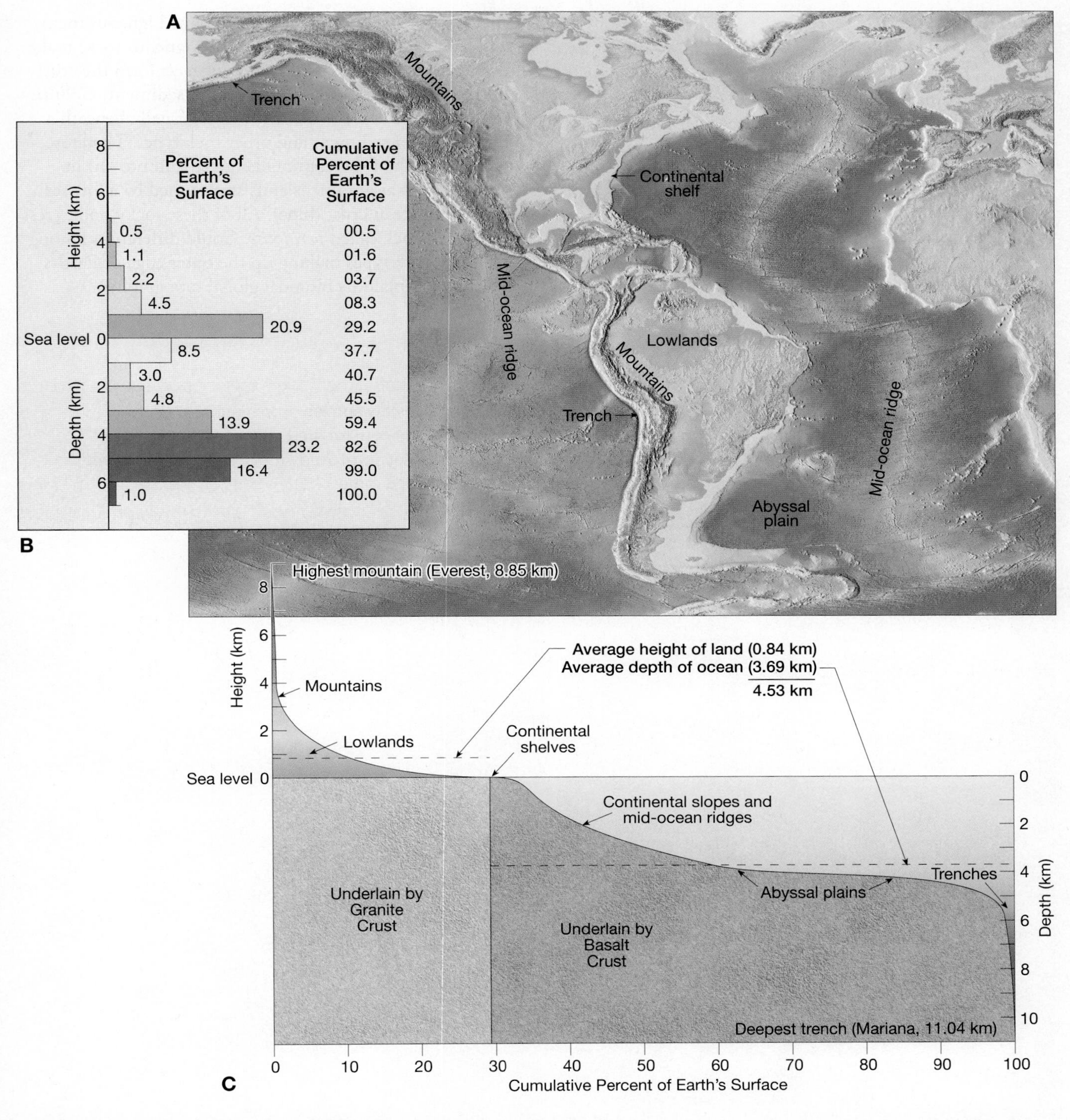

FIGURE 1.11 Global topography of Earth. A. Portion of Earth with ocean water removed, based on satellite-based radar and laser technologies. B. Histogram of global topography. C. Hypsometric curve (or hypsographic curve) of Earth's global topography. (Refer to text for discussion.)

Global Isostasy

The average elevation of the continents is about 0.84 km above sea level (+0.84 km), but the average elevation of the ocean basins is 3.87 km below sea level (−3.87 km). Therefore the difference between the average continental and ocean basin elevations is 4.71 km! If the continents did not sit so much higher than the floor of the ocean basins, then Earth would have no dry land and there would be no humans. What could account for this elevation difference? One clue may be the difference between crustal granite and basalt in relation to mantle peridotite.

Granite (light-colored, coarse-grained igneous rock) and basalt (dark-colored, fine-grained igneous rock) make up nearly all of Earth's crust. *Basaltic rocks* form the crust of the oceans, beneath a thin veneer of sediment. *Granitic rocks* form the crust of the continents, usually beneath a thin veneer of sediment and other rock types. Therefore, you can think of the continents (green and brown) in **FIGURE 1.11A** as granitic islands surrounded by a low sea of basaltic ocean crust (blue). All of these rocky bodies rest on mantle rock called *peridotite*. Could differences among the three rock types making up the outer edge of Earth's geosphere explain its bimodal global topography?

ACTIVITY 1.1 Geologic Inquiry

Name: ____________________ **Course/Section:** ____________________ **Date:** ____________

A. Carefully observe and analyze each part of **FIGURE 1.1**. Describe below what you see in each part. Be sure to record qualitative data (e.g., names, colors, shapes, textures, relationships of what you see) and quantitative data (e.g., amounts, sizes) in your descriptions. Be prepared to share your descriptions (data) with others.

1.1A: Aster satellite images of Escondida mining region, Chile:
1.1B: Ground view of Escondida open pit mine:
1.1C: Boulders in Escondida open pit mine:
1.1D: Minerals of Escondida open pit mine:
1.2E: Coin:
1.1F: Circuit board:
1.1G: Copper atoms:

B. How is everything in **FIGURE 1.1** related?

C. According to the Minerals Education Coalition, every American born in 2012 will consume 978 pounds (80.7 kg) of copper in his/her lifetime (**www.mineralseducationcoalition.org**). Analyze the three parts of **FIGURE 1.1** listed below, and do your best to answer the questions based on what you observe.

1.1A: Aster satellite images of Escondida mining region, Chile. How could geologists use these images, at this scale of observation, to find new sources of copper ore?
1.1B: Ground view of Escondida open pit mine. How can geologists and miners locate copper ore when they view the mine at this scale of observation?
1.1D: Minerals of Escondida open pit mine. What must be done with these ore minerals to provide you with the copper you need?

D. REFLECT & DISCUSS Analyze **FIGURE 1.1A**, ASTER satellite images of Chile's Escondida Mine and vicinity. This is primarily a mine for copper ore minerals, from which copper is extracted. Some silver and gold are also extracted from the same ore. The ore is mined from large open pits. Notice how these pits appear in the images.

1. Imagine that you are a geologist who has been hired by Escondida Mine to find the best location for a new pit. Which location, A, B, or C, is probably the best site for a new pit? What evidence and critical thinking process leads you to this conclusion?

2. What plan of scientific investigation would you carry out to see if the location you chose above is actually a good source for more copper ore?

ACTIVITY 1.2 Spheres of Matter, Energy, and Change

Name: ______________________ **Course/Section:** ______________________ **Date:** ____________

A. Complete the table below.

State of Matter	Sphere	What is the main source of energy that powers the sphere (Sun or geothermal energy)?	Give examples of named parts of this sphere that you have personally encountered.
GAS: What is a gas?	What sphere is made mostly of gases?		
LIQUID: What is a liquid?	What sphere is made mostly of liquid water?		
	What subsphere of Earth in Figure 1.5 is a mostly liquid rock?	Geothermal energy	Not encountered by humans.
SOLID: What is a solid?	What subsphere is made mostly of water ice?		
	What sphere is made mostly of solid rock (besides water ice)?		
SOLIDS, LIQUIDS, AND GASES	What sphere consists of living parts containing solids, liquids, and gases?		

B. Study the processes of change in **FIGURE 1.6**, then complete the table below as done for deposition.

Process of Change	Sphere(s) involved in the process or product	Give an example of how you observed the process happening or how you encountered the result of the process	What caused the process to happen?
Deposition	atmosphere geosphere	I saw frost crystals on cold metal surfaces and windows of my car last winter.	The temperature of the metal was so cold that water vapor in the air formed ice crystals on contact with the metal.
	hydrosphere geosphere atmosphere	At the seashore, sand covered up my feet as waves crashed onto the beach where I was standing.	Wind caused waves. Waves carried the sand. When waves broke, they lost energy and the sand settled out.
Evaporation			
Condensation			
Decomposition reaction			
Dissolution			
Chemical precipitation			

C. Many of Earth's physical environments and ecosystems (communities of organisms and the physical environments in which they live) occur at the boundary between, or at intersection among, two or more spheres.

1. Add the following environments and ecosystems to the correct field of the Venn diagram below.

- Surface of a leaf
- Beach
- Phytoplankton floating at the surface of the ocean
- Surface of an ocean, lake, stream
- Moldy brick basement walls
- Surface of a glacier
- Soil
- Lava flowing over a forest
- Bottom of ocean, lake, stream
- Seafloor rock with attached oysters

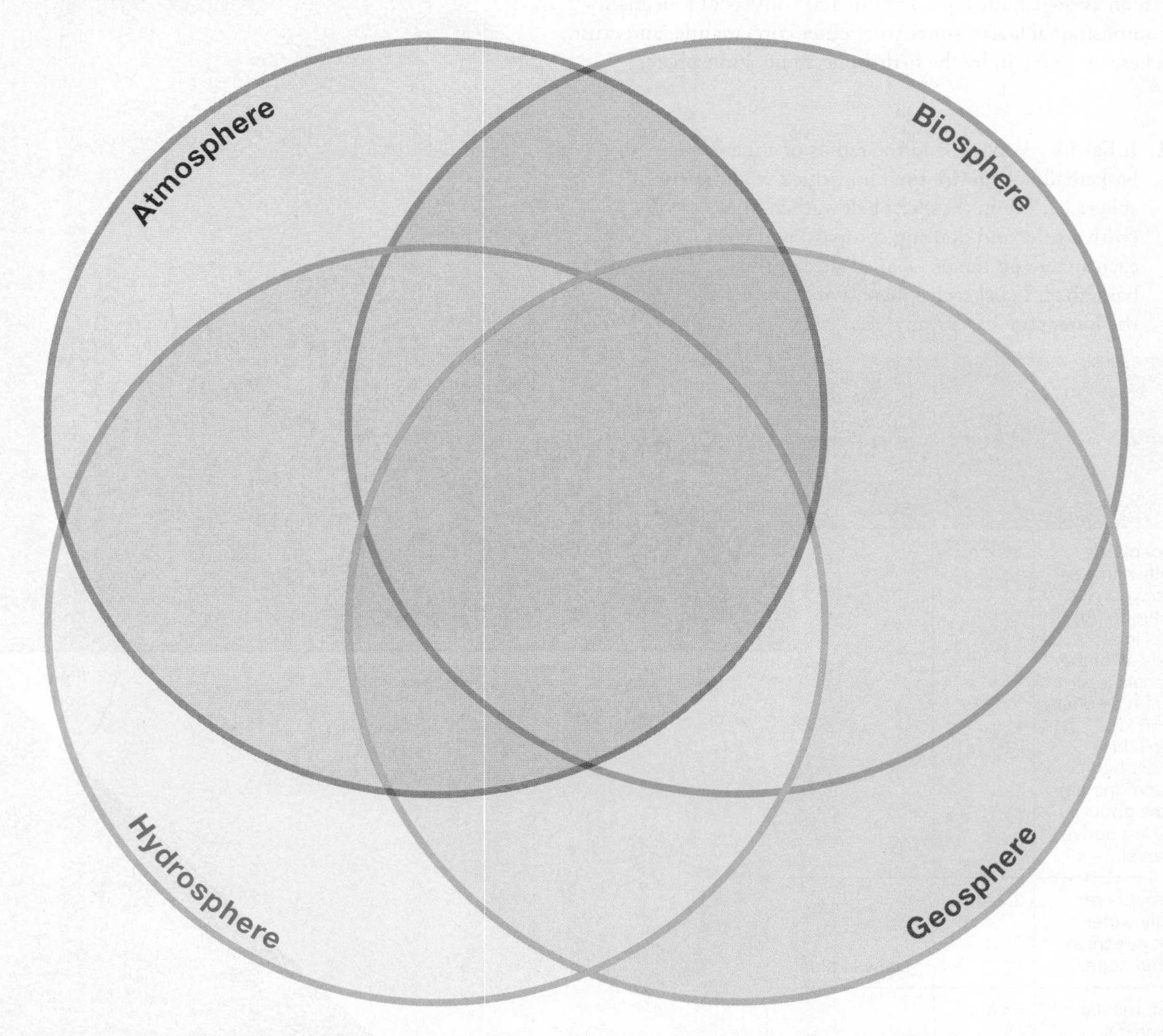

2. In the four sphere fields above (labeled atmosphere, hydrosphere, geosphere, and biosphere) add an "S" for solar energy and a "G" for geothermal energy to indicate which kind of energy *primarily* powers it. Where two of the fields overlap, write an "SS" or "SG" to indicate the sum of energies that power the field. Use the same convention with three letters to indicate the sum of energy sources where three fields overlap.

E. **REFLECT & DISCUSS** Do you think that most change on Earth occurs within individual systems, at boundaries between two systems, or at the intersections of more than two systems? Why?

ACTIVITY 1.3 Modeling Earth Materials and Processes

Name: ____________________ **Course/Section:** ____________________ **Date:** ________

A. Geoscientists conceptualize Earth as a dynamic system of interacting material spheres (subsystems). The rocky body of Earth (geosphere) has an average radius of 6371 km and consists of four main compositional layers: inner core, outer core, mantle, and crust. These are overlain by the hydrosphere and atmosphere.

1. If Earth's geosphere had the radius of a men's basketball (119 mm), then how thick would each sphere be? Fill in the chart below, then draw (with a ruler and drafting compass) and label each sphere on the pie-shaped slice of this basketball. Label each sphere. For example, the inner core has already been done.

SPHERE	ACTUAL THICK-NESS	THICKNESS IN MM, IF THE GEOSPHERE IS THE SIZE OF A BASKET-BALL
Atmosphere: mostly nitrogen (N), oxygen (O), and argon (Ar) gases in air. Nearly all of the materials in air occur in a sphere just 16 km (10 mi) thick (troposphere). "Space" (no air) begins about 1000 km above sea level.	16 km	
Hydrosphere: mostly water (H_2O, ocean) in a liquid state.	3.7 km	Draw in blue!
Crust: mostly oxygen (O), silicon (Si), aluminum (Al), and iron (Fe).	25 km	
Mantle: mostly oxygen (O), silicon (Si), magnesium (Mg), and iron (Fe) in a solid state.	2900 km	
Outer Core: mostly iron (Fe) and nickel (Ni) in a liquid state.	2250 km	
Inner Core: mostly iron (Fe) in a solid state	1196 km	22.3 mm

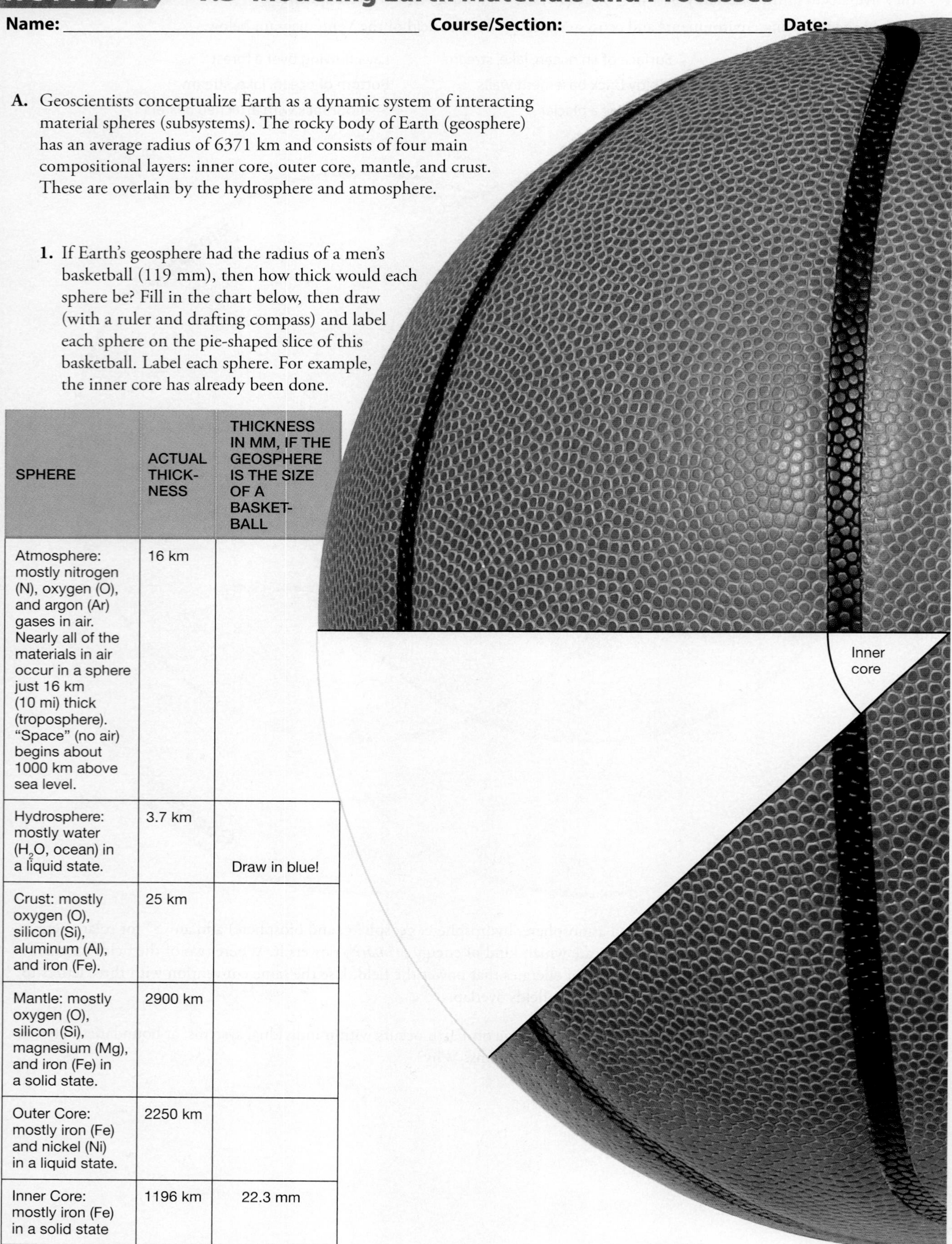

2. Recall that Earth's actual average radius is 6371 km (6,371,000 meters) and that the radius of the basketball is only 119 mm (0.119 meters). Calculate the fractional scale (show your work) and ratio scale of the basketball model.

Fractional scale: **Ratio scale:**

B. MODELING LANDSLIDE HAZARDS

1. Place a ruler flat on a table in front of you. Place a coin in the center of the ruler. What happens if you lift one end of the ruler?

2. The coin did not slide off of the ruler at the very second you started to lift one end of the ruler. Why?

3. Why did the coin start sliding when it did?

4. **REFLECT & DISCUSS** Landslides are sudden downslope movements of rock, soil, and mud that occur in every country of the world. The U.S. Geological Survey warns that landslides are a major geologic hazard because they happen in all 50 states and U.S. territories and cause \$1–2 billion in damages and more than 25 fatalities in the U.S. on average each year. Describe below how you would modify your ruler and coin model and use it to study what factors can trigger a landslide?

ACTIVITY 1.4 Measuring and Determining Relationships

Name: ______________________ **Course/Section:** ______________________ **Date:** ____________

A. Make the following unit conversions using the Mathematical Conversions chart on page xii.

1. 10 mi = ______________ km **3.** 16 km = ______________ m **5.** 25.4 mL = ______________ cm^3

2. 1 ft = ______________ m **4.** 25 m = ______________ cm **6.** 1.3 liters = ______________ cm^3

B. Write these numbers using scientific notation

1. 6,555,000,000 = ______________ **2.** 0.000001234 = ______________

C. Using a ruler, draw a line segment that has a length of exactly 1 cm (1 centimeter). A line occupies only one dimension of space, so a line that is 1 cm long is 1 cm^1.

D. Using a ruler, draw a square area that has a length of exactly 1 cm and a width of exactly 1 cm. An area occupies two dimensions of space, so a square that is 1 cm long and 1 cm wide is 1 cm^2 of area ($1 \text{ cm} \times 1 \text{ cm} = 1 \text{ cm}^2$).

E. Using a ruler, draw a cube that has a length of 1 cm, width of 1 cm, and height of 1 cm. This cube made of centimeters occupies three dimensions of space, so it is 1 cm^3 (1 cubic centimeter) of volume.

F. Explain how you could use a small graduated cylinder and a gram balance to determine the density of water (ρ_{water}) in g/cm^3. Then use your procedures to calculate the density of water as exactly as you can. Show your data and calculations.

G. Obtain a small lump of clay (grease-based modeling clay) and determine its density (ρ_{clay}) in g/cm^3. There is more than one way to do this, so develop and apply a procedure that makes the most sense to you. Explain the procedure that you use, show your data, and show your calculations.

H. Reconsider your answers to items **F** and **G** and the fact that modeling clay sinks in water.

1. Why does modeling clay sink in water?

2. What could you do to a lump of modeling clay to get it to float in water? Try your hypothesis and experiment until you get the clay to float.

I. **REFLECT & DISCUSS** How is the distribution of Earth's spheres related to their relative densities? Why?

J. RATES:

1. Some geologists infer that Grand Canyon in Arizona is about 6 million years old. Its greatest depth is 1.6 km.

   a. At what rate (in mm/year) is the canyon being cut into the geosphere? Show your work, and give your final answer in scientific notation

   b. Based on the rate of canyon cutting that you just calculated above, how many millimeters has the Grand Canyon deepened during your lifetime? Show your work.

2. The geosphere is energized mostly by geothermal heat (heat originating in Earth's core). Therefore, it is much hotter deep inside Earth than it is near the surface. The **geothermal gradient** is the rate of temperature increase with depth from Earth's surface. The deepest mine on Earth is located on the African continent (Carletonville, South Africa: **FIGURE 1.5**). It is 3.9 km deep, and rocks at that depth are 60°C. Assuming that rocks at the surface of the mine are 0°C, what is the geothermal gradient at the mine? Show your work.

K. SINGLE-LINE GRAPH:
The amount of CO_2 in the atmosphere has been monitored at Mauna Loa Observatory, Hawaii since 1959. Below is a chart of NOAA (U.S. National Oceanic and Atmospheric Administration) data from the observatory, showing how the concentration of CO2 in ppmv (parts per million volume) has changed per decade since 1962. Plot the data onto the graph as neatly and perfectly as possible, then draw a best fit line through the points.

Annual Average Concentration of Atmospheric Carbon Dioxide at Mauna Loa, Hawaii	
Year	CO_2 (ppmv)
1962	318
1972	327
1982	341
1992	356
2002	373
2012	393

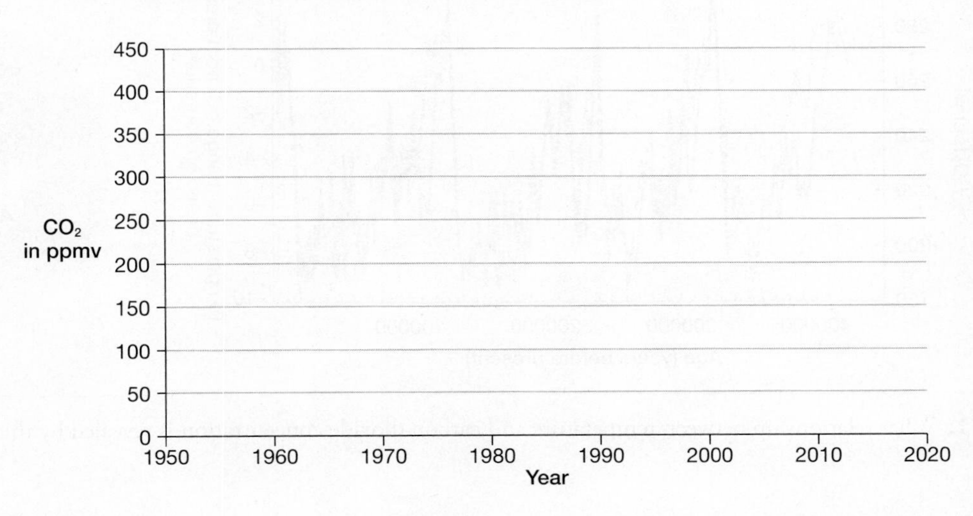

1. What does the graph show?

2. What are two ways you can tell from the graph that the concentration of CO_2 has increased since 1962?

L. BAR GRAPH: Using the data from part L, calculate the average rate of increase in CO_2 concentration per year for the time intervals 1962–1972, 1972–1982, 1982–1992, 1992–2002, and 2002–2012. Then plot the results as a bar graph.

Average Yearly Rate of Increase in the Concentration of Atmospheric Carbon Dioxide at Mauna Loa, Hawaii

Time interval	Rate of increase per year
1962–1972	
1972–1982	
1982–1992	
1992–2002	
2002–2012	

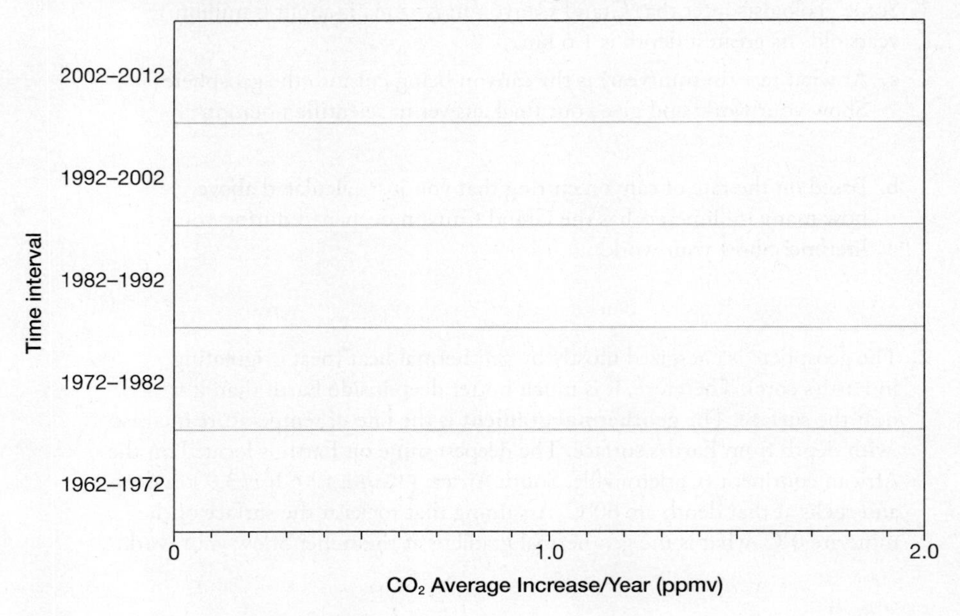

M. TWO-LINE GRAPH: Two-line graph of data obtained by analysis of a core of ice from Vostok Station, Antarctica (from NOAA: U.S. National Oceanic and Atmospheric Administration). Blue line shows how temperature at Vostok has changed in degrees Celsius from present (0 is the present temperature). Red line shows how the concentration of carbon dioxide (in ppmv: parts per million volume) has changed at Vostok over the same interval of time.

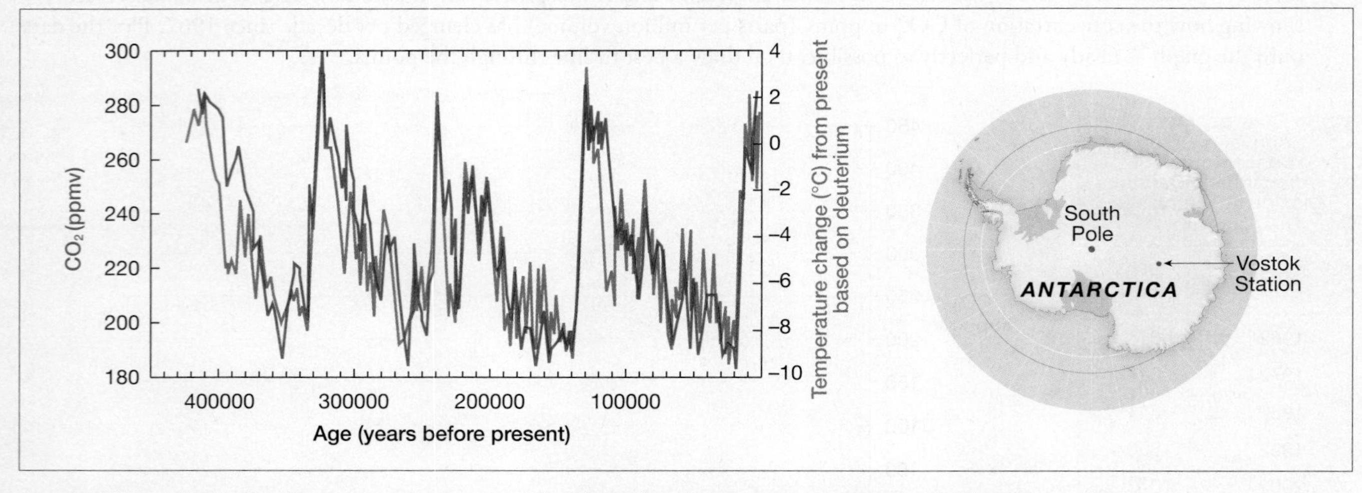

1. What relationship between temperature and carbon dioxide concentration is revealed by this graph?

N. REFLECT & DISCUSS What do you predict will happen to Earth's atmospheric temperature in the future? How do the graphs above (Parts K, L, and M) help you to answer this question?

ACTIVITY 1.5 Density, Gravity, and Isostasy

Name: ______________________ **Course/Section:** ______________________ **Date:** ____________

A. Obtain one of the wood blocks provided at your table. Determine the density of the wood block (ρ_{wood}) in g/cm^3. Show your calculations.

B. Float the same wood block in a bowl of water (like **FIGURE 1.10A**) and mark the equilibrium line (waterline).

1. Measure and record H_{block} (total height of the wood block) in cm: ____________ cm
2. Measure and record H_{below} (height of the wood block that is submerged below the water line) in cm: ____________ cm
3. Measure and record H_{above} (height of the wood block that is above the water line) in cm: ____________ cm

C. Write an isostasy equation (mathematical model) that expresses how the density of the wood block (ρ_{wood}) compared to the density of the water (ρ_{water}) is related to the height of the wood block that floats *below* the equilibrium line (H_{below}). [*Hint:* Recall that the wood block achieves isostatic equilibrium (motionless balanced floating) when it displaces a volume of water that has the same mass as the entire wood block. For example, if the wood block is 80% as dense as the water, then only 80% of the wood block will be below the equilibrium line (water line). Therefore, the portion of the wood block's height that is below the equilibrium line (H_{below}) is equal to the total height of the wood block (H_{block}) times the ratio of the density of the wood block (ρ_{wood}) to the density of water (ρ_{water}).

D. Change your answer in Part C to an equation (mathematical model) that expresses how the density of the wood block (ρ_{wood}) compared to the density of the water (ρ_{water}) is related to the height of the wood block that floats *above* the equilibrium line (H_{above}).

E. The density of water ice (in icebergs) is 0.917 g/cm^3. The average density of (salty) ocean water is 1.025 g/cm^3.

1. Use your isostasy equation for (H_{below}) (Question C) to calculate how much of an iceberg is submerged below sea level. Show your work.
2. Use your isostasy equation for (H_{above}) (Question D) to calculate how much of an iceberg is exposed above sea level. Show your work.
3. Notice the graph paper grid overlay on the picture of an iceberg in **FIGURE 1.10B**. Use this grid to determine and record the cross-sectional area of this iceberg that is below sea level and the cross-sectional area that is above sea level (by adding together all of the whole boxes and fractions of boxes that overlay the root of the iceberg or the exposed top of the iceberg). Use this data to calculate what proportion of the iceberg is below sea level (the equilibrium line) and what proportion is above sea level. How do your results compare to your calculations in Questions E1 and E2?
4. What will happen as the top of the iceberg melts?

F. **REFLECT & DISCUSS** Clarence Dutton proposed his isostasy hypothesis to explain how some ancient shorelines have been elevated to where they now occur on the slopes of adjacent mountains. Use *your* understanding of isostasy and icebergs to explain how this may happen.

ACTIVITY 1.6 Isostasy and Earth's Global Topography

Name: ______________________ **Course/Section:** ______________________ **Date:** ___________

A. *As exactly as you can,* weigh (grams) and determine the volume (by water displacement, **FIGURE 1.9**) of a sample of basalt. Add your data to the basalt density chart below. Calculate the density of your sample of basalt to tenths of a g/cm^3. Then determine the average density of basalt using all ten lines of sample data in the basalt density chart.

B. *As exactly as you can,* weigh (grams) and determine the volume (by water displacement, **FIGURE 1.9**) of a sample of granite. Add your data to the granite density chart below. Calculate the density of your sample of granite to tenths of a g/cm^3. Then determine the average density of granite using all ten lines of sample data in the granite density chart.

A. BASALT DENSITY CHART

Basalt Sample Number	Sample Weight (g)	Sample Volume (cm^3)	Sample Density (g/cm^3)
1	40.5	13	3.1
2	29.5	10	3.0
3	46.6	15	3.0
4	31.5	10	3.2
5	37.6	12	3.1
6	34.3	11	3.1
7	78.3	25	3.1
8	28.2	9	3.1
9	55.6	18	3.1
10			

Average density of basalt = ______________

B. GRANITE DENSITY CHART

Granite Sample Number	Sample Weight (g)	Sample Volume (cm^3)	Sample Density (g/cm^3)
1	32.1	12	2.7
2	27.8	10	2.8
3	27.6	10	2.8
4	31.1	11	2.8
5	58.6	20	2.9
6	62.1	22	2.8
7	28.8	10	2.9
8	82.8	30	2.8
9	52.2	20	2.6
10			

Average density of granite = ______________

C. Seismology (the study of Earth's structure and composition using earthquake waves), mantle xenoliths, and laboratory experiments indicate that the upper mantle is peridotite rock. The peridotite has an average density of about 3.3 g/cm^3 and is capable of slow flow. Seismology also reveals the thicknesses of crust and mantle layers.

1. Seismology indicates that the average thickness of basaltic ocean crust is about 5.0 km. Use the average density of basalt (from part A above) and your isostasy equation (Activity 1.5, item D) to calculate how high (in kilometers) basalt floats in the mantle. Show your work.

2. Seismology indicates that the average thickness of granitic continental crust is about 30.0 kilometers. Use the average density of granite (from part B above) and your isostasy equation (Activity 1.5, item D) to calculate how high (in kilometers) granite floats in the mantle. Show your work.

3. What is the difference (in km) between your answers in C1 and C2?

4. How does this difference between C1 and C2 compare to the actual difference between the average height of continents and average depth of oceans on the hypsographic curve (**FIGURE 1.11C**)?

D. REFLECT & DISCUSS Reflect on all of your work in this laboratory. Explain why Earth has a bimodal global topography.

E. REFLECT & DISCUSS How is a mountain like the iceberg in **FIGURE 1.10B**?

F. REFLECT & DISCUSS Clarence Dutton was not the first person to develop the concept of a floating crust in equilibrium balance with the mantle, which he called *isostasy* in 1889. Two other people proposed floating crust (isostasy) hypotheses in 1855 (See illustration below). John Pratt (a British physicist and Archdeacon of Calcutta) studied the Himalaya Mountains and hypothesized that floating blocks of Earth's crust have different densities, but they all sink to the same *compensation level* within the mantle. The continental blocks are higher because they are less dense. George Airy (a British astronomer and mathematician) hypothesized that floating blocks of Earth's crust have the same density but different thicknesses. The continental blocks are higher because they are thicker. Do you think that one of these two hypotheses (Pratt vs. Airy) is correct, or would you propose a compromise between them? Explain.

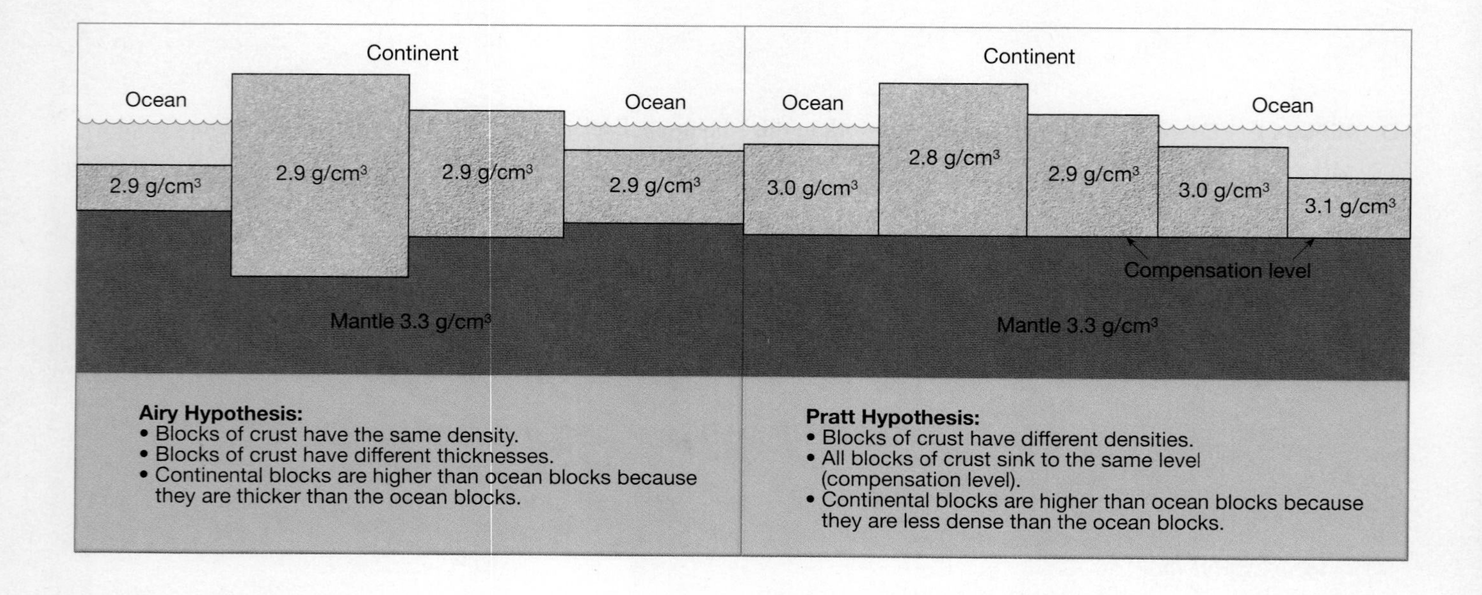

PRE-LAB VIDEO

LABORATORY 2

Plate Tectonics and the Origin of Magma

CONTRIBUTING AUTHORS
Edward A. Hay • *De Anza College*
Cherukupalli E. Nehru • *Brooklyn College (CUNY)*
C. Gil Wiswall • *West Chester University of Pennsylvania*

View of a rift valley, looking north across Thingvelir National Park, Iceland. Cliffs in the foreground (west side of rift) are part of the North American Plate. Hills in the background are across the rift valley, on the Eurasian Plate. The plates are diverging (moving apart). (Photo by Ragnar Sigurdsson/Arctic/Alamy)

BIG IDEAS

Tectonics is the study of global processes that create and deform lithosphere. Plate tectonics is the theory that Earth's lithosphere is broken into dozens of plates (thin curved pieces). The plates are created and destroyed, move about, and interact in ways that cause earthquakes and create major features of the continents and ocean basins (like volcanoes, mountain belts, ocean ridges, and trenches).

FOCUS YOUR INQUIRY

THINK About It | Is the lithosphere beneath your feet really moving?

ACTIVITY 2.1 Plate Motion Inquiry Using GPS Time Series *(p. 43)*

THINK About It | What causes plate tectonics?

ACTIVITY 2.2 Is Plate Tectonics Caused by a Change in Earth's Size? *(p. 45)*

ACTIVITY 2.3 Lava Lamp Model of Earth *(p. 46)*

THINK About It | How are plate boundaries identified? How and at what rates does plate tectonics affect Earth's surface?

ACTIVITY 2.4 Paleomagnetic Stripes and Seafloor Spreading *(p. 47)*

ACTIVITY 2.5 Atlantic Seafloor Spreading *(p. 49)*

ACTIVITY 2.6 Using Earthquakes to Identify Plate Boundaries *(p. 50)*

ACTIVITY 2.7 San Andreas Transform-Boundary Plate Motions *(p. 50)*

THINK About It | What are hot spots, and how do they help us explain plate tectonics?

ACTIVITY 2.8 Hot Spots and Plate Motions *(p. 50)*

THINK About It | How and where does magma form?

ACTIVITY 2.9 The Origin of Magma *(p. 51)*

Introduction

Looking at a world map, you may have noticed how edges of the continents seem like they could fit together like pieces of a jigsaw puzzle. You are not alone. Francis Bacon wrote about the fit of the continents in 1620, and geographer Antonio Snider-Pellegrini made maps in 1858 showing how the American and African continents may have fit together in Earth's past.

Continental Drift Hypothesis

In 1915, Alfred Wegener matched up all of the continents and published a **Continental Drift Hypothesis**—that all continents were once part of a single supercontinent (Pangea), parts of which drifted apart to form the smaller modern continents. A *hypothesis* is a tentative idea that must be tested repeatedly to verify or falsify its validity. Little did Wegener know, but his hypothesis would start a 50-year scientific investigation! Testing the hypothesis would require years of *verification* (finding data that supports the hypothesis), *falsification* (finding data that suggests the hypothesis must be false), and comparison with competing hypotheses. But the scientific process would yield a unifying view of Earth's rocky body (geosphere).

Shrinking Earth Hypothesis

When Wegener proposed that the continents had drifted apart, most scientists were skeptical. Wegener presented no evidence of a natural process that would force continents apart, and "anti-drift" scientists argued that it was impossible for continents to drift or plow through solid oceanic rocks. They favored a hypothesis of Earth involving stationary landforms that could rise and fall but not drift sideways. They also reasoned that Earth was cooling from an older semi-molten state, so it must be shrinking. The **Shrinking Earth Hypothesis** suggested that continents must be moving together, not drifting apart. If so, then Earth's crust must be shrinking into less space, and flat rock layers in ocean basins are being squeezed and folded between the continents (as observed in the Alps, Himalayan Mountains, and Appalachian Mountains).

Expanding Earth Hypothesis

Two other German scientists, Bernard Lindemann (in 1927) and Otto Hilgenberg (in 1933), independently evaluated the Continental Drift and Shrinking Earth Hypotheses. Both men agreed with Wegener's notion that the continents had split apart from a supercontinent (Pangea), but they proposed a new **Expanding Earth Hypothesis** (which they developed and published separately) to explain how it was possible. According to this hypothesis, Earth was once much smaller (about 60% of its modern size) and covered entirely by granitic crust. As Earth expanded, the granitic crust split apart into the shapes of the modern continents and basaltic ocean crust was exposed between them (and covered by ocean).

Seafloor Spreading Hypothesis

During the 1960s more data emerged in favor of Wegener's Continental Drift Hypothesis. For example, geologists found that it was not only the shapes (outlines) of the continents that matched up like pieces of a Pangea jig-saw puzzle. Similar bodies of rock and the patterns they make at Earth's surface also matched up like a picture on the puzzle pieces. Abundant studies also revealed that ocean basins were generally younger than the continents. An American geologist, Harry Hess, even developed a **Seafloor-Spreading Hypothesis** to explain this. He hypothesized that seafloor crust is created along mid-ocean ridges above regions of upwelling magma from Earth's mantle. As old seafloor crust moves from the elevated mid-ocean ridges to the trenches, new magma rises and fills fractures along the mid-ocean ridge. This creates new crust while old crust at the trenches begins descending back into the mantle.

Harry Hess' hypothesis was supported by studies showing that although Earth's rocky body (geosphere) has distinct compositional layers (inner core, outer core, mantle, and crust), it can also be divided into layers that have distinct physical behaviors. Two of these physical layers are the lithosphere and asthenosphere. The **lithosphere** (Greek *lithos* = rock) is a physical layer of rock that is composed of Earth's brittle crust and brittle uppermost mantle, called *lithospheric mantle.* It normally has a thickness of 70–150 km but has an average thickness of about 100 km. The lithosphere rests on the **asthenosphere** (Greek *asthenos* = weak), a physical layer of the mantle about 100–250 km thick that has plastic (ductile) behavior. It tends to flow rather than fracture. The lithosphere is broken into fragments called **lithospheric plates** (**FIGURE 2.1**), which rest and move upon the weak asthenosphere. Zones of abundant earthquake and volcanic activity are concentrated along the unstable boundaries (**plate boundaries**) between the lithospheric plates, and many of the plate boundaries are visible as linear features on Earth's surface. By the end of the 1960s, this new view of Earth had emerged, and it has become the unifying theory of geology. Unlike a hypothesis, which is only a tentative idea to be tested and evaluated, a *theory* is an idea that is widely accepted because it has been well tested, evaluated, and verified. Theories evolve from the testing and evaluation of one or more hypotheses.

Plate Tectonics Theory

According to the **Plate Tectonics Theory**, Earth's lithosphere is broken into dozens of plates (flat pieces) that move about and interact in ways that cause earthquakes and create major features of the continents and ocean basins (like volcanoes, mountain belts, ocean ridges, and trenches). *Plates are created and spread apart* along **divergent boundaries** such as mid-ocean ridges and continental rifts (**FIGURE 2.2**),

On which tectonic plate do you live?

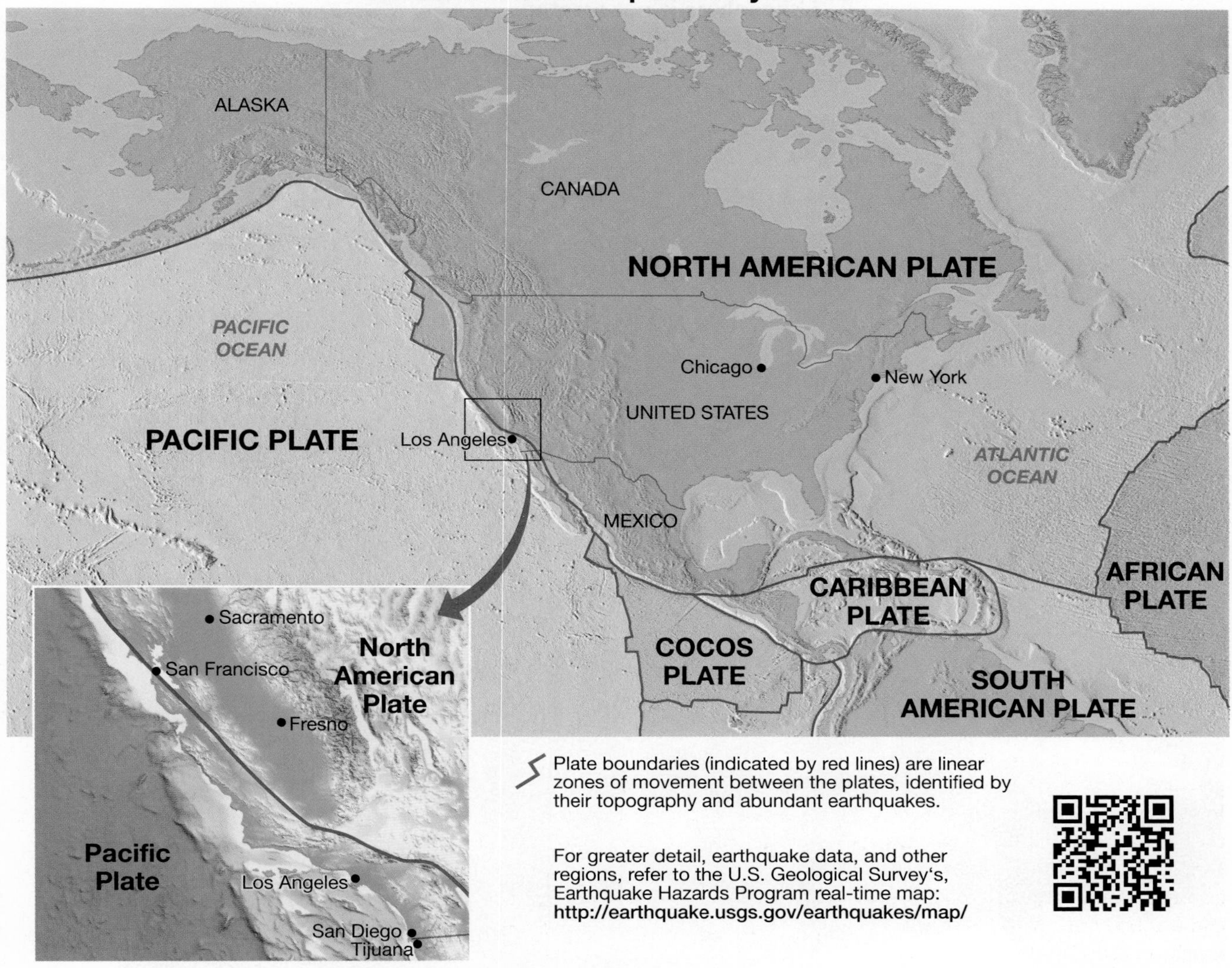

FIGURE 2.1 Lithospheric plates associated with North America. The plate boundaries are identified as linear zones of abundant earthquakes; whereas the plates tend to have few concentrated zones of earthquake activity.

where magma rises up between plates that are spreading apart. The magma cools to form new rock on the edges of both plates. *Plates are destroyed* along **convergent boundaries,** where the edge of one plate *subducts* (descend beneath the edge of another plate) back into the mantle (**FIGURE 2.2**). When the ocean edge of one plate subducts beneath the ocean edge of another, an island arc forms (like the Aleutian Islands of Alaska). When the ocean edge of one plate subducts beneath the continental edge of another plate, then a mountainous volcanic arc forms on the continent's edge (as in the Andes and Cascades Mountains). When the continental edges of two plates converge, the crust of both plates crumples into large folds and merge to form a tall mountain belt (like the Alpine-Himalayan Mountain Belt) that is eventually destroyed by weathering and erosion of the mountains). *Plates slide past one another* along **transform boundaries**, where plates are neither created nor destroyed (**FIGURE 2.2**).

The Theory of Plate Tectonics has become a unifying theory of geology in the way that the Theory of Evolution unifies all elements of biology. It explains such things as the origin and distribution of earthquakes and volcanoes, how the main features of oceans and continents form, how continents seem to drift about, patterns in the distribution of fossil and living organisms, and the origin and distribution of Earth materials and hazards. It ties together all of the sub-disciplines of geology.

Calculating Rates—The Math You Need

Several of the activities in this laboratory require you to calculate rates. You can review and learn more about calculating rates (including practice problems) at this site featuring The Math You Need, When You Need It math tutorials for students in introductory geoscience courses: **http://serc.carleton.edu/mathyouneed/rates/index.html**

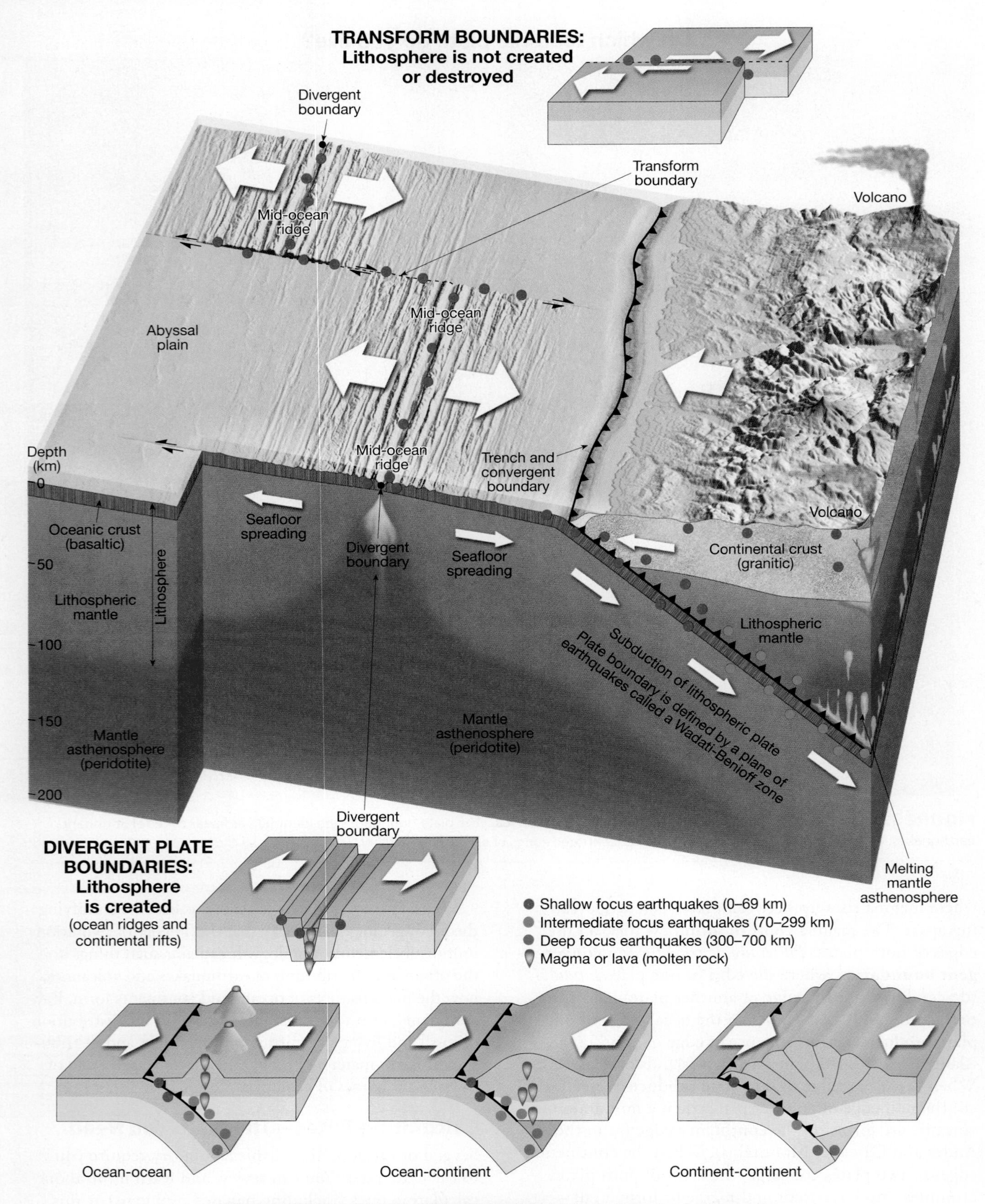

FIGURE 2.2 Three kinds of plate boundaries: divergent, convergent, and transform boundaries. White arrows indicate motions of the lithospheric plates. Half arrows on the transform fault boundary indicate relative motion of the two blocks on either side of the fault. The focus of an earthquake is the exact location where an earthquake occurred (shallow, intermediate, or deep). Water in subducted plates can lower the melting point of rock just above them at intermediate depths and lead to formation of volcanoes.

ACTIVITY

2.1 Plate Motion Inquiry Using GPS Time-Series

THINK About It | **Is the lithosphere beneath your feet really moving?**

OBJECTIVE Use NASA GPS (Global Positioning System) time series to determine the vector direction and rate of lithospheric plate motion where you live.

PROCEDURES

1. **Before you begin**, read about the GPS time series below. Also, this is **what you will need** to do the activity:
   ____ calculator, ruler
   ____ protractor (optional), cut from GeoTools Sheet 4 at the back of the manual.
   ____ Activity 2.1 Worksheets (pp. 55–56) and pencil
2. **Answer every question on the worksheet in a way that makes sense to you** and be prepared to compare your work and inferences with others.

GPS—Global Positioning System

The Global Positioning System (GPS) is a technology used to make *precise* (exact) and *accurate* (error free) measurements of the location of points on Earth. It is used for geodesy—the science of measuring changes in Earth's size and shape, and the position of objects, over time.

The GPS technology is based on a constellation of about 30 satellites that take just 12 hours to orbit Earth. They are organized among six circular orbits (20,200 km, or 12,625 mi above Earth) so that a minimum of six satellites will be in view to users anywhere in the world at any time. The GPS constellation is managed by the United States Air Force for operations of the Department of Defense, but it is free for anyone to use anywhere in the world. Billions of people rely on GPS daily.

How GPS Works

Each GPS satellite communicates simultaneously with fixed ground-based Earth stations and other GPS satellites, so it knows exactly where it is located relative to the center of Earth and Universal Time Coordinated (UTC, also called Greenwich Mean Time). Each GPS satellite also transmits its own radio signal on a different channel, which can be detected by a fixed or handheld GPS receiver. If you turn on a handheld GPS receiver in an unobstructed outdoor location, then the receiver immediately acquires (picks up) the radio channel of the strongest signal it can detect from a GPS satellite. It downloads the navigational information from that satellite channel, followed by a second, third, and so on. A receiver must acquire and process radio transmissions from at least four GPS satellites to triangulate a determination of its exact position and elevation—this is known as a fix. But a fix based on more than four satellites is more accurate. In North America and Hawaii, the accuracy of the GPS constellation is also augmented by WAAS (Wide Area Augmentation System) satellites operated by the Federal Aviation Administration. WAAS uses ground-based reference stations to measure small variations in GPS satellites signals and correct them. The corrections are transmitted up to geostationary WAAS satellites, which broadcast the corrections back to WAAS-enabled GPS receivers on Earth.

GPS Accuracy

The more channels a GPS receiver has, the faster and more accurately it can process data from the most satellites. Most 12-channel GPS receivers are accurate to within 9 meters of your precise location. Comparable WAAS-enabled GPS receivers are accurate to within 3 meters. The best GPS receivers have millimeter accuracy and can be used to measure things like the movement of lithospheric plates over years of time.

Using Gps to Study Lithospheric Plate Motion

NASA compiles geodetic information from receivers located at more than 2000 GPS reference stations throughout the world. At each GPS station, there is a rigid GPS monument (concrete and steel structure) attached firmly to bedrock (solid rock that is beneath the soil and a part of the lithosphere) or a building that is anchored in bedrock. A GPS receiver antenna is firmly attached to the monument, so it will not move unless its bedrock anchor moves. The bedrock may move in response to volcanic activity or earthquakes. It will also move as the lithospheric plate (that it is a part of) moves.

GPS Time Series

The exact location and elevation of each NASA GPS station has been monitored over time, as a **time series** (a series of observations made over time, FIGURE 2.3), by the California Institute of Technology's Jet Propulsion Laboratory (under NASA contract). Because each GPS station is anchored in bedrock of the lithosphere, the time series data provide data on movement (if any) of the lithospheric plate to which it is attached. Plate motion is determined using one graph of time series data for how the GPS station changed its latitude (position north or south) and another graph of time series data for how the GPS station changed its longitude (position east or west). The average rates of motion, in mm/yr or cm/yr, are provided with the

HOW TO PLOT AN ABSOLUTE PLATE MOTION VECTOR FROM NASA GPS TIME-SERIES AND CALCULATE VELOCITY OF THE PLATE MOTION

A. Choose a data location (geodetic position) at http://sideshow.jpl.nasa.gov/post/series.html and left-click on the green dot to open a small time-series box. Double-click on the box to enlarge. The time-series data in this example are for Sydney, Australia.

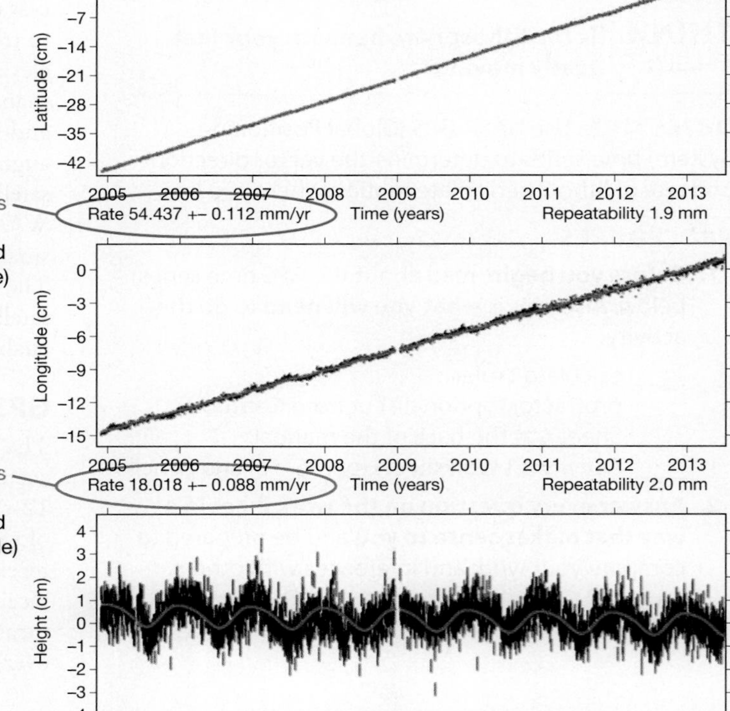

B. Record Latitude vector direction and velocity (be sure to note mm/yr or cm/yr).

- Positive value indicates North Latitude
- Negative value indicates South Latitude

Rate = 54.4 mm/yr
Positive value indicates North Latitude
(a negative value would indicate South Latitude)

C. Record Longitude vector direction and velocity (be sure to note mm/yr or cm/yr).

- Positive value indicates East Latitude
- Negative value indicates West Latitude

Rate = 18.0 mm/yr
Positive value indicates East Longitude
(a negative value would indicate West Longitude)

D. Plot the vector data on a Plate Motion Plotter like the one used here. In this example:

- The plotter has no numerical scale, so a determine the scale you wish to use (in this case 10 mm/yr scale was noted in black on the graph).
- Plot line (red) for 54.4 mm/yr North Latitude.
- Plot line (blue) for 18.0 mm/yr East Longitude.
- Draw line (black arrow) from the origin (center point in the graph), through the point where the Latitude (red) and Longitude (blue) lines intersect, and on through the compass edge of the plotter. Read the direction that the GPS station is moving (North 15.5 degrees East).

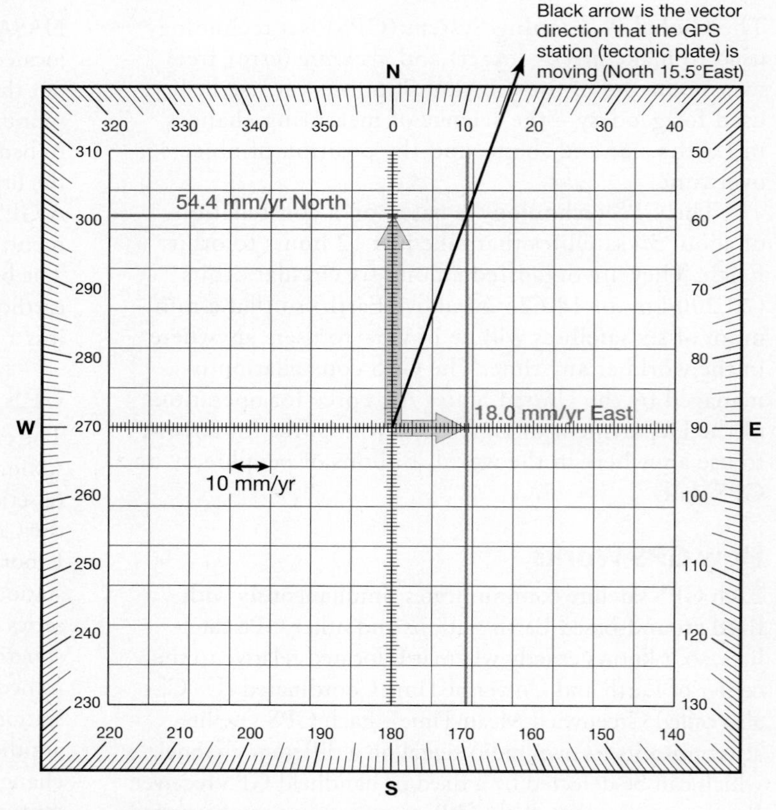

E. Calculate the velocity that the GPS station is moving as follows:

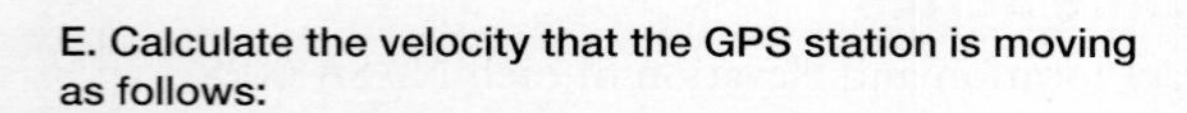

$$\text{Velocity} = \sqrt{(\text{Latitude velocity})^2 + (\text{Longitude velocity})^2}$$

In this example:

$$\text{Velocity} = \sqrt{(54.4\ \text{mm/yr})^2 + (18.0\ \text{mm/yr})^2}$$

$$= 57.3\ \text{mm/yr}$$

F. Sydney, Australia sits on a lithospheric plate that is moving North 15.5° East, at a velocity of 57.3 mm/yr.

FIGURE 2.3 How to determine direction and velocity of lithospheric plate motion using JPL-NASA GPS time-series data. (From the California Institute of Technology's, Jet Propulsion Laboratory. http://slideshow.jpl.nasa.gov/post/series.html)

time series graph. You will need to combine the latitude and longitude data to determine the overall direction and rate of plate motion for the years that the data were collected. See **FIGURE 2.3**.

ACTIVITY

2.2 Is Plate Tectonics Caused by a Change in Earth's Size?

THINK About It | **What causes plate tectonics?**

OBJECTIVE Are Plate Tectonics Caused by a Change in Earth's Size?—Analyze and interpret Earth's tectonic forces and plate boundaries to infer if plate tectonics is caused by a change in Earth's size.

PROCEDURES

1. **Before you begin**, read below about what could cause plate tectonics and study **FIGURE 2.5**. Also, this is **what you will need** to do the activity:
   ____ calculator, ruler
   ____ Activity 2.2 Worksheets (pp. 57–58) and pencil
2. **Answer every question on the worksheet in a way that makes sense to you** and be prepared to compare your work and inferences with others.

What Causes Plate Tectonics?

Recall that geoscientists have historically tried to understand the cause of Earth's oceans, mountains, and global tectonics by questioning if Earth could be shrinking, expanding, or staying the same in size. This question can be evaluated by studying Earth's natural forces and faults in relation to those that you might predict to occur if the size of Earth were changing. By comparing your predictions to observations of the kinds of strains and faults actually observed, it is possible to determine if Earth's size is changing and infer whether a change in Earth's size could cause plate tectonics.

Earth Forces and the Faults They Produce

Three kinds of directed force (stress) can be applied to a solid mass of rock and cause it to deform (strain) by bending or even faulting (**FIGURE 2.4**). **Compression** compacts a block of rock and squeezes it into less space. This can cause *reverse faulting* (also known as thrust faulting) in which the hanging wall block is forced up the footwall block in opposition to the pull of gravity (**FIGURE 2.4**). **Tension** (also called *dilation*) pulls a block of rock apart and increases its length. This can cause *normal faulting*, in which gravity pulls the hanging wall block down and forces it to slide down off of the footwall block (see **FIGURE 2.4**). **Shear** offsets a block of rock from side to side and may eventually tear it apart into two blocks of rock that slide past each other along

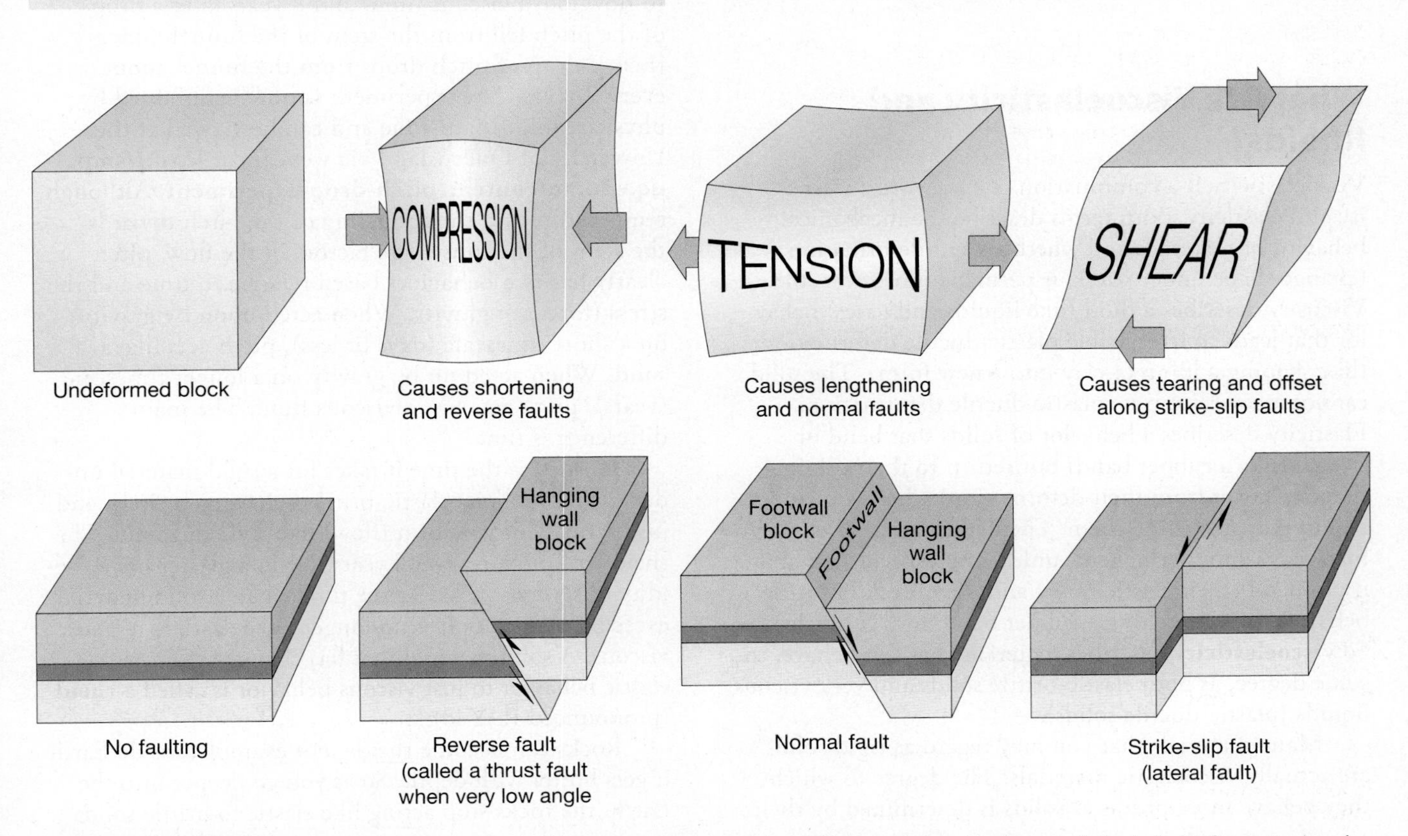

FIGURE 2.4 Stress and strain. Three kinds of stress (applied force, as indicated by arrows) cause characteristic strain (deformation) and faulting types.

a lateral or *strike-slip fault* (**FIGURE 2.4**). Plate tectonic forces can be understood by how the lithosphere is strained and faulted.

ACTIVITY

2.3 Lava Lamp Model of Earth

THINK About It | **What causes plate tectonics?**

OBJECTIVE Investigate viscoelasticity and rheidity, then evaluate a lava lamp model of mantle convection.

PROCEDURES

1. **Before you begin**, read about viscoelasticity, rheids, mantle convection, and seismic tomography below. Also, this is **what you will need** to do the activity:
   ____ blue and red colored pencils or pens
   ____ plastic ruler or popsicle stick
   ____ Activity 2.3 Worksheets (pp. 59–60) and pencil
   ____ Silly Putty™ and Lava lamp (provided in lab)
2. **Answer every question on the worksheet in a way that makes sense to you** and be prepared to compare your work and inferences with others.

What Are Viscoelasticity and Rheids?

Viscoelasticity is a combination of two words: viscosity and elasticity. Both terms describe the mechanical behavior of materials and whether their deformation (change shape under stress) is permanent or reversible. Viscosity describes a fluid (like liquids and gases) behavior that leads to irreversible plastic-ductile deformation (like shaping a lump of clay into a new form). The solid cannot recover from its plastic-ductile deformation. Elasticity describes a behavior of solids that bend or stretch (like a rubber band) but return to their original shape (recover from their deformation) when you stop bending or stretching them. The deformation of elastic materials is not permanent, unless you exceed their ability to bend (their elastic limit) and they break (a brittle behavior, like stretching a rubber band until it breaks). So **viscoelasticity** describes materials that can behave, to some degree, as both elastic-brittle solids and very viscous liquids (plastic-ductile solids).

Many materials that you may regard as rigid solids are actually viscoelastic materials. The degree to which they behave more or less as solids is determined by their level of internal energy (*temperature*) and/or amount of *stress* (force, pressure) acting upon them over a given amount of *time*. To understand how temperature affects viscoelastc materials, think about steel. Red hot steel has a great amount of thermal energy, so it has more of a plastic-ductile behavior (can easily be hammered into new shapes) than cold steel. Cold steel is less plastic-ductile and more brittle than red hot steel. The effects of stress and time are more variable. Using lots of stress like hard hammer strikes, you can hammer a small ball of red hot steel into a flat disk in seconds. Alternatively, you can press on the steel with giant cold steel rolling pins over minutes of time to roll the steel into a flat disk. The effect is the same; the difference is the amount of stress and the amount of time it is applied. One study of these effects has been carried out at the University of Queensland since 1927.

Physicist Thomas Parnell has studied the properties of pitch, a black solid material derived from naturally occurring asphalt. Unless heated, pitch normally behaves as a brittle solid with little elasticity, so it shatters when struck with a hammer. But at the same temperature, under the stress of gravity, Parnell found that Pitch also behaves as a viscous fluid. In 1927, he heated some pitch to make it a liquid, and then poured it into a funnel with a sealed stem at its base. There, it solidified into its rigid solid form. He waited 3 years, and then cut one the stem of the funnel to see what would happen. After several years it began to flow like a viscous fluid, and 8 years later a drop of the pitch fell from the stem of the funnel. Since then, a drop of pitch drops from the funnel about every 9 years. The experiment is now maintained by physicist John Mainstone and can be viewed at the University of Queensland via webcam at **http://smp.uq.edu.au/content/pitch-drop-experiment**. Although, temperature and the funneling of the pitch towards the stem of the funnel are factors in the flow, pitch clearly has two behaviors based relative to time and the stress (force) of gravity. When acted upon by gravity on a short timescale (days or less), pitch acts like a solid. When acted up by gravity on a longer timescale (years), pitch acts like a viscous fluid. The main difference is time.

Rheidity is the time it takes for a solid material under stress to lose its elastic-brittle behavior, entirely, and just permanently deform (flow) like a viscous fluid. The rheidty of pitch is several years. Ice in a glacier has a rheidity of several weeks. At the time when a solid material exceeds its rheidity it is no longer viscoelastic—it is just viscous. A solid material that has changed from viscoelastic behavior to just viscous behavior is called a **rheid** (pronounced RAY-id).

Rocks can become rheids. For example, inside Earth it gets hotter with depth. So as you go deeper into the Earth, the rocks stop acting like elastic to brittle solids (rigid rocks) and change their behavior to that of viscous, plastic-ductile solids. The zone where this happens even has a name—the *brittle-ductile transition zone.* So mantle rocks are rheids—they flow and permanently deform (change shape) like viscous fluids. Rocks of the lithosphere and rock samples in your hand are not. They are viscoelastic solids that behave much more like elastic (to brittle) solids than viscous fluids.

Mantle Convection

While much is known about plate tectonics, and the plates have been identified and named (**FIGURE 2.5**), there has been uncertainty about how mantle rocks beneath the asthenosphere may influence this process. In the 1930s, an English geologist named Arthur Holmes speculated that the mantle may contain convection cells in which there is a circular flow of material like in boiling pot of soup. He proposed that such flow could carry continents about the Earth like a giant conveyor belt. This idea was also adapted in the 1960s by Harry Hess, who hypothesized that mantle flow is the driving mechanism of plate tectonics. New technologies provide an opportunity to evaluate this hypothesis. For example, seismic tomography now provides sound evidence that processes at least 660 kilometers deep inside the mantle may have dramatic effects on plate tectonics at the surface.

Seismic Tomography

Earth's mantle is nearly 3000 km thick and occurs between the crust and the molten outer core. Although mantle rocks behave like a brittle solid on short timescales associated with earthquakes, they seem to become rheids and flow like a viscous fluid on longer timescales of hundreds to thousands of years. Geologists use a technique called *seismic tomography* to detect this mantle flow.

The word *tomography* (Greek: *tomos* = slice, *graphe* = drawing) refers to the process of making drawings of slices through an object or person. Geologists use seismic tomography to view slices of Earth's interior similar to the way that medical technologists view slices of the human body. The human body slices are known as CAT (computer axial tomography) scans and are constructed using X-rays to penetrate and image the human body. The tomography scans of Earth's interior are constructed using seismic waves to penetrate and image the body of Earth.

In seismic tomography, geologists collect data on the velocity (rate and direction) of many thousands of seismic waves as they pass through Earth. The waves travel fastest through rocks that are the densest and presumed to be coolest. The waves travel slower through rocks that are less dense and presumed to be warmer. When a computer is used to analyze all of the data, from all directions, it is possible to generate seismic tomography images of Earth. These images can be viewed individually or combined to form three-dimensional perspectives. The computer can also assist in false coloring seismic tomography images to show bodies of mantle rock that are significantly warmer or cooler than the rest of the mantle (**FIGURE 2.6**).

ACTIVITY 2.4 Paleomagnetic Stripes and Seafloor Spreading

THINK About It — **How are plate boundaries identified? How and at what rates does plate tectonics affect Earth's surface?**

OBJECTIVE Analyze paleomagnetic stripes and infer how seafloor spreading is related to Cascade Range volcanoes.

PROCEDURES

1. **Before you begin**, read about Earth's magnetism and paleomagnetism below. Also, this is **what you will need** to do the activity:
   ____ calculator, ruler
   ____ Activity 2.4 Worksheets (pp. 61–62) and pencil
2. **Answer every question on the worksheet in a way that makes sense to you** and be prepared to compare your work and inferences with others.

Earth's Magnetism and Paleomagnetism

If you drop a pen, it falls to the floor. The pen is under the influence of Earth's gravity—an invisible force field that pulls everything and everyone towards the center of our planet. But Earth has another force field that is not so obvious, its magnetic field. It is as though a giant bar magnet resides inside Earth, giving our planet both a magnetic north pole and a magnetic south pole. Invisible lines of the magnetic force field arc out through space from the south magnetic pole, travel around the outside of Earth at its equator, then arc back into the north magnetic pole. The strength of a magnetic field is measured in units called teslas, and a microtesla is a millionth of a tesla. Small magnets used to hold notes on refrigerator doors have a strength of about 50,000 microteslas, and Earth's magnetic field strength ranges from just 30 microteslas at the equator to about 50 microteslas at the poles. Therefore, a refrigerator magnet

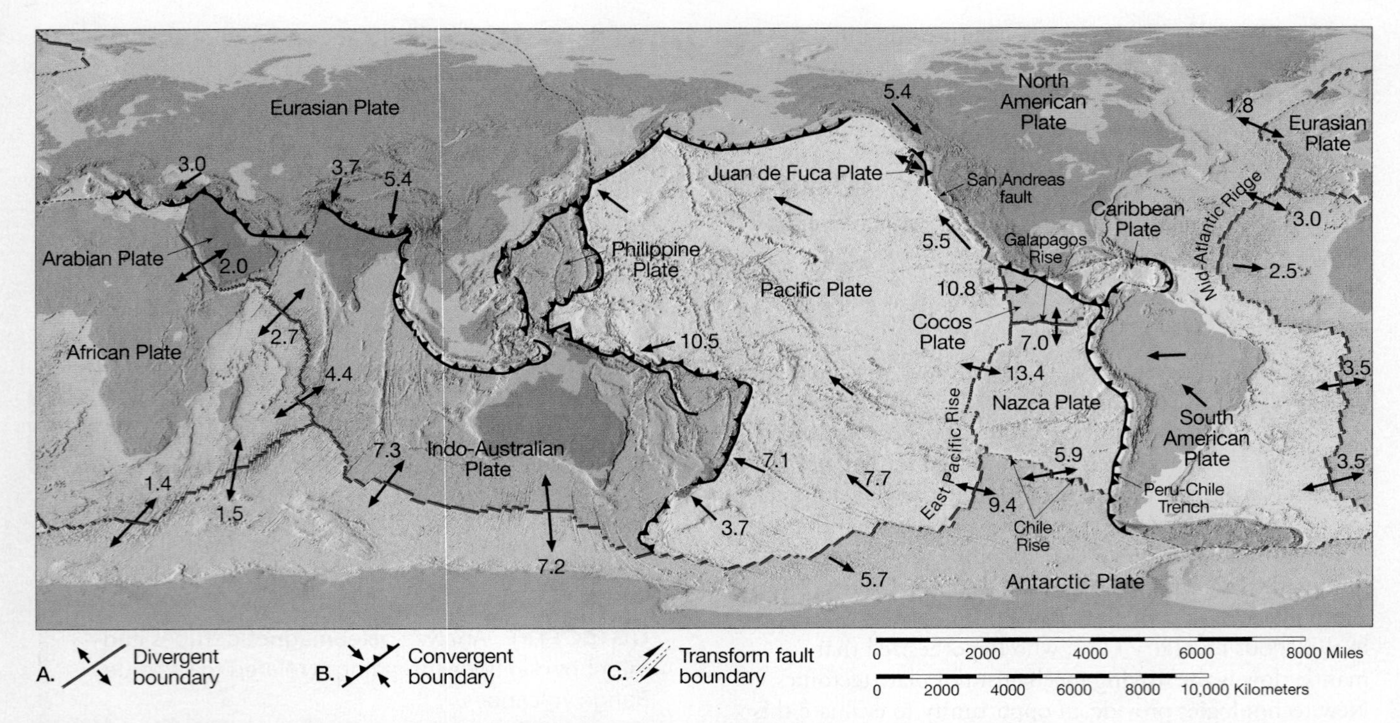

FIGURE 2.5 Earth's lithospheric plates and their boundaries. Numerals indicate relative (in relation to each other) rates of plate motion in centimeters per year (cm/yr); not the actual (absolute) rates of individual plate motion that you may have studied in Activity 2.1. Divergent plate boundaries (red) occur where two adjacent plates form and move apart (diverge) from each other. Convergent plate boundaries (hachured with triangular "teeth") occur where two adjacent plates move together. Transform fault plate boundaries (dashed) occur along faults where two adjacent plates slide past each other. Refer back to **FIGURE 2.2** for another perspective of the three kinds of plate boundaries.

is three orders of magnitude (1000 times) stronger than Earth's magnetic field. Even so, you can use the tiny magnetic needle in a compass to detect Earth's magnetic field. Magnetic compass needles are not attracted to the geographic North Pole. Instead, they are attracted to the magnetic north pole, which is located in the Arctic Islands of Northern Canada, about 700 km (450 mi) from the geographic North Pole.

Paleomagnetism

Tiny crystals of iron-rich minerals, such as magnetite (Fe_3O_4), acquire and retain the directional signature of Earth's magnetic field when they form. This ancient magnetism is called **paleomagnetism**. Magnetic mineral crystals lose this magnetism if heated above the *Curie Point* of about 580°C. Only when they cool below the Curie Point do mineral crystals acquire and retain the signature of Earth's magnetic field for the time and place where they cooled. This happens when volcanic lava cools and crystallizes below the Curie Point.

Magnetic Reversals

When geologists first started detecting paleomagnetism in layers of cooled lava (volcanic rock) stacked one atop the other, they discovered that Earth's magnetic field has not always been the same. It has undergone periodic **reversals**. During times of **normal polarity**, the north-seeking end of a compass needle (and tiny iron-bearing mineral crystals in volcanic rock) points in the direction of Earth's present north magnetic pole. But during times of **reversed polarity**, the north-seeking end of a compass needle points in the opposite direction (geographic south).

Magnetic Anomalies and Paleomagnetic Stripes

Magnetic anomalies are deviations from the average strength of the magnetic field in a given area. Areas of higher than average strength are positive anomalies, and areas of less than average strength are negative anomalies. In the 1950s the U.S. Coast and Geodetic Survey scanned the ocean for magnetic anomalies and discovered that rocks of the sea floor contained alternating striped patterns of high and low magnetic anomalies, called **paleomagnetic stripes**. They also discovered that the pattern of paleomagnetic stripes was symmetrical on opposite sides of mid-ocean ridges. In 1963, geologists Fred Vine, Drummond Matthews, and Lawrence Morley discovered that the symmetrical pattern of paleomagnetic stripes in seafloor rocks was the result of two processes: the formation of seafloor and reversals of Earth's magnetic field. They proposed that as volcanoes

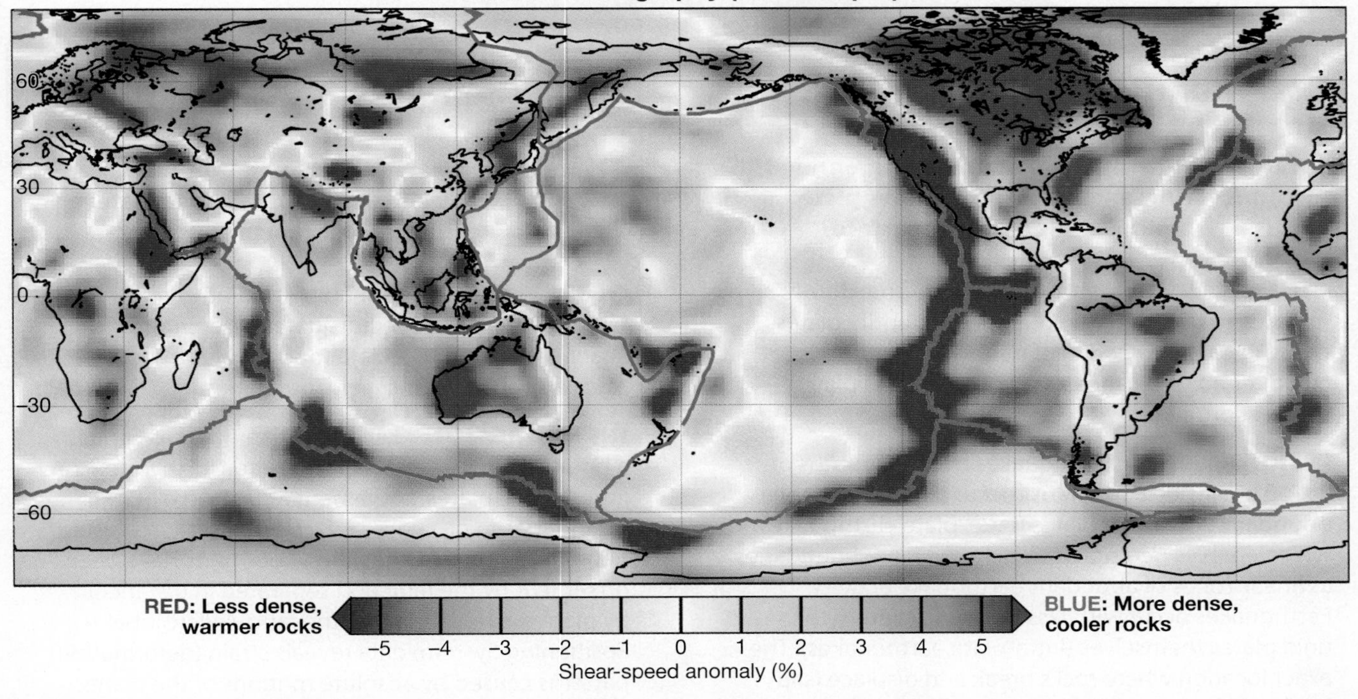

FIGURE 2.6 Seismic tomography image (horizontal slice) of Earth's mantle at a depth of 80 km. Note scale indicating the percent change in shear speed of the seismic waves. Red false coloring indicates slower shear speed in warmer rock that is less dense and ascending. Blue false coloring indicates faster shear speed in cooler rock that is static or descending. (Image constructed by Sergei Lebedev, Dublin Institute for Advanced Studies, and Rob D. van der Hilst, MIT/IRIS Consortium, based on a large global data set retrieved from IRIS (Incorporated Research Institutions for Seismology). Used with permission.)

erupted along a mid-ocean ridge, the lava cooled below the Curie Point and recorded reversals of Earth's paleomagnetic field. Rocks formed during times of normal polarity now have magnetic signatures that add to the modern field strength and create a positive anomaly. Rocks formed during times of reversed polarity have magnetic signatures that oppose the modern field and create a negative anomaly. The symmetrical pattern of paleomagnetic stripes developed as new crust was formed and magnetized and older crust moved down and spread away from both sides of the ridge under the influence of gravity (a process that Harry Hess called **seafloor spreading**).

Polar Reversal Time Scale

Radiometric dating of volcanic rocks containing paleomagnetism enables geologists to time the pattern of Earth's paleomagnetic field reversals. The **polar reversal time scale** is a column of named *chrons* (time intervals) of normal and reversed polarity, from oldest at the bottom to youngest at the top, combined with a time scale in millions of years. We are living in the Brunhes normal chron, which began 0.78 million years ago, after the Matayama reversed chron. Most chrons are also subdivided into named subchrons. In Activity 2.4, you will use a simpler version of a polar reversal time scale in which the time intervals have been color coded (instead of named) so they are easy to recognize on sight.

ACTIVITY

2.5 Atlantic Seafloor Sreading

THINK About It **How are plate boundaries identified? How and at what rates does plate tectonics affect Earth's surface?**

OBJECTIVE Infer how fracture zones and shapes of coastlines provide clues about how and when North America and Africa were once part of the same continent.

INTRODUCTION If you did Activity 2.4, then you have already studied sea floor spreading about the Gorda and Juan de Fuca Ridges off the northwest coast of the United States. This activity is an investigation of seafloor spreading about the Mid-Atlantic Ridge.

PROCEDURES

1. This is **what you will need** to do the activity:
   ____ red and blue pencils or pens
   ____ calculator, ruler
   ____ Activity 2.5 Worksheet (p. 63) and pencil
2. **Answer every question on the worksheet in a way that makes sense to you** and be prepared to compare your work and inferences with others.

ACTIVITY

2.6 Using Earthquakes to Identify Plate Boundaries

THINK About It **How are plate boundaries identified? How and at what rates does plate tectonics affect Earth's surface?**

OBJECTIVE Apply earthquake data from South America to define plate boundaries, identify plates, construct a cross section of a subduction zone, and infer how volcanoes may be related to plate subduction.

INTRODUCTION Earthquakes occur at depths of 0–700 km inside Earth. Most occur along the mobile boundaries between plates (inter-plate earthquakes), which enables geologists to map the plate boundaries as linear zones of abundant earthquake activity. Earthquakes occur with less frequency within the rigid plates themselves (intra-plate earthquakes). The exact location where rocks break and displace (slip past one another) to make an earthquake is called the **focus** of the earthquake. Shallow focus earthquakes (0–69 km deep) are the most common kind of earthquakes and occur at all three main kinds of plate boundaries (divergent, convergent, and transform: FIGURE 2.2). Intermediate (70–299 km deep) and deep focus earthquakes (300–700 km deep) occur mostly in *Wadati-Benioff zones* of earthquake activity associated with plate subduction at convergent plate boundaries (FIGURE 2.2).

PROCEDURES

1. This is **what you will need** to do the activity:
   ____ red pencil or pen
   ____ ruler
   ____ Activity 2.6 Worksheets (pp. 64–65) and pencil
2. **Answer every question on the worksheet in a way that makes sense to you** and be prepared to compare your work and inferences with others.

ACTIVITY

2.7 San Andreas Transform-Boundary Plate Motions

THINK About It **How are plate boundaries identified? How and at what rates does plate tectonics affect Earth's surface?**

OBJECTIVE Analyze maps of geology and GPS-based plate motion vectors data to determine absolute and relative plate motions along the San Andreas Fault to transform boundaries.

INTRODUCTION California's San Andreas Fault is a boundary between the Pacific and North American lithospheric plates. Movement along the plate boundary can be characterized and measured using geologic maps that show how rock units have been *offset* (cut by the fault and separated in distance) by plate movement along the fault. GPS (Global Positioning System) data reveals strain (deformation) patterns caused by absolute motions of the plates.

PROCEDURES

1. **Before you begin**, read about the GPS time series below. Also, this is **what you will need** to do the activity:
   ____ calculator, ruler
   ____ Activity 2.7 Worksheets (pp. 66–67) and pencil
2. **Answer every question on the worksheet in a way that makes sense to you** and be prepared to compare your work and inferences with others.

ACTIVITY

2.8 Hot Spots and Plate Motions

THINK About It **What are hot spots, and how do they help us explain plate tectonics?**

OBJECTIVE Determine rates and directions of plate motions as they have moved over hot spots.

PROCEDURES

1. **Before you begin**, read about plate tectonics and hot spots below. Also, this is **what you will need** to do the activity:
   ____ calculator, ruler
   ____ Activity 2.8 Worksheets (pp. 68–69) and pencil
2. **Answer every question on the worksheet in a way that makes sense to you** and be prepared to compare your work and inferences with others.

Evaluating Plate Tectonics and Hot Spots

The Plate Tectonics Model is widely applied by geoscientists to help explain many regional and global features of the geosphere. Another regional feature of Earth is hot spots, centers of volcanic activity that persist in a stationary location for tens-of-millions of years. Geologists think they are either a) the result of long-lived narrow *plumes* of hot rock rising rapidly from Earth's mantle (like a stream of heated lava rising in a lava lamp), or b) the slow melting of a large mass of hot mantle rock in the upper mantle that persists for a long interval of geologic time.

The Hawaiian Hot Spot and Pacific Plate Motion

As a lithospheric plate migrates across a stationary hot spot, a volcano develops directly above the hot spot. When the plate slides on, the volcano that was over the hot spot becomes dormant, and over time, it migrates many kilometers from the hot spot. Meanwhile, a new volcano arises as new lithosphere passes over the hot spot. The result is a string of volcanoes, with one end of the line located over the hot spot and quite active, and the other end distant and inactive. In between is a succession of volcanoes that are progressively older with distance from the hot spot. The Hawaiian Islands and Emperor Seamount chain (**FIGURE 2.7**) are thought to represent such a line of volcanoes that formed over the Hawaiian hot spot.

ACTIVITY

2.9 The Origin of Magma

THINK About It | **How and where does magma form?**

OBJECTIVE Apply physical and graphical models of rock melting to infer how magma forms in relavtion to pressure, temperature, water, and plate tectonics.

PROCEDURES

1. **Before you begin**, read about the origin of magma below. Also, this is **what you will need** to do the activity:
   ____ Materials provided in lab: hot plate, aluminum foil, sugar cubes, water, dropper
   ____ calculator, ruler
   ____ Activity 2.9 Worksheets (pp. 70–72) and pencil
2. **Answer every question on the worksheet in a way that makes sense to you** and be prepared to compare your work and inferences with others.

The Origin of Magma

If you have watched videos of the fountains and rivers of lava produced by Kilauea volcano in Hawaii, you may have wondered how much of Earth's interior is made up of melted rock, or magma. Seismic studies indicate that nearly all of Earth's mantle and crust are solid rock—not magma. Therefore, except for some specific locations where active volcanoes occur, there is no reservoir or layer of magma beneath Earth's surface just waiting to erupt. On a global scale, the volume of magma that feeds active volcanoes is actually very small. What, then, are the special conditions that cause these rare bodies of upper mantle and lower crust magma to form?

Magma generally forms in three plate tectonic settings (divergent plate boundaries, convergent plate boundaries, and hot spots). Its origin (rock melting) is also influenced by underground temperature, underground pressure (lithostatic pressure), and the kind of minerals that comprise underground rocks.

Temperature (T)

Rocks are mostly masses of solid mineral crystals. Therefore, some or all of the mineral crystals must melt to form magma. According to the Kinetic Theory, a solid mineral crystal will melt if its kinetic energy (motion of its atoms and molecules) exceeds the attractive forces that hold together its orderly crystalline structure. Heating a crystal is the most obvious way to melt it. If enough heat energy is applied to the crystal, then its kinetic energy level may rise enough to cause melting. The specific temperature at which crystals of a given mineral begin to melt is the mineral's **melting point.**

Partial Melting. All minerals have different melting points. So when heating a rock comprised of several different kinds of mineral crystals, one part of the rock (one kind of mineral crystal) will melt before another part (another kind of mineral crystal). Geologists call this **partial melting** of rock. But where would the heat come from to begin melting rocks below the ground?

Geothermal Gradient. Unless you live near a volcano or hot spring, you probably are not aware of Earth's body heat. But South African gold miners know all about it. The deeper they mine, the hotter it gets. In the deepest mine (**FIGURE 1.6**), 3.8 kilometers below ground, rock temperatures are 60 °C (140 °F) and the mine must be air conditioned. This gradient of increasing temperature with depth is called the **geothermal gradient.** This gradient also varies between ocean crust and continental crust, but the global average for all of Earth's crust is about 25 °C (77 °F) per kilometer. In other words, rocks located 1 kilometer below your house are about 25 °C warmer than the foundation of your house. If the geothermal gradient continued at this rate through the mantle, then the mantle would eventually melt at depths of 100–150 kilometers. Seismology shows that this does not occur, so temperature is not the only factor that determines whether a rock melts or remains solid. Pressure is also a factor.

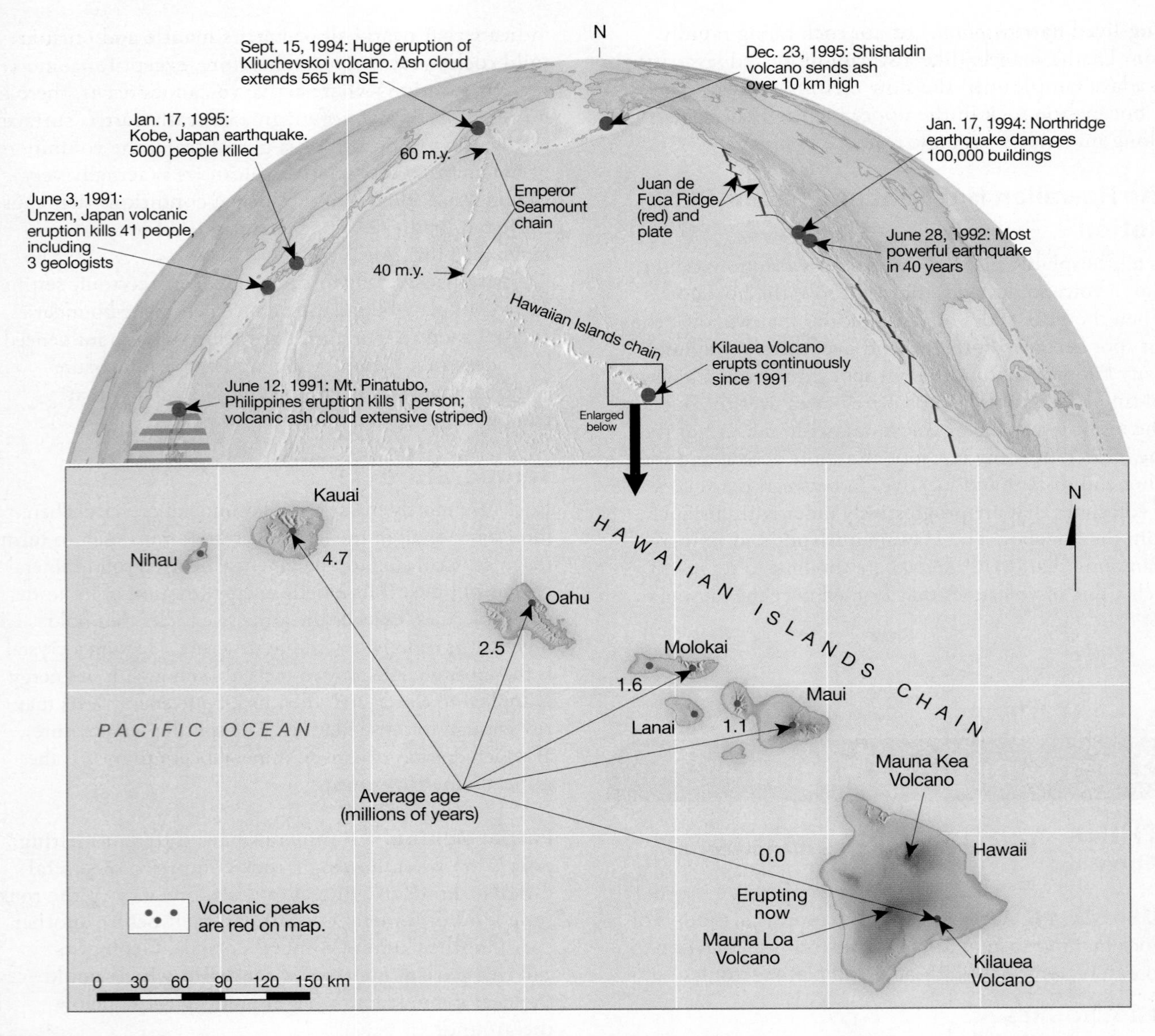

FIGURE 2.7 Effects of a hot spot on the Pacific seafloor. Top map shows the northern Pacific Ocean, adjacent landmasses, some notable geologic hazards (natural disasters), the Hawaiian Islands chain, and the Emperor Seamount chain. Lower map shows details of the Hawaiian Islands chain, including locations of volcanic peaks.

Pressure (P)

When you press your hand against something like a bookshelf, you can apply all of your body *weight* against the surface *area* under your hand. Therefore, **pressure** is expressed as amount of weight applied per unit of area. For example, imagine that you weigh 100 pounds and that your hand is 5 inches long and 4 inches wide. If you exert all of your weight against a wall by leaning against the wall with one hand, then you are exerting 100 pounds of weight over an area of 20 square inches (5 inches × 4 inches × 20 square inches). This means that you are exerting 5 pounds of pressure per square inch of your hand.

Confining Pressure. Atoms and molecules of air (atmosphere) are masses of matter that are pulled by gravity toward the center of Earth. But they cannot reach Earth's center because water, rocks, and your body are in their path. As a result, the weight of the air presses against surfaces of water, rocks, and your body. If you stand at sea level, then your body is confined by 14.7 pounds of weight pressing on every square inch of your body (14.7 lbs/in^2). This is called atmospheric **confining pressure.** Scientists also refer to this as one *atmosphere* (1 atm) of pressure.

You do not normally feel one atmosphere of confining pressure, because your body exerts the same pressure to keep you in equilibrium (balance) with your surroundings.

But if you ever dove into the deep end of a swimming pool, then you experienced the confining pressure exerted by the water plus the confining pressure of the atmosphere. The deeper you dove, the more pressure you felt. It takes 10 m (33.9 ft) of water to exert another 1 atm of confining pressure on your body.

Rocks are about three times denser than water, so it takes only about 3.3 m of rock to exert a force equal to that of 10 m of water or the entire thickness of the atmosphere! 100 m of rock exert a confining pressure of about 30 atm, and 1 km (1000 m) of rock exerts a confining pressure of about 300 atm. At 300 atm/km, a rock buried 5 km underground is confined by 1500 atm of pressure!

Decompression Melting. The confining pressure under kilometers of rock is so great that a mineral crystal cannot melt at its "normal" melting point observed on Earth's surface. The pressure confines the atoms and molecules and prevents them from flowing apart. More heat is required to raise the kinetic energy level of atoms and molecules in the crystal enough to melt the crystal. Consequently, an increase in confining pressure causes an increase in the melting point of a mineral. Reducing confining pressure lowers the melting point of a mineral. This means that if a mineral is already near its melting point, and its confining pressure decreases enough, then it will melt. This is called **decompression melting**.

Pressure-Temperature (P-T) Diagrams

Geologists understand that rock melting (the origin of magma) is related to both temperature and pressure. Therefore, they heat and pressurize rock samples under controlled conditions in geochemical laboratories to determine how rock melting is influenced by specific combinations of both pressure and temperature. Samples are pressurized and heated to specific P-T points to determine if they remain solid, undergo partial melting, or melt completely. The data are then plotted as specific points on a **pressure-temperature (P-T) diagram** such as the one in **FIGURE 2.8** for mantle peridotite. Mantle

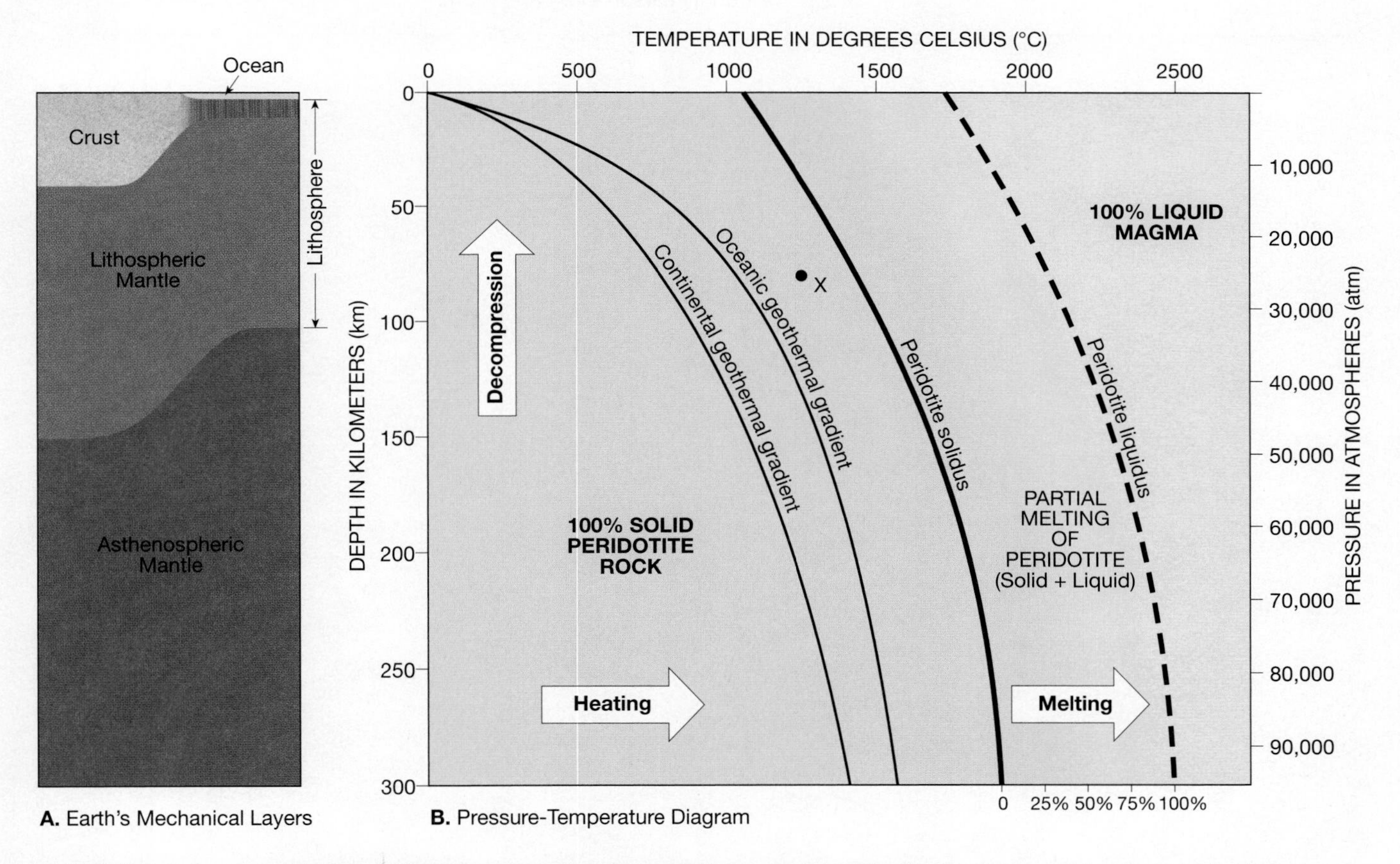

FIGURE 2.8 Pressure-Temperature diagram relative the geosphere. A. The physical layers of the geosphere vary in physical properties, such as melting point, depending on the temperature and pressure. B. The pressure-temperature (P-T) diagram shows the environmental conditions that exist across the physical layers shown in A. The diagram shows how P-T conditions affect peridotite rock (made of olivine, pyroxene, amphibole, and garnet mineral crystals). At P-T points below (to the left of) the *peridotite solidus*, all mineral crystals in the rock remain solid. At P-T points above (to the right of) the *peridotite liquidus*, all mineral crystals in the rock melt to liquid. At P-T points between the solidus and liquidus, the rock undergoes partial melting—one kind of mineral at a time, so solid and liquid are present. Continental and oceanic geothermal gradients are curves showing how temperature normally varies according to depth below the continents and ocean basins. Temperatures along both of these geothermal gradients are too cool to begin partial melting of peridotite. Both gradients occur below (to the left of) the peridotite solidus (1 atm 5 about 1 bar).

peridotite is made of olivine, pyroxene, amphibole, and garnet mineral crystals. Therefore, this diagram also shows the combined effects of pressure and temperature on a rock made of several different minerals. At P-T points below (to the left of) the *solidus,* all mineral crystals in the rock remain solid. At P-T points above (to the right of) the *liquidus,* all mineral crystals in the rock melt to liquid. At P-T points between the solidus and liquidus, the rock undergoes partial melting—one kind of mineral at a time. Therefore, a P-T diagram also reveals stability fields for states (phases) of matter. In this case (see **FIGURE 2.8**), there are stability fields for solid, solid + liquid, and liquid.

Notice that lines for the continental and oceanic geothermal gradients are also plotted on **FIGURE 2.8**. They show how temperature normally varies according to depth below the continents and ocean basins. Temperatures along both of these geothermal gradients are not great enough to begin melting peridotite. Both gradients occur along temperatures below (to the left of) the peridotite solidus.

ACTIVITY 2.1 Plate Motion Inquiry Using GPS Time Series

Name: ____________________ **Course/Section:** ____________________ **Date:** __________

A. Analyze **FIGURE 2.1**. On what lithospheric plate do you live? ____________________

B. Go to the **JPL-NASA GPS Time Series website** at **http://sideshow.jpl.nasa.gov/post/series.html**. The map displays GPS stations as small green dots with a yellow line. The yellow line points away from the green dot in the direction that the GPS station (and lithospheric plate to which it is anchored) is moving. Notice that you can scroll in on the map to enlarge it and reveal more GPS stations. Find the GPS station that is the closest to where you live. Record the station name, then complete the Plate Motion Plotter below for the station by following the directions in **FIGURE 2.3**.

1. GPS Station Name:

2. Latitude vector direction (North or South) and velocity: ____________________

3. Longitude vector direction (East or West) and velocity: ____________________

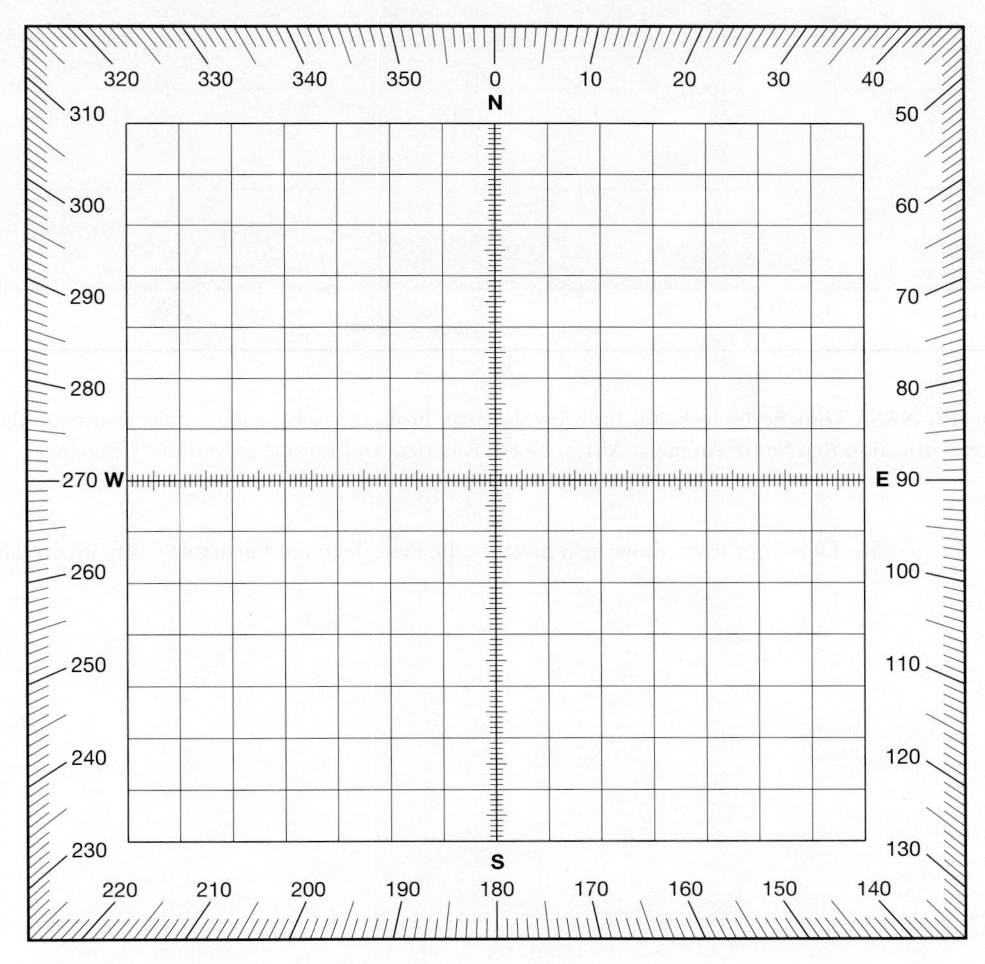

4. In what direction is this GPS station moving?

5. At what velocity is this GPS station (and lithospheric plate to which it is attached) moving?

6. Return to the **JPL-NASA Time Series website,** and click on "Geodetic Positions and Velocities" above the map. Scroll down to the name of your station, and record its current coordinate position in latitude and longitude below. For latitude, notice that a positive number indicates degrees North Latitude, and a negative number indicates degrees South Latitude. For longitude, a positive number indicates degrees East Longitude, and a negative number indicates West Longitude.

   Current Latitude: ______________________ Current Longitude: ______________________

7. Use the latitude and longitude coordinates to plot the location of your GPS station on the map below as a dot. Then add an arrow to show the direction that the station is moving (from #4 above). Beside the arrow, write the velocity that the GPS station (and lithospheric plate) is moving (from part 5 above).

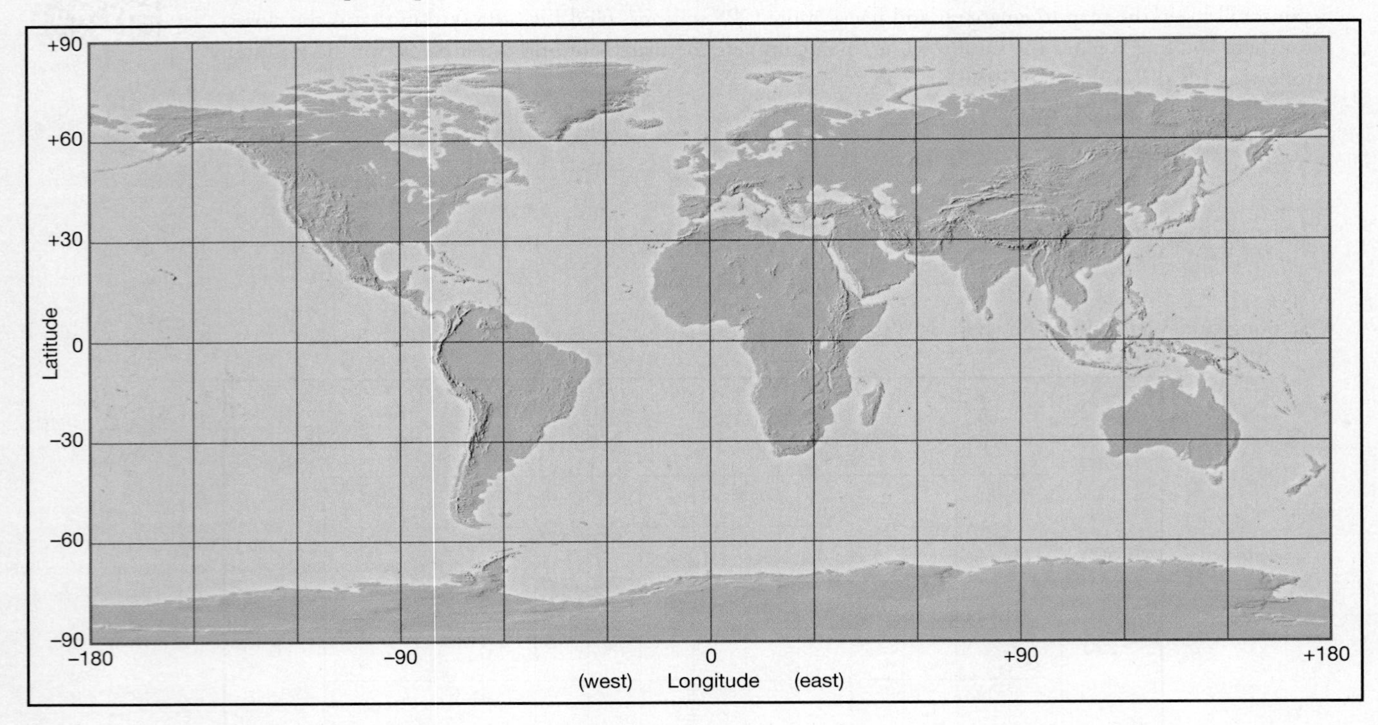

C. Return to the **JPL-NASA Time Series website**, and view the map. From the website map, draw arrows on the map above to show the general direction that South America, Africa, North America, and Europe are currently moving.

D. REFLECT & DISCUSS Does your work above help to verify the Plate Tectonic Theory or falsify it? Explain.

ACTIVITY 2.2 Is Plate Tectonics Caused by a Change in Earth's Size?

Name: ______________________ **Course/Section:** ______________________ **Date:** ____________

A. Recall that geoscientists have historically tried to understand the cause of oceans, mountains, and global tectonics by questioning if Earth could be shrinking or expanding in size. This question can be evaluated by studying Earth's natural forces and faults in relation to those that you might predict to occur if the size of Earth were changing. Analyze **FIGURE 2.4** to see how three kinds of faults are caused by three kinds of stress (applied force). To the right of this diagram, predict what kind of faulting you would mostly expect to find in Earth's lithosphere if Earth were expanding or contracting (shrinking).

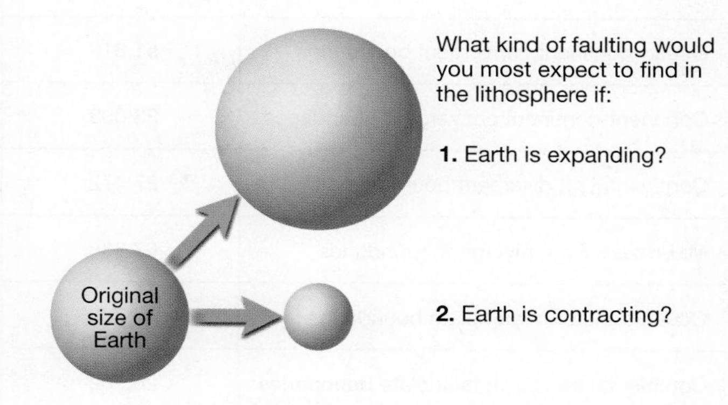

What kind of faulting would you most expect to find in the lithosphere if:

1. Earth is expanding?

2. Earth is contracting?

B. Refer to **FIGURES 2.2** and **2.4**. Fill in the table below to indicate what kind of stress and faulting characterizes each kind of plate boundary.

Plate Boundary Type	Main Stress (applied force)	Main Fault Type
Divergent		
Convergent		
Transform		

C. Refer to **FIGURE 2.5**, a map showing the distribution of Earth's lithospheric plates and three main kinds of plate boundaries: divergent (red), convergent (hachured), and transform (dashed). Also note in **FIGURE 2.2** that there are two kinds of divergent boundaries (continental rifts, mid-ocean ridges) and three kinds of convergent boundaries (ocean-ocean, ocean-continent, and continent-continent).

1. Geologist Peter Bird conducted a classical study of Earth's lithospheric plates and plate boundaries in 2003. He found that the lithosphere is actually broken into 52 plates, many of which are too small to see on a world map. So the typical global map of about 13 main plates shown in **FIGURE 2.5** and most textbooks is quite generalized. He also identified the total combined lengths of all of the different kinds of plate boundaries, as summarized in the two left-hand columns of the table on the back of this page. Complete the two right-hand columns of the table.

Lithospheric Plate Boundary Type	Total Length (kilometers)	Total Length of the three main plate boundary types	Percentage of each of the three main plate boundary types
Ocean-ocean convergent boundaries	17,449	Convergent	Convergent
Ocean-continent convergent boundaries	51,310		
Continent-continent convergent boundaries	23,003		
Continental rift divergent boundaries	27,472	Divergent	Divergent
Mid-ocean ridge divergent boundaries	67,338		
Ocean transform fault plate boundaries	47,783	Transform	Transform
Continental transform fault plate boundaries	26,132		

2. Do you think Earth's size is increasing (expanding), decreasing (shrinking), or staying about the same in size? Justify your answer by citing evidence from your work above.

3. Peter Bird also calculated that Earth's lithosphere is being created at a rate of 3.4 km^2/yr and being destroyed at a rate of 3.4 km^2/yr. At these rates, and the fact that Earth's surface area is 510,000,000 km^2, how long would it take to recycle Earth's entire lithosphere? Show your work.

D. REFLECT & DISCUSS Do you think that plate tectonics is being caused by a change in Earth's size, or something else? If something else, then what do you suggest? Explain.

ACTIVITY 2.3 Lava Lamp Model of Earth

Name: ______________________ **Course/Section:** ______________________ **Date:** ____________

A. Earthquake shear waves travel through both the crust and mantle of Earth. Such waves cannot travel through fluids (liquids, gases), so rocks of both crust and mantle are mostly solid rock, not liquid rock. Yet Plate Tectonic Theory states that lithospheric plates are made of rigid, stiff bodies of elastic-brittle rock (crust and lithospheric mantle) that rest on asthenosphere and deeper parts of the mantle made of weak, ductile rock that flows like soft plastic or a viscous (thick) fluid. Explore how solid rock can be rigid and stiff in the lithosphere but soft and fluid-like in the asthenosphere.

1. Obtain a piece of Silly Putty™ from your instructor. Perform the following tests on it, check the boxes to indicate whether the test results characterize Silly Putty™ as a solid or a liquid, and answer the two questions.

Test	Behaves like a solid	Behaves like a liquid (fluid)
1. Roll the Silly Putty™ into a ball and bounce it on the table.		
2. Hold opposite ends of the mass of Silly Putty™, and pull it apart slowly.		
3. Hold opposite ends of the mass of Silly Putty™, and pull it apart as fast as you can.		
4. Roll the Silly Putty™ into a ball, then press down on it with your thumb.		
5. Roll the Silly Putty™ into a ball, and allow it to sit for 2-3 minutes, or longer.		

Under what conditions of pressure and time does Silly Putty™ behave like a solid?	Under what conditions of pressure and time does Silly Putty™ behave like a liquid?

2. What is a rheid? Is Silly Putty™ a rheid?

3. REFLECT & DISCUSS How does your research on Silly Putty™ help explain how rocks may behave in the lithosphere and beneath the lithosphere?

B. A "lava lamp" is inactive when the light is off, but a lighted lava lamp is dynamic and ever changing. Observe the rising and sinking motion of the lava-like wax in a lighted lava lamp.

1. Describe the motions of the "lava" that occur over one full minute of time, starting with lava at the bottom of the lamp and its path through the lamp.

2. What causes the "lava" to move from the base of the lamp to the top of the lamp? (Be as specific and complete as you can.)

3. What causes the "lava" to move from the top of the lamp to the base of the lamp? (Be as specific and complete as you can.)

4. What is the name applied to this kind of cycle of change? (*Hint:* Refer to **FIGURE 1.6** on page 13.)

C. Observe the seismic tomography image in **FIGURE 2.6**: a slice through Earth's mantle at a depth of 350 kilometers. Unlike the lava lamp that you viewed in a vertical profile from the side of the lamp, this image is a horizontal slice of Earth's mantle viewed from above. This image is also false colored to show where rocks are significantly warmer and less dense (colored red) versus cooler and more dense (colored blue).

1. How is Earth's mantle like a lava lamp?

2. How is Earth's mantle different from a lava lamp?

D. Compare the tectonic plates and plate boundaries in **FIGURE 2.5** to the red and blue regions of the seismic tomography image in **FIGURE 2.6**.

1. Under what kind of plate tectonic feature do the warm, less dense rocks (red) occur most often?

2. Under what kind of plate tectonic feature do the cool, more dense rocks (blue) occur most often?

E. **REFLECT & DISCUSS** Based on your work in **B–D**, draw a vertical cross section (vertical slice) of Earth that shows how mantle convection may be related to plate tectonics. Include and label the following features in your drawing: mid-ocean ridge (divergent plate boundary), lithospheric plate(s) with ocean crust, subduction zone (convergent plate boundary), lithospheric plate with continental crust, arrows to indicate the convection motion of the mantle. Use colored pencils to show where the mantle rocks in your vertical cross section would be red and blue like the false colored mantle rocks in **FIGURE 2.6**.

ACTIVITY 2.4 Paleomagnetic Stripes and Seafloor Spreading

Name: ______________________ **Course/Section:** ______________________ **Date:** ____________

A. Analyze the seafloor part of the map below, just off the Pacific Coast, west of California, Oregon, Washington, and southwest Canada. The colored bands are seafloor magnetic anomalies. Colored bands are rocks with a positive (+) magnetic anomaly, so they have normal polarity, like now. The white bands are rocks with a negative (-) magnetic anomaly, so they have reversed polarity. Different colors indicate the ages of the rocks, in millions of years as shown in the polar reversal time scale provided.

1. Using a pencil, draw a line on the sea floor to show where new ocean crust and lithosphere is forming now (zero millions of years old). Using FIGURE 2.2 as a guide, label the segments of your line that are **Juan de Fuca Ridge** and **Gorda Ridge** (divergent plate boundaries). Then label the segments of your pencil line that are **transform fault** plate boundaries. **Add half-arrows to the transform fault boundaries** to show the relative motion of the rocks.

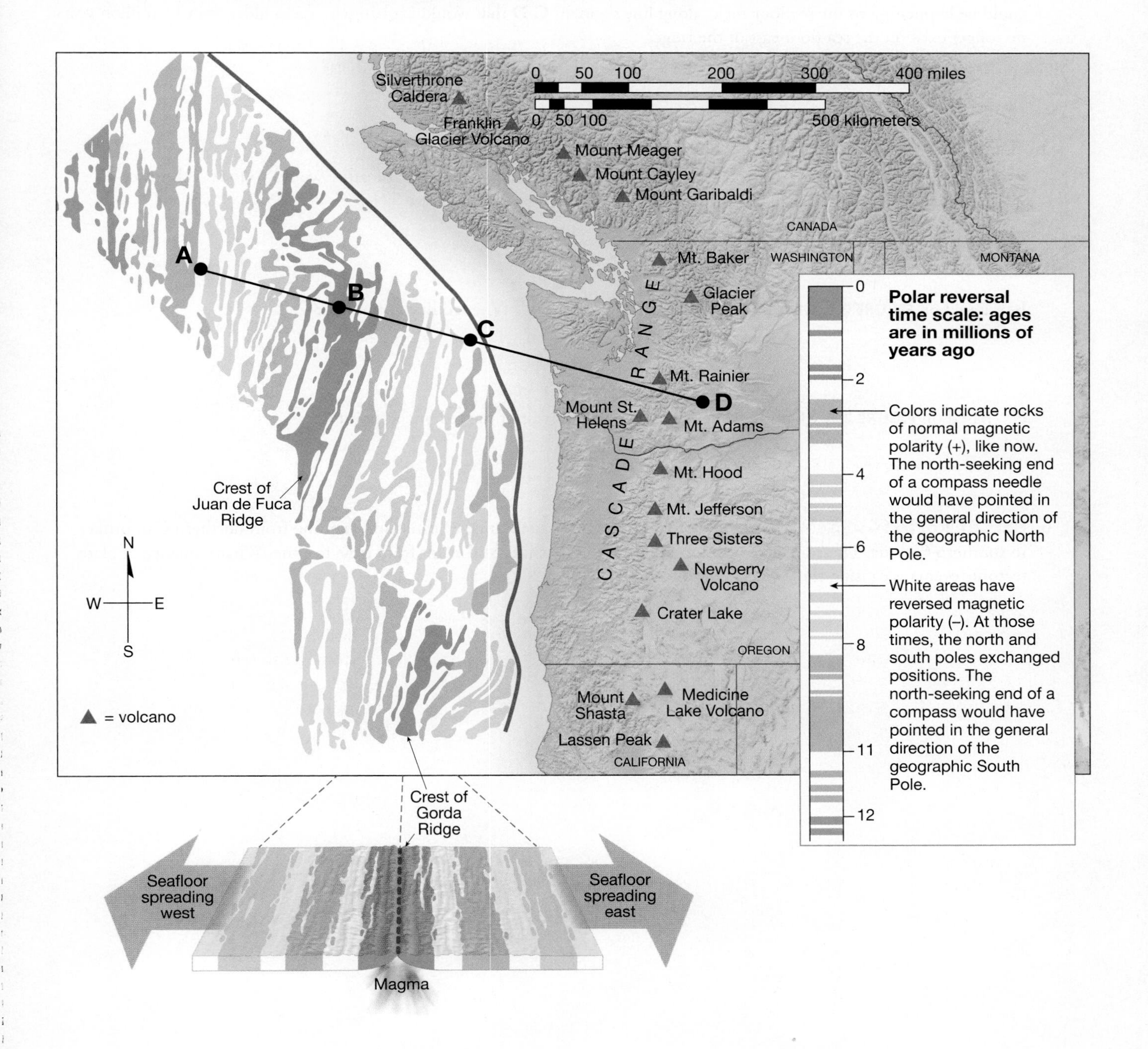

2. What has been the average rate and direction of seafloor spreading in cm per year (cm/yr) west of the Juan de Fuca Ridge, from **B** to **A**? Show your work.

3. What has been the average rate and direction of seafloor spreading in cm per year (cm/yr) east of the Juan de Fuca Ridge, from **B** to **C**? Show your work.

4. Notice that rocks older than 11 million years are present west of the Juan de Fuca Ridge, but not east of the ridge. What could be happening to the seafloor rocks along line segment **C-D** that would explain why rocks older than 11 million years no longer exist on the sea floor east of the ridge?

5. Notice the red line running through point **C:**

   a. If you could take a submarine to view the sea floor along this line, then what feature would you see? (Hint: see **FIGURE 2.2.**)

   b. Based on **FIGURE 2.1**, what lithospheric plate is located east of the red line at point **C**?

   c. Based on **FIGURE 2.1**, what lithospheric plate is located west of the red line at point **C**?

6. **REFLECT & DISCUSS** Notice the line of volcanoes (Cascade Range volcanic arc) running from northern California to southern Canada. These are active volcanoes, meaning that they still erupt from time to time. What sequence of plate tectonic events is causing these volcanoes to form?

ACTIVITY 2.5 Atlantic Seafloor Spreading

Name: ______________________ **Course/Section:** ______________________ **Date:** ____________

A. The map below shows the ages of seafloor basalt (the actual floor of the ocean, beneath the modern mud and sand) between North America and Africa.

1. Draw a red line on the map to show the exact location of the divergent plate boundary between the North American Plate and the African and Eurasian Plates. Refer to **FIGURE 2.5** for assistance as needed.

2. Draw two blue lines on the map to show the exact position of two different transform fault plate boundaries. Refer to **FIGURE 2.2** for assistance as needed.

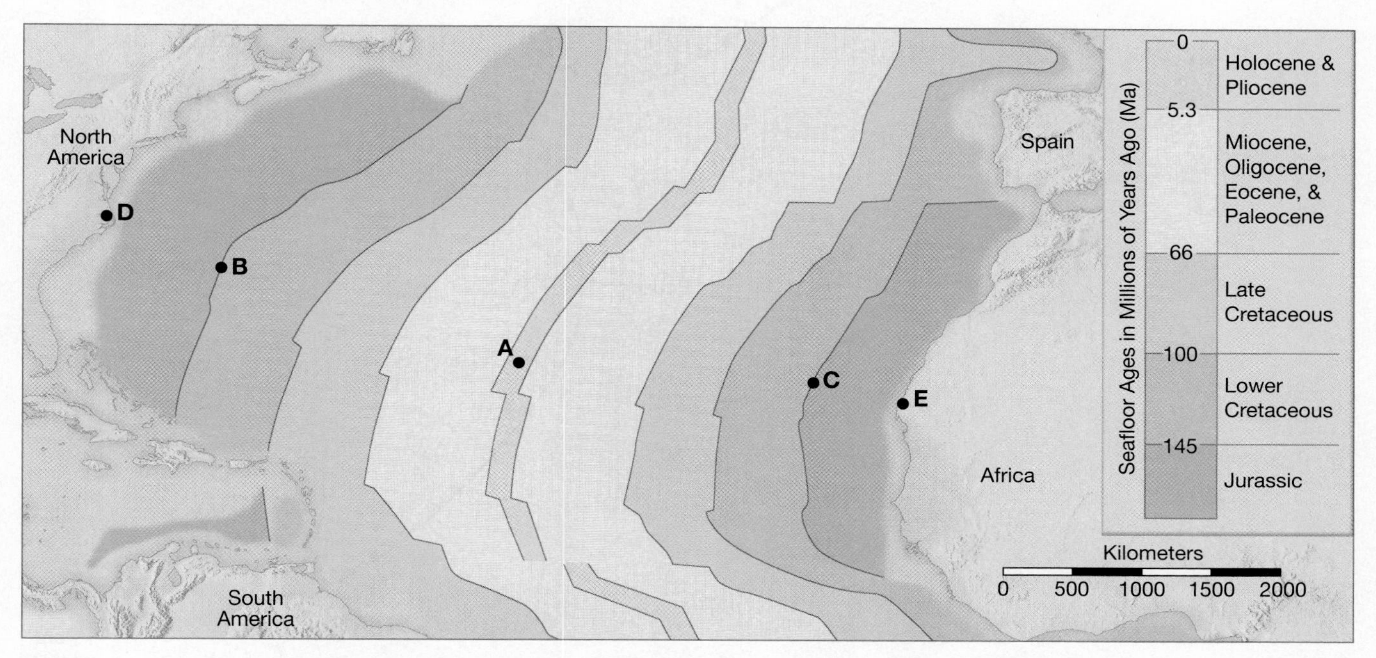

B. Notice that points B and C were together 145 million years ago, but did the sea floor spread apart at exactly the same rate on both sides of the mid-ocean ridge? How can you tell?

C. How far apart are points B and C today, in kilometers? ______________ km

1. Calculate the average rate, in km per million years, that points B and C have moved apart over the past 145 million years. Show your work.

2. Convert your answer above from km per million years to mm per year.

D. REFLECT & DISCUSS Based on your answer in **C1** above, how many millions of years ago and in what geologic period of time were Africa and North America part of the same continent? Show your work.

E. REFLECT & DISCUSS Based on your answer in **C2** above, how far in meters have Africa and North America moved apart since the United States was formed in 1776?

ACTIVITY 2.6 Using Earthquakes to Identify Plate Boundaries

Name: ______________________ **Course/Section:** ______________________ **Date:** ____________

A. Refer to **FIGURE 2.2** for background on how the depth of earthquake foci is related to plate tectonics. On the map below, use a red colored pencil or pen to draw lines (as exactly as you can) that indicate where plate boundaries occur at Earth's surface. Then label the East Pacific Rise, Galapagos Rise, Chile Rise, and all of the plates (refer to **FIGURE 2.5**).

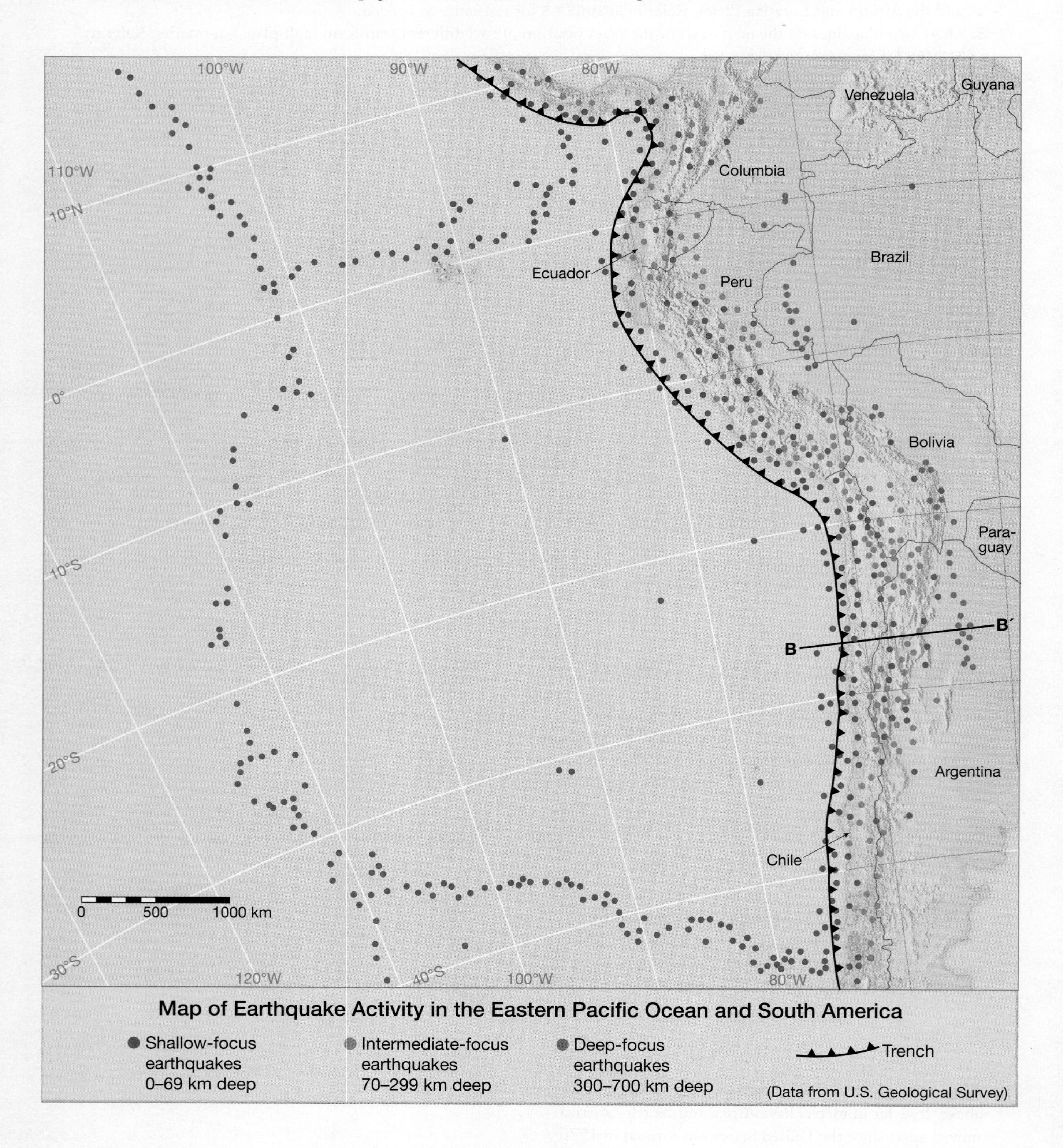

B. Notice line B–B' on the map in Part **A** and the fact that shallow, intermediate, and deep earthquakes occur along it. Volcanoes also occur at Earth's surface along this line. Plot the locations of earthquake foci (depth of earthquake vs. its location east or west of the trench) on the cross section below using data in the accompanying table (provided by the U.S. Geological Survey). For volcanoes, draw a small triangle on the surface (depth of zero).

Location East or West of Trench	Depth of Earthquake (or volcano location)	Location East or West of Trench	Depth of Earthquake (or volcano location)	Location East or West of Trench	Depth of Earthquake (or volcano location)
200 km West	20 km	220 km East	30 km	410 km East	150 km
160 km West	25 km	250 km East	volcano	450 km East	50 km
60 km West	10 km	260 km East	120 km	450 km East	150 km
30 km West	25 km	300 km East	volcano	470 km East	180 km
0 (trench)	20 km	300 km East	110 km	500 km East	30 km
10 km East	40 km	330 km East	volcano	500 km East	160 km
20 km East	30 km	330 km East	40 km	500 km East	180 km
50 km East	60 km	330 km East	120 km	540 km East	30 km
51 km East	10 km	350 km East	volcano	590 km East	20 km
55 km East	30 km	390 km East	volcano	640 km East	10 km
60 km East	20 km	390 km East	40 km	710 km East	30 km
80 km East	70 km	390 km East	140 km	780 km East	530 km
100 km East	10 km	410 km East	volcano	800 km East	560 km
120 km East	80 km	410 km East	25 km	820 km East	610 km
200 km East	110 km	410 km East	110 km	880 km East	620 km

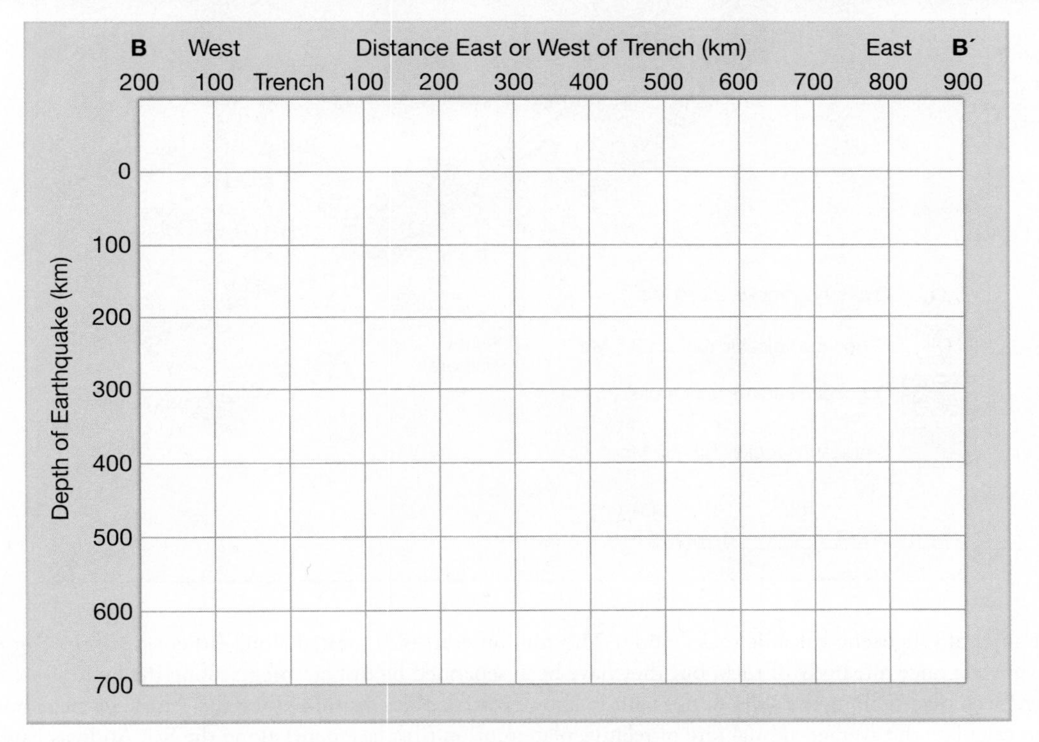

1. What kind of plate boundary is shown in your cross section? ____________________
2. Draw a line in the cross section to show the probable top surface of the subducting plate.
3. Label the part of your cross section that probably represents earthquakes in the lithosphere.
4. At what depth does magma probably originate here just above the subducting plate: ____________________ km
   How can you tell?

5. **REFLECT & DISCUSS** What is the deepest earthquake plotted on your cross section? Do you think there is a lower limit below which earthquakes are not likely to occur? Explain your answer. (Hint: Think about how pressure and temperature influence the behavior of rock in the upper mantle).

ACTIVITY 2.7 San Andreas Transform-Boundary Plate Motions

Name: ______________________ **Course/Section:** ______________________ **Date:** ____________

Study the geologic map of southern California below, showing the position of the famous San Andreas Fault, a transform plate boundary between the North American Plate (east side) and the Pacific Plate (west side). It is well known to all who live in southern California that plate displacements along the fault cause frequent earthquakes, which place humans and their properties at risk.

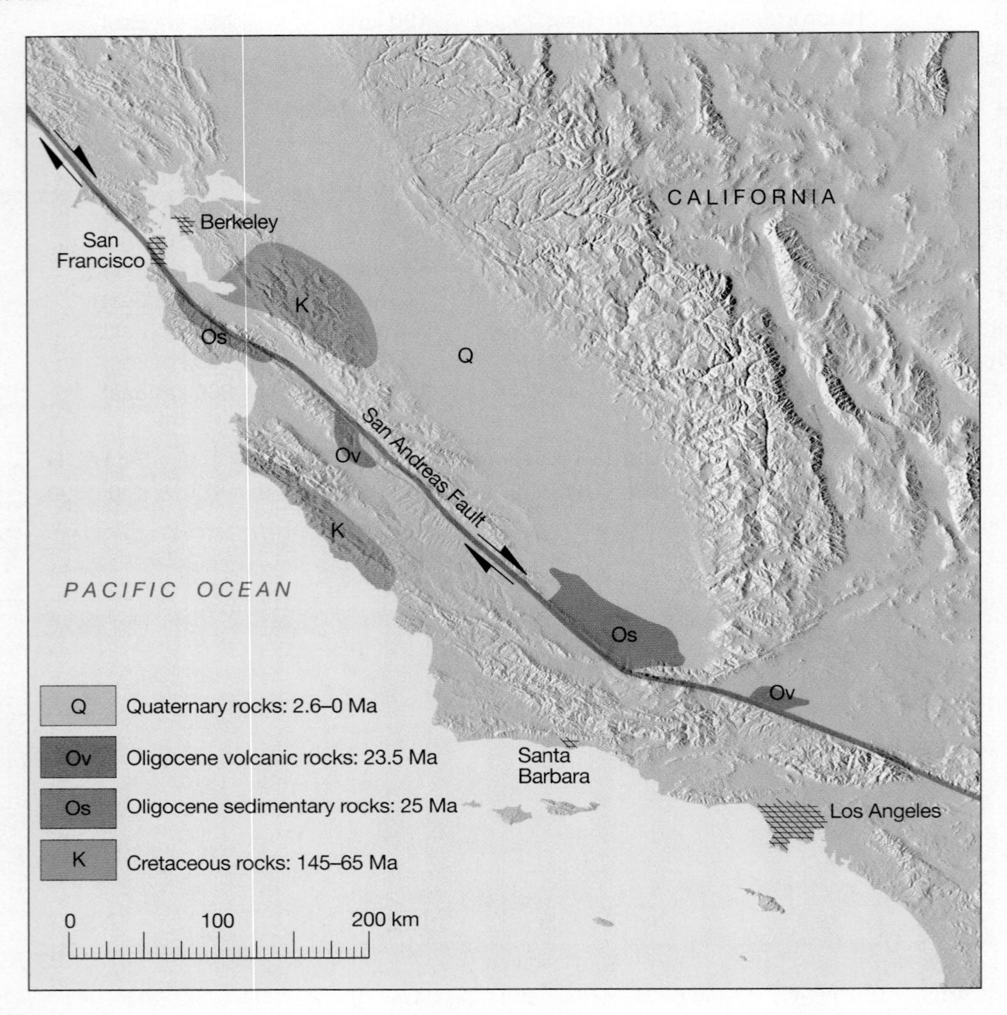

A. The two bodies of Oligocene volcanic rocks (about 23.5 million years old) located along either side of the San Andreas Fault (map above) were once one body of rock, but they have been separated by displacements along the fault. Note that half arrows have been placed along the sides of the fault to show **relative plate motion** along the transform plate boundary here.

1. You can calculate the average annual rate of relative plate motion (displacement) along the San Andreas Fault by measuring how much the Oligocene volcanic rocks have been offset by the fault and by assuming that these rocks began separating soon after they formed. What is the average rate of fault displacement in centimeters per year (cm/yr)? Show your work.

2. An average displacement of about 5 m (16 ft) along the San Andreas Fault was associated with the devastating 1906 San Francisco earthquake that killed people and destroyed properties. Assuming that all displacement along the fault was produced by Earth motions of this magnitude, how often must such earthquakes have occurred in order to account for the total displacement? Show your work.

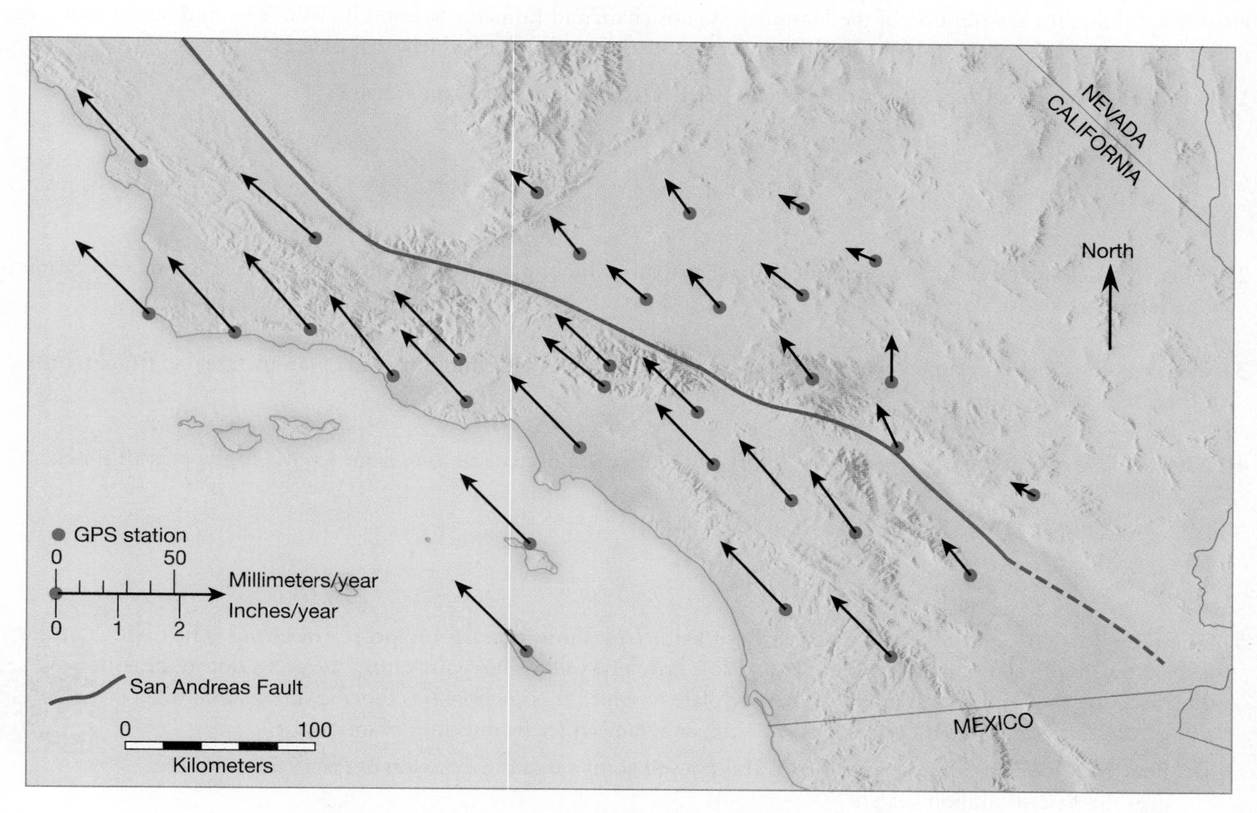

B. The above map shows some Global Positioning System (GPS) reference stations and observations from the JPL-NASA GPS Time Series website at **http://sideshow.jpl.nasa.gov/post/series.html**. Length of the arrows indicates **absolute plate motion**, the direction and rate that the plate is moving in mm/yr at the GPS station (which is attached to bedrock of the plate).

1. Notice that both plates are moving northwest here. Estimate in cm/year how much faster the Pacific Plate is moving than the North American Plate. ____________________ cm/yr
2. Add half arrows along the San Andreas Fault to show the relative movement between the two plates.

C. REFLECT & DISCUSS What is the difference between absolute plate motion and relative plate motion?

ACTIVITY 2.8 Hot Spots and Plate Motions

Name: ______________________ **Course/Section:** ______________________ **Date:** ____________

As a lithospheric plate migrates across a stationary hot spot, a volcano develops directly above the hot spot. When the plate slides past the hot spot the overlying volcano becomes dormant, and over time, it migrates many kilometers from the hot spot. Meanwhile, a new volcano arises as a new part of the plate passes over the hot spot. The result is a string of volcanoes, with one end of the line located over the hot spot and quite active, and the other end distant and inactive. In between is a succession of volcanoes that are progressively older with distance from the hot spot.

A. **FIGURE 2.7** shows the distribution of the Hawaiian Islands chain and Emperor Seamount chain. The numbers indicate age in millions of years old or ago (Ma), obtained from the basaltic igneous rock of which each island is composed.

1. In general, how is the Emperor Seamount chain related to the Hawaiian Islands chain?

2. What was the rate in centimeters per year (cm/yr) and direction of plate motion of the 2300 km long Emperor Seamount Chain from 20 to 40 Ma?

3. What was the rate in centimeters per year (cm/yr) and direction of plate motion in the Hawaiian region from 4.7 to 1.6 Ma?

4. What was the rate in centimeters per year (cm/yr) and direction of plate motion from 1.6 Ma to the present time?

5. Go to the JPL-NASA GPS Time Series website at **http://sideshow.jpl.nasa.gov/post/series.html**. The map displays each GPS station as a green dot and yellow line. The yellow line points from the green dot to the direction that the GPS station (and lithospheric plate to which it is anchored) is moving. Look at Hawaii. JPL-NASA GPS station "NPOC" is located there, and you can see its direction of motion.

a. How does the current motion of NPOC on Hawaii compare to the direction of Pacific Plate motion over the past 40 million years?

b. The NPOC GPS station on Hawaii, and the Pacific Plate to which it is attached, has the following motion: Latitude vector direction & velocity: +1.4825 mm/yr, Longitude vector direction & velocity: –5.1612 mm/yr
Using the formula in **FIGURE 2.3**, Part E, what is the current velocity (in mm/yr) of Pacific Plate motion at Hawaii? Show your work.

B. **REFLECT & DISCUSS** Based on all of your work above, explain how the direction and rate of Pacific Plate movement changed over the past 60 million years.

C. The map below shows the distribution of volcanic calderas (sub-circular depressions caused by repeated volcanic explosions and the associated faulting, tilting, and collapse of the crust) in Wyoming, Idaho, and Nevada. Geologist, Mark Anders mapped and dated these calderas. He also discovered that all of them are now inactive except for the one located directly over the Yellowstone National Park region. The hot springs, geysers, and earthquakes at Yellowstone indicate that the site is still volcanically active.

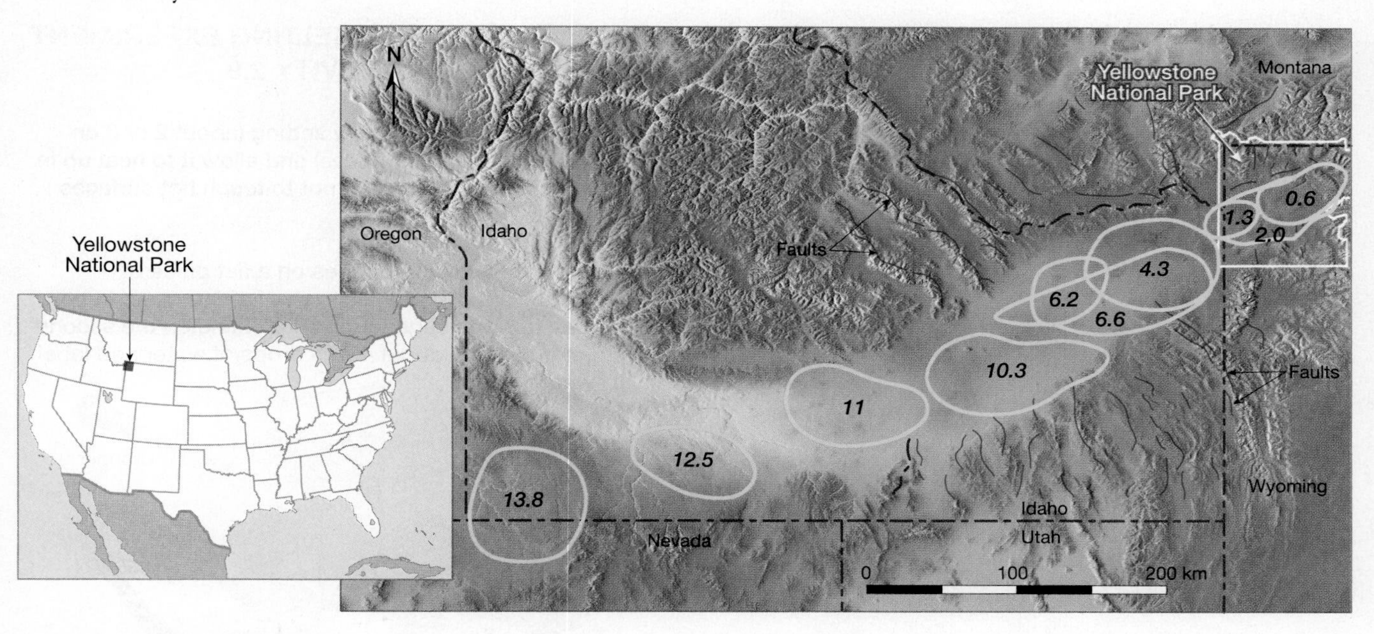

1. Do you think that Yellowstone National Park could be located over a hot spot? Why?

2. Based on the map, what was the rate in centimeters per year (cm/yr) and direction of North American Plate motion at Yellowstone since 10.3 Ma?

   Add an arrow (vector) and rate label to the USA map above to show this movement.

3. **REFLECT & DISCUSS** How do hot spots help us understand plate tectonic processes and rates?

ACTIVITY 2.9 The Origin of Magma

Name: ______________________ **Course/Section:** ______________________ **Date:** ____________

A. Examine the pressure-temperature (P-T) diagram for mantle peridotite in **FIGURE 2.8**, and locate point **X**. This point represents a mass of peridotite buried 80 km underground.

1. According to the continental geothermal gradient, rocks buried 80 km beneath a continent would normally be heated to what temperature?

2. According to the oceanic geothermal gradient, rocks buried 80 km beneath an ocean basin would normally be heated to what temperature?

3. Is the peridotite at point **X** a mass of solid, a mixture of solid and liquid, or a mass of liquid? How do you know?

4. What would happen to the mass of peridotite at point **X** if it were heated to 1750 °C?

5. What would happen to the mass of peridotite at point **X** if it were heated to 2250 °C?

B. At its current depth, the peridotite at point **X** in **FIGURE 2.8** is under about 25,000 atm of pressure.

1. At what depth and pressure will this peridotite begin to melt if it is uplifted closer to Earth's surface and its temperature remains the same?

Depth: ________ Pressure: ________

2. What is the name applied to this kind of melting?

3. Name a process that could uplift mantle peridotite to start it melting in this way, and name a specific plate tectonic setting where this may be happening now. (*Hint:* Study **FIGURES 2.2, 2.5,** and **2.6.**)

PROCEDURES FOR MELTING EXPERIMENT IN ACTIVITY 2.9

1. Turn the hot plate on a low setting (about 2 or 3 on most commercial hot plates) and allow it to heat up in a safe location (be careful not to touch hot surfaces directly).

2. Next, place two sugar cubes on a flat piece of aluminum foil or in aluminum foil baking cups. Label (on the foil) one sugar cube "dry." Moisten the second sugar cube with about 4 or 5 drops of water and label it "wet."

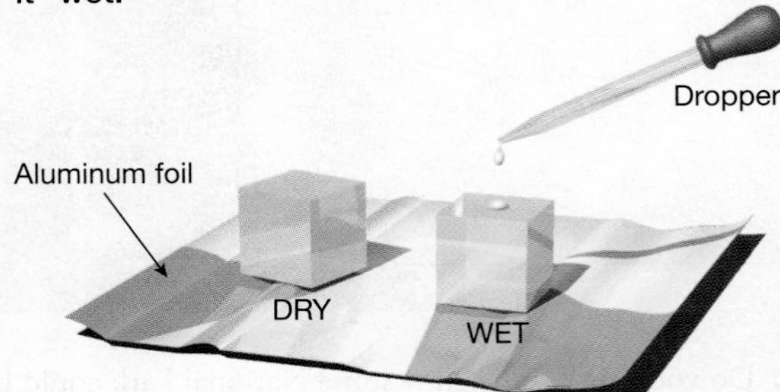

3. Carefully place the aluminum foil with the labeled sugar cubes onto the hot plate and observe what happens. When one of the sugar cubes begins to melt, use crucible tongs and/or hot pads to remove the foil and sugar cubes from the hot plate and avoid burning the sugar.

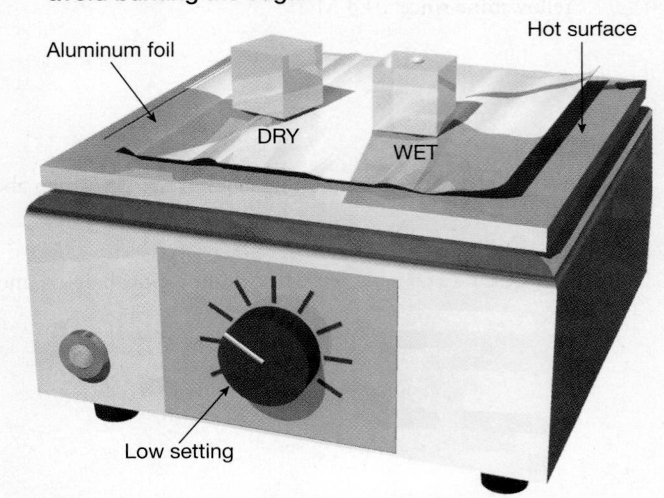

4. Turn off and un-plug the hot plate as soon as you finish #3 above. Be careful not to touch the hot surface!

FIGURE 2.9 Procedures for melting experiment in Activity 2.9. Be careful not to burn yourself on the hot surface or in molten sugar. Do not create a hazard by burning the sugar until it smokes or catches fire.

C. REFLECT & DISCUSS Based on your answers above, what are two environmental changes that can cause the peridotite at point **X** (see **FIGURE 2.8**) to begin partial melting?

D. Obtain the materials shown in **FIGURE 2.9**. Turn the hot plate on a low setting (about 3 on most commercial hot plates) and allow it to heat up in a safe location. Next place two sugar cubes on a flat piece of aluminum foil. Label (on the foil) one sugar cube "dry." Moisten the second sugar cube with a few drops of water, and label it "wet." Carefully place the aluminum foil (with the sugar cubes) onto the hot plate and observe what happens. (*Note:* Turn off the hot plate when one cube begins to melt.)

1. Which sugar cube melted first?

2. The rapid melting that you observed in the moistened sugar cube is called "flux melting," because flux is a material that promotes (speeds up) melting. What was the flux?

3. How would the P-T diagram in **FIGURE 2.9** change if all of the peridotite in the diagram was "wet" peridotite?

4. In what specific kind of plate tectonic setting could water enter Earth's mantle and cause flux melting of mantle peridotite? (*Hint:* **FIGURE 2.3**)

E. REFLECT & DISCUSS Examine this cross section of a plate boundary.

1. What kind of plate boundary is this?

2. Name the specific process that led to the formation of magma in this cross section.

3. Describe the sequence of plate tectonic and magma generating processes that led to formation of the volcanoes (oceanic ridge) in this cross section.

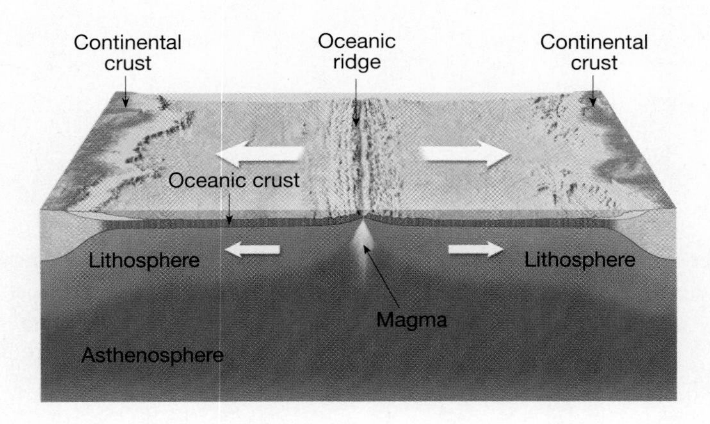

F. **REFLECT & DISCUSS** Examine this cross section of a plate boundary.

1. What kind of plate boundary is this?

2. Name the specific process that led to the formation of magma in this cross section.

3. Describe the sequence of plate tectonic and magma generating processes that led to formation of the volcanoes (continental volcanic arc) in this cross section.

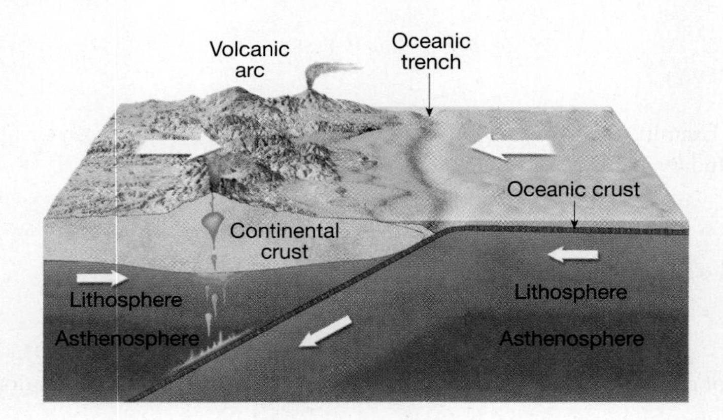

LABORATORY

Mineral Properties, Identification, and Uses

CONTRIBUTING AUTHORS
Jane L. Boger • *SUNY, College at Geneseo*
Philip D. Boger • *SUNY, College at Geneseo*
Roseann J. Carlson • *Tidewater Community College*
Charles I. Frye • *Northwest Missouri State University*
Michael F. Hochella, Jr. • *Virginia Polytechnic Institute*

Bingham Canyon Mine, southwest of Salt Lake City, Utah. It is primarily a copper mine, but gold, silver, and other metals have also been extracted from the ore here for over a century. (Michael Collier)

BIG IDEAS

Minerals comprise rocks and are described and classified on the basis of their physical and chemical properties. Every person depends on minerals and elements refined from them, but the supply of minerals is nonrenewable, and the magnitude of their use may be unsustainable.

FOCUS YOUR INQUIRY

THINK About It What are minerals and crystals, and how are they related to rocks and elements?

ACTIVITY 3.1 Mineral and Rock Inquiry *(p. 74)*

THINK About It How and why do people study minerals?

ACTIVITY 3.2 Mineral Properties *(p. 77)*

ACTIVITY 3.3 Determining Specific Gravity (SG) *(p. 86)*

THINK About It How and why do people study minerals? How do you personally depend on minerals and elements extracted from them?

ACTIVITY 3.4 Mineral Identification and Uses *(p. 88)*

THINK About It How do you personally depend on minerals and elements extracted from them? How sustainable is your personal dependency on minerals and elements extracted from them?

ACTIVITY 3.5 The Mineral Dependency Crisis *(p. 89)*

THINK About It How sustainable is your personal dependency on minerals and elements extracted from them?

ACTIVITY 3.6 Urban Ore *(p. 99)*

ACTIVITY

3.1 Mineral and Rock Inquiry

THINK About It What are minerals and crystals, and how are they related to rocks and elements?

OBJECTIVE Analyze rock samples, and infer how minerals are related to and distinguished from rocks, crystals, and chemical elements.

PROCEDURES

1. **Before you begin**, do not look up definitions and information. Use your current knowledge, and complete the worksheet with your current level of ability. Also, this is **what you will need** to do the activity:

   ____ Activity 3.1 Worksheet (p. 101) and pencil
2. **Then answer every question on the worksheet in a way that makes sense to you** and be prepared to compare your ideas with others.
3. **After you complete the worksheet**, read about minerals and rocks below and be prepared to discuss your observations, interpretations, and inferences with others.

Minerals and Rocks

Many people think of minerals as the beautiful natural crystals mined from the rocky body of Earth and displayed in museums or mounted in jewelry. But table salt, graphite in pencil leads, and gold nuggets are also minerals.

What Are Minerals?

According to geologists, **minerals** are inorganic, naturally occurring solids that have a definite chemical composition, distinctive physical properties, and crystalline structure. In other words, each mineral

- occurs in the solid, rocky body of Earth, where it formed by processes that are inorganic (not involving life).
- has a definite chemical composition of one or more chemical elements that can be represented as a chemical formula (like NaCl for halite, FeS_2 for pyrite, and Au for pure "native gold").
- has physical properties (like hardness, how it breaks, and color) that can be used to identify it.
- has crystalline structure—an internal patterned arrangement or geometric framework of atoms that can be revealed by external crystal faces (**FIGURES 3.1A, B**), the way a mineral breaks (**FIGURE 3.2B**), and in atomic-resolution images (**FIGURE 3.2C**).

A few "minerals," such as limonite (rust) and opal (**FIGURE 3.3**) never form crystals, so they do not have crystalline structure. They are mineral-like materials (*mineraloids*) rather than true minerals. And even though all true minerals normally form by inorganic processes, some organisms make them as shells or other parts of their bodies. These so-called *biominerals* are of obvious organic origin (made by plants and animals). Examples include aragonite mineral crystals in clam shells and tiny magnetite crystals in the human brain. People make *cultured* mineral crystals in laboratories. Their chemical and physical properties are identical to naturally-formed mineral crystals, but they are not true minerals because they are *synthetic* (man-made, not natural).

How Are Minerals Classified?

Geologists have identified and named thousands of different kinds of minerals, but they are often classified into smaller groups according to their importance, use, or chemistry. For example, a group of only about twenty are known as **rock-forming minerals,** because they are the minerals that make up most of Earth's crust. Another group is called the **industrial minerals**, because they are the main non-fuel raw materials used to sustain industrialized societies like ours. Some industrial minerals are used in their raw form, such as quartz (quartz sand), muscovite (used in computer chips), and gemstones. Most are refined to obtain specific elements such as iron, copper, and sulfur. All minerals are also classified into the following chemical classes:

- **Silicate minerals** are composed of pure silicon dioxide (SiO_2, called quartz) or silicon-oxygen ions $(SiO_4)^{4-}$ combined with other elements. Examples are olivine: $(Fe, Mg)_2SiO_4$, potassium feldspar: $KAlSi_3O_8$, and kaolinite: $Al_2(Si_4O_{10})(OH)_8$.
- **Oxide minerals** contain oxygen (O^{2-}) combined with a metal (except for those containing silicon, which are silicate minerals). Examples are hematite: Fe_2O_3, magnetite: Fe_2O_3, and corundum: Al_2O_3.
- **Hydroxide minerals** contain hydroxyl ions $(OH)^-$ combined with other elements (except for those containing silicon, which are silicate minerals). Examples are goethite: FeO(OH) and limonite: $FeO(OH) \cdot nH_2O$.
- **Sulfide minerals** contain sulfur ions (S^{2-}) combined with metal(s) and no oxygen. Examples are pyrite: FeS_2, galena: PbS, and sphalerite: ZnS. When they are scratched or crushed, one can usually smell the sulfur in these minerals.
- **Sulfate minerals** contain sulfate ions $(SO_4)^{2-}$ combined with other elements. Examples include gypsum: $CaSO_4 \cdot H_2O$ and barite: $BaSO_4$.
- **Carbonate minerals** contain carbonate ions $(CO_3)^{2-}$ combined with other elements. Examples include calcite: $CaCO_3$ and dolomite: $CaMg(CO_3)_2$. These minerals react with acid, the way baking soda (which is the mineral named nahcolite and the chemical compound named sodium bicarbonate: $NaHCO_3$) reacts with acetic acid (CH_3COOH) in vinegar. Geologists use dilute hydrochloric acid (HCl) to detect carbonate minerals because the reaction makes larger bubbles. If a mineral reacts with the dilute HCl, then it is a carbonate mineral.

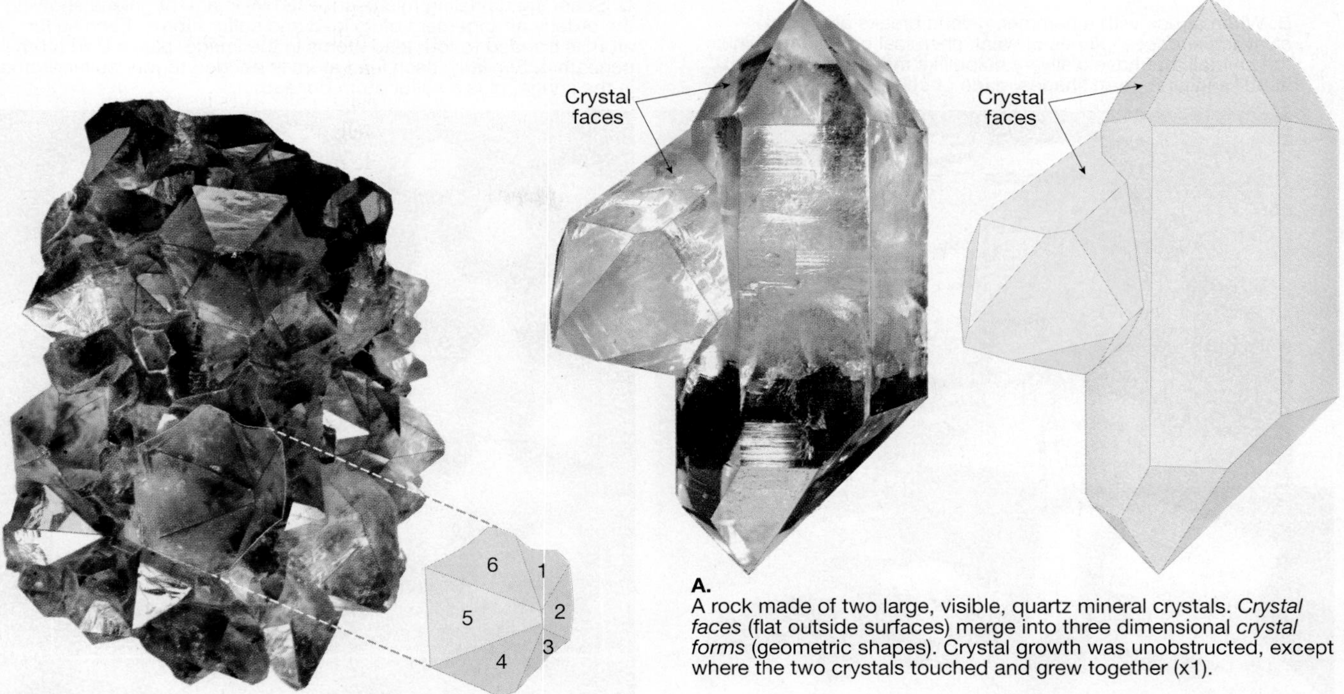

Top view: Crystal growth was unobstructed so crystal faces are developed (x1).

A.
A rock made of two large, visible, quartz mineral crystals. *Crystal faces* (flat outside surfaces) merge into three dimensional *crystal forms* (geometric shapes). Crystal growth was unobstructed, except where the two crystals touched and grew together (x1).

Side view: Deformed crystal faces among crowded intergrown crystals (x1).

B.
Rock made of many quartz mineral crystals. Note how crystal growth was obstructed as the sides of many crystals grew together (side view), but tips of the crystals (top view) grew unobstructed into six-sided pyramids. Iron impurity gives the purple amethyst variety of quartz its color.

C.
Crystal growth of the calcite mineral crystals in this rock (marble) was obstructed in every direction. The crystals grew together as a dense mass of odd-shaped crystals instead of perfect crystal forms.

Thin section (x30). The layers of agate are made of long intergrown quartz mineral crystals.

D.
Slice of rock (agate) cut with a diamond saw and polished. The layers are made of quartz mineral crystals that are *cryptocrystalline* (not visible in hand sample). They can only be seen in a thin section (thin transparent slice of the rock mounted on a glass slide) magnified with a microscope to 30 times larger than their actual size (x30).

FIGURE 3.1 Minerals and rocks. Most rocks are made of one or more mineral crystals.

B. When struck with a hammer, galena breaks along flat *cleavage surfaces* (planes of weak chemical bonding within the crystal) that have a silvery color, like metal, and intersect at 90° angles to form shapes made of cubes.

C. Scanning tunneling microscope (STM) image of galena showing the orderly arrangement of its lead and sulfur atoms. Each sulfur atom is bonded to four lead atoms in the image, plus a lead atom beneath it. Similarly, each lead atom is bonded to four sulfur atoms in the image, plus a sulfur atom beneath it.

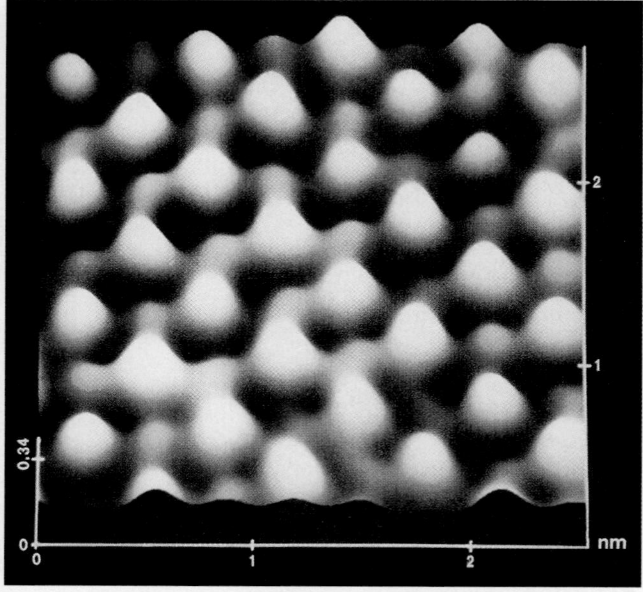

Blue = S (sulfur) atoms, Orange = Pb (lead) atoms
nm = nanometer = 1 millionth of a millimeter

A. Galena mineral crystals form cubic shapes that tarnish to a dull gray color.

FIGURE 3.2 Crystal shape, cleavage, and atomic structure. Galena is lead sulfide—PbS. It is an ore mineral from which lead (Pb) and sulfur (S) are extracted. (STM image by C.M. Eggleston, University of Wyoming)

- **Halide minerals** contain a halogen ion (F^-, Cl^-, Br^-, or I^-) combined with a metal. Examples are halite: NaCl and fluorite: CaF_2.
- **Phosphate minerals** contain phosphate ions $(PO_4)^{3-}$ combined with other elements. An example is apatite: $Ca_5F(PO_4)_3(OH, F, Cl)$.
- **Native elements** are elements in pure form, not combined with different elements. Examples include graphite: C, copper: Cu, sulfur: S_2, gold: Au, and silver: Ag.

How Are Minerals Related to Rocks?

Most **rocks** are aggregates of one or more mineral crystals. For example, mineral crystals comprise all of the rocks in FIGURE 3.1. Notice that you can easily detect the mineral crystals in FIGURES 3.1A and 3.1B by their flat **faces**, which are an external feature of the internal geometric framework of their atoms. However, the crystals in many rocks have grown together in such a crowded way that few faces are visible (FIGURES 3.1C). Some rocks are also **cryptocrystalline**, made of crystals that are only visible under a microscope (FIGURE 3.1D).

Earth is sometimes called the "third rock" (rocky planet) from the Sun, because it is mostly made of rocks. But rocks are generally made of one or more minerals, which are the natural materials from which every inorganic item in our industrialized society has been manufactured. Therefore, minerals are the physical foundation of both our rocky planet and our human societies.

Opal is a residue of hydrated silicon dioxide that forms light-colored translucent masses like this. Notice its lack of crystals and cleavage. This "precious" opal has been polished to enhance its internal flashes of color.

Limonite forms dull powdery yellow-brown to dense dark brown masses like this. Notice its lack of crystals and cleavage. It is a residue of hydrated iron oxide and/or hydrated iron oxyhydroxide that you know as rust.

FIGURE 3.3 Mineraloids. Opal and limonite are naturally-occurring inorganic materials, but they are *amorphous* (non-crystalline; they never form crystals). This makes them *mineraloids* (amorphous mineral-like materials), rather true minerals, but they are described, identified, and listed as minerals.

ACTIVITY

3.2 Mineral Properties

THINK About It | **How and why do people study minerals?**

OBJECTIVE Analyze and describe the physical and chemical properties of minerals.

PROCEDURES

1. **Before you begin**, read the following background information. This is **what you will need**:
   - ____ Activity 3.2 Worksheets (pp. 102–103) and pencil
   - ____ set of mineral samples (obtained as directed by your instructor)
   - ____ set of mineral analysis tools (obtained as directed by your instructor)
   - ____ cleavage goniometer cut from GeoTools Sheet 1 at the back of the manual
2. **Then follow your instructor's directions** for completing the worksheets.

What Are a Mineral's Chemical and Physical Properties?

The **chemical properties** of a mineral are its characteristics that can only be observed and measured when or after it undergoes a chemical change due to reaction with another material. This includes things like if or how it tarnishes (reacts with air or water) and whether or not it reacts with acid. For example, calcite and other carbonate (CO_3-containing) minerals react with acid, and native copper tarnishes to a dull brown or green color when it reacts with air or water.

The **physical properties** of a mineral are its characteristics that can be observed (and sometimes measured) without changing its composition. This includes things like how it looks (color, luster, clarity) before it tarnishes or weathers by reacting with air or water, how well it resists scratching (hardness), how it breaks or deforms under stress (cleavage, fracture, tenacity), and the shapes of its crystals. For example, quartz crystals are hard to scratch, glassy, and transparent, while talc is easily scratched, opaque, and feels greasy.

In this activity, you will use the properties of color and clarity (before and after tarnishing), crystal form, luster (before and after tarnishing), streak, hardness, cleavage, and fracture to describe mineral samples. Additional properties—such as tenacity, reaction with acid, magnetic attraction, specific gravity, striations, and exsolution lamellae—can also be helpful in analyzing particular minerals.

Color and Clarity. A mineral's **color** is usually its most noticeable property and may be a clue to its identity. Minerals normally have a typical color, like gold. A rock made up of one color of mineral crystals is usually made up of one kind of mineral, and a rock made of more than one color of mineral crystals is usually made up of more than one kind of mineral. However, there are exceptions, like the agate in FIGURE 3.1D. It has many colors, but they are simply *varieties* (var.)—different colors—of the mineral quartz. This means that a mineral cannot be identified solely on the basis of its color. The mineral's other properties must also be observed, recorded, and used collectively to identify it. Most minerals also tend to exhibit one color on freshly broken surfaces and a different color on tarnished or weathered surfaces. Be sure to note this difference, if present, to aid your identification.

Mineral crystals may vary in their **clarity**: degree of transparency or their ability to transmit light. They may be *transparent* (clear and see-through, like window glass), *translucent* (foggy, like looking through a steamed-up shower door), or *opaque* (impervious to light, like concrete and metals). It is good practice to record not only a mineral's color, but also its clarity. For example, the crystals in FIGURE 3.1B are purple in color and have transparent to translucent clarity. Galena mineral crystals (FIGURE 3.2) are opaque.

Crystal Forms and Mineral Habits. The geometric shape of a crystal is its **crystal form**. Each form is bounded by flat **crystal faces** that intersect at specific angles and in symmetrical relationships (FIGURE 3.1A and B). The crystal faces are the outward reflection of the way that atoms or groups of atoms bonded together in a three-dimensional pattern as the crystal grew under specific environmental conditions. There are many named crystal forms (FIGURE 3.4). Combinations of two or more crystals can also form named patterns, shapes, or twins (botryoidal, dendritic, radial, fibrous: FIGURE 3.4). A mass of mineral crystals lacking a distinctive pattern of crystal growth is called *massive*.

Development of Crystal Faces. The terms euhedral, subhedral, and anhedral describe the extent to which a crystal's faces and form are developed. *Euhedral crystals* have well developed crystal faces and clearly defined and recognizable crystal forms (FIGURE 3.1A). They develop only if a mineral crystal is unrestricted as it grows. This is rare. It is more common for mineral crystals to crowd together as they grow, resulting in a massive network of intergrown crystals with deformed crystal faces and odd shapes or imperfect crystal forms (FIGURE 3.1B). *Subhedral* crystals are imperfect but have enough crystal faces that their forms are recognizable. *Euhedral* crystals have no crystal faces, so they have no recognizable crystal form (FIGURE 3.1C). Most of the laboratory samples of minerals that you will analyze do not exhibit their crystal forms because they are small broken pieces of larger crystals. But whenever the form or system of crystals in a mineral sample can be detected, then it should be noted and used as evidence for mineral identification.

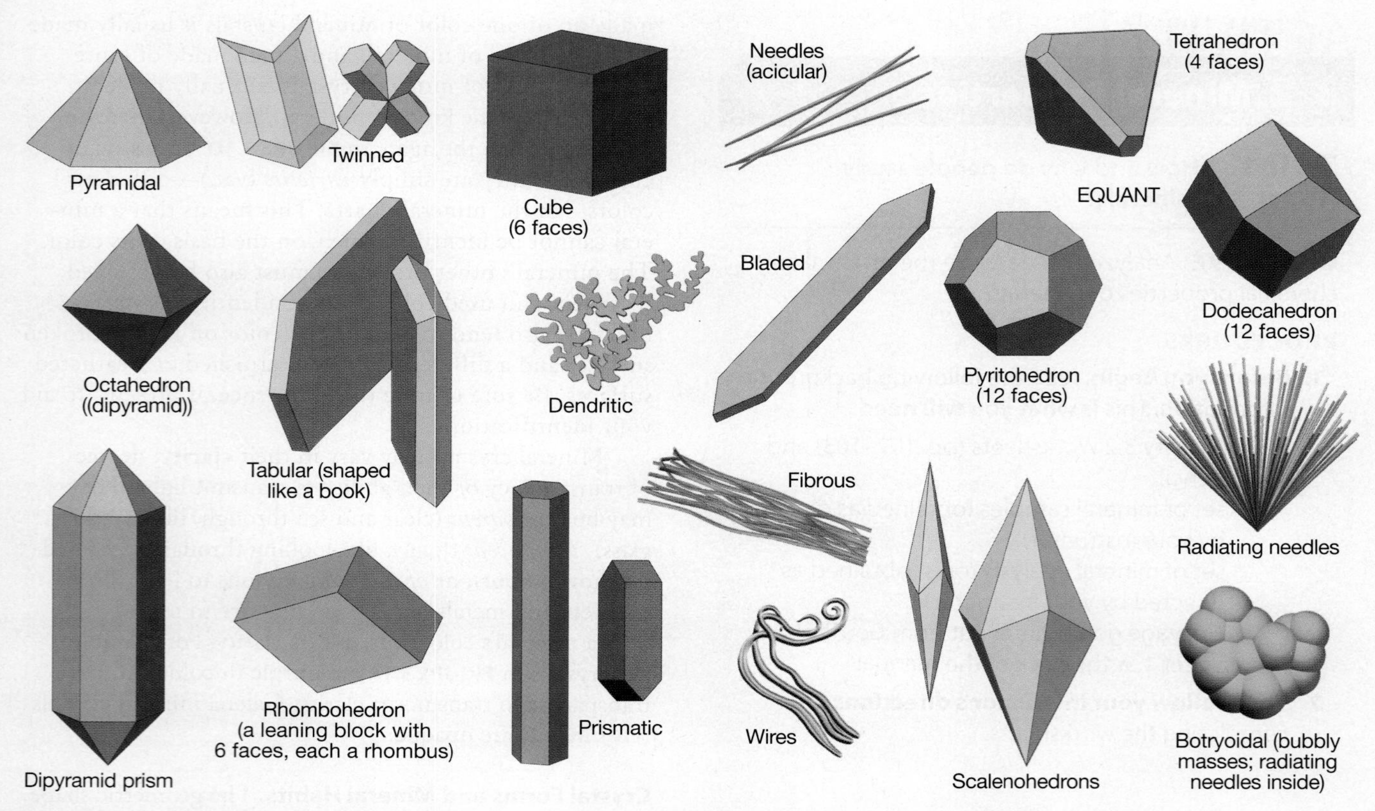

FIGURE 3.4 Crystal forms and combinations. *Crystal form* is the geometric shape of a crystal, and is formed by intersecting flat outer surfaces called *crystal faces*. Combinations of two or more crystals can form patterns, shapes, or twins that also have names. *Massive* refers to a combination of mineral crystals so tightly inter-grown that their crystal forms cannot be seen in hand sample.

Crystal Systems. Each specific crystal form can be classified into one of six *crystal systems* (FIGURE 3.5) according to the number, lengths, and angular relationships of imaginary geometric axes along which its crystal faces grew. The crystal systems comprise 32 classes of crystal forms, but only the common crystal forms are illustrated in FIGURE 3.5.

Mineral Habit. A mineral's **habit** is the characteristic crystal form(s) or combinations (clusters, coatings, twinned pairs) that it habitually makes under a given set of environmental conditions. Pyrite forms under a variety of environmental conditions so it has more than one habit. Its habit is cubes, pyritohedrons, octahedrons, or massive (FIGURE 3.4).

Luster. A mineral's **luster** is a description of how light reflects light from its surfaces. Luster is of two main types—metallic and nonmetallic—that vary in intensity from bright (very reflective, shiny, polished) to dull (not very reflective, not very shiny, not polished). For example, if you make a list of objects in your home that are made of metal (e.g., coins, knives, keys, jewelry, door hinges, aluminum foil), then you are already familiar with metallic luster. Yet the metallic objects can vary from bright (very reflective—like polished jewelry, the polished side of aluminum foil, or new coins) to dull (non-reflective—like unpolished jewelry or the unpolished side of aluminum foil).

Metallic Luster. Minerals with a **metallic luster (M)** reflect light just like the metal objects in your home—they have opaque, reflective surfaces with a silvery, gold, brassy, or coppery sheen (FIGURES 3.2B, 3.6A, 3.7A).

Nonmetallic Luster. All other minerals have a **nonmetallic luster (NM)**—a luster unlike that of the metal objects in your home (FIGURES 3.1, 3.2A, 3.3). The luster of nonmetallic minerals can also be described with the more specific terms below:

- Vitreous—very reflective luster resembling freshly broken glass or a glossy photograph
- Waxy—resembling the luster of a candle
- Pearly—resembling the luster of a pearl
- Earthy (dull)—lacking reflection, like dry soil
- Greasy—resembling the luster of grease, oily

Tarnish and Submetallic Luster. Most metallic minerals will normally tarnish (chemically weather) to a more dull nonmetallic luster, like copper coins. Notice how the exposed metallic copper crystals in FIGURE 3.6 and the galena crystals in FIGURE 3.2A have tarnished to a nonmetallic luster. Always observe freshly broken surfaces of a mineral (e.g., FIGURE 3.2B) to determine whether it has a metallic or nonmetallic luster. It is also useful to note a mineral's luster on fresh versus tarnished

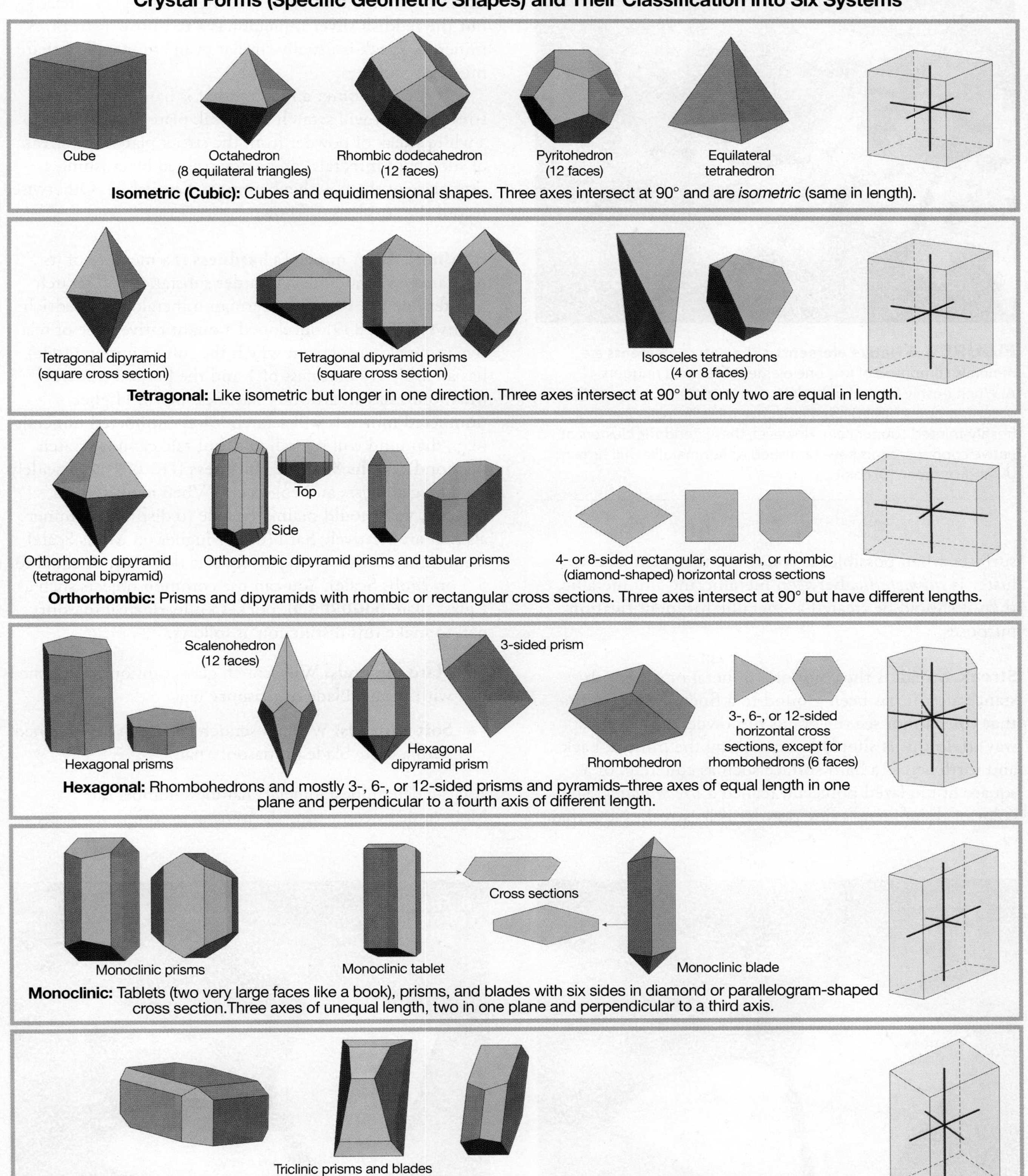

FIGURE 3.5 Crystal systems. Each specific crystal form can be classified into one of six *crystal systems* (major groups) according to the number, lengths, and angular relationships of imaginary geometric axes along which its crystal faces grew (red lines in the right-hand models of each system above). Only the common crystal forms of each class are illustrated and named above.

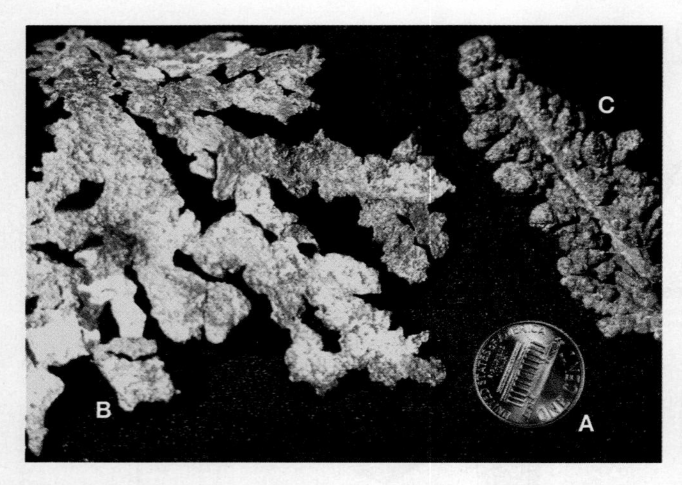

FIGURE 3.6 Native elements. The native elements are minerals composed of just one element, like gold nuggets. A. When freshly formed or broken, native copper (Cu, naturally-occurring pure copper) has a reflective metallic luster like this freshly-minted copper coin. However, these dendritic clusters of native copper crystals have tarnished to nonmetallic dull brown (A) and/or green (B) colors.

surfaces when possible. If you think that a mineral's luster is *submetallic*, between metallic and nonmetallic, then it should be treated as metallic for identification purposes.

Streak. Streak is the color of a mineral or other substance after it has been ground to a fine powder (so fine that you cannot see the grains of powder). The easiest way to do this is simply by scratching the mineral back and forth across a hard surface such as concrete, or a square of unglazed porcelain (called a *streak plate*). The color of the mineral's fine powder is its streak. Note that the brassy mineral in FIGURE 3.7 has a dark gray streak, but the reddish silver mineral has a red-brown streak. A mineral's streak is usually similar even among all of that mineral's varieties.

If you encounter a mineral that is harder than the streak plate, it will scratch the streak plate and make a white streak of powder from the streak plate. The streak of such hard minerals can be determined by crushing a tiny piece of them with a hammer (if available). Otherwise, record the streak as unknown.

Hardness (H). A mineral's **hardness** is a measure of its resistance to scratching. A harder substance will scratch a softer one (FIGURE 3.8). German mineralogist Friedrich Mohs (1773–1839) developed a quantitative scale of relative mineral hardness on which the softest mineral (talc) has an arbitrary hardness of 1 and the hardest mineral (diamond) has an arbitrary hardness of 10. Higher-numbered minerals will scratch lower-numbered minerals (e.g., diamond will scratch talc, but talc cannot scratch diamond). **Mohs Scale of Hardness** (FIGURE 3.9) is widely used by geologists and engineers. When identifying a mineral, you should mainly be able to distinguish minerals that are relatively hard (6.0 or higher on Mohs Scale) from minerals that are relatively soft (less than or equal to 5.5 on Mohs Scale). You can use common objects such as a glass plate (FIGURE 3.9), pocket knife, or steel masonry nail to make this distinction as follows.

- **Hard minerals:** Will scratch glass; cannot be scratched with a knife blade or masonry nail.
- **Soft minerals:** Will not scratch glass; can be scratched with a knife blade or masonry nail.

You can determine a mineral's hardness number on Mohs Scale by comparing the mineral to common objects

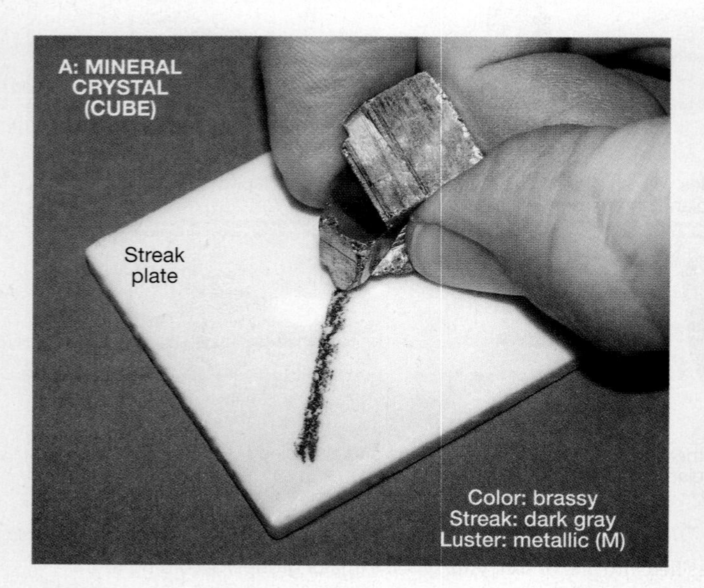

FIGURE 3.7 Streak tests. Determine a mineral's streak (color in powdered form) by scratching it across a streak plate with significant force, then blowing away larger pieces of the mineral to reveal the color of the powder making the streak. If you do not have a streak plate, then determine the streak color by crushing or scratching part of the sample to see the color of its powdered form.

shown in **FIGURE 3.9** or pieces of the minerals in Mohs Scale. Commercial *hardness kits* contain a set of all of the minerals in **FIGURE 3.9** or a set of metal scribes of known hardnesses. When using such kits to make hardness comparisons, remember that the harder mineral/object is the one that scratches, and the softer mineral/object is the one that is scratched.

Cleavage and Fracture. Cleavage is the tendency of some minerals to break (*cleave*) along flat, parallel surfaces (**cleavage planes**) like the flat surfaces on broken pieces of galena (**FIGURE 3.2B**). Cleavage planes are surfaces of weak chemical bonding (attraction) between repeating, parallel layers of atoms in a crystal. Each different set of parallel cleavage planes is referred to as a *cleavage direction*. Cleavage can be described as excellent, good, or poor (**FIGURE 3.10**). An *excellent cleavage* direction reflects light in one direction from a set of obvious, large, flat, parallel surfaces. A *good cleavage* direction reflects light in one direction from a set of many small, obvious, flat, parallel surfaces. A *poor cleavage* direction reflects light from a set of small, flat, parallel surfaces that are difficult to detect. Some of the light is reflected in one direction from the small cleavage surfaces, but most of the light is scattered randomly by fracture surfaces separating the cleavage surfaces.

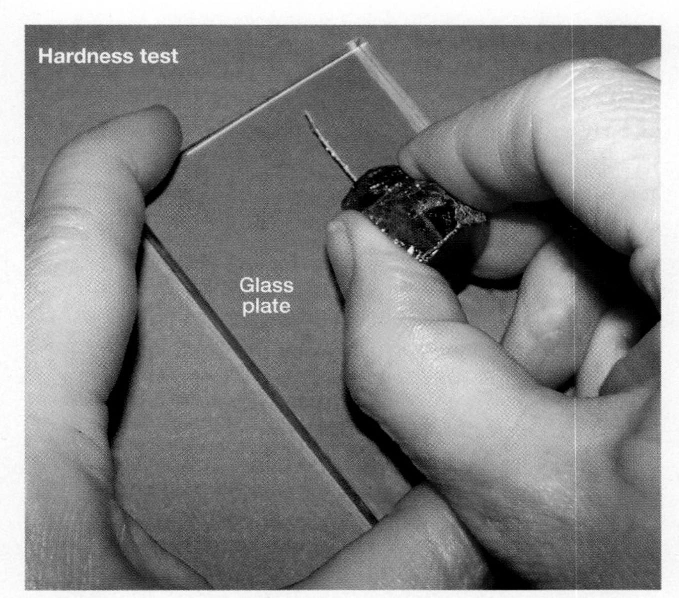

FIGURE 3.8 Hardness test. You can test a mineral's hardness (resistance to scratching) using a glass plate, which has a hardness of 5.5 on Mohs Scale of Hardness (**FIGURE 3.9**). Be sure the edges of the glass have been dulled. If not, then wrap the edges in masking tape or duct tape. Hold the glass plate firmly against a flat table top, then forcefully try to scratch the glass with the mineral sample. A mineral that scratches the glass is a *hard* mineral (i.e., harder than 5.5). A mineral that does not scratch the glass is a *soft* mineral (i.e., less than or equal to 5.5).

	Mohs Scale of Hardness*	Hardness of Some Common Objects (Harder objects scratch softer objects)
HARD	10 Diamond	
	9 Corundum	
	8 Topaz	
	7 Quartz	
		6.5 Streak plate
	6 Orthoclase Feldspar	
		5.5 Glass, Masonry nail, Knife blade
SOFT	5 Apatite	
		4.5 Wire (iron) nail
	4 Fluorite	
		3.5 Brass (wood screw, washer)
	3 Calcite	2.9 Copper coin (penny)
		2.5 Fingernail
	2 Gypsum	
	1 Talc	

* A scale for measuring relative mineral hardness (resistance to scratching).

FIGURE 3.9 Mohs Scale of Hardness (resistance to scratching). *Hard minerals* have a Mohs hardness number greater than 5.5, so they scratch glass and cannot be scratched with a knife blade or masonry (steel) nail. *Soft minerals* have a Mohs hardness number of 5.5 or less, so they do not scratch glass and are easily scratched by a knife blade or masonry (steel) nail. A mineral's hardness number can be determined by comparing it to the hardness of other common objects or minerals of Mohs Scale of Hardness.

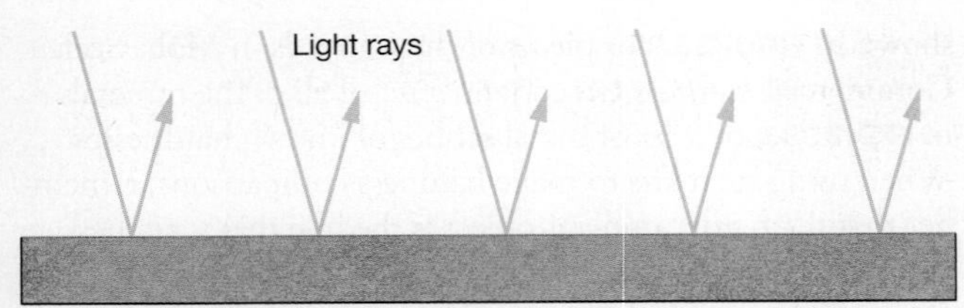

A. Cleavage excellent or perfect (large, parallel, flat surfaces)

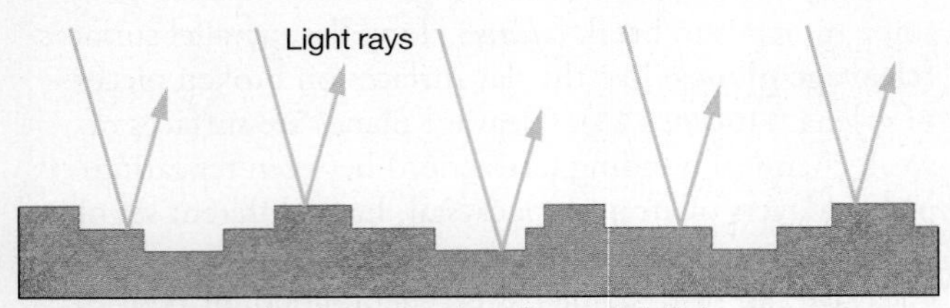

B. Cleavage good or imperfect (small, parallel, flat, stair-like surfaces)

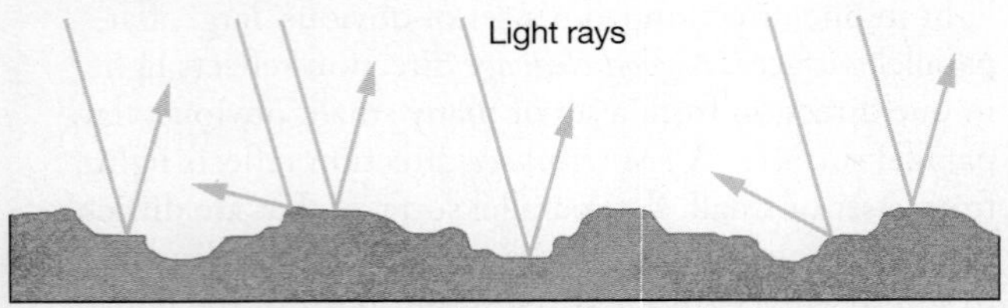

C. Cleavage poor (a few small, flat surfaces difficult to detect)

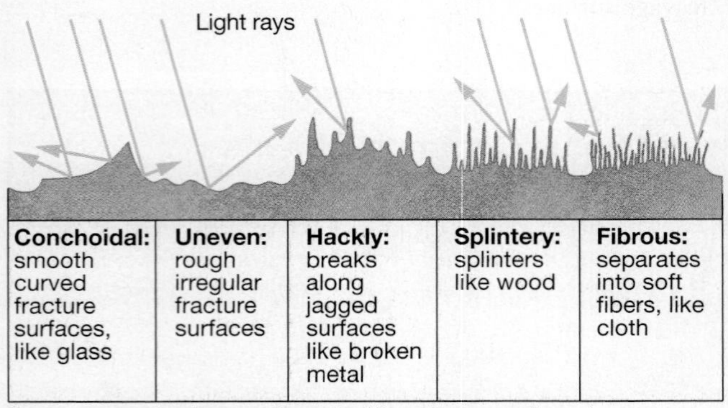

D. Fractures (broken surfaces lacking cleavage planes)

FIGURE 3.10 Recognizing cleavage and fracture. Illustrated cross sections of mineral samples to show degrees of development of cleavage—the tendency for a mineral to break along one or more sets of parallel, planar, reflective surfaces called *cleavage planes*. If a broken piece of a mineral crystal is rotated in bright light, its cleavage planes will be revealed by periodic flashes of light from one large, or many small, flat parallel surfaces. If no such reflective flashes of light occur, then the mineral sample has no cleavage. *Fracture* refers to any break in a mineral that does not occur along a cleavage plane. Therefore, fracture surfaces are normally not flat and they never occur in parallel sets.

Fracture refers to any break in a mineral that does not occur along a cleavage plane. Therefore, fracture surfaces are normally not flat and they never occur in parallel sets. Fracture can be described as *uneven* (rough and irregular, like the milky quartz in FIGURE 3.11B), *splintery* (like splintered wood), or *hackly* (having jagged edges, like broken metal). Pure quartz (FIGURE 3.11A) and mineraloids like opal (FIGURE 3.3) tend to fracture like glass—along ribbed, smoothly curved surfaces called *conchoidal fractures*.

Cleavage Direction. Cleavage planes are parallel surfaces of weak chemical bonding (attraction) between repeating parallel layers of atoms in a crystal, and more than one set of cleavage planes can be present in a crystal. Each different set has an orientation relative to the crystalline structure and is referred to as a **cleavage direction** (FIGURE 3.12). For example, muscovite (FIGURE 3.13) has one excellent cleavage direction and splits apart like pages of a book (book cleavage). Galena (FIGURE 3.2) breaks into small cubes and shapes made of cubes, so it has three cleavage directions developed at right angles to one another. This is called cubic cleavage (FIGURE 3.12).

Cleavage Direction in Pyriboles. Minerals of the pyroxene (e.g., augite) and amphibole (e.g., hornblende) groups generally are both dark-colored (dark green to black), opaque, nonmetallic minerals that have two good

A: Pure quartz (var. rock crystal) is colorless, transparent, nonmetalic, and has conchoidal fracture (like glass).

(x1)

(x1)

B: Milky quartz forms when the quartz has microscopc fluid inclusions, usually water. It has an irregular (rough, uneven) fracture.

FIGURE 3.11 Fracture in quartz—SiO_2 (silicon dioxide). These hand samples are broken pieces of quartz mineral crystals so no crystal faces are present. Note the absence of cleavage and the presence of conchoidal (like glass) to uneven fracture.

Number of Cleavages and Their Directions	Name and Description of How the Mineral Breaks	Shape of Broken Pieces (cleavage directions are numbered)	Illustration of Cleavage Directions
No cleavage (fractures only)	No parallel broken surfaces; may have conchoidal fracture (like glass)	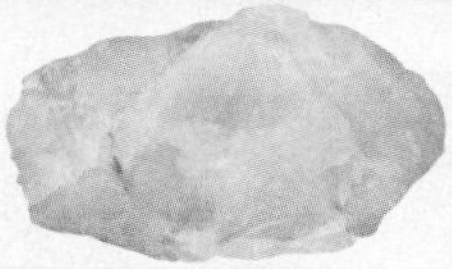 Quartz	None (no cleavage)
1 cleavage	**Basal (book) cleavage** "Books" that split apart along flat sheets	 1 Muscovite, biotite, chlorite (micas)	
2 cleavages intersect at or near 90°	**Prismatic cleavage** Elongated forms that fracture along short *rectangular* cross sections	Orthoclase 90° (K-spar) 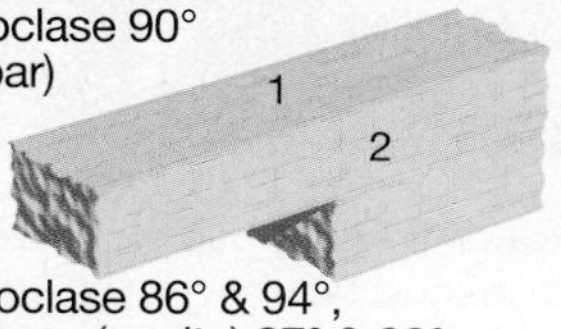 1 2 Plagioclase 86° & 94°, pyroxene (augite) 87° & 93°	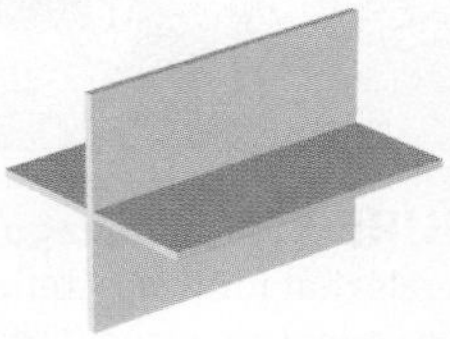
2 cleavages do not intersect at 90°	**Prismatic cleavage** Elongated forms that fracture along short *parallelogram* cross sections	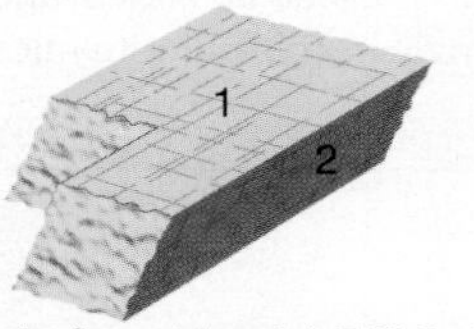 1 2 Amphibole (hornblende) 56° & 124°	
3 cleavages intersect at 90°	**Cubic cleavage** Shapes made of cubes and parts of cubes	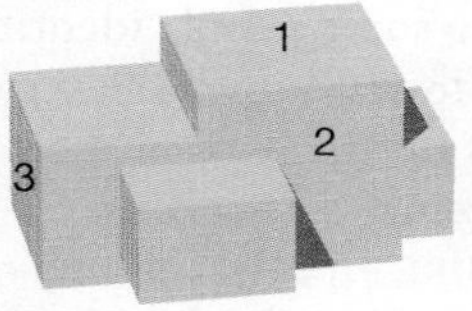 1 2 3 Halite, galena	
3 cleavages do not intersect at 90°	**Rhombohedral cleavage** Shapes made of rhombohedrons and parts of rhombohedrons	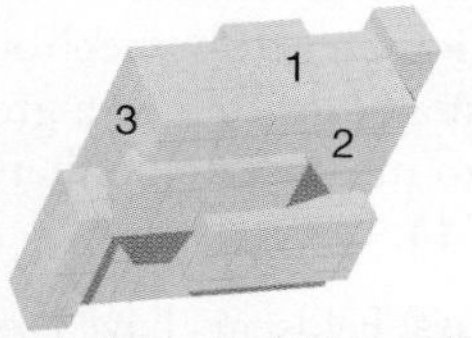 1 3 2 Calcite and dolomite 75° & 105°	
4 main cleavages intersect at 71° and 109° to form octahedrons, which split along hexagon-shaped surfaces; may have secondary cleavages at 60° and 120°	**Octahedral cleavage** Shapes made of octahedrons and parts of octahedrons	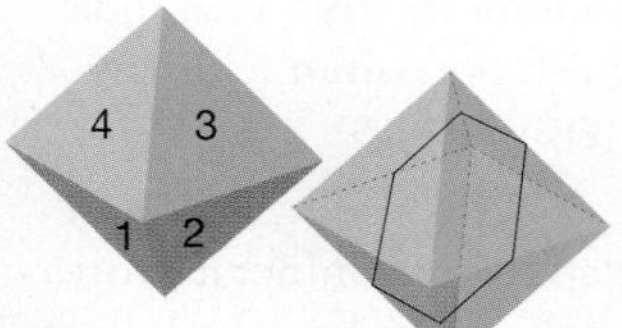 4 3 1 2 Fluorite	
6 cleavages intersect at 60° and 120°	**Dodecahedral cleavage** Shapes made of dodecahedrons and parts of dodecahedrons	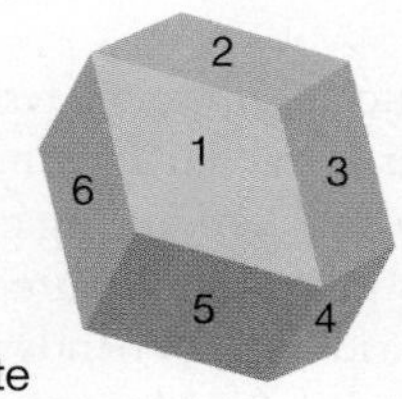 2 1 3 6 5 4 Sphalerite	

FIGURE 3.12 Cleavage in minerals.

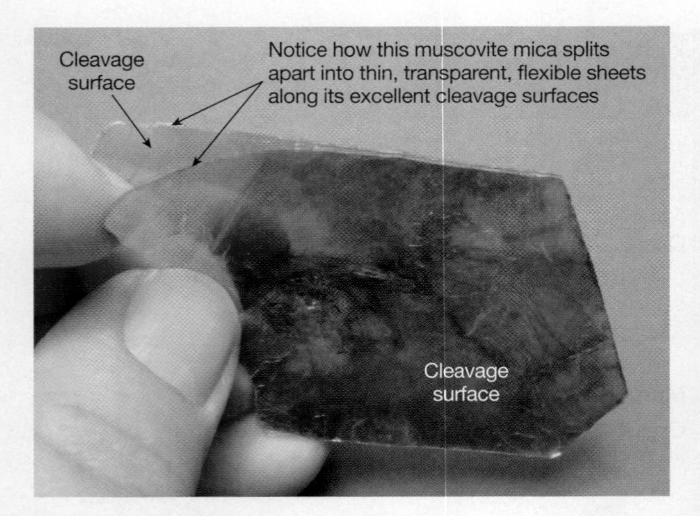

FIGURE 3.13 Cleavage in mica. Mica is a group of silicate minerals that form very reflective (vitreous) tabular crystals with one excellent cleavage direction. The crystals split easily into thin sheets, like pages of a book. This is called *book cleavage*. Muscovite mica is usually silvery brown in color. Biotite mica is always black.

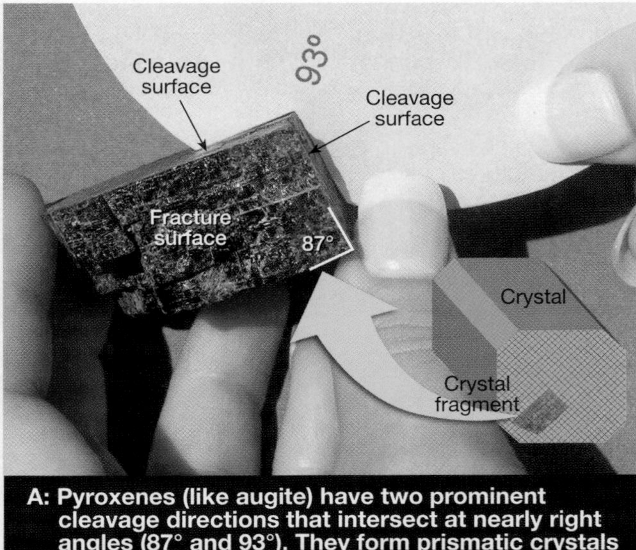

FIGURE 3.14 Cleavage in pyroxenes and amphiboles. Pyroxenes and amphiboles are two groups of dark colored silicate minerals with many similar properties. The main feature that distinguishes them is their cleavage.

cleavage directions. The two groups of minerals are sometimes difficult to distinguish, so some people identify them collectively as *pyriboles*. However, pyroxenes can be distinguished from amphiboles on the basis of their cleavage. The two cleavages of pyroxenes intersect at 87° and 93°, nearly at right angles (**FIGURE 3.14A**). The two cleavages of amphiboles intersect at angles of 56° and 124° (**FIGURE 3.14B**). These angles can be measured in hand samples using the cleavage goniometer from GeoTools Sheet 1 at the back of this manual. Notice how a green cleavage goniometer was used to measure angles between cleavage directions in **FIGURE 3.14**.

Cleavage Direction in Feldspars. Feldspars have two excellent to good cleavage directions, plus uneven fracture (**FIGURE 3.15**). The cleavage goniometer from GeoTools Sheet 1 can be used to distinguish potassium feldspar (orthoclase) from plagioclase (**FIGURE 3.15**).

Other Properties. There are additional mineral properties, too numerous to review here. However, the following other properties are typical of specific minerals or mineral groups:

Tenacity is the manner in which a substance resists breakage. Terms used to describe mineral tenacity include *brittle* (shatters like glass), *malleable* (like modeling clay or gold; can be hammered or bent permanently into new shapes), *elastic* or *flexible* (like a plastic comb; bends but returns to its original shape), and *sectile* (can be carved with a knife).

Reaction to acid differs among minerals. Cool, dilute hydrochloric acid (1–3% HCl) applied from a dropper bottle is a common "acid test." All of the so-called *carbonate minerals* (minerals with a chemical composition including carbonate, CO_3) will effervesce ("fizz") when a drop of such dilute HCl is applied to one of their freshly exposed surfaces (**FIGURE 3.16**). Calcite ($CaCO_3$) is the most commonly encountered carbonate mineral and effervesces in the acid test. Dolomite $[Ca,Mg(CO_3)_2]$ is

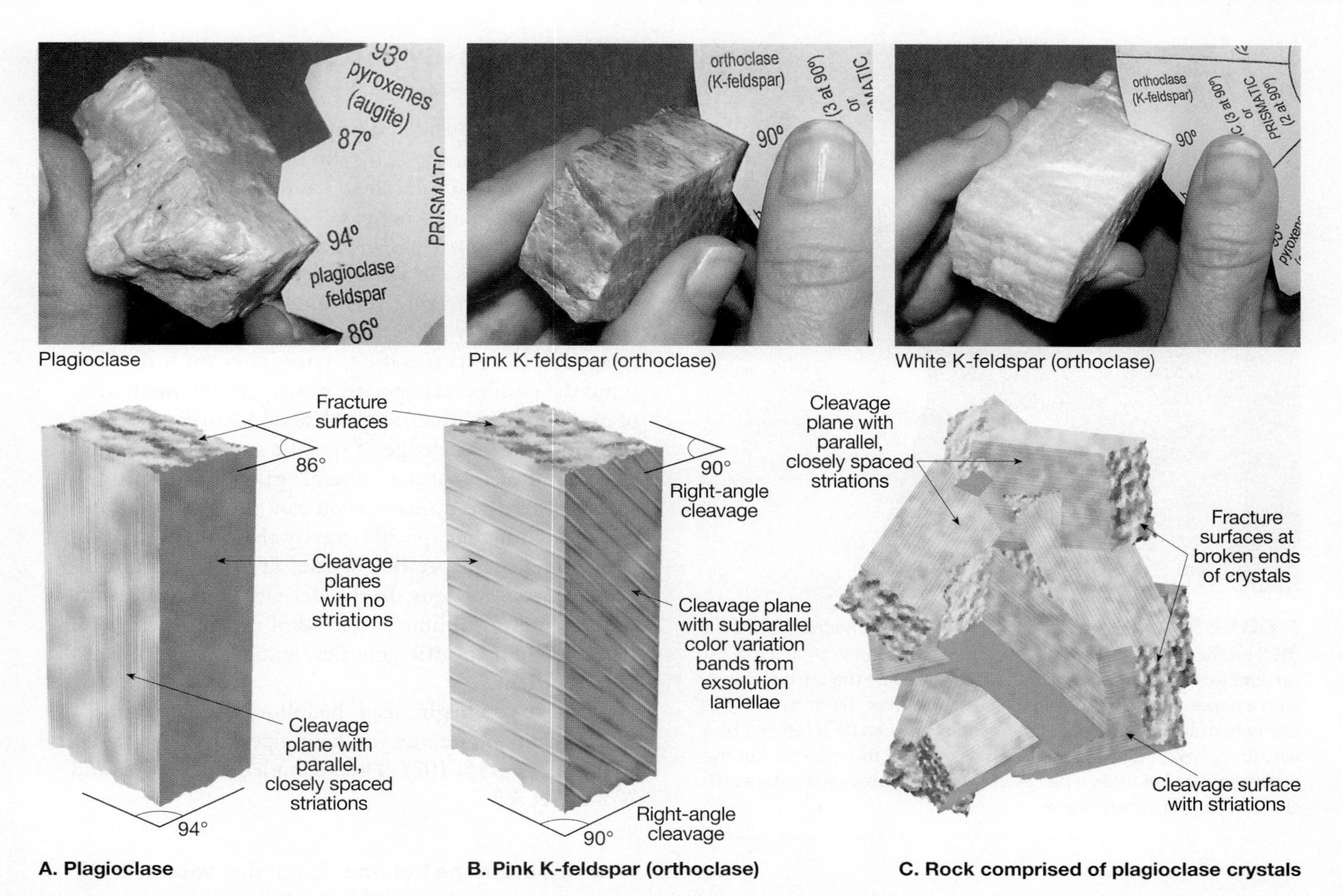

FIGURE 3.15 Common feldspars. Note how the cleavage goniometer can be used to distinguish potassium feldspar (K-spar, orthoclase) from plagioclase. The K-spar or orthoclase (Greek, *ortho*—right angle and *clase*—break) has perfect right-angle (90°) cleavage. Plagioclase (Greek, *plagio*—oblique angle and *clase*—break) does not. **A.** Plagioclase often exhibits *hairline striations* on some of its cleavage surfaces. They are caused by *twinning*: microscopic intergrowths between symmetrically-paired microcrystalline portions of the larger crystal. **B.** K-par (orthoclase) crystals may have intergrowths of thin, discontinuous, *exsolution lamellae*. They are actually microscopic layers of plagioclase that form as the K-spar cools, like fat separates from soup when it is refrigerated. **C.** Hand sample of a rock that is an aggregate of intergrown plagioclase mineral crystals. Individual mineral crystals are discernible within the rock, particularly the cleavage surfaces that have characteristic hairline striations.

another carbonate mineral that resembles calcite, but it will fizz in dilute HCl only if the mineral is first powdered. (It can be powdered for this test by simply scratching the mineral's surface with the tip of a rock pick, pocket knife, or nail.) If HCl is not available, then undiluted vinegar can be used for the acid test. It contains acetic acid (but the effervescence will be much less violent).

Striations are straight "hairline" grooves on the cleavage surfaces or crystal faces of some minerals. This can be helpful in mineral identification. For example, you can use the striations of plagioclase feldspar (**FIGURE 3.15A**) to distinguish it from potassium feldspar (K-feldspar, **FIGURE 3.15B**). *Plagioclase feldspar* has faint hairline striations on surfaces of one of its two cleavage directions. In contrast, *K-feldspar* (orthoclase) sometimes has internal *exsolution lamellae,* which are faint streaks of plagioclase that grew inside of it.

Magnetism influences some minerals, such as magnetite. The test is simple: check to see if the mineral is attracted to a magnet. Lodestone is a variety of magnetite that is itself a natural magnet. It will attract steel paperclips. Some other minerals may also be weakly attracted to a magnet (e.g., hematite, bornite, and pyrrhotite).

Specific Gravity (SG). Density is a measure of an object's mass (weighed in grams, g) divided by its volume (in cubic centimeters, cm^3). **Specific gravity** is the ratio of the density of a substance divided by the density of water. Since water has a density of 1 g/cm^3 and the units cancel out, specific gravity is the same number as density but without any units. For example, the mineral quartz has a density of 2.65 g/cm^3 so its specific gravity is 2.65 (i.e., SG = 2.65). **Hefting** is an easy way to judge the specific gravity of one mineral relative to another. This is done by holding a piece of the first mineral in one hand and holding an equal-sized piece of the second mineral in your other hand. Feel the difference in weight between the two samples (i.e., heft the samples). The sample that feels heavier has a higher specific gravity than the other. Most metallic minerals have higher specific gravities than nonmetallic minerals.

FIGURE 3.16 Acid test. Place a drop of weak hydrochloric acid (HCl) on the sample. If it effervesces (reacts, bubbles), then it is a carbonate (CO_3– containing) mineral. Please wipe the sample dry with a paper towel after doing this test! Note that the mineral in this example occurs in several different colors and can be scratched by a wire (iron) nail. The yellow sample is a crystal of this mineral, but the other samples are broken pieces of crystals that reveal the mineral's characteristic cleavage angles.

ACTIVITY

3.3 Determining Specific Gravity (SG)

THINK About It | **How and why do people study minerals?**

OBJECTIVE Measure the volume and mass of minerals, calculate their specific gravities, and use the results to identify them.

PROCEDURES

1. **Before you begin,** read the following background information. Your instructor will provide laboratory equipment, but this is **what you will need** to bring to lab:

   ___ Activity 3.3 Worksheet (p. 104) and pencil
   ___ calculator

2. **Then follow your instructor's directions** about where to obtain laboratory equipment and mineral samples, how to work safely, and how to complete the worksheet.

Why Are Density and Specific Gravity Important?

Have you ever considered buying silver coins as an investment? If so, then you should be wary of deceptive sales. For example, there have been reports of less valuable silver-plated copper coins marketed as pure silver coins. Copper has a specific gravity of 8.94, which is very close to silver's specific gravity of 9.32. So, even experienced buyers cannot tell a solid silver coin from a silver-plated copper coin just by hefting it to approximate its specific gravity. They must determine the coin's exact specific gravity as one method of ensuring its authenticity. Mineral identification is also aided by knowledge of specific gravity. If you heft same-sized pieces of the minerals galena (lead sulfide, an ore of lead) and quartz, you can easily tell that one has a much higher specific gravity than the other. But the difference in specific gravities of different minerals is not always so obvious. In this activity you will learn how to measure the volume and mass of mineral samples, calculate their specific gravities, and use the results to identify them.

Before you begin, read the following background information and be sure you have a pencil, eraser, and Worksheet 3.3 (p. 102). **Then** complete the activity and Worksheet 3.3.

How to Determine Volume. Recall that **volume** is the amount of space that an object takes up. Most mineral samples have odd shapes, so their volumes cannot be calculated from linear measurements. Their volumes must be determined by measuring the volume of water they displace. This is done in the laboratory with a *graduated cylinder* (FIGURE 3.17), an instrument used to measure volumes of fluid (fluid volume). Most graduated cylinders are graduated in metric units called milliliters (mL or ml), which are thousandths of a liter. *You should also note that 1 mL (1 ml) of fluid volume is exactly the same as 1 cm^3 of linear volume.*

Procedures for determining the volume of a mineral sample are provided in FIGURE 3.17. Note that when you pour water into a glass graduated cylinder, the surface of the liquid is usually a curved *meniscus*, and the volume is read at the bottom of its concave surface. In most plastic graduated cylinders, however, there is no meniscus. The water level is flat and easy to read.

If you slide a mineral sample into a graduated cylinder full of water (so no water splashes out), then it takes up space previously occupied by water at the bottom of the graduated cylinder. This displaced water has nowhere to go except higher into the graduated cylinder. Therefore, the volume of the mineral sample is exactly the same as the volume of fluid (water) that it displaces.

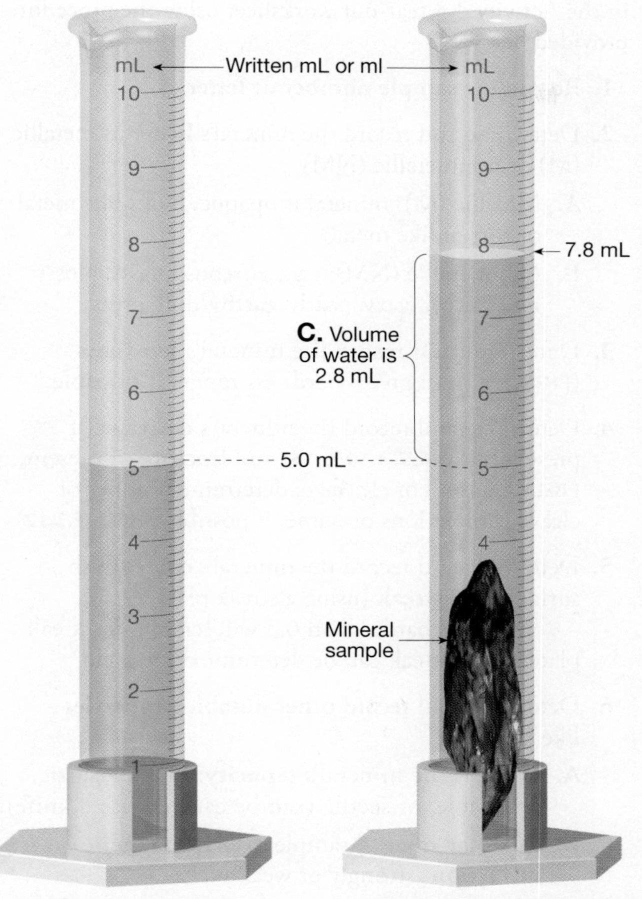

PROCEDURES

A. Place water in the bottom of a graduated cylinder. Add enough water to be able to totally immerse the mineral sample. It is also helpful to use a dropper bottle or wash bottle and bring the volume of water (before adding the mineral sample) up to an exact graduation mark like the 5.0 mL mark above. Record this starting volume of water.

B. Carefully slide the mineral sample down into the same graduated cylinder, and record the ending volume of the water (7.8 mL in the above example).

C. Subtract the starting volume of water from the ending volume of water to obtain the displaced volume of water. In the above example: 7.8 mL − 5.0 mL = 2.8 mL (2.8 mL is the same as 2.8 cm^3). This volume of displaced water is the volume of the mineral sample.

FIGURE 3.17 How to determine volume of a mineral sample.

How to Determine Mass. Earth materials do not just take up space (volume). They also have a mass of atoms that can be weighed. You will use a gram balance to measure the **mass** of materials (by determining their weight under the pull of Earth's gravity). The gram (g) is the basic unit of mass in the metric system, but instruments used to measure grams vary from triple-beam balances to spring scales to digital balances (page viii). Consult with your laboratory instructor or other students to be sure that you understand how to read the gram balance provided in your laboratory.

How to Calculate Density and Specific Gravity. Every material has a *mass* that can be weighed and a *volume* of space that it occupies. However, the relationship between a material's mass and volume tends to vary from one kind of material to another. For example, a bucket of rocks has much greater mass than an equal-sized bucket of air. Therefore a useful way to describe an object is to determine its mass per unit of volume, called **density**. *Per* refers to division, as in miles *per* hour (distance divided by time). So density is the measure of an object's mass divided by its volume (density = mass ÷ volume). Scientists and mathematicians use the Greek character rho (ρ) to represent density. Also, the gram (g) is the basic metric unit of mass, and the cubic centimeter is the basic unit of metric volume (cm^3), so density (ρ) is usually expressed in grams per cubic centimeter (g/cm^3). For example:

$$\frac{\text{Mineral sample weighs 44.0 grams}}{\text{Mineral sample takes up 11.0 ml of volume}}$$

$$= \frac{44.0 \text{ g}}{11.0 \text{ cm}^3} = 4.00 \text{ g/cm}^3 = \rho$$

Specific gravity (SG) is the ratio of the density of a substance divided by the density of water. Since water has a density of 1 g/cm^3 and the units cancel out, specific gravity is the same number as density but without any units. In the example above, the specific gravity of the mineral sample would be 4.00 (i.e., SG = 4.00). The mineral quartz has a density of 2.65 g/cm^3 so its specific gravity is 2.65 (i.e., SG = 2.65).

Calculating Density and Specific Gravity—The Math You Need

You can learn more about calculating density and specific gravity at this site featuring The Math You Need, When You Need It math tutorials for students in introductory geoscience courses: **http://serc.carleton.edu/mathyouneed/density/index.html**

ACTIVITY

3.4 Mineral Identification and Uses

THINK About It **How and why do people study minerals? How do you personally depend on minerals and elements extracted from them?**

OBJECTIVE Identify common minerals on the basis of their properties and assess how you depend on them.

PROCEDURES

1. **Before you begin**, read the introduction and mineral identification procedures below. Your instructor will provide laboratory equipment, but this is **what you will need** to bring to lab:

   ___ Activity 3.4 Worksheets (pp. 105–108) and pencil

2. **To complete the activity, follow your instructor's directions** about where to obtain a set of mineral analysis tools and mineral samples and any additional procedures for completing Worksheet 3.4 (which you will also need for Activity 3.5).
3. **When you have completed your worksheets**, reflect on how you depend on each of the minerals that you identified. What did you learn about how you depend on minerals? Be prepared to discuss this question and your mineral identifications. Save your Activity 3.4 worksheets to complete Activity 3.5.

Introduction

You are expected to learn how to identify common minerals on the basis of their properties and assess how you depend on them. The ability to identify minerals is one of the most fundamental skills of an Earth scientist. It also is fundamental to identifying rocks, for you must first identify the minerals comprising them. Only after minerals and rocks have been identified can their origin, classification, and alteration be adequately understood. Mineral identification is based on your ability to describe mineral properties using identification charts (FIGURES 3.18–3.20) and a Mineral Database (FIGURE 3.21). The database also lists the chemical composition and some common uses for each mineral. Some minerals, like halite (table salt) and gemstones are used in their natural state. Others are valuable as *ores*—materials from which specific chemical *elements* (usually metals) can be extracted at a profit.

Mineral Identification Procedures

Obtain a set of mineral samples and analysis tools according to your instructor's instructions. For each sample, fill in the Activity 3.3 tear-out worksheet using the procedures provided below.

1. Record the **sample number or letter.**
2. Determine and record the mineral's **luster** as metallic (M) or nonmetallic (NM)
   - **A.** Metallic (M): mineral is opaque, looks like metal or sort of like metal)
   - **B.** Nonmetallic (NM): e.g., vitreous (glassy, glossy reflection), waxy, pearly, earthy/dull, greasy
3. Determine and record the mineral's **hardness** (FIGS. 3.8, 3.9): give a hardness range, if possible.
4. Determine and record the mineral's **cleavage** (if present, FIGURES. 3.10–3.16) and **fracture** (if present, FIGURE 3.10). For cleavage, determine number of cleavage directions or name, if possible (FIGURE 3.12).
5. Determine and record the mineral's **color** (fresh surface) and **streak** (using a streak plate).

   Minerals harder than 6.5 will scratch the streak plate, so no streak can be determined for them.
6. Determine and record **other notable properties** like these:
   - **A.** What is the mineral's **tenacity**: brittle, elastic, malleable, or sectile (can be carved with a knife)?
   - **B.** Does the mineral sample display **magnetic attraction** (strongly or weakly)?
   - **C.** Does the mineral sample display a **reaction with acid** (dilute HCl)?
   - **D.** If crystals are visible, then what is their **crystal form?**
   - **E.** Does the mineral sample have **striations** on cleavage surfaces or crystal faces or **exsolution lamellae** (FIGURE 3.15)?
   - **F.** Estimate **specific gravity** (SG) as low, intermediate, or high.
   - **G.** Does the mineral sample have any unique diagnostic properties like **smell** when scratched or during acid test?
7. Use mineral identification figures to identify the **name of the mineral.**
   - **A.** If the mineral is opaque and metallic or submetallic, follow steps 1–5 in FIGURE 3.18.
   - **B.** If the mineral is light colored and nonmetallic, then follow steps 1–4 in FIGURE 3.19.
   - **C.** If the mineral is dark colored and nonmetallic, then follow steps 1–4 in FIGURE 3.20.

8. Use the Mineral Database (FIGURE 3.21) and FIGURE 3.22 to determine and record the mineral's **chemical composition** and help you determine **how you personally depend on the mineral** (including commodities refined from it). For more information about specific minerals or elements, you can refer to the U.S. Geological Survey's Mineral Commodity Summaries (**http://minerals.usgs.gov/minerals/pubs/mcs/**).

ACTIVITY 3.5 The Mineral Dependency Crisis

THINK About It **How do you personally depend on minerals and elements extracted from them? How sustainable is your dependency on minerals and elements extracted from them?**

OBJECTIVE Evaluate your personal and U.S. dependency on minerals.

PROCEDURES

1. **Before you begin**, read the background information below and on page 99. Your instructor will provide laboratory equipment, but this is **what you will need** to bring to lab:
   ___ Activity 3.5 Worksheet (p. 109) and pencil
   ___ Activity 3.4 Worksheets that you already completed
2. **Then refer to FIGURE 3.22, and follow your instructor's directions** about how to complete the Activity 3.5 worksheet.

Mineral Dependency

Did you know that some of the minerals used to make your cell phone and fluorescent light bulbs are quite rare nonrenewable resources? Many high-tech products depend on such nonrenewable mineral resources, yet many are either not mined within the United States or are mined here only in small quantities. The locations where they can be economically extracted in the United States have already been mined or are too small to be developed. Of particular concern are minerals mined as ores for *rare earth elements*, a group of 17 elements used in products like fluorescent light bulbs, flat screen televisions, cell phones, computers, solar panels, wind turbines, hybrid cars, cameras, DVDs, rechargeable batteries, magnets, medical equipment, night-vision goggles, missile systems, and medical equipment. China currently produces nearly all of the world's supply of rare earth elements, and the United States produces almost none. This has created what is widely known as the "rare earth crisis," and a shortage of rare earth elements used to make fluorescent light bulbs has become widely known as the "phosphor crisis." Yet the United States also relies on foreign supplies of many other minerals and elements extracted from them. Has the United States entered an unsustainable level of mineral dependency?

U.S. Net Import Reliance on Non-fuel Mineral Resources

Commodities are natural materials that people buy and sell, because they are required to sustain our wants and needs. Three classes are: agricultural products, energy resources, and non-fuel mineral resources. The **non-fuel mineral resources** include *rocks, minerals* used in their unrefined state or as ore from which specific elements can be profitably refined, and chemical *elements* extracted from ores. The U.S. Geological Survey (USGS) has determined that the United States was the world's largest user of non-fuel mineral resources in 2012 (about 12,000 pounds, or 11.3 metric tons, per person each year). To sustain its needs, the U.S. imported some of the minerals (and elements already extracted from them) that it needed. FIGURE 3.22 shows the 2012 U.S. net import reliance (expressed as a percent) on some selected minerals and elements refined from them. The United States exports some of the same non-fuel mineral resources that it imports, so net import reliance is the total of U.S. production and imports, minus the percentage of exports. A net import reliance of 80% means that 80% of the resource is imported. FIGURE 3.22 does not include all of the rare earth elements. Also not shown are minerals and elements for which the U.S. is less than 5% import reliant (or a net exporter).

USGS Mineral Resources Data System (MRDS)

Recall that commodities are natural materials that people buy and sell, because they are required to sustain our wants and needs. Three classes are: agricultural products, energy resources, and non-fuel mineral resources. The U.S. Geological Survey (USGS) divides the non-fuel mineral resources into two groups. **Nonmetallic mineral resources** are mostly rocks made of unrefined minerals (such as rock salt) and rocks (gravel, granite, marble). **Metallic mineral resources** are *ores* (rocks or minerals from which chemicals, usually metals, can be extracted at a profit) and chemical *elements* that have already been extracted from ore minerals. The **USGS Mineral Resources Data System (MRDS)** is a global database of both kinds of mineral resources and where they have been found and/or processed.

METALLIC AND SUBMETALLIC (M) MINERAL IDENTIFICATION

STEP 1: What is the mineral's hardness?	STEP 2: Does the mineral have cleavage?	STEP 3: What is the mineral's streak?	STEP 4: Match the mineral's physical properties to other characteristic properties below.	STEP 5: Mineral name. Find out more about it in the mineral database (Fig.3.21).
HARD (H > 5.5) Scratches glass Not scratched by masonry nail or knife blade	Cleavage absent, poor, or not visible	Dark gray to black	Color silvery gold; Tarnishes brown; H 6–6.5; Brittle; conchoidal to uneven fracture; Crystals: cubes (may be striated), pyritohedrons, or octahedrons; Distinguished from chalcopyrite, which is soft	Pyrite
			Silvery dark gray to black; Tarnishes gray or rusty yellow-brown; Strongly attracted to a magnet and may be magnetized; H 6–6.5; Crystals: octahedrons	Magnetite
HARD or SOFT		Yellow-brown	Color submetallic silvery brown; Tarnishes to dull and earthy yellow-brown to brown rust colors; H 1–5.5; More commonly occurs in its nonmetallic yellow to brown forms (H 1–5)	Limonite
		Brown	Color silvery black to black; Tarnishes gray to black; H 5.5–6; May be weakly attracted to a magnet; Crystals: octahedrons	Chromite
		Red to red-brown	Color steel gray, reddish-silver, to glittery bright silver (var. specular); Both metallic varieties have the characteristic red-brown streak; May be attracted to a magnet; H 5–6; Also occurs in nonmetallic, dull to earthy, red to red-brown forms	Hematite
SOFT (H ≤ 5.5) Does not scratch glass Scratched by masonry nail or knife blade	Cleavage good to excellent	Dark gray to black	Color bright silvery gray; Tarnishes dull gray; Brittle: breaks into cubes and shapes made of cubes; H 2.5; Crystals: cubes or octahedrons; Feels heavy for its size because of high specific gravity	Galena
		White to pale yellow-brown	Color silvery yellow-brown, silvery red, or black with submetallic to resinous luster; Tarnishes brown or black; H 3.5–4.0; smells like rotten eggs when scratched, powdered, or in acid test	Sphalerite
	Cleavage absent, poor, or not visible	Dark gray to black	Color bright silvery gold; Tarnishes bronze brown brassy gold, or iridescent blue-green and red; H 3.5–4.0; Brittle; uneven fracture; Crystals: tetrahedrons	Chalcopyrite
			Color characteristically brownish-bronze; Tarnishes bright iridescent purple, blue, and/or red, giving It its nickname "peacock ore"; May be weakly attracted to a magnet; H 3; Usually massive, rare as cubes or dodecahedrons	Bornite
			Color opaque brassy to brown-bronze; Tarnishes dull brown, may have faint iridescent colors; Fracture uneven to conchoidal; No cleavage; Attracted to a magnet; H 3.5–4.5; Usually massive or masses of tiny crystals; Resembles chalcopyrite, which is softer and not attracted to a magnet	Pyrrhotite
			Color dark silvery gray to black; Can be scratched with your fingernail; Easily rubs off on your fingers and clothes, making them gray; H 1–2	Graphite
		Yellow-brown	Metallic or silky submetallic luster, Color dark brown, gray, or black; H 5–5.5; Forms layers of radiating microscopic crystals and botryoidal masses	Goethite
		Copper	Color copper; Tarnishes dull brown or green; H 2.5–3.0; Malleable and sectile; Hackly fracture; Usually forms dendritic masses or nuggets	Copper (native copper)
		Gold	Color yellow gold; Does not tarnish; Malleable and sectile; H 2.5–3.0; Forms odd-shaped masses, nuggets, or dendritic forms	Gold (native gold)
		Silvery white	Color silvery white to gray; Tarnishes gray to black; H 2.5–3.0; Malleable and sectile; Forms dendritic masses, nuggets, or curled wires	Silver (native silver)

FIGURE 3.18 **Identification chart for opaque minerals with metallic or submetallic luster (M) on freshly broken surfaces.**

DARK TO MEDIUM-COLORED NONMETALLIC (NM) MINERAL IDENTIFICATION			
STEP 1: What is the mineral's hardness?	**STEP 2: What is the mineral's cleavage?**	**STEP 3: Compare the mineral's physical properties to other distinctive properties below.**	**STEP 4: Find mineral name(s) and check the mineral database for additional properties (Figure 3.21).**
HARD (H > 5.5) Scratches glass Not scratched by masonry nail or knife blade	Cleavage excellent or good	Translucent to opaque dark gray; blue-gray, or black; May have silvery iridescence; 2 cleavages at nearly 90° and with striations; H 6	Plagioclase feldspar
		Translucent to opaque brown, gray, green, or red; 2 cleavages at nearly right angles; Exsolution lamellae; H 6	Potassium feldspar (K-spar)
		Green to black; Vitreous luster; H 5.5–6.0; 2 cleavages at about 124° and 56° plus uneven fracture; Usually forms long blades and masses of needle-like crystals	Actinolite (amphibole)
		Dark gray to black; Vitreous luster; H 5.5–6.0; 2 cleavages at about 124° and 56° plus uneven fracture; Forms long crystals that break into blade-like fragments	Hornblende (amphibole)
		Dark green to black; Dull to vitreous luster; H 5.5–6.0; two cleavages at nearly right angles (93° and 87°) plus uneven fracture; Forms short crystals with squarish cross sections; Breaks into blocky fragments	Augite (pyroxene)
	Cleavage absent, poor, or not visible	Transparent or translucent gray, brown, or purple; Greasy luster; Massive or hexagonal prisms and pyramids; H 7	Quartz Smoky quartz (black/brown var.), Amethyst (purple var.)
		Gray, black, or colored (dark red, blue, brown) hexagonal prisms with flat striated ends; H 9	Corundum Emery (black impure var.), Ruby (red var.) Sapphire (blue var.)
		Opaque red-brown or brown; Luster waxy; Cryptocrystalline; H 7	Jasper (variety of quartz)
		Transparent to translucent dark red to black; Equant (dodecahedron) crystal form or massive; H 7	Garnet
		Opaque gray; Luster waxy; Cryptocrystalline; H 7	Chert (gray variety of quartz)
		Opaque black; Luster waxy; Cryptocrystalline; H 7	Flint (black variety of quartz)
		Black or dark green; Long striated prisms; H 7–7.5	Tourmaline
		Olive green, Transparent or translucent; No cleavage; Usually has many cracks and conchoidal to uneven fracture; Single crystals or masses of tiny crystals resembling green granulated sugar or aquarium gravel; The crystals have vitreous (glassy) luster	Olivine
		Opaque dark gray to black; Tarnishes gray to rusty yellow-brown; Cleavage absent; Strongly attracted to a magnet; May be magnetized; H 6–6.5	Magnetite
		Opaque green; Poor cleavage; H 6–7	Epidote
		Opaque brown prisms and cross-shaped twins; H 7	Staurolite
SOFT (H ≤ 5.5) Does not scratch glass Scratched by masonry nail or knife blade	Cleavage excellent or good	Yellow-brown, brown, or black; vitreous to resinous luster (may also be submetallic); Dodecahedral cleavage; H 3.5–4.0; Rotten egg smell when scratched or powdered	Sphalerite
		Purple cubes or octahedrons; Octahedral cleavage; H 4	Fluorite
		Black short opaque prisms; Splits easily along 1 excellent cleavage into thin sheets; H 2.5–3	Biotite (black mica)
		Green short opaque prisms; Splits easily along 1 excellent cleavage into thin sheets; H 2–3	Chlorite
	Cleavage absent, poor, or not visible	Opaque rusty brown or yellow-brown; Massive and amorphous; Yellow-brown streak; H 1–5.5	Limonite
		Rusty brown to red-brown, may have shades of tan or white; Earthy and opaque; Contains pea-sized spheres that are laminated internally; H 1–5; Pale brown streak	Bauxite
		Deep blue; Crusts, small crystals, or massive; Light blue streak; H 3.5–4	Azurite
		Opaque green or gray-green; Dull or silky masses or asbestos; White streak; H 2–5	Serpentine
		Opaque green in laminated crusts or massive; Streak pale green; Effervesces in dilute HCl; H 3.5–4	Malachite
		Translucent or opaque dark green; Can be scratched with your fingernail; Feels greasy or soapy; H 1	Talc
		Transparent or translucent green, brown, blue, or purple; Brittle hexagonal prisms; Conchoidal fracture; H 5	Apatite
		Opaque earthy brick red to dull red-gray, or gray; H 1.5–5; Red-brown streak; Magnet may attract the gray forms	Hematite

FIGURE 3.19 Identification chart for dark to medium-colored minerals with nonmetallic (NM) luster on freshly broken surfaces.

LIGHT-COLORED NONMETALLIC (NM) MINERAL IDENTIFICATION

STEP 1: What is the mineral's hardness?	STEP 2: What is the mineral's cleavage?	STEP 3: Compare the mineral's physical properties to other distinctive properties below.	STEP 4: Find mineral name(s) and check the mineral database for additional properties (Figure 3.21).
HARD (H > 5.5) Scratches glass Not scratched by masonry nail or knife blade	Cleavage excellent or good	White or pale gray; 2 good cleavages at nearly 90° plus uneven fracture; May have striations; H 6	Plagioclase feldspar
		Orange, pink, pale brown, green, or white; H 6; 2 good cleavages at 90° plus uneven fracture; exsolution lamellae	Potassium feldspar
		Pale brown, white, or gray; Long slender prisms; 1 excellent cleavage plus fracture surfaces; H 6–7	Sillimanite
		Blue, very pale green, white, or gray; Crystals are blades; H 4–7	Kyanite
	Cleavage absent, poor, or not visible	Gray, white, or colored (dark red, blue, brown) hexagonal prisms with flat striated ends; H 9	Corundum vars. ruby (red), sapphire (blue)
		Colorless, white, gray, or other colors; Greasy luster; Massive or hexagonal prisms and pyramids; Transparent or translucent; H 7	Quartz: vars. rose (pink), rock crystal (colorless), milky (white), citrine (amber)
		Opaque gray or white; Luster waxy; H 7	Chert (variety of quartz)
		Colorless, white, yellow, light brown, or pastel colors; Translucent or opaque; Laminated or massive; Cryptocrystalline; Luster waxy; H 7	Chalcedony (variety of quartz)
		Pale green to yellow; Transparent or translucent; H 7; No cleavage; Usually has many cracks and conchoidal to uneven fracture; Single crystals or masses of tiny crystals resembling green or yellow granulated sugar or aquarium gravel; Crystals vitreous (glassy)	Olivine
SOFT (H ≤ 5.5) Does not scratch glass Scratched by masonry nail or knife blade	Cleavage excellent or good	Colorless, white, yellow, green, pink, or brown; 3 excellent cleavages; Breaks into rhombohedrons; Effervesces in dilute HCl; H 3	Calcite
		Colorless, white, gray, creme, or pink; 3 excellent cleavages; Breaks into rhombohedrons; Effervesces in dilute HCl only if powdered; H 3.5–4	Dolomite
		Colorless or white with tints of brown, yellow, blue, black; Short tabular crystals and roses; Very heavy; H 3–3.5	Barite
		Transparent, colorless to white; H 2, easily scratched with your fingernail; White streak; Blade-like crystals or massive	Gypsum var. selenite
		Colorless, white, gray, or pale green, yellow, or red; Spheres of radiating needles; Luster silky; H 5–5.5	Natrolite (zeolite)
		Colorless, white, yellow, blue, brown, or red; Cubic crystals; Breaks into cubes; Salty taste; H 2.5	Halite
		Colorless, purple, blue, gray, green, yellow; Cubes with octahedral cleavage; H 4	Fluorite
		Colorless, yellow, brown, or red-brown; Short opaque prisms; Splits along 1 excellent cleavage into thin flexible transparent sheets; H 2–2.5	Muscovite (white mica)
	Cleavage absent, poor, or not visible	White, gray or yellow; Earthy to pearly; massive form; H 2, easily scratched with your fingernail; White streak	Gypsum var. alabaster
		White to gray; Fibrous form with silky or satiny luster; H 2, easily scratched with your fingernail	Gypsum var. satin spar
		Yellow crystals or earthy masses; Luster greasy; H 1.5–2.5; Smells like rotten eggs when powdered	Sulfur (Native sulfur)
		Opaque pale blue to blue-green; Conchoidal fracture; H 2-4; Massive or amorphous earthy crusts; Very light blue streak	Chrysocolla
		Opaque green, yellow, or gray; Dull or silky masses or asbestos; White streak; H 2–5	Serpentine
		Opaque white, gray, green, or brown; Can be scratched with fingernail; Greasy or soapy feel; H 1	Talc
		Opaque earthy white to very light brown masses of "white clay"; H 1–2; Powdery to greasy feel	Kaolinite
		Mostly pale brown to tan or white; Earthy and opaque; Contains pea-sized spheres that are laminated internally; H 1–5; Pale brown to white streak	Bauxite
		Colorless to white, orange, yellow, blue, gray, green, or red; May have internal play of colors; H 5.0–5.5; Amorphous; Often has many cracks; Conchoidal fracture	Opal
		Colorless or pale green, brown, blue, white, or purple; Brittle hexagonal prisms; Conchoidal fracture; H 5	Apatite

FIGURE 3.20 Identification chart for light-colored minerals with nonmetallic (NM) luster on freshly broken surfaces.

MINERAL DATABASE (Alphabetical Listing)					
Mineral	**Luster and Crystal System**	**Hardness**	**Streak**	**Distinctive Properties**	**Some Uses**
ACTINOLITE (amphibole)	Nonmetallic (NM) Monoclinic	5.5–6	White	Color dark green or pale green; Forms needles, prisms, and asbestose fibers; Good cleavage at 56° and 124°; SG = 3.1	Green gem varieties are the gemstone "nephrite jade"; asbestos products
AMPHIBOLE: See HORNEBLENDE and ACTINOLITE					
APATITE $Ca_5F(PO_4)_3$ calcium fluorophosphate	Nonmetallic (NM) Hexagonal	5	White	Color pale or dark green, brown, blue, white, or purple; Sometimes colorless; Transparent or opaque; Brittle; Conchoidal fracture; Forms hexagonal prisms; SG = 3.1–3.4	Used mostly to make fertilizer, pesticides; Transparent varieties sold as gemstones
ASBESTOS: fibrous varieties of AMPHIBOLE and SERPENTINE					
AUGITE (pyroxene) calcium ferromagnesian silicate	Nonmetallic (NM) Monoclinic	5.5–6	White to pale gray	Color dark green to brown or black; Forms short, 8-sided prisms; Two good cleavages that intersect at 87° and 93° (nearly right angles); SG = 3.2–3.5	Ore of lithium, used to make lithium batteries, ovenware glazes, high temperature grease, and to treat depression
AZURITE $Cu_3(CO_3)_2(OH)_2$ copper carbonate hydroxide	Nonmetallic (NM) Monoclinic	3.5–4	Light blue	Color a distinctive deep blue; Forms crusts of small crystals, opaque earthy masses, or short and long prisms; Brittle; Effervesces in dilute HCl; SG = 3.7–3.8	Ore of copper used to make pipes, electrical wire, coins, ammunition, bronze, brass; added to vitamin pills for healthy hair and skin; Gemstone
BARITE $BaSO_4$ barium sulfate	Nonmetallic (NM) Orthorhombic	3–3.5	White	Colorless to white, with tints of brown, yellow, blue, or red; Forms short tabular crystals and rose-shaped masses (Barite roses); Brittle; Cleavage good to excellent; Very heavy, SG = 4.3–4.6	Ore of barium, used to harden rubber, make fluorescent lamp electrodes, and in fluids used to drill oil/gas wells
BAUXITE Mixture of aluminum hydroxides	Nonmetallic (NM) No visible crystals	1–3	White	Brown earthy rock with shades of gray, white, and yellow; Amorphous; Often contains rounded pea-sized structures with laminations; SG = 2.0–3.0	Ore of aluminum used to make cans, foil, airplanes, solar panels; Ore of gallium used to make LED bulbs and liquid crystal displays in cell phones, computers, flat screen televisions
BIOTITE MICA ferromagnesian potassium, hydrous aluminum silicate $K(Mg,Fe)_3\ (Al,Si_3O_{10})(OH,F)_2$	Nonmetallic (NM) Monoclinic	2.5–3	Gray-brown to white	Color black, green-black, or brown-black; Cleavage excellent; Forms very short prisms that split easily into very thin, flexible sheets; SG = 2.7–3.1	Used for fire-resistant tiles, rubber, paint
BORNITE Cu_5FeS_4 copper-iron sulfide	Metallic (M) Isometric	3	Dark gray to black	Color brownish bronze; Tarnishes bright purple, blue, and/or red; May be weakly attracted to a magnet; H 3; Cleavage absent or poor; Forms dense brittle masses; Rarely forms crystals	Ore of copper, used to make pipes, electrical wire, coins, ammunition, bronze, brass; added to vitamin pills for healthy hair and skin
CALCITE $CaCO_3$ calcium carbonate	Nonmetallic (NM) Hexagonal	3	White	Usually colorless, white, or yellow, but may be green, brown, or pink; Opaque or transparent; Excellent cleavage in 3 directions not at 90°; Forms prisms, rhombohedrons, or scalenohedrons that break into rhombohedrons; Effervesces in dilute HCl; SG = 2.7	Used to make antacid tablets, fertilizer, cement; Ore of calcium
CHALCEDONY SiO_2 cryptocrystalline quartz	Nonmetallic (NM) No visible crystals	7	White*	Colorless, white, yellow, light brown, or other pastel colors in laminations; Often translucent; Conchoidal fracture; Luster waxy; Cryptocrystalline; SG = 2.5–2.8	Used as an abrasive; Used to make glass, gemstones (agate, chrysoprase)

*Streak cannot be determined with a streak plate for minerals harder than 6.5. They scratch the streak plate.

FIGURE 3.21 Mineral Database. This is an alphabetical list of minerals and their properties and uses.

MINERAL DATABASE (Alphabetical Listing)					
Mineral	**Luster and Crystal System**	**Hardness**	**Streak**	**Distinctive Properties**	**Some Uses**
CHALCOPYRITE $CuFeS_2$ copper-iron sulfide	Metallic (M) Tetragonal	3.5–4	Dark gray	Color bright silvery gold; Tarnishes bronze brown, brassy gold, or iridescent blue-green and red; Brittle; No cleavage; Forms dense masses or elongate tetrahedrons; SG = 4.1–4.3	Ore of copper, used to make pipes, electrical wire, coins, ammunition, bronze, brass; added to vitamin pills for healthy hair and skin
CHERT SiO_2 cryptocrystalline quartz	Nonmetallic (NM) No visible crystals	7	White*	Opaque gray or white; Luster waxy; Conchoidal fracture; SG = 2.5–2.8	Used as an abrasive; Used to make glass, gemstones
CHLORITE ferromagnesian aluminum silicate $(Mg,Fe,Al)_6(Si,Al)_4O_{10}(OH)_8$	Nonmetallic (NM) Monoclinic	2–2.5	White	Color dark green; Cleavage excellent; Forms short prisms that split easily into thin flexible sheets; Luster bright or dull; SG = 2–3	Used as a "filler" (to take up space and reduce cost) in plastics for car parts, appliances; Massive pieces carved into art sculptures
CHROMITE $FeCr_2O_4$ iron-chromium oxide	Metallic (M) Isometric	5.5–6	Dark brown	Color silvery black to black; Tarnishes gray to black; No cleavage; May be weakly attracted to a magnet; Forms dense masses or granular masses of small crystals (octahedrons)	Ore of chromium for chrome, stainless steel, mirrors, yellow and green paint pigments and ceramic glazes, and pills for healthy metabolism and cholesterol levels
CHRYSOCOLLA $CuSiO_3 \cdot 2H_2O$ hydrated copper silicate	Nonmetallic (NM) Orthorhombic	2–4	Very light blue	Color pale blue to blue-green; Opaque; Forms cryptocrystalline crusts or may be massive; Conchoidal fracture; Luster shiny or earthy; SG = 2.0–4.0	Ore of copper, used to make pipes, electrical wire, coins, ammunition, bronze, brass; added to vitamin pills for healthy hair and skin; Gemstone
COPPER (NATIVE COPPER) Cu copper	Metallic (M) Isometric	2.5–3	Copper	Color copper; Tarnishes brown or green; Malleable; No cleavage; Forms odd-shaped masses, nuggets, or dendritic forms; SG = 8.8–9.0	Ore of copper, used to make pipes, electrical wire, coins, ammunition, bronze, brass; added to vitamin pills for healthy hair and skin
CORUNDUM Al_2O_3 aluminum oxide	Nonmetallic (NM) Hexagonal	9	White*	Gray, white, black, or colored (red, blue, brown, yellow) hexagonal prisms with flat striated ends; Opaque to transparent; Cleavage absent; SG = 3.9–4.1 H 9	Used for abrasive powders to polish lenses; gemstones (red ruby, blue sapphire); emery cloth
DOLOMITE $CaMg(CO_3)_2$ magnesian calcium carbonate	Nonmetallic (NM) Hexagonal	3.5–4	White	Color white, gray, creme, or pink; Usually opaque; Cleavage excellent in 3 directions; Breaks into rhombohedrons; Resembles calcite, but will effervesce in dilute HCl only if powdered; SG = 2.8–2.9	Ore of magnesium used to make paper; lightweight frames for jet engines, rockets, cell phones, laptops; pills for good brain, muscle, and skeletal health
EPIDOTE complex silicate	Nonmetallic (NM) Monoclinic	6–7	White*	Color pale or dark green to yellow-green; Massive or forms striated prisms; Cleavage poor, SG = 3.3–3.5	Used as a green gemstone
FELDSPAR: See PLAGIOCLASE (Na-Ca Feldspars) and POTASSIUM FELDSPAR (K-Spar)					
FLINT SiO_2 cryptocrystalline quartz	Nonmetallic (NM) No visible crystals	7	White*	Color black to very dark gray; Opaque to translucent; Conchoidal fracture; Crypto-crystalline; SG = 2.5–2.8	Used as an abrasive; Used to make glass; Black gemstone
FLUORITE CaF_2 calcium fluoride	Nonmetallic (NM) Isometric	4	White	Colorless, purple, blue, gray, green, or yellow; Cleavage excellent; Crystals usually cubes; Transparent or opaque; Brittle; SG = 3.0–3.3	Ore of fluorine used in fluoride toothpaste, refrigerant gases, rocket fuel

*Streak cannot be determined with a streak plate for minerals harder than 6.5. They scratch the streak plate.

FIGURE 3.21 (continued)

MINERAL DATABASE (Alphabetical Listing)					
Mineral	**Luster and Crystal System**	**Hardness**	**Streak**	**Distinctive Properties**	**Some Uses**
GALENA PbS lead sulfide	Metallic (M) Isometric	2.5	Gray to dark gray	Color bright silvery gray; Tarnishes dull gray; Forms cubes and octahedrons; Brittle; Cleavage good in three directions, so breaks into cubes; SG = 7.4–7.6	Ore of lead for television glass, auto batteries, solder, ammunition; May be an ore of bismuth (an impurity) used as a lead substitute in pipe solder and fishing sinkers; May be an ore of silver (an impurity) used in jewelry, electrical circuit boards
GARNET complex silicate	Nonmetallic (NM) Isometric	7	White*	Color usually red, black, or brown, sometimes yellow, green, pink; Forms dodecahedrons; Cleavage absent but may have parting; Brittle; Translucent to opaque; SG = 3.5–4.3	Used as an abrasive; Red gemstone
GOETHITE FeO(OH) iron oxide hydroxide	Metallic (M) Orthorhombic	5–5.5	Yellow-brown	Color dark brown to black; Tarnishes yellow-brown; Forms layers of radiating microscopic crystals; SG = 3.3–4.3	Ore of iron for iron and steel used in machines, buildings, bridges, nails, tools, file cabinets; Added to pills and foods to aid hemoglobin production in red blood cells
GOLD (NATIVE GOLD) Au pure gold	Metallic (M) Isometric	2.5–3.0	Gold-yellow	Color gold to yellow-gold; Does not tarnish; Ductile, malleable and sectile; Hackly fracture; SG = 19.3; No cleavage; Forms odd-shaped masses, nuggets, and dendritic forms	Ductile and malleable metal used for jewelry; Electrical circuitry in computers, cell phones, car air bags; Heat shields for satellites
GRAPHITE C carbon	Metallic (M) Hexagonal	1	Dark gray	Color dark silvery gray to black; Forms flakes, short hexagonal prisms, and earthy masses; Greasy feel; Very soft; Cleavage excellent in 1 direction; SG = 2.0–2.3	Used for pencils, anodes (negative ends) of most batteries, synthetic motor oil, carbon steel, fishing rods, golf clubs
GYPSUM $CaSO_4 \cdot 2H_2O$ hydrated calcium sulfate	Nonmetallic (NM) Monoclinic	2	White	Colorless, white, or gray; Forms tabular crystals, prisms, blades, or needles (satin spar variety); Transparent to translucent; Very soft; Cleavage good; SG = 2.3	Plaster-of-paris, wallboard, drywall, art sculpture medium (alabaster)
HALITE NaCl sodium chloride	Nonmetallic (NM) Isometric	2.5	White	Colorless, white, yellow, blue, brown, or red; Transparent to translucent; Brittle; Forms cubes; Cleavage excellent in 3 directions, so breaks into cubes; Salty taste; SG = 2.1–2.6	Table salt, road salt; Used in water softeners and as a preservative; Sodium ore
HEMATITE Fe_2O_3 iron oxide	Metallic (M) or Nonmetallic (NM) Hexagonal	1–6	Red to red-brown	Color silvery gray, reddish silver, black, or brick red; Tarnishes red; Opaque; Soft (earthy) and hard (metallic) varieties have same streak; Forms thin tabular crystals or massive; May be attracted to a magnet; SG = 4.9–5.3	Red ochre pigment in paint and cosmetics. Ore of iron for iron and steel used in machines, buildings, bridges, nails, tools, file cabinets; Added to pills and foods to aid hemoglobin production in red blood cells
HORNBLENDE (amphibole) calcium ferromagnesian aluminum silicate	Nonmetallic (NM) Monoclinic	5.5–6.0	White to pale gray	Color dark gray to black; Forms prisms with good cleavage at 56° and 124°; Brittle; Splintery or asbestos forms; SG = 3.0–3.3	Fibrous varieties used for fire-resistant clothing, tiles, brake linings
JASPER SiO_2 cryptocrystalline quartz	Nonmetallic (NM) No visible crystals	7	White*	Color red-brown, or yellow; Opaque; Waxy luster; Conchoidal fracture; Cryptocrystalline; SG = 2.5–2.8	Used as an abrasive; Used to make glass, gemstones
KAOLINITE $Al_4(Si_4O_{10})(OH)_8$ aluminum silicate hydroxide	Nonmetallic (NM) Triclinic	1–2	White	Color white to very light brown; Commonly forms earthy, microcrystalline masses; Cleavage excellent but absent in hand samples; SG = 2.6	Used for pottery, clays, polishing compounds, pencil leads, paper
K-SPAR: See POTASSIUM FELDSPAR					
KYANITE $Al_2(SiO_4)O$ aluminum silicate oxide	Nonmetallic (NM) Triclinic	4–7	White*	Color blue, pale green, white, or gray; Translucent to transparent; Forms blades; SG = 3.6–3.7	High temperature ceramics, spark plugs

*Streak cannot be determined with a streak plate for minerals harder than 6.5. They scratch the streak plate.

MINERAL DATABASE (Alphabetical Listing)					
Mineral	**Luster and Crystal System**	**Hardness**	**Streak**	**Distinctive Properties**	**Some Uses**
LIMONITE $Fe_2O_3 \cdot nH_2O$ hydrated iron oxide and/or $FeO(OH) \cdot nH_2O$ hydrated iron oxide hydroxide	Metallic (M) or Nonmetallic (NM) Amorphous	1–5.5	Yellow-brown	Color yellow-brown to dark brown; Tarnishes yellow to brown; Amorphous masses; Luster dull or earthy; Hard or soft; SG = 3.3–4.3	Yellow ochre pigment in paint and cosmetics. Ore of iron for iron and steel used in machines, buildings, bridges, nails, tools, file cabinets; Added to pills and foods to aid hemoglobin production in red blood cells
MAGNETITE Fe_3O_4 iron oxide	Metallic (M) or Nonmetallic (NM) Isometric	6–6.5	Dark gray	Color silvery gray to black; Opaque; Forms octahedrons; Tarnishes gray; No cleavage; Attracted to a magnet and can be magnetized; SG = 5.0–5.2	Ore of iron for iron and steel used in machines, buildings, bridges, nails, tools, file cabinets; Added to pills and foods to aid hemoglobin production in red blood cells
MALACHITE $Cu_2CO_3(OH)_2$ copper carbonate hydroxide	Nonmetallic (NM) Monoclinic	3.5–4	Green	Color green, pale green, or gray-green; Usually in crusts, laminated masses, or microcrystals; Effervesces in dilute HCl; SG = 3.6–4.0	Ore of copper, used to make pipes, electrical wire, coins, ammunition, bronze, brass; added to vitamin pills for healthy hair and skin; Gemstone
MICA: See BIOTITE and MUSCOVITE					
MUSCOVITE MICA potassium hydrous aluminum silicate $KAl_2(Al,Si_3O_{10})(OH,F)_2$	Nonmetallic (NM) Monoclinic	2–2.5	White	Colorless, yellow, brown, or red-brown; Forms short opaque prisms; Cleavage excellent in 1 direction, can be split into thin flexible transparent sheets; SG = 2.7–3.0	Computer chip substrates, electrical insulation, roof shingles, Cosmetics with a satiny sheen
NATIVE COPPER: See COPPER					
NATIVE GOLD: See GOLD					
NATIVE SILVER: See SILVER					
NATIVE SULFUR: See SULFUR					
NATROLITE (ZEOLITE) $Na_2(Al_2Si_3O_{10}) \cdot 2H_2O$ hydrous sodium aluminum silicate	Nonmetallic (NM) Orthorhombic	5–5.5	White	Colorless, white, gray, or pale green, yellow, or red; Forms masses of radiating needles; Silky luster; SG = 2.2–2.4	Used in water softeners
OLIVINE $(Fe,Mg)_2SiO_4$ ferromagnesian silicate	Nonmetallic (NM) Orthorhombic	7	White*	Color pale or dark olive-green to yellow, or brown; Forms short crystals that may resemble sand grains; Conchoidal fracture; Cleavage absent; Brittle; SG = 3.3–3.4	Green gemstone (peridot); Ore of magnesium used to make paper; lightweight frames for jet engines, cell phones, laptops; pills for good brain, muscle, and skeletal health
OPAL $SiO_2 \cdot nH_2O$ hydrated silicon dioxide	Nonmetallic (NM) Amorphous	5–5.5	White	Colorless to white, orange, yellow, brown, blue, gray, green, or red; may have play of colors (opalescence); Amorphous; Cleavage absent; Conchoidal fracture; SG = 1.9–2.3	Gemstone
PLAGIOCLASE FELDSPAR $NaAlSi_3O_8$ to $CaAl_2Si_2O_8$ calcium-sodium aluminum silicate	Nonmetallic (NM) Triclinic	6	White	Colorless, white, gray, or black; May have iridescent play of color from within; Translucent; Forms striated tabular crystals or blades; Cleavage good in two directions at nearly 90°; SG = 2.6–2.8	Used to make ceramics, glass, enamel, soap, false teeth, scouring powders
POTASSIUM FELDSPAR $KAlSi_3O_8$ potassium aluminum silicate	Nonmetallic (NM) Monoclinic	6	White	Color orange, brown, white, green, or pink; Forms translucent prisms with subparallel exsolution lamellae; Cleavage excellent in two directions at nearly 90°; SG = 2.5–2.6	Used to make ceramics, glass, enamel, soap, false teeth, scouring powders
PYRITE ("fool's gold") FeS_2 iron sulfide	Metallic (M) Isometric	6–6.5	Dark gray	Color silvery gold; Tarnishes brown; H 6–6.5; Cleavage absent to poor; Brittle; Forms opaque masses, cubes (often striated), or pyritohedrons; SG = 4.9–5.2	Ore of sulfur for matches, gunpowder, fertilizer, rubber hardening (car tires), fungicide, insecticide, paper pulp processing
PYRRHOTITE FeS iron sulfide	Metallic (M) Monoclinic	3.5–4.5	Dark gray to black	Color brassy to brown-bronze; Tarnishes dull brown, sometimes with faint iridescent colors; Fracture uneven to conchoidal; No cleavage; attracted to a magnet; SG = 4.6	Ore of iron and sulfur; Impure forms contain nickel and are used as nickel ore; the nickel is used to make stainless steel

*Streak cannot be determined with a streak plate for minerals harder than 6.5. They scratch the streak plate.

MINERAL DATABASE (Alphabetical Listing)

Mineral	Luster and Crystal System	Hardness	Streak	Distinctive Properties	Some Uses
PYROXENE: See AUGITE					
QUARTZ SiO_2 silicon dioxide	Nonmetallic (NM) Hexagonal	7	White*	Usually colorless, white, or gray but uncommon varieties occur in all colors; Transparent to translucent; Luster greasy; No cleavage; Forms hexagonal prism and pyramids; SG = 2.6–2.7 Some quartz varieties are: • var. flint (opaque black or dark gray) • var. smoky (transparent gray) • var. citrine (transparent yellow-brown) • var. amethyst (purple) • var. chert (opaque gray) • var. milky (white) • var. jasper (opaque red or yellow) • var. rock crystal (colorless) • var. rose (pink) • var. chalcedony (translucent, waxy luster)	Used as an abrasive; Used to make glass, gemstones
SERPENTINE $Mg_6Si_4O_{10}(OH)_8$ magnesium silicate hydroxide	Nonmetallic (NM) Monoclinic	2–5	White	Color pale or dark green, yellow, gray; Forms dull or silky masses and asbestos forms; No cleavage; SG = 2.2–2.6	Fibrous varieties used for fire-resistant clothing, tiles, brake linings
SILLIMANITE $Al_2(SiO_4)O$ aluminum silicate	Nonmetallic (NM) Orthorhombic	6–7	White	Color pale brown, white, or gray; One good cleavage plus fracture surfaces; Forms slender prisms and needles; SG = 3.2	High-temperature ceramics
SILVER (NATIVE SILVER) Ag pure silver	Metallic (M) Isometric	2.5–3.0	White to silvery white	Color silvery white to gray; Tarnishes dark gray to black; Ductile, malleable and sectile; Hackly fracture; No cleavage; Forms nuggets, curled wires, and dendritic forms; SG = 10.5	Ductile and malleable metal used for jewelry and silverware; Electrical circuit boards for computers and cell phones; Photographic film
SPHALERITE ZnS zinc sulfide	Metallic (M) or Nonmetallic (NM) Isometric	3.5–4	White to pale yellow-brown	Color silvery yellow-brown, dark red, or black; Tarnishes brown or black; Dodecahedral cleavage excellent to good; Smells like rotten eggs when scratched/powdered; Forms misshapen tetrahedrons or dodecahedrons; SG = 3.9–4.1	Ore of zinc for brass, galvanized steel and roofing nails, skin-healing creams, pills for healthy immune system and protein production: Ore of Indium (an impurity) used to make solar cells
STAUROLITE iron magnesium zinc aluminum silicate	Nonmetallic (NM) Monoclinic	7	White to gray*	Color brown to gray-brown; Tarnishes dull brown; Forms prisms that interpenetrate to form natural crosses; Cleavage poor; SG = 3.7–3.8	Gemstone crosses called “fairy crosses”
SULFUR (NATIVE SULFUR) S sulfur	Nonmetallic (NM) Orthorhombic	1.5–2.5	Pale yellow	Color bright yellow; Forms transparent to translucent crystals or earthy masses; Cleavage poor; Luster greasy to earthy; Brittle; SG = 2.1	Used for matches, gunpowder, fertilizer, rubber hardening (car tires), fungicide, insecticide, paper pulp processing
TALC $Mg_3Si_4O_{10}(OH)_2$ hydrous magnesian silicate	Nonmetallic (NM) Monoclinic	1	White	Color white, gray, pale green, or brown; Forms cryptocrystalline masses that show no cleavage; Luster silky to greasy; Feels greasy or soapy (talcum powder); Very soft; SG = 2.7–2.8	Used as a “filler” (to take up space and reduce cost) in plastics for car parts, appliances; Massive pieces carved into art sculptures
TOURMALINE complex silicate	Nonmetallic (NM) Hexagonal	7–7.5	White*	Color usually opaque black or green, but may be transparent or translucent green, red, yellow, pink or blue; Forms long striated prisms with triangular cross sections; Cleavage absent; SG = 3.0–3.2	Crystals used in radio transmitters; gemstone
ZEOLITE: A group of calcium or sodium hydrous aluminum silicates. See NATROLITE.					

*Streak cannot be determined with a streak plate for minerals harder than 6.5. They scratch the streak plate.

2012 U.S. NET IMPORT RELIANCE ON SELECTED NON-FUEL MINERAL COMMODITIES			
COMMODITY (Element, Ore, or Raw Mineral)	**ORE MINERAL or RAW MINERAL**	**Percent Import Reliance**	**WHAT IS THIS COMMODITY USED FOR?**
Fluorine ore (F): fluorspar	**Fluorite**	**100**	Fluorine is used in fluoride toothpaste, fluorocarbon refrigerant gases and fire extinguishers, and fluoropolymer plastics that coat non-stick fry pans and insulate wiring in cell phones, laptops, and airplanes.
Graphite (C)	**Graphite**	**100**	Used to make carbon steel, pencils, carbon fiber reinforced plastics in car bodies, and negative ends of most batteries (including those in all cell phones, power tools, computers, and hybrid/electric vehicles).
Indium metal (In)	**Sphalerite** with In as an impurity	**100**	Indium is used to make solar cells, and liquid-crystal displays (LCDs) in cell phones, computers, and flat-screen television sets.
Mica (sheet)	**Muscovite**	**100**	Muscovite is used in heating elements of hair dryers and toasters, joint compound, and cosmetics with a satiny or glittery sheen.
Quartz crystal (industrial)	**Quartz** var. rock crystal	**100**	Crystals of cultured pure quartz are used to make quartz watches and the frequency controls and timers in every computer and cell phone.
Niobium metal (Nb, "columbium")	**Columbite** (in "coltan")	**100**	Niobium is used to make high-strength non-corrosive steel alloys (for jet engines, power plants) and arc welding rods, plus electrical insulation coatings in cell phones, computers, and electronic games.
Tantalum metal (Ta)	**Tantalite** (in "coltan")	**100**	Tantalum is used to make "tantalum capacitors" that buffer the flow of electricity between a battery and electronic parts in the circuits of cell phones, laptops, iPods, and most other electrical devices.
Gallium metal (Ga)	**Bauxite** is Ga ore	**99**	Gallium is used to make light-emitting diode (LED) bulbs and liquid-crystal displays (LCDs) in things like cell phones, computers, and flat-screen television sets.
Vanadium metal (V)	**Magnetite** with V as an impurity	**96**	Vanadium is used for cutting tools; mixed with iron to make lightweight shock-resistant steel for car axles and gears, springs, and cutting tools.
Bismuth metal (Bi)	**Galena** with Bi as an impurity	**92**	Bismuth is used as a nontoxic replacement for lead (in ceramic glazes, fishing sinkers, food processing equipment, plumbing, and shot for hunting) and in antidiarrheal medications.
Barium metal ore (Ba)	**Barite**	**80**	Barium (Ba) is widely used to make capacitors (that store energy) and memory cells in cell phones and other portable electronic devices.
Zinc metal (Zn)	**Sphalerite** is an ore of Zn	**72**	Zinc is used to make alloys like brass, skin-healing creams, and galvanized (rust-proof) steel and roofing nails; added to vitamin pills for a healthy immune system and to aid protein production.
Chromium metal (Cr)	**Chromite** is an ore of Cr	**70**	Chromium is used to make stainless steel, yellow and green ceramic glazes and paints, and military camouflage paints; added to vitamin pills for healthy metabolism and lower cholesterol levels.
Garnet (industrial)	**Garnet**	**65**	Industrial garnet is used as an abrasive in things like sandpaper and sandblasting.
Silver metal (Ag)	**Native silver; Galena** with Ag as an impurity	**57**	Silver is used to make jewelry and silverware, photographic film, and solder on electrical circuit boards of computers and cell phones.
Nickel metal (Ni)	**Pyrrhotite** contains Ni as an impurity	**49**	Nickel is used to make rechargeable batteries (Ni-Cd) for portable electronic devices, screw-end caps of light bulbs, and stainless steel.
Magnesium metal (Mg)	**Dolomite** and **Olivine** are Mg ores	**46**	Magnesium is used to make strong, lightweight frames for jet engines and rockets, lightweight cell phone and laptop cases, and incendiary flares and bombs; added to vitamin pills to aid good brain and muscle function and strengthen bones.
Tungsten metal (W)	**Wolframite** is W ore	**42**	Tungsten is a dense metal that makes cell phones and pagers vibrate (by attaching it to an electric motor spinning off center); also used for light bulb filaments, golf clubs, and tungsten carbide cutting tools.
Copper metal (Cu)	**Azurite, Bornite, Chalcopyrite, Chrysocolla,** and **Malachite** are Cu ores	**35**	Copper is used to make copper pipes; electrical wire for homes, businesses, electric motors, and circuit boards in cell phones and other electrical devices. Hybrid cars contain about 100 pounds (45 kg) of copper. Added to vitamin pills for healthy hair and skin.
Aluminum (Al)	**Bauxite is Al ore**	**20**	Aluminum is a lightweight silvery metal used to make drink cans, foil, airplanes, and solar panels.
Salt	**Halite**	**19**	Used as table salt, road salt (to melt snow), in water softeners, and as a food preservative.
Sulfur (S)	**Native Sulfur; Pyrite** is a S ore	**19**	Used to make matches, gunpowder, fertilizer, fungicide, insecticide, and harden rubber (car tires).
Gypsum	**Gypsum**	**12**	Used to make Plaster-of-Paris, drywall and for art (alabaster).
Iron metal (Fe), **Steel**	**Geothite, Limonite, Magnetite,** and **Hematite** are Fe ores	**11**	Iron and steel are used to construct machines, buildings, bridges, nails, bolts, tools, file cabinets; iron is added to vitamin pills to aid hemoglobin production in red blood cells for oxygen transport.
Cement	**Calcite**	**7**	Calcite is processed into cement, which is used to make concrete.

FIGURE 3.22 Selected Net Non-Fuel Mineral Resource Imports by the United States in 2012. This list includes only some of the mineral resources reported in *USGS Mineral Commodity Summaries 2013*. Net import reliance is the total of U.S. production and imports, minus the percentage of exports. (Adapted from USGS Mineral Commodity Summaries, 2013)

What Are Conflict Minerals?

Niobium (Nb) and tantalum (Ta) are two rare metallic mineral resources (metal elements). Niobium (formerly called columbium) is mixed with iron to make high-strength non-corrosive steel alloys (for rockets, jet engines, chemical pipelines, nuclear power plants) and superconducting magnets for medical MRI scanners. Tantalum is used to make alloys with high heat tolerance for electronic microchips (for small electrical devices like cell phones) and non-irritating steel for surgical steel tools. Niobium is refined from the mineral columbite, and tantalum is refined from the mineral tantalite. The two minerals commonly occur together as an ore called "coltan." Unfortunately, a primary source of coltan is the Democratic Republic of Congo (DRC) in Africa, where rebel groups use forced labor to mine the coltan and engage in extreme violence against women and children. The United States regards all coltan from DRC as a "conflict mineral." It is illegal to import any coltan, niobium, or tantalum into the United States if it originated in the DRC or another conflict-mineral site.

ACTIVITY 3.6 Urban Ore

THINK About It **How sustainable is your dependency on minerals and elements extracted from them?**

OBJECTIVE Evaluate the prospect of recycling products and mining discarded products to extract their metals.

PROCEDURES

1. **Before you begin,** read the background information ahead. You must also obtain access to a computer that has Internet access. This is **what you will need** to bring to lab:

   ___ Activity 3.6 Worksheet (p. 110) and pencil
   ___ calculator

2. **Then follow your instructor's directions** about how to complete the worksheet.

How Ores and Precious Metals Are Weighed

Did you know that it is common practice in the United States to weigh gold and its ore using different systems of measurement? Within the United States, mining companies use an avoirdupois system to weigh bulk amounts of rock like gold *ore*—material (usually rocks or minerals) from which chemicals can be extracted at a profit. If they sell the gold ore to another country, then its weight is quoted using the metric system (page xiii at the front of the manual). However, once gold is extracted from its ore, its weight is quoted everywhere in the world using the troy system.

Grams Are Metric; Ounces Are Not

A gram is a metric unit of weight (mass) equal to one thousandth of a kilogram—roughly the mass of a paper clip. Any metric scale or balance can be used to measure the weight (mass) of an object in the metric unit called a gram. However, there are no metric ounces. Ounces are used in the avoirdupois and troy systems of measurement. An avoirdupois ounce is 28.349523125 grams, and a troy ounce is 31.1034768 grams. Also, note that fluid ounces are units of volume, not weight.

What Is Avoirdupois?

Avoirdupois (avdp) is pronounced in English as "aver-due-pois," with the "pois" as in poison. It refers to a system of weights (masses) that is widely used in the United States, and parts of Canada and Great Britain, to weigh everything except precious metals, gems, and pharmaceuticals (drugs). It is based on a 28.35-gram *ounce* (oz), 16-avdp-ounce *pound* (lb), and 2000-pound *short ton*. The concept of a pound originated in the Roman Empire, developed into a 16-ounce pound in Europe by the year 1300, and was the system of weights adopted by the 13 British Colonies that became the United States of America. However, the United States developed the 2000-pound "short ton" that is still used today (instead of the British 2240-pound "long ton").

What Are Troy Weights?

Troy (t) refers to an old British system of weights (masses) that is now used globally to quote the weights of precious metals (gold and silver) and gems. It is based on a 31.1-gram *troy ounce* (ozt) and a 12-troy-ounce *troy pound* (lbt). Because some gems and jewelry pieces are small, jewelers often express their weight in troy "pennyweight" (abbreviated "DWT"). One pennyweight is equal to 1/20th of a troy ounce (0.05 ozt) or 1.555 grams.

Unit Conversion—The Math You Need

You can learn more about unit conversion (including problems on converting weights) at this site featuring The Math You Need, When You Need It math tutorials for students in introductory geoscience courses: **http://serc.carleton.edu/mathyouneed/units/index.html**

ACTIVITY 3.1 Mineral and Rock Inquiry

Name: ______________________ **Course/Section:** ______________________ **Date:** ____________

A. All of the samples below are rocks from Earth's crust. Record how many crystals you see in each sample (Write 1, 2, 3, or many). Then make a numbered list of how many different kinds of minerals are in the sample and describe each one in your own words. Complete parts **B** and **C**.

How many **crystals** do you see in this sample? _____
List the number of different **minerals** in the sample and give a description of each one.

How many **crystals** do you see in this sample? _____
List the number of different **minerals** in the sample and give a description of each one:

How many **crystals** do you see in this sample? _____
List the number of different **minerals** in the sample and give a description of each one:

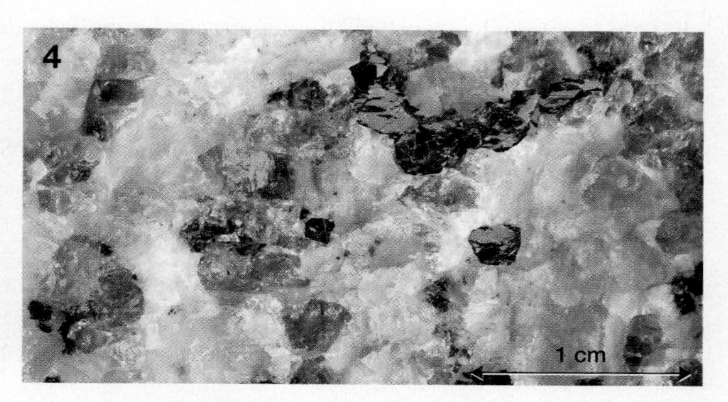

How many **crystals** do you see in this sample? _____
List the number of different **minerals** in the sample and give a description of each one:

B. Which of these samples seems to have crystals of a valuable chemical element? _______ What element? _______

C. **REFLECT & DISCUSS** Based on your observations in this activity—what is a rock, and how are rocks related to minerals and crystals?

ACTIVITY 3.2 Mineral Properties

Name: ______________________ **Course/Section:** ______________________ **Date:** ____________

A. Indicate whether the luster of each of the following materials looks metallic (M) or nonmetallic (NM):

1. a mirror: ________ **2.** butter: ________ **3.** ice: ________ **4.** a rusty nail: ________

B. What is the streak color (i.e., color in powdered form) of each of the following substances?

1. salt: ____________ **2.** wheat: ____________ **3.** pencil lead: ____________

C. What is the crystal form (**FIGURE 3.4**) of the:

1. quartz in **FIGURE 3.1B**? ____________ **2.** native copper in **FIGURE 3.6**? ____________

D. Look up quartz in the Mineral Database (**FIGURE 3.21**, page 93) to find a list of the varieties (var.) of quartz. Then identify each quartz variety below, and write its name beneath the image.

var. ____________ var. ____________ var. ____________ var. ____________

E. A mineral can be scratched by a masonry nail or knife blade but not by a wire (iron) nail (**FIGURE 3.9**).

1. Is this mineral hard or soft? ____________

2. What is the hardness number of this mineral on Mohs Scale? ________

3. What mineral on Mohs Scale has such a hardness? ____________

F. A mineral can scratch calcite, and it can be scratched by a wire (iron) nail.

1. What is the hardness number of this mineral on Mohs Scale? ________

2. Which mineral on Mohs Scale has this hardness? ____________

G. The brassy, opaque, metallic mineral in **FIGURE 3.7A** is the same as the mineral in **FIGURE 3.8**. What is this mineral's hardness, and how can you tell?

H. Analyze the mineral samples and figure caption in **FIGURE 3.16**.

1. What is this mineral's hardness (give a number or range of numbers)? ____________

2. Very carefully cut out the cleavage goniometer from GeoTools Sheet 1 at the back of this manual. Be sure to cut the angles as exactly as possible. **Sketch the characteristic shape** that this mineral breaks into. Using the cleavage goniometer, measure the angles between flat flat cleavage surfaces of this mineral in **FIGURE 3.16**, and record the angles here:

What is the name of this kind of cleavage?

I. A mineral sample weighs 27 grams and takes up 10.4 cubic centimeters of space. What is the SG (specific gravity) of this mineral? Show your work.

J. Analyze these two photomicrographs of ice crystals (snowflakes) by William Bentley.

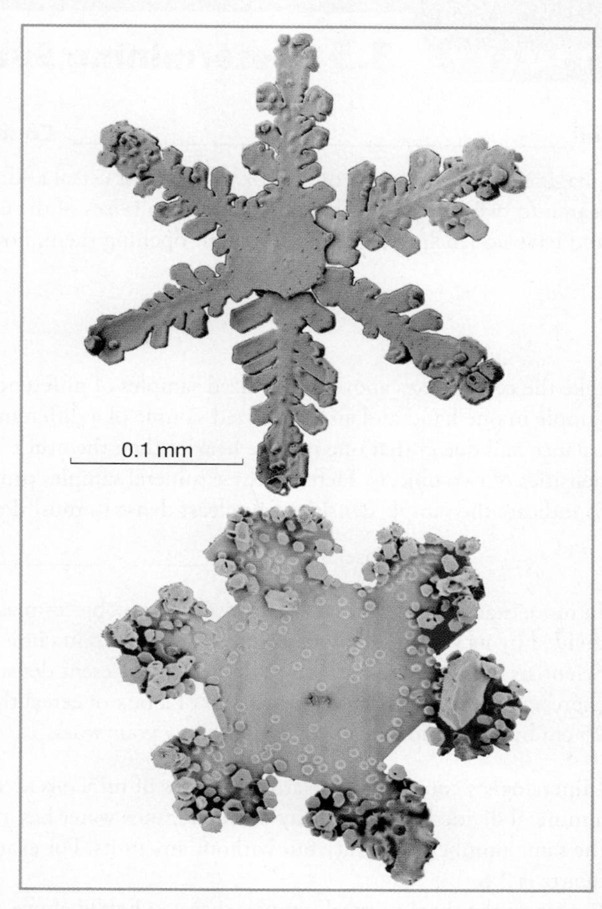

1. Based on **FIGURE 3.4**, what is the crystal form of the top crystal?

2. Notice that the crystals are symmetrical, but not exactly. Imperfections are common in crystals, but their underlying crystal form can still be detected. To what crystal system in **FIGURE 3.5** do ice crystals belong? How can you tell?

3. **REFLECT & DISCUSS** The habit of snowflakes (crystals of water ice) includes a variety of different crystal forms. Why don't all snowflakes have the same crystal form?

K. Analyze each crystalline household material pictured below and identify which crystal system it belongs to. (Use a hand lens or microscope to observe actual samples of the materials if they are available.)

1. Sucrose (table sugar) belongs to the ______________________________ crystal system.
   How can you tell?

2. Epsomite (epsom salt) belongs to the ______________________________ crystal system.
   How can you tell?

3. Halite (table salt) belongs to the ______________________________ crystal system.
   How can you tell?

4. **REFLECT & DISCUSS** Which of these crystalline household materials (sucrose, epsomite, or halite) cannot be a mineral? Why not?

ACTIVITY 3.3 Determining Specific Gravity (SG)

Name: ______________________ **Course/Section:** ______________________ **Date:** ____________

A. Imagine that you want to buy a box of breakfast cereal and get the most cereal for your money. You have narrowed your search to two brands of cereal that are sold in boxes of the exact same size and price. The boxes are made of opaque cardboard and have no labeling of weight. Without opening them, how can you tell which box contains the most cereal?

B. Like the cereal boxes above, equal-sized samples of different minerals often have different weights. If you hold a mineral sample in one hand and an equal-sized sample of a different mineral in the other hand, then it is possible to act like a human balance and detect that one may be heavier than the other. This is called **hefting**, and it is used to estimate the relative densities of two objects. Heft the three mineral samples provided to you, then write sample numbers/letters on the lines below to indicate the sample densities from least dense to most dense.

(Least dense) ____________ ____________ ____________ (Most dense)

C. In more exact terms, **density** is a measure of an object's mass (weighed in grams, g) divided by its volume (how much space it takes up in cubic centimeters, cm^3). Scientists use the Greek character rho (ρ) to represent density, which is always expressed in g/cm^3. What is the density of a box of cereal that is 20 cm by 25 cm by 5 cm and weighs 0.453 kg? Show your work.

D. Mineralogists compare the relative densities of minerals according to their **specific gravity (SG)**: the ratio of the density of a mineral divided by the density of water. Since water has a density of 1 g/cm^3, and the units cancel out, specific gravity is the same number as density but without any units. For example, the density of quartz is 2.6 g/cm^3, so the specific gravity of quartz is 2.6.

Return to the three mineral samples that you hefted above, and do the following:

1. First (while they are still dry), determine and record the mass (weight) of each sample in grams.
2. Use the water displacement method to measure and record the volume of each sample (**FIGURE 3.15**). Recall that one fluid milliliter (mL or ml on the graduated cylinder) equals one cubic centimeter.
3. Calculate the specific gravity of each sample.
4. Identify each sample based on the list of specific gravities of some common minerals.

Sample	Mass in Grams (g)	Volume in Cubic cm (cm^3)	Specific Gravity (SG)	Mineral Name

SG OF SOME MINERALS	
2.1	Sulfur
2.6–2.7	Quartz
3.0–3.3	Fluorite
3.5–4.3	Garnet
4.4–4.6	Barite
4.9–5.2	Pyrite
7.4–7.6	Galena
8.8–9.0	Native copper
10.5	Native silver
19.3	Native gold

D. REFLECT & DISCUSS Were your data and calculations accurate enough to be useful in identifying the samples? If not, how could they be made more accurate?

Name: ______________________ Course/Section: ______________________ Date: ______________

ACTIVITY 3.4 Mineral Analysis, Identification, and Uses

Sample Letter or Number	MINERAL DATA CHART						
	Luster*	Hardness	Color / Streak	Cleavage / Fracture	Other notable properties; tenacity, magnetic attraction, reaction with acid, specific gravity, smell, etc	Name (Fig. 3.18, 3.19, or 3.20) and chemical composition (Fig. 3.21)	How do you depend on this mineral or elements from it? (Fig. 3.21)

*M = metallic or submetallic, NM = nonmetallic

Name: ______ Course/Section: ______ Date: ______

MINERAL DATA CHART

Sample Letter or Number	Luster*	Hardness	Color / Streak	Cleavage / Fracture	Other notable properties; tenacity, magnetic attraction, reaction with acid, specific gravity, smell, etc	Name (Fig. 3.18, 3.19, or 3.20) and chemical composition (Fig. 3.21)	How do you depend on this mineral or elements from it? (Fig. 3.21)

*M = metallic or submetallic, NM = nonmetallic

Name: ______________________ Course/Section: ______________________ Date: ____________

MINERAL DATA CHART

Sample Letter or Number	Luster*	Hardness	Color / Streak	Cleavage / Fracture	Other notable properties; tenacity, magnetic attraction, reaction with acid, specific gravity, smell, etc	Name (Fig. 3.18, 3.19, or 3.20) and chemical composition (Fig. 3.21)	How do you depend on this mineral or elements from it? (Fig. 3.21)

*M = metallic or submetallic, NM = nonmetallic

Name: ______________ Course/Section: ______________ Date: ______________

MINERAL DATA CHART

Sample Letter or Number	Luster*	Hardness	Color / Streak	Cleavage / Fracture	Other notable properties; tenacity, magnetic attraction, reaction with acid, specific gravity, smell, etc	Name (Fig. 3.18, 3.19, or 3.20) and chemical composition (Fig. 3.21)	How do you depend on this mineral or elements from it? (Fig. 3.21)

*M = metallic or submetallic, NM = nonmetallic

ACTIVITY 3.5 The Mineral Dependency Crisis

Name: ______________________ **Course/Section:** ______________________ **Date:** ____________

A. Refer to the list of selected net non-fuel mineral resource Imports by the United States in 2012 (**FIGURE 3.22**).

1. Based on **FIGURE 3.22**, complete the table below.

Element used to make cell phones	**Mineral ore(s)** from which it is extracted	How is the element used to make cell phones?

2. Would the United States be able to manufacture cell phones if a world crisis prevented it from importing minerals and elements? _____yes _____ no What evidence from **FIGURE 3.22** supports your answer?

B. Refer to **FIGURE 3.22** and your uses of minerals recorded in your completed Worksheet 3.4. How would your lifestyle change if the U.S. could no longer import the minerals and elements that it imported 100% of in 2012?

C. REFLECT & DISCUSS Do you think that it will be possible in the future for the United States to sustain its 2012 levels of net import reliance for the selected commodities and minerals in **FIGURE 3.22**? Explain your answer.

ACTIVITY 3.6 Urban Ore

Name: ______________________ **Course/Section:** ______________________ **Date:** __________

A. Recall that "ore" is a material (usually rocks or minerals) from which chemicals can be extracted at a profit.

1. More than half of the gold mined in the U.S. is from mines in northern Nevada. These mines produced an average of 3.2 grams of gold per 2000-pound short ton of ore in 2012.

a. Search the Internet for "New York spot gold price" in U.S. dollars (USD) per ounce, and enter it here. Note that ounces of gold are always quoted in troy ounces (ozt), but some people incorrectly report it as "oz."

NY spot gold price: ______________________

b. There are 31.1 grams (g) in one troy ounce (ozt). How many troy ounces of gold are extracted from one short ton of average Nevada ore? Show your work.

c. What is the gold worth (in USD) from one ton of the Nevada ore? Show your work.

d. It costs about $640 to extract 1 troy ounce of Nevada gold from the mine. So how much does it cost to mine and extract the gold from one short ton of the Nevada ore? Show your work.

e. Based on your answers in **c** and **d** above, what is the current average profit per ton of gold ore from northern Nevada?

B. The average iPhone 5 contains 0.0012 grams of gold and weighs 3.951 ounces (avdp).

1. There are 16.00 ounces in one avoirdupois pound and 2000 pounds in one short ton (avoirdupois ton). How many iPhone 5 cell phones are there in one short ton? Show your work.

2. Based on your work above, and the fact that there are about 0.0012 grams of gold in one iPhone 5, how many grams of gold are there in one short ton of iPhone 5 cell phones? Show your work.

3. There are 31.1 grams in one troy ounce. How many troy ounces of gold are there in 1 short ton of iPhone 5s? Show your work.

4. Based on the New York spot gold price in USD per troy ounce) that you determined above (part **A1a**), what is the current value of the gold in one short ton of iPhone 5 cell phones?

C. REFLECT & DISCUSS What materials besides cell phones could the U.S. recycle or mine as "urban ore" for metals noted in **FIGURE 3.22**, and what impact would this have on the environment and the ability of the U.S. to sustain its need for metals and mineral ores?

LABORATORY 4

Rock-Forming Processes and the Rock Cycle

El Capitan, Yosemite, California, is a large body of igneous rock that formed when magma cooled underground. The rock weathered from above it to reveal the igneous rock, which is crumbling to form sediment. (Henrik Lehnerer/Gamma/Glow Images)

BIG IDEAS

Rocks can be classified as igneous, sedimentary, or metamorphic on the basis of their present composition and texture and how they formed.The rock cycle model explains how all rocks can be formed, deformed, transformed, melted, and reformed as a result of environmental factors and natural processes that affect them.

FOCUS YOUR INQUIRY

THINK About It What are rocks made of, and where and how do they form?

ACTIVITY 4.1 Rock Inquiry *(p. 112)*

ACTIVITY 4.2 What Are Rocks Made Of? *(p. 113)*

ACTIVITY 4.3 Rock-forming Minerals *(p. 117)*

THINK About It How are a rock's composition and texture used to classify it as igneous, sedimentary, or metamorphic?

ACTIVITY 4.4 What Is Rock Texture? *(p. 117)*

THINK About It How are rocks formed, deformed, transformed, melted, and reformed as a result of environmental factors and natural processes of the rock cycle?

ACTIVITY 4.5 Rocks and the Rock Cycle Model *(p. 119)*

Introduction

Rocks are the solid materials that make up most of Earth, our Moon, and the other rocky planets of our solar system. You might say that the largest rock on Earth is its entire rocky body, the geosphere. And like all bodies, it is a complex system of interacting parts and processes. Bedrock sticks out of the ground at "outcrops," where it is physically and chemically broken down into loose rock frgaments, mineral particles, and chemical residues. But at the same time, new bedrock is forming from cooling lava at volcanoes. The bodily function of the geosphere is its cycle of rock creation, deformation, and destruction—the "rock cycle." Every rock is part of the cycle and contains a story of where it has been on its path through the cycle. So the next time you see a rock, look closely for clues about its origin and the story it has to tell.

ACTIVITY

4.1 Rock Inquiry

THINK About It | **What are rocks made of, and where and how do they form?**

OBJECTIVE Analyze rock samples and infer where and how they formed?

PROCEDURES

1. **Before you begin**, do not look up definitions and information. Use your current knowledge, and complete the worksheet with your current level of ability. Also, this is **what you will need** to do the activity:
   ____ pencil(s) with eraser
   ____ Activity 4.1 Worksheet (p. 121)
2. **Analyze both rocks, and complete the worksheet in a way that makes sense to you.**
3. **After you complete the worksheet**, read about rocks and the rock cycle below and be prepared to discuss your observations, interpretations, and inferences with others.

Rocks and the Rock Cycle

Most rocks are solid aggregates of mineral grains (particles), either mineral crystals or clasts (broken pieces) of mineral crystals and rocks (e.g., pebbles, gravel, sand, and silt). There are, however, a few notable rocks that are not made of mineral grains. For example, *obsidian* is a rock made of volcanic glass, and *coal* is a rock made of plant fragments.

Three Main Groups of Rocks

Rock-forming materials come from Earth's mantle (as molten rock called *magma* while underground and *lava* when it erupts to the surface), space (meteorites), organisms (parts of plants and animals), or the fragmentation and chemical decay of mineral crystals and other rocks. Environmental changes and processes affect these materials and existing rocks in ways that produce three main rock groups (FIGURE 4.1):

- **Igneous rocks** form when magma or lava cool to a solid form—either glass or masses of tightly intergrown mineral crystals. The crystals are large if they had a long time to grow in a slowly cooling magma, and they are small if they formed quickly in a rapidly cooling lava.
- **Sedimentary rocks** form mostly when mineral crystals and clasts (broken pieces, fragments) of plants, animals, mineral crystals, or rocks are compressed or naturally cemented together. They also form when mineral crystals precipitate from water to form a rocky mass such as *rock salt* or cave stalactites.
- **Metamorphic rocks** are rocks deformed or changed from one form to another (transformed) by intense heat, intense pressure, and/or the action of hot fluids. This causes the rock to recrystallize, fracture, change color, and/or flow. As the rock flows, the flat layers are folded and the mineral crystals are aligned like parallel needles or scales.

The Rock Cycle

All rocks are part of a system of rock-forming processes, materials, and products that is often portrayed in a conceptual model called the **rock cycle** (FIGURE 4.2). The rock cycle model explains how all rocks can be formed, deformed, transformed, melted, and reformed as a result of environmental factors and natural processes that affect them.

Igneous Processes. An idealized path (broad purple arrows) of rock cycling and redistribution of matter is illustrated in FIGURE 4.2, starting with igneous processes. If magma (from the mantle or lower crust) cools, then it solidifies into igneous rocks that are masses of glass or aggregates of intergrown mineral crystals.

Sedimentary Processes. If these igneous rocks are uplifted, then sedimentary processes force other changes to occur. The igneous rocks are **weathered** (fragmented into grains, chemically decayed to residues, or even dissolved), **eroded** (worn away) and **transported** (moved to a new place), and later deposited to form **sediment** (an accumulation of chemical residues and fragmented rocks, mineral crystals, plants, or animals). Meteorites (dust and rocks from space) may be incorporated into the sediment. Sediment is **lithified** (hardened) into sedimentary rock as it compacts under its own weight or gets naturally cemented with crystals precipitated from water.

Metamorphic Processes. If the sedimentary rock is subjected to metamorphic processes (intense heat, intense pressure, or the chemical action of hot fluids), then it will *deform* (fold, fracture, or otherwise change its shape) and *transform* (change color, density, composition, and/or general form) to metamorphic rock. And if the heat is great enough, then the metamorphic rock will melt (an igneous process) to form another body of magma that will begin the cycle again.

Multiple Pathways Through the Rock Cycle. Of course, not all rocks undergo change along such an idealistic path. There are *at least* three changes that each rock could undergo. The arrows in FIGURE 4.2

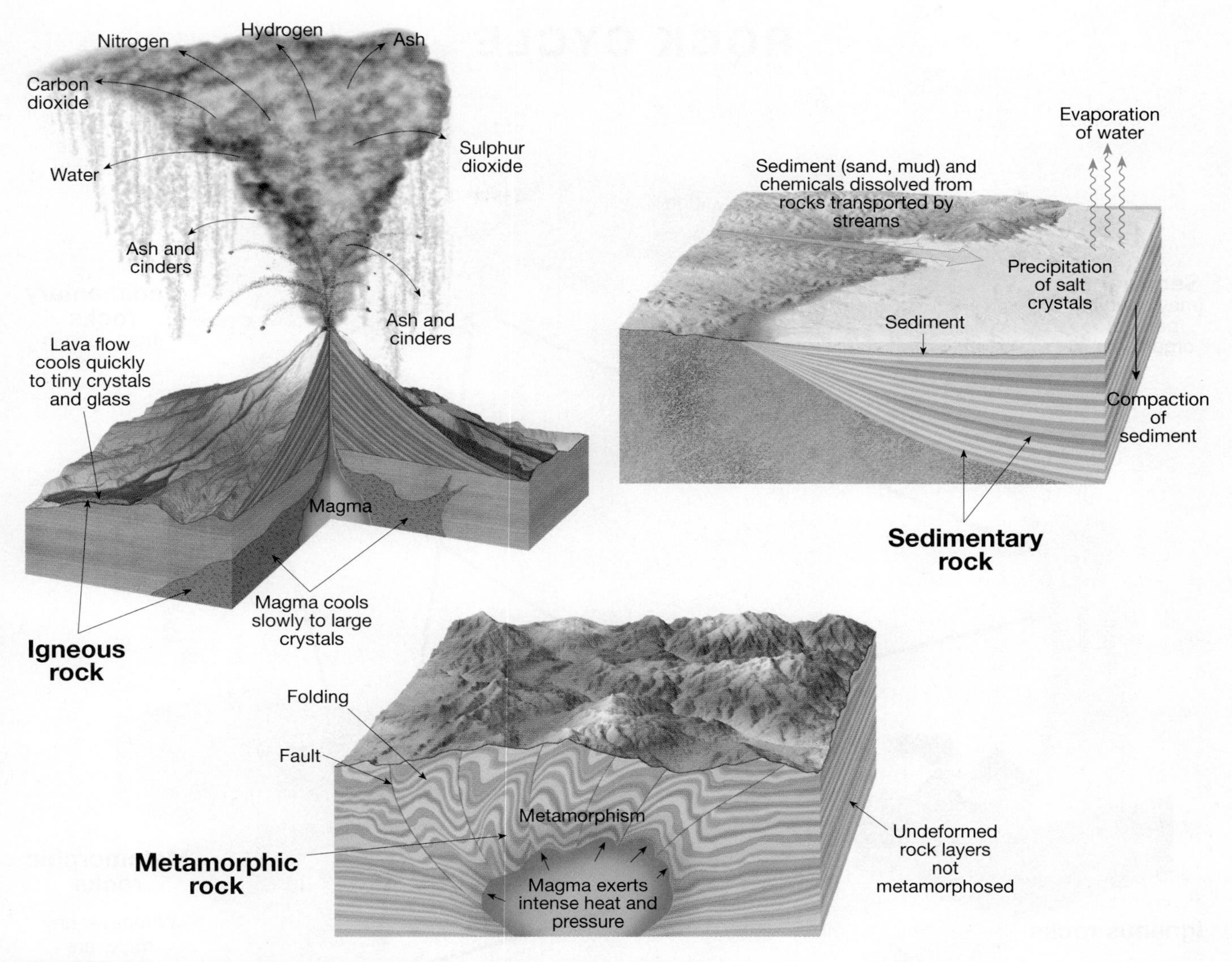

FIGURE 4.1 **The origin of igneous, sedimentary, and metamorphic rocks.**

show that any rock from one group can be transformed to either of the other two groups *or* recycled within its own group. Igneous rock can be (1) weathered and eroded to form sediment that is lithified to form sedimentary rock; (2) transformed to metamorphic rock by intense heat, intense pressure, and/or hot fluids; or (3) re-melted, cooled, and solidified back into another igneous rock. Sedimentary rock can be (1) melted, cooled, and solidified into an igneous rock; (2) transformed to metamorphic rock by intense heat, intense pressure, and/or hot fluids; or (3) weathered and eroded back to sediment that is lithified back into another sedimentary rock. Metamorphic rock can be (1) weathered and eroded to form sediment that is lithified into sedimentary rock; (2) melted, cooled, and solidified into igneous rock; or (3) re-metamorphosed into a different type of metamorphic rock by intense heat, intense pressure, or hot fluids.

ACTIVITY 4.2 What Are Rocks Made Of?

THINK About It | **What are rocks made of, and where and how do they form?**

OBJECTIVE Analyze rock samples and describe what they are made of.

PROCEDURES

1. **Before you begin**, read about rock composition below. Also, this is **what you will need:**
   ____ pencil with eraser
   ____ Activity 4.2 Worksheet (p. 122)
2. **Then follow your instructor's directions** for completing the worksheets.

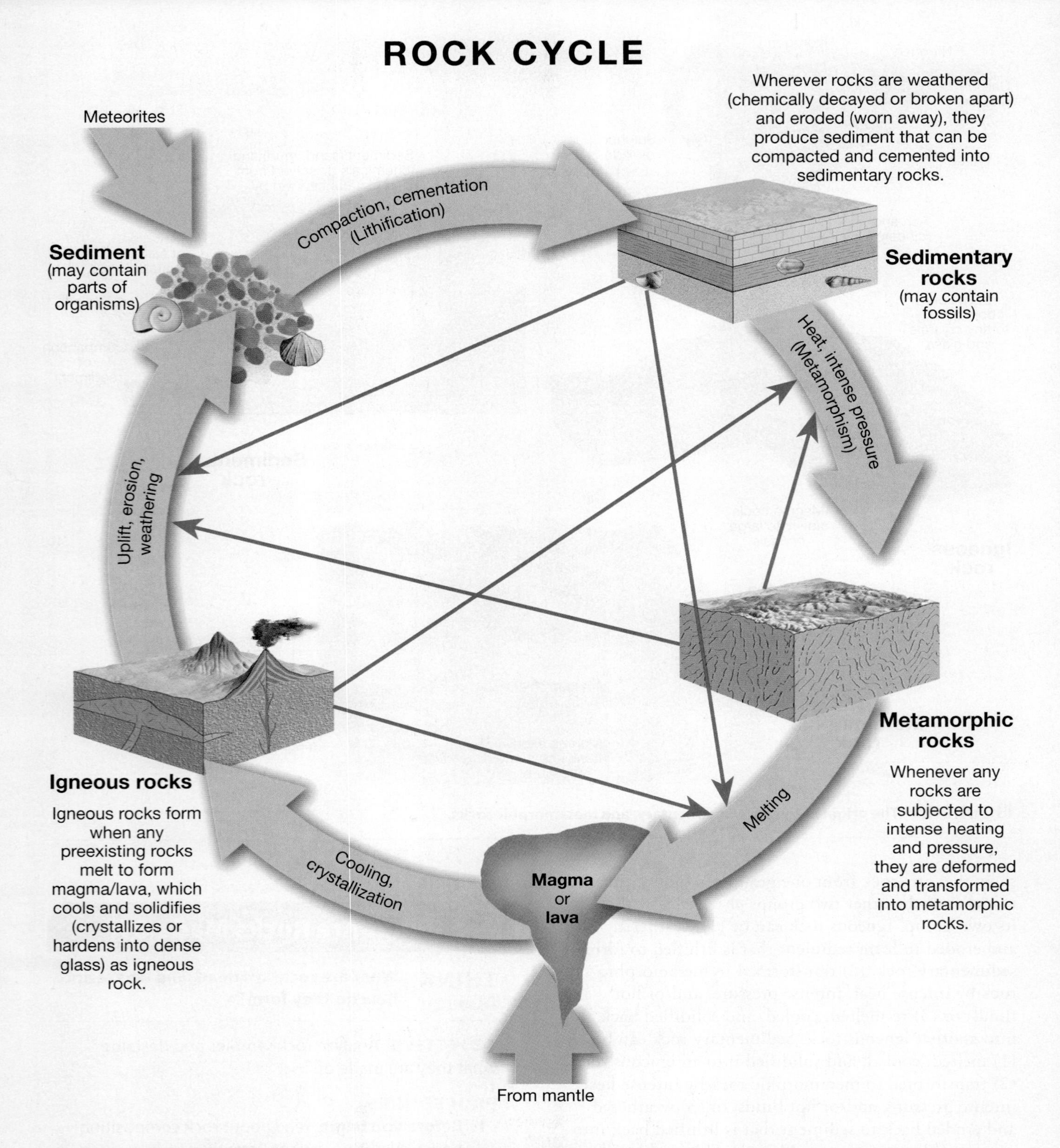

FIGURE 4.2 The Rock Cycle—a conceptual model of how all rocks can be formed, transformed, destroyed, and re-formed as a result of environmental factors and natural processes that affect them. Environmental changes and processes affect these materials and existing rocks in ways that produce three main rock groups. Arrows show that a rock from one group can be transformed to either of the other two groups, or it can be recycled within its own group. An *idealized rock cycle path* is shown by the broad large arrows. But there are *at least* two other changes that each rock could undergo.

Rock Composition

Composition of a rock refers to what it is made of. *Chemical composition* refers to the chemical elements that make up the rock. This determines how the rock will react with materials of different composition, such as whether or not it will react with and decay (tarnish, dissolve, chemically disintegrate) in air or water. It also determines rock color. For example, ferromagnesian-rich rocks (iron- and magnesium-rich rocks) generally have a dark color and ferromagnesian-poor rocks generally have a light color. But the chemical elements in a rock are normally bonded together in tangible materials like minerals that, in turn, make up most rocks. So the *physical composition* of rocks is a description of what visible materials they are made of, in whole or part. It is your job as a geologist, using your eyes and simple tools (like a hand magnifying lens), to describe and identify what physical materials are made of.

Volcanic Glass

Glass is an amorphous (containing no definite form; not crystalline) solid that forms by cooling molten (liquefied by heat) materials like melted rock (lava) or quartz sand (the main ingredient that is melted to make window glass). Volcanic glass (obsidian) looks and breaks just like window glass, except that it is usually dark colored. But how does it form? When a volcano erupts, and lava is erupted onto Earth's surface, it begins to cool. If the lava is fluid enough (has low viscosity), and stays liquefied long enough, then its elements and molecules will bond together and form mineral crystals. But if the lava is too viscous, and cools too quickly, then mineral crystals do not form and the solid material that remains is volcanic glass.

Grains in Rocks

Most rocks are made of **grains**—mineral crystals or other hard, visible particles. To view the grains in a rock hand sample, start with your own eyes and look closely. If you cannot see or identify the grains, then also try using a hand lens. Most geologists use a 10x hand lens, meaning that objects viewed through the lens appear ten times larger than in real life. Here is a list of the kinds of grains that comprise most rocks. You should look for the following:

- **Mineral grains.** Mineral crystals are the most common kind of grains in rocks. There are thousands of kinds of minerals, but twenty or fewer make up the bulk of most rocks and are known as rock-forming minerals (**FIGURE 4.3**). Whenever possible, try to identify and record what kind(s) of mineral crystals are present in any rock that you analyze. Also try to determine if the mineral crystals are *in situ* or not. **In situ mineral grains** are present in the rock where they originally formed. Examples are the intergrown mineral crystals in an igneous rock that formed from cooling of lava or magma and intergrown halite crystals in rock salt that formed in an evaporating sea. The intergrown mineral crystals lock together to form the rock. Also, *in situ* mineral crystals are usually arranged randomly, and they may be engulfed in glass or a mass of smaller intergrown crystals. **Detrital mineral grains** are not *in situ*—they are not intergrown, and do not lock together to form the rock. This is because they were removed from the place or rock where they originally formed and were transported by wind, water, ice, organisms, and/or gravity to a new place. There, they may become or have already become part of another rock. Most detrital mineral grains are clasts (see below), such as quartz pebbles.

- **Clasts.** Physical weathering is the cracking, crushing, and wearing away of Earth materials. The cracking and crushing causes big rocks, animal shells, and plants to be fragmented into broken pieces called **clasts** (from the Greek *klastós,* meaning broken in pieces). Plant fragments and shells or bones that have been separated or broken are often singled out as **bioclasts**. Broken mineral crystals are **detrital mineral grains** (described above), and broken pieces of rock are called **rock fragments**. Similarly, geologists have names for size classes of clasts (gravel, sand, silt, clay).

- **Gravel, sand, silt, and clay.** These terms are often used to describe what a rock or other feature is made of. For example, there is sand in a sandbox, and sandstone is made of sand. But the terms are actually names for size classes of clasts (called Wentworth size classes after C.K. Wentworth, who devised the scale in 1922). **Gravel** is a mass of grains that are mostly larger than 2 mm (like aquarium gravel, pebbles, cobbles, and boulders). **Sand** is a mass of grains that are mostly 1/16 to 2 mm in diameter (like sand in a sandbox or making up a sandy beach). **Silt** is finer than sand, so much that you can barely see and feel the grains. The grains are generally too small to identify with a hand lens or your unaided eye, so geologists refer to them collectively as silt. **Clay** is even finer than silt. If you ever played with pottery clay, then you know that it can dry on your hands as a light-colored slippery powder. You can tell it is there, but grains are too small to feel or see individually (even with a hand lens). Thus, geologists refer to these microscopic grains collectively as clay.

SOME COMMON ROCK-FORMING MINERALS

Mineral	Description	Occurs in Igneous Rocks	Occurs in Sedimentary Rocks	Occurs in Metamorphic Rocks
Augite pyroxene	Very dark green to brown or dark gray, hard mineral (Hardness 5.5 – 6.0) with two cleavages about 90 degees apart.	as *in situ* crystals		
Biotite mica	Glossy black mineral that easily splits into thin transparent sheets along its excellent cleavage. Hardness 2.5 – 3.0.	as *in situ* crystals		as foliated *in situ* crystals
Calcite	Usually colorless, yellow, white, or amber. Breaks along three excellent cleavages (none at 90 degrees) to form rhombohedrons (leaning blocks). Hardness 3. Reacts with dilute hydrochloric acid (HCl).		as *in situ* crystals	as *in situ* crystals
Chlorite	Green mica-like mineral that splits into thin glossy transparent sheets along its excellent cleavage. Hardness 2.0 – 2.5. Occurs in large crystals or fine-grained masses.		as microscopic crystals that color the rock green	as *in situ* crystals
Garnet	Red to black rounded crystals with no cleavage. Very hard (Hardness 7).	Rarely, as *in situ* crystals	as detrital clasts	as *in situ* crystals
Gypsum	Colorless, white, or gray mineral. Easily scratched (Hardness 2.0), even with a fingernail.		as *in situ* crystals	
Halite	Colorless, white, yellow, gray cubes, that break into cubic shapes because they have three excellent cleavages 90 degrees apart. Brittle. Hardness 2.5.		as *in situ* crystals	
Hornblende amphibole	Dark gray to black, hard mineral (Hardness 5.5 – 6.0). Breaks along glossy cleavage surfaces about 56 and 124 degrees apart.	as *in situ* crystals		as *in situ* crystals
Kaolinite	Earthy white, gray or very light brown clayey masses that leave powder on your fingers. Very fine-grained. No visible crystals. Hardness 1 – 2.		as *in situ* earthy masses	
Muscovite mica	Colorless, brown, yellow, or white minerals that easily splits into transparent thin sheets along its excellent cleavage. Hardness 2.0 – 2.5.	as *in situ* crystals	tiny flakes as clasts	as foliated *in situ* crystals
Olivine	Pale to dark olive green or yellow mineral with no cleavage. Very hard (Hardness 7). Crystals may resemble sand grains. Brittle.	as *in situ* crystals		as *in situ* crystals
Plagioclase feldspar	Usually white to pastel gray, but may be colorless or black with iridescent play of colors. Exhibits fracture surfaces and two good cleavages. Cleavage surfaces may have thin striations. Hardness 6.	as *in situ* crystals	as detrital clasts	as *in situ* crystals
Potassium feldspar (orthoclase)	Usually pink-orange or pale brown, may be white. Usually has internal discontinuous streaks (exsolution lamellae). Exhibits fracture surfaces and two good cleavages. Hardness 6.	as *in situ* crystals	as detrital clasts	as *in situ* crystals
Quartz	Usually transparent to translucent gray or milky white, may be colorless. No cleavage. Breaks along uneven fractures or curved conchoidal fractures (like glass). Very hard (Hardness 7).	as *in situ* crystals	as *in situ* crystals and commonly as detrital clasts	as *in situ* crystals
Serpentine	Usually pale to dark green, opaque masses with no visible crystals or cleavage. Usually scratches easily (Hardness 2 – 5).			as *in situ* masses

In situ ("in place") mineral grains are present in the rock where they originally formed. Examples are mineral crystals newly formed from cooling of lava or magma (in igneous rocks), crystals newly formed or recyrstallized in rock or hot watery solutions under conditions of intense heat and pressure or (in metamorphic rocks), or as newly formed crystals precipitated from evaporating surface or ground water (in sedimentary rocks).
Detrital mineral grains are not in situ. They did not form where they are now found, are not intergrown, and do not lock together to form the rock. They were removed from the place or rock where they originally formed and were moved by wind, water, ice, organisms, and/or gravity to a new place. Examples are pebbles and sand grains in sedimentary and metamorphic rocks.
Foliated mineral grains are flat or blade-like crystals that have been aligned and layered, like scales on a fish, during metamorphism.

FIGURE 4.3 Some common rock-forming minerals.

ACTIVITY

4.3 Rock-forming Minerals

THINK About It **What are rocks made of, and where and how do they form?**

OBJECTIVE Analyze and identify samples of some common rock-forming minerals.

PROCEDURES

1. Before you begin, read about Rock Composition (page 116 and FIGURE 4.3). Also, this is **what you will need:**
   ___ pencil with eraser
   ___ Activity 4.3 Worksheet (p. 123)
   ___ optional: a set of mineral samples (obtained as directed by your instructor)
   ___ optional: a set of mineral analysis tools (obtained as directed by your instructor)
2. **Then follow your instructor's directions** for completing the worksheets.

ACTIVITY

4.4 What Is Rock Texture?

THINK About It **How are a rock's composition and texture used to classify it as igneous, sedimentary, or metamorphic?**

OBJECTIVE Determine textures of rocks and classify them based on their composition and texture.

PROCEDURES

1. **Before you begin**, read about rock textures below, and how rock composition and texture can be used to determine if a rock is igneous, sedimentary, or metamorphic. Also, this is **what you will need:**
   ___ pencil with eraser
   ___ Activity 4.4 Worksheets (pp. 124–126)
   ___ optional: a set of rock samples (obtained as directed by your instructor)
2. **Then follow your instructor's directions** for completing the worksheets.

Rock Texture

Another very important property of rocks is **texture**—a description of the grains and other parts of a rock and their size, shape, and arrangement. Carefully review the textures below. You will need to identify them in rock samples on Worksheet 4.2.

- **Glassy** refers to rocks that have no visible grains, and break along wavy, curved glossy surfaces—just like a broken glass bottle. An example is *obsidian*, a dense dark-colored volcanic glass. Another example is anthracite coal (hard coal), which has a glassy texture but is not really glass. It is made of plant fragments and parts that are so small and compacted together, that rock just looks glassy. Coal is also opaque in hand sample and less dense than true glass.
- **Fine-grained** refers to rocks made mostly of grains that are barely visible and too small to identify even when magnified with a hand lend (grains generally < 1 mm in diameter).
- **Coarse-grained** refers to rocks made mostly of grains that are visible and large enough to identify with either a hand lens or your unaided eyes (grains generally > 1 mm in diameter).
- **Vesicular** refers to rocks with round or oval holes, called *vesicles*, that resemble the holes in a sponge or Swiss cheese. The holes are bubbles of volcanic gases that bubbled through the lava that cooled to make the rock before the bubbles could escape. Some volcanic rocks have just scattered vesicles, but *pumice* is a rock containing so many vesicles that it floats in water. The vesicles in pumice are tiny glassy bubbles with sharp edges where they break, so pumice is used as an abrasive in polishes and as a cosmetic exfoliant bar to soften skin (pumice stones, Lava™ soap).
- **Crystalline** texture refers to fine- and coarse-grained rocks in which the grains are intergrown mineral crystals that glitter when rotated in bright light. (The light reflects off the flat crystal faces or cleavage surfaces like tiny mirrors.) The crystals may be **heterogranular** (a mixture of two or more significantly different sizes) or **equigranular** (all about the same size). The crystals may also be randomly arranged or else *foliated*—a metamorphic texure in which mineral grains have been aligned or layered, causing the rock to break or reflect light in a specific direction like the layered scales on a fish.
- **Clastic** texture means that the rock is mostly made of clasts (fragments; broken pieces) of minerals or other rocks (a rock made mostly of plant fragments or broken or separated bones and shells is called **bioclastic**). The clasts may be **angular**—freshly broken with sharp corners and edges, or **rounded**—having corners and edges worn down from transportation and grain abrasion. Recall that the terms gravel, sand, silt, and clay are also textural terms. **Gravelly** rocks are made mostly of gravel (grains larger than 2 mm; equal to or coarser-grained than aquarium gravel). **Sandy** rocks are mostly made of sand (grains 1/16 to 2 mm in diameter, coarse-grained like sand in a sandbox or a sandy beach). Silty and clayey rocks are fine-grained. The **silty** rocks are mostly made of grains that you

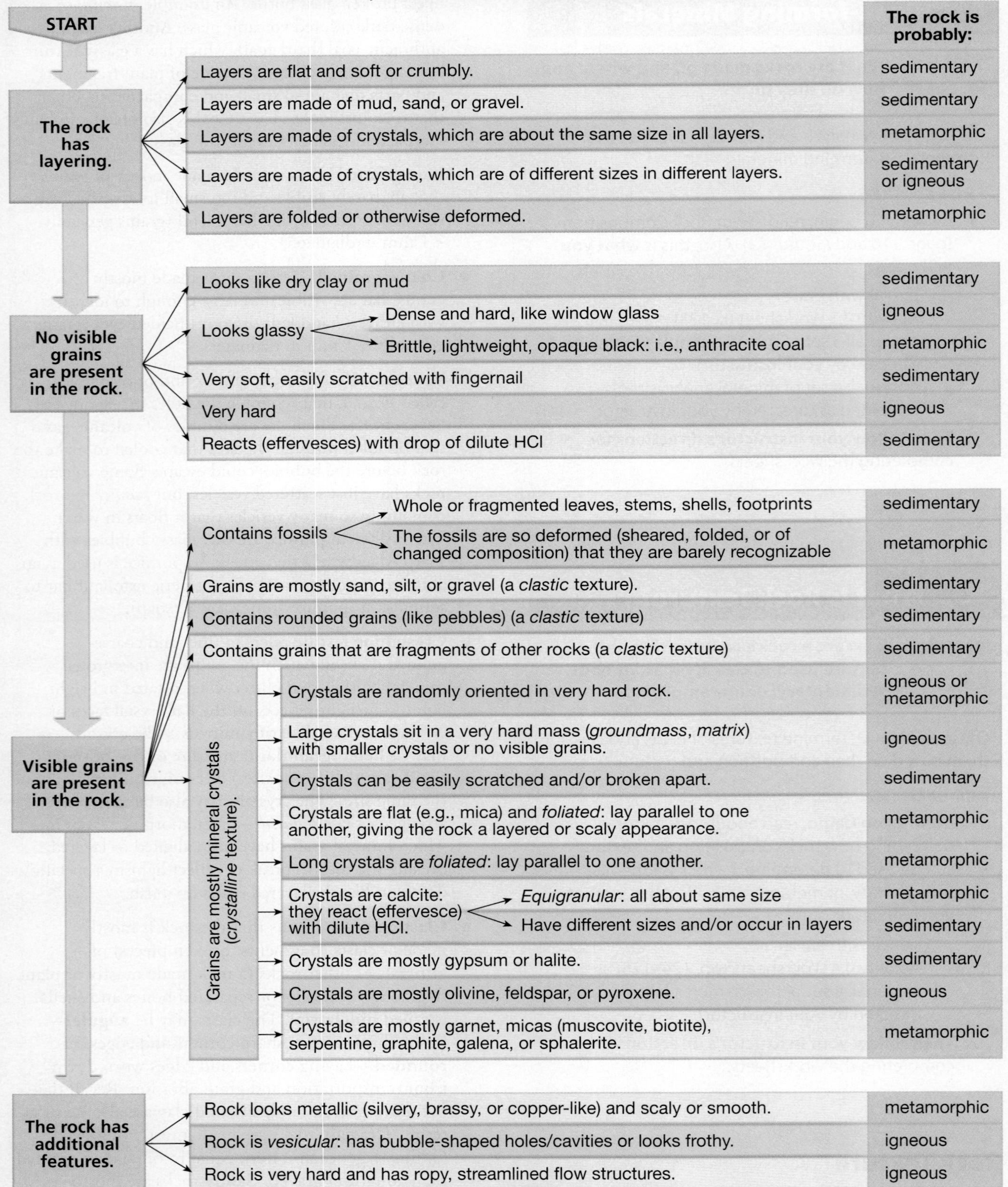

FIGURE 4.4 **Flowchart for classification of rocks as igneous, sedimentary, or metamorphic.**

can barely see and feel but are too small to identify with a hand lens or your unaided eye. **Clayey** rocks are mostly made of clay, which has grains too small to feel or see (even with a hand lens).

- **Layered** texture. Some rocks have grains arranged in layers that can be observed at more than one scale: over a region, in an outcrop, or in a hand sample. Sedimentary rocks generally have **flat layers** made of either clastic grains (gravel, sand, silt, clay, shells, plant fragments) or crystals of gypsum, halite, or calcite. Metamorphic rock layers are generally not flat-lying and **foliated** (a metamorphic texure described above in which mineral grains have been aligned or layered, causing the rock to break or reflect light in a specific direction like the layered scales on a fish). Metamorphic rock layers may also be **folded,** like you would fold a napkin. If the folds are smooth and unbroken, then the rock must have been soft and ductile (due to high thermal energy) when it was folded. The foliation is due to directed pressure and shearing during metamorphism. Brittle rocks do not fold easily. They tend to fracture (break, form a clastic texture) and move apart along faults.

ACTIVITY

4.5 Rocks and the Rock Cycle Model

THINK About It **How are rocks formed, deformed, transformed, melted, and reformed as a result of environmental factors and natural processes of the rock cycle?**

OBJECTIVE Analyze and classify rocks, infer how they formed, and predict how they may change according to the rock cycle.

PROCEDURES

1. Before you begin, read about The Rock Cycle (page 114 and FIGURE 4.2). Also, this is **what you will need:**

   ___ pencil with eraser

   ___ Activity 4.5 Worksheets (pp. 127–128)

2. **Then follow your instructor's directions** for completing the worksheets using Figure 4.5.

Rock Classification

All rocks are classified as igneous, sedimentary, or metamorphic, based on their properties of composition and texture and how they formed. Some properties are characteristic of more than one rock type. For example, igneous, sedimentary, and metamorphic rocks all can be dark, light, or made of mineral particles. Therefore, it is essential to classify a rock based on more than one of its properties.

Igneous Composition and Texture

Recall that igneous rocks form when molten rock (rock liquefied by heat and pressure in the mantle) cools to a solid form (FIGURE 4.1). Molten rock exists both below Earth's surface (where it is called *magma*) and at Earth's surface (where it is called *lava*). Igneous rocks can have various textures, including crystalline (heterogranular), glassy, or vesicular (bubbly). They commonly contain mineral crystals of olivine, pyroxene, or feldspars. Igneous rocks from cooled lava flows may have ropy, streamlined shapes or layers (from repeated flows of lava). Igneous rocks usually lack fossils and organic grains.

Sedimentary Composition and Texture

Recall that sedimentary rocks form in two ways (FIGURE 4.1). **Lithification** is the hardening of sediment—masses of loose Earth materials such as clasts (rock fragments, detrital mineral grains, pebbles, gravel, sand, silt, mud, shells, plant fragments) and products of chemical decay (clay, rust). **Precipitation** produces mineral crystals that collect as *in situ* aggregates, such as the rock salt that remains when ocean water evaporates. The lithification process occurs as layers of sediments are **compacted** (pressure-hardened) or **cemented** (glued together by tiny crystals precipitated from fluids in the pores of sediment).

Thus, most sedimentary rocks are layered and have a **clastic** texture (i.e., are made of grains called *clasts*—fragments of rocks, mineral crystals, shells, and plants—usually rounded into pebbles, gravel, sand, and mud). The sedimentary grains are arranged in layers due to sorting by wind or water. Sedimentary rocks may also include **fossils**—bones, impressions, tracks, or other evidence of ancient life.

The crystalline sedimentary rocks are layered aggregates of crystals precipitated from water. This includes the icicle-shaped stalactites that hang from the roofs of caves. Common minerals of these precipitated sedimentary rocks include calcite, dolomite, gypsum, or halite.

Metamorphic Composition and Textures

Recall that metamorphic rocks are rocks that have been deformed and transformed by intense heat, intense pressure, or the chemical action of hot fluids (FIGURE 4.1). Therefore, metamorphic rocks have textures indicating significant deformation (folds, extensive fractures, faults, and foliation). Fossils, if present, also are deformed (stretched or compressed). Metamorphic rocks often contain garnet, tourmaline, or foliated layers of mica. Serpentine, epidote, graphite, galena, and sphalerite occur only in metamorphic rocks. Metamorphism can occur over large regions, or in thin "contact" zones (like burnt crust on a loaf of bread) where the rock was in contact with magma or other hot fluids.

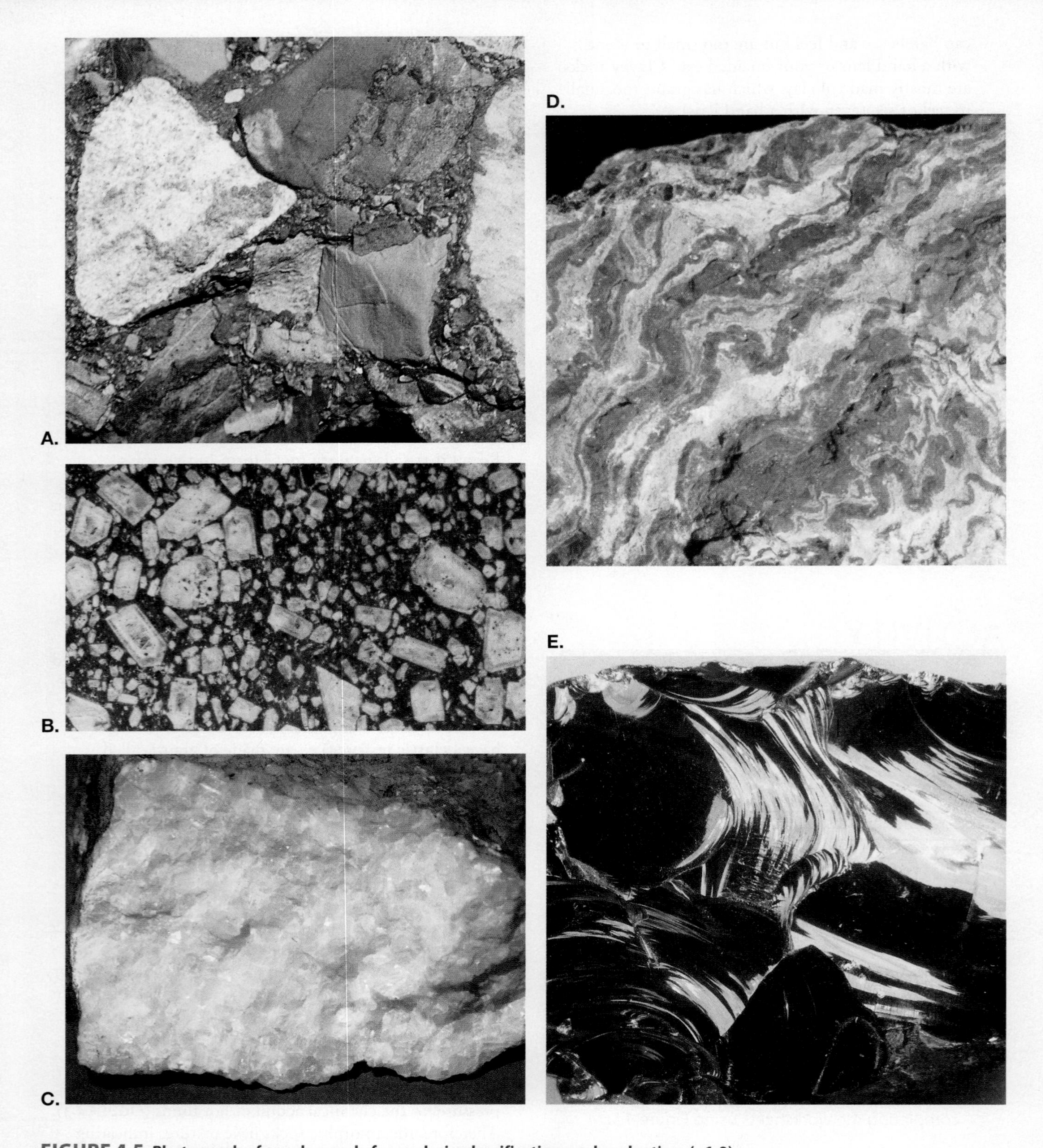

FIGURE 4.5 Photograph of a rock sample for analysis, classification, and evaluation. (×1.0)

ACTIVITY 4.1 Rock Inquiry

Name: ______________________ **Course/Section:** ______________________ **Date:** __________

A. REFLECT & DISCUSS Describe the rock below, where it may have formed, and how it may have formed.

B. REFLECT & DISCUSS Describe the rock below, where it may have formed, and how it may have formed.

ACTIVITY 4.2 What Are Rocks Made Of?

Name: ____________________ **Course/Section:** ________________ **Date:** __________

A. What is the difference between *in situ* mineral grains and detrital mineral grains?

B. REFLECT & DISCUSS Rocks are made of the materials listed below and described on page 115. Below each sample, write the name of every kind of material it contains from the list below. (All samples are x1.) Be prepared to compare your observations with the observations of the other geologists.

Mineral grains (*in situ*)	Clasts (detrital minerals)	Gravel	Silt	Glass
Mineral grains (detrital)	Clasts (rock fragments)	Sand	Clay	Bioclasts

1

2

3

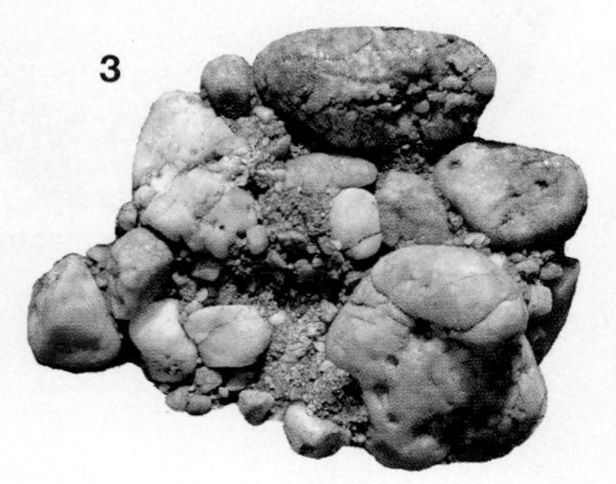

4

5

6

7

8

9

ACTIVITY 4.3 Rock-forming Minerals

Name: ______________________ Course/Section: ______________________ Date: ______________

A. REFLECT & DISCUSS Refer to **FIGURE 4.3** (page 116) and identify each rock-forming mineral below. Write its name below the picture. Be prepared to compare your observations with the observations of the other geologists. (All samples x1.)

1

2

3

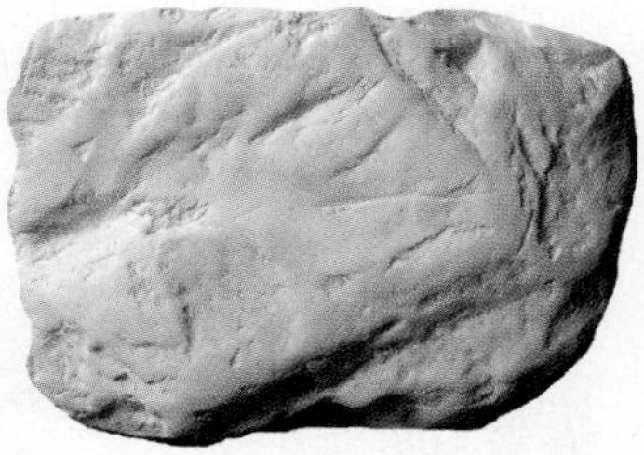

4

5

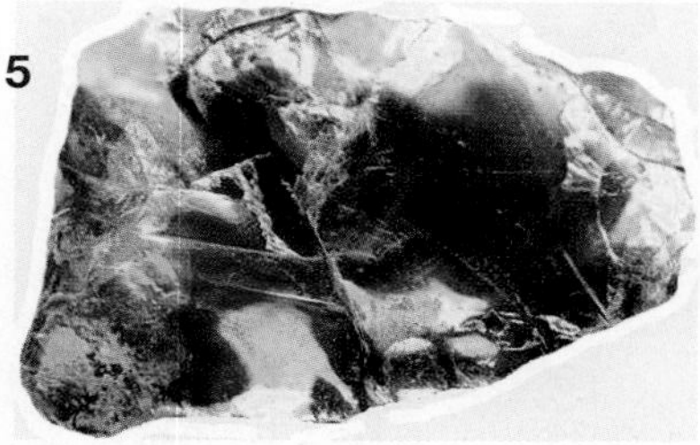

6

7

8

9

10

11

12

13

14

15

ACTIVITY 4.4 What Is Rock Texture?

Name: ______________________ **Course/Section:** ______________________ **Date:** ____________

A. Review the following list of textures on page 117. Below each sample, write the name of every one of these textures it contains. Be prepared to compare your observations with the observations of the other geologists. (All samples x1.)

Glassy	Fine-grained	Crystalline (heterogranular)	Clastic (angular)	Layered (flat)
Vesicular	Coarse-grained	Crystalline (equigranular)	Clastic (rounded)	Layered (foliated)
			Bioclastic	Layered (folded)
			Clastic (gravely, sandy, silty, clayey)	

1

2

3

4

5

6

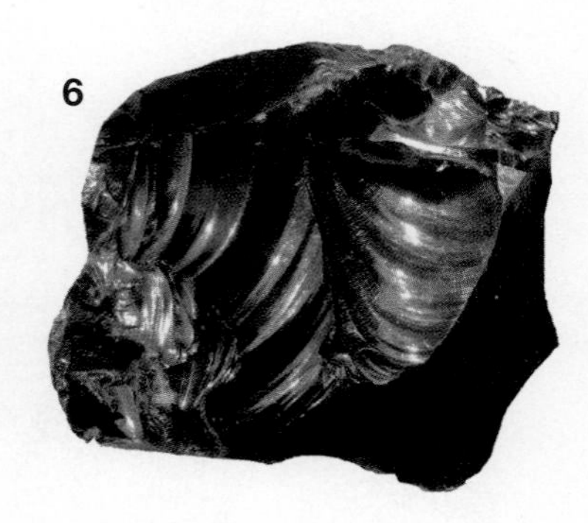

7

8

9

10

11

12

B. REFLECT & DISCUSS Based on the texture(s) and composition of each of the rocks in part A, tell whether you think it is igneous, sedimentary, or metamorphic (in its current state) and why?

1.

2.

3.

4.

5.

6.

7.

8.

9.

10.

11.

12.

ACTIVITY 4.5 Rocks and the Rock Cycle Model

Name: ______________________ **Course/Section:** ______________________ **Date:** ____________

A. On the rock cycle below, color arrows orange if they indicate a process leading to formation of igneous rocks, brown if they indicate a process leading to formation of sedimentary rocks, and green if they indicate a process leading to formation of metamorphic rocks. Place check marks in the table to indicate what rock group(s) is/are characterized by each of the processes and rock properties.

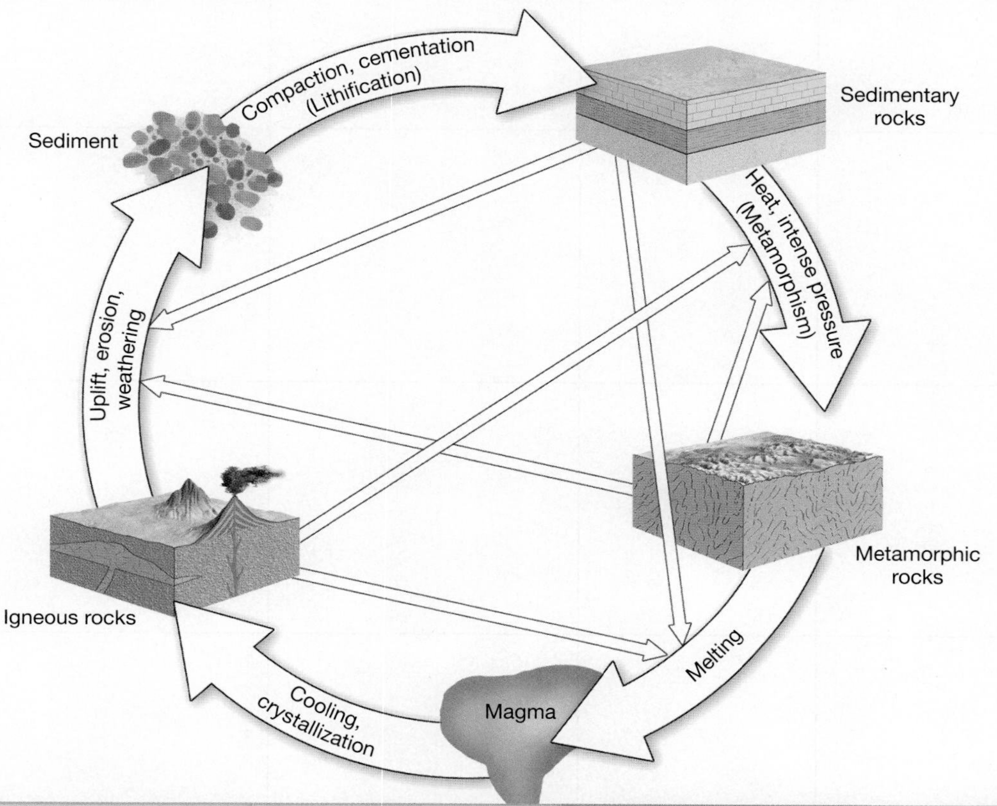

Processes and Rock Properties	Igneous	Sedimentary	Metamorphic
lithification of sediment			
intense heating (but no melting)			
crystals precipitate from water			
solidification of magma/lava			
melting of rock			
compaction of sediment			
cementation of grains			
folding of rock			
crystalline			
foliated			
common fossils			

B. **FIGURE 4.5** contains photographs of five rocks (**a–e**). For each photograph, record the following information in the chart on the next page:

1. In column one (left-hand column), note the figure number of the rock sample photograph to be analyzed.
2. In column two (blue), list the rock properties that you can observe in the sample.
3. In column three (pink), classify the rock as igneous, sedimentary, or metamorphic.
4. In column four (yellow), describe, as well as you can, how the rock may have formed.
5. On the rock cycle diagram above, write the figure number of the photograph/rock sample to show where it fits in the rock cycle model.
6. In column five (green), predict from the rock cycle (**FIGURE 4.2**) three different changes that the rock could undergo next if left in a natural setting.

Sample	ROCK PROPERTIES (grain types, textures)	ROCK CLASSIFICATION (igneous, sedimentary, metamorphic) Figure 4.4	HOW DID THE ROCK FORM?	WHAT ARE THREE CHANGES THE ROCK COULD UNDERGO? (according to the rock cycle model, Figure 4.2)
Figure 4.5a				1. 2. 3.
Figure 4.5b				1. 2. 3.
Figure 4.5c				1. 2. 3.
Figure 4.5d				1. 2. 3.
Figure 4.5e				1. 2. 3.

C. **REFLECT & DISCUSS** Starting with a sedimentary rock, describe a series of processes that could transform the rock into each of the other two kinds of rock and back into a sedimentary rock.

LABORATORY 5

Igneous Rocks and Processes

CONTRIBUTING AUTHORS
Harold E. Andrews • *Wellesley College*
James R. Besancon • *Wellesley College*
Claude E. Bolze • *Tulsa Community College*
Margaret D. Thompson • *Wellesley College*

Explosive volcanic eruptions like this one eject partially molten volcanic bombs that become rounded and cool as they fly through the air. (Superstock)

BIG IDEAS

Igneous rocks form wherever magma or lava cool to a solid state. The composition and texture of igneous rock samples, and the shapes of bodies of igneous rock, can be used to classify them and infer their origin. Lava and igneous rock-forming processes can be observed at volcanoes, which occur along lithospheric plate boundaries and hot spots, are linked to underground bodies of magma, and can pose hazards to humans.

FOCUS YOUR INQUIRY

THINK About It What do igneous rocks look like? How can they be classified into groups?

ACTIVITY 5.1 Igneous Rock Inquiry *(p. 130)*

THINK About It What are igneous rocks composed of? How is composition used to classify and interpret igneous rocks?

ACTIVITY 5.2 Minerals That Form Igneous Rocks *(p. 130)*
ACTIVITY 5.3 Estimate Rock Composition *(p. 131)*

THINK About It What are igneous rock textures? How is texture used to classify and interpret igneous rocks?

ACTIVITY 5.4 Glassy and Vesicular Textures of Igneous Rocks *(p. 133)*
ACTIVITY 5.5 Crystalline Textures of Igneous Rocks *(p. 134)*

THINK About It How are rock composition and texture used to classify, name, and interpret igneous rocks?

ACTIVITY 5.6 Rock Analysis, Classification, and Origin *(p. 135)*
ACTIVITY 5.7 Thin Section Analysis and Bowen's Reaction Series *(p. 135)*
ACTIVITY 5.8 Analysis and Interpretation of Igneous Rocks *(p. 141)*

THINK About It How can the shapes of bodies of igneous rock be used to classify them and infer their origin?

ACTIVITY 5.9 Geologic History of Southeastern Pennsylvania *(p. 142)*

Introduction

Right now, there are more than a hundred volcanoes erupting or threatening to erupt on continents and islands around the world. Some pose direct threats to humans. Others pose indirect threats, such as earthquakes and episodic melting of glaciers. In the oceans, deep under water and far from direct influence on humans, there are likely hundreds more volcanoes. The exact number is unknown, because they are erupting at places on the sea floor that humans rarely see.

Most of the world's volcanoes occur along its 260,000 kilometers of linear boundaries between lithospheric plates. The rest are largely associated with hot spots. All of the volcanoes overlie bodies of molten (hot, partly or completely melted) rock called **magma**, which is referred to as **lava** when it reaches Earth's surface at the volcanoes. In addition to their liquid rock portion, or *melt*, magma and lava contain dissolved gases (e.g., water, carbon dioxide, sulfur dioxide) and solid particles. The solid particles may be pieces of rock that have not yet melted and/or mineral crystals that may grow in size or abundance as the magma cools. **Igneous rocks** form when magma or lava cool to a solid state. The bodies of igneous rock may be as large as those in Yosemite Park, where bodies of magma cooled underground to form batholiths of igneous rock, tens of kilometers in diameter. They may be as small as centimeter-thick layers of volcanic ash, which is composed of microscopic fragments of igneous rock (mostly volcanic glass pulverized by an explosive volcanic eruption).

ACTIVITY 5.1 Igneous Rock Inquiry

THINK About It **What are igneous rocks composed of, and how can they be classified into groups?**

OBJECTIVE Analyze and describe samples of igneous rock, then infer how they can be classified into groups.

PROCEDURES

1. **Before you begin**, do not look up definitions and information. Use your current knowledge, and complete the worksheet with your current level of ability. Also, this is **what you will need** to do the activity:
   ____ Activity 5.1 Worksheet (p. 143) and pencil
   ____ optional: a set of igneous rock samples (obtained as directed by your instructor)
2. **Analyze the rocks, and complete the worksheet in a way that makes sense to you.**
3. **After you complete the worksheet**, be prepared to discuss your observations, interpretations, and inferences with others.

ACTIVITY 5.2 Minerals That Form Igneous Rocks

THINK About It **What are igneous rocks composed of? How is composition used to classify and interpret igneous rocks?**

OBJECTIVE Identify samples of eight minerals that form most igneous rocks and categorize them as mafic or felsic.

PROCEDURES

1. **Before you begin**, read Mafic and Felsic Rock-Forming Minerals below. Also, this is **what you will need:**
   ____ Activity 5.2 Worksheet (p. 144) and pencil
   ____ optional: a set of mineral samples (obtained as directed by your instructor)
   ____ optional: a set of mineral analysis tools (obtained as directed by your instructor)
2. **Then follow your instructor's directions** for completing the worksheet.

Mafic and Felsic Rock-Forming Minerals

There are eight silicate minerals that form most igneous rocks. This is because silicon and oxygen are the most common elements in magma and lava. The silicon and oxygen naturally forms silicon-oxygen tetrahedra, in which one silicon atom shares electrons with four oxygen atoms (FIGURE 5.1). This creates a silicon-oxygen tetrahedron (four-pointed pyramid) with four electrons too many, so each oxygen atom also shares an electron with another adjacent silicon atom. The simplest ratio of silicon to oxygen is 1:2, written SiO_2 and called **silica**.The mineral quartz is a crystalline form of pure silica. However, with the abundance of other chemicals in magma and lava, silicon-oxygen tetrahedra often bond with other kinds of metal atoms to make the other silicate minerals commonly found in igneous rocks. Although each one has its own unique properties that can be used to identify it, the minerals are also categorized into two chemical groups.

Felsic Minerals

The name *felsic* refers to feldspars (*fel-*) and other silica-rich (*-sic*) minerals. The common felsic minerals in igneous rocks are gray translucent *quartz*, light gray opaque *plagioclase feldspar*, pale-orange to pink opaque *potassium feldspar*, and glossy pale-brown to silvery-white *muscovite.* They are all light colored because their chemical composition lacks iron and magnesium.

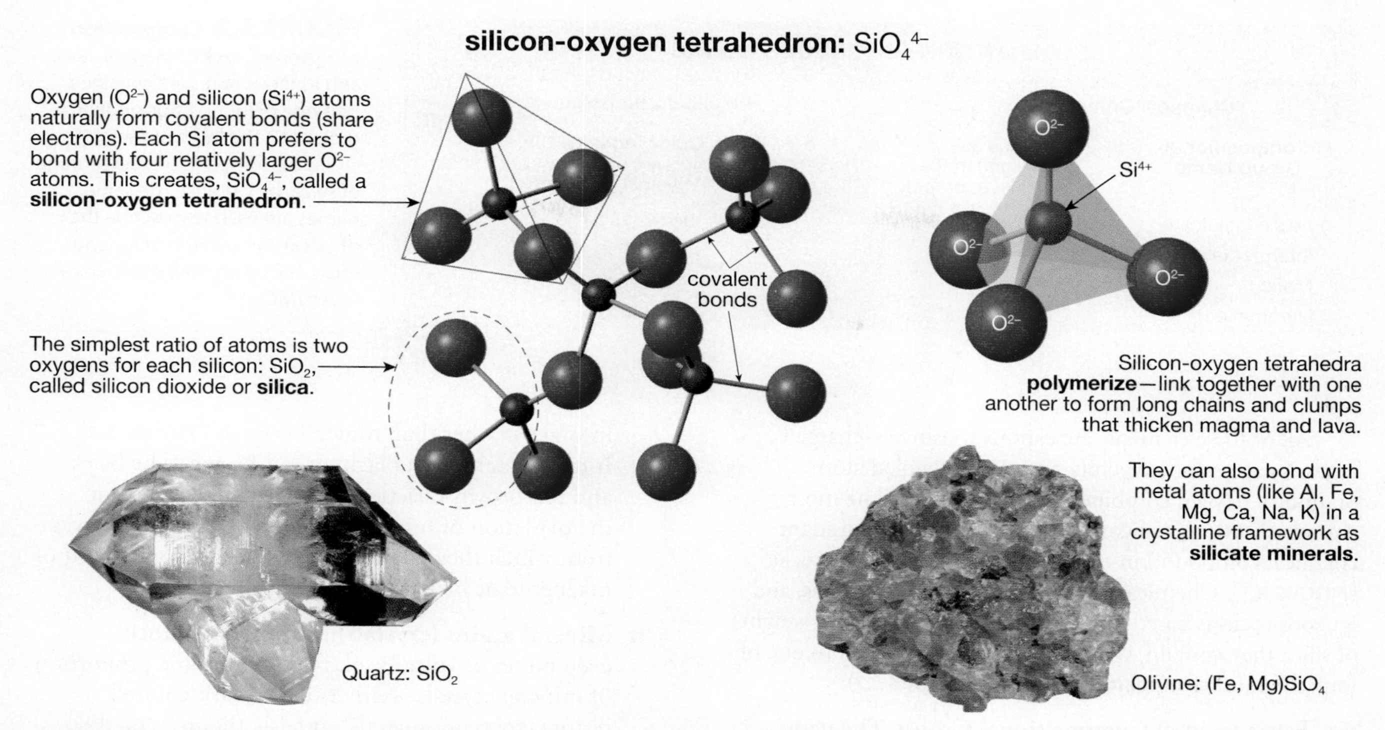

FIGURE 5.1 Silica and silicate minerals. Silicon (Si) and oxygen (O) are, by far, the most abundant chemical elements in magma, lava, and igneous rocks. They form silica (SiO_2), which thickens magma and lava (makes it more viscous) and bonds togther, alone (as quartz), or with metal atoms to make silicate minerals besides quartz in igneous rocks.

Mafic Minerals

The name *mafic* refers to minerals with magnesium (*ma-*) and iron (*-fic*) in their chemical formulas, so they are also called *ferromagnesian* minerals. They get their dark color from the abundant proportion of iron and magnesium in their chemical composition. The common mafic minerals in igneous rocks are glossy black *biotite*, dark gray to black *amphibole*, dark green to green-gray *pyroxene*, and olive-green *olivine*.

ACTIVITY

5.3 Estimate Rock Composition

THINK About It **What are igneous rocks composed of? How is composition used to classify and interpret igneous rocks?**

OBJECTIVE Determine the compositional group of an igneous rock using methods of visual estimation and point counting.

PROCEDURES

1. **Before you begin**, read about Composition of Igneous Rocks and How to Assign Rock Samples to Chemical Groups (p. 145). Also, this is **what you will need:**

   ____ Activity 5.3 Worksheet (p. 145) and pencil
2. **Then follow your instructor's directions** for completing the worksheet.

Composition of Igneous Rocks

Composition of a rock refers to what it is made of. *Chemical compositon* refers to the chemical elements that make up the rock. This determines how the rock will react with materials of different compositon, such as whether or not it will react with and decay (tarnish, dissolve, chemically disintegrate) in air or water. It also determines rock color. For example, ferromagnesian-rich rocks (iron- and magnesium-rich rocks) generally have a dark color and ferromagnesian-poor rocks generally have a light color. But the chemical elements in a rock are normally bonded together in tangible materials like minerals that, in turn, make up most rocks. So the *phsyical composition* of rocks is a description of what visible materials they are made of, in whole or part. It is your job as a geologist, using your eyes and simple tools (like a hand magnifying lens), to describe and identify what physical materials igneous rocks are made of.

Chemical Composition—Four Groups

Magmas, lavas, and igneous rocks are composed mostly of the same eight elements that characterize the average composition of Earth's crust. They are oxygen (O), silicon (Si), aluminum (Al), iron (Fe), magnesium (Mg), calcium (Ca), sodium (Na), and potassium (K).

COMPOSITION OF IGNEOUS ROCKS		
Chemical Composition		**Physical Composition**
Compositional Group Name	Silica % (by weight) in the magma, lava, or rock	**Mafic Color Index (MCI):** Percent of mafic (green, dark gray, and black) mineral crystals in the rock
Felsic (acidic)	above 65%	below 15%
Intermediate	54 – 64%	16 – 45%
Mafic	45 – 53%	46 – 85%
Ultramafic	below 45%	above 85%

FIGURE 5.2 Composition of igneous rocks. Magma, lava, and igneous rocks are classified into one of four compositional groups on the basis of their chemical composition (percentage of silica, by weight). The same names are used to describe the physical compositon of igneous rocks, based on their mafic color index (MCI).

All of these elements are cations (positively-charged atoms), except for oxygen (a negatively-charged atom, or anion); oxygen combines with the cations. The most abundant cation is silicon, so silica is the most abundant chemical compound in magmas, lavas, and igneous rocks (FIGURE 5.1). Chemical classification of magmas, lavas, and igneous rocks is based on the amount (percentage by weight) of silica they contain, which is used to assign them to one of four chemical **compositonal groups** (FIGURE 5.2):

- **Felsic (acidic) Compositional Group**. The name *felsic* refers to feldspars (*fel-*) and other silica-rich (*-sic*) minerals, but it is now also used (in place of "acidic") to decsribe magmas, lavas, and igneous rocks containing more than 60% silica.
- **Mafic (basic) Compositional Group.** The name *mafic* refers to minerals with magnesium (*ma-*) and iron (*-fic*) in their chemical formulas (also called *ferromagnesian* minerals), but it is now also used (in place of "basic") to describe magmas, lavas, and igneous rocks containing 45–53% silica.
- **Ultramafic (ultrabasic) Compositional Group.** As the name implies, this term was originally used to describe igneous rocks made almost entirely of mafic minerals. However, it now also is used (in place of "ultrabasic") to describe magmas, lavas, and igneous rocks containing less than 45% silica.
- **Intermediate Compositional Group**. This name refers to magmas, lavas, and igneous rocks that contain 54–64 % silica; a composition between mafic and felsic.

Physical Composition

The visible materials that comprise igneous rocks include volcanic glass and **grains**—mineral crystals and other hard discrete particles.

- **Volcanic glass.** Glass is an amorphous (containing no definite form; not crystalline) solid that forms by cooling viscous molten materials like melted rock (magma, lava) or quartz sand (the main ingredient that is melted to make window glass. Volcanic glass (obsidian) looks and breaks just like window glass, and it is transparent to translucent when held up to a light. It is mostly associated with felsic rocks, because they have a high percentage of silica that can polymerize into glass rather than mineral crystals (FIGURE 5.1). It may be tan, gray, black, or red-brown. The black and red-brown varieties get their dark color from the oxidation of minute amounts of iron in the lavas from which they cooled. It takes just a tiny amount of magnetite or hematite to darken the glass.
- **Mineral grains (crystals).** Most igneous rocks, even pieces of volcanic glass, contain some proportion of mineral crystals—either mafic (dark-colored ferromagnesian minerals) or felsic (light-colored silica-rich minerals). If you have not read Mafic and Felsic Rock-Forming Minerals on page 130, then you should do so now.
- **Pyroclasts (tephra).** *Pyroclasts* (from Greek meaning "fire broken") are rocky materials that have been fragmented and/or ejected by explosive volcanic eruptions (FIGURE 5.3). They include *volcanic ash* fragments (pyroclasts < 2 mm), *lapilli* or *cinders* (pyroclasts 2–64 mm), and *volcanic bombs* or *blocks* (pyroclasts > 64 mm). A mass of pyroclastic debris is called *tephra.*
- **Xenoliths.** Magma is physically contained within the walls of bedrock (crust, mantle) through which it moves. Fragments of the wall rock occasionally break free and become incorporated into the magma. When the magma cools, the fragments of wall rock are contained within the younger igneous rock as xenoliths.

How to Assign Rock Samples to Chemical Groups

The process of chemically anaylzing rocks to determine their proportions of specific elements is generally time consuming and expensive. Therefore, geologists have devised methods of hand sample analysis that enable them to assign igneous rocks to their compositional groups.

Using a Visual Estmation of Percent Chart

You can estimate the abundance of any mineral or other type of grain in a rock by using a Visual Estimation of Percent Chart provided at the back of the manual (GeoTools Sheets 1 and 2). The percentage of the circle that is black is noted on the charts (5%, 15%, 45%, 85%) for both small and large visible grains. The charts on GeoTools Sheet 2 are transparent, so you can lay them directly onto the rock.

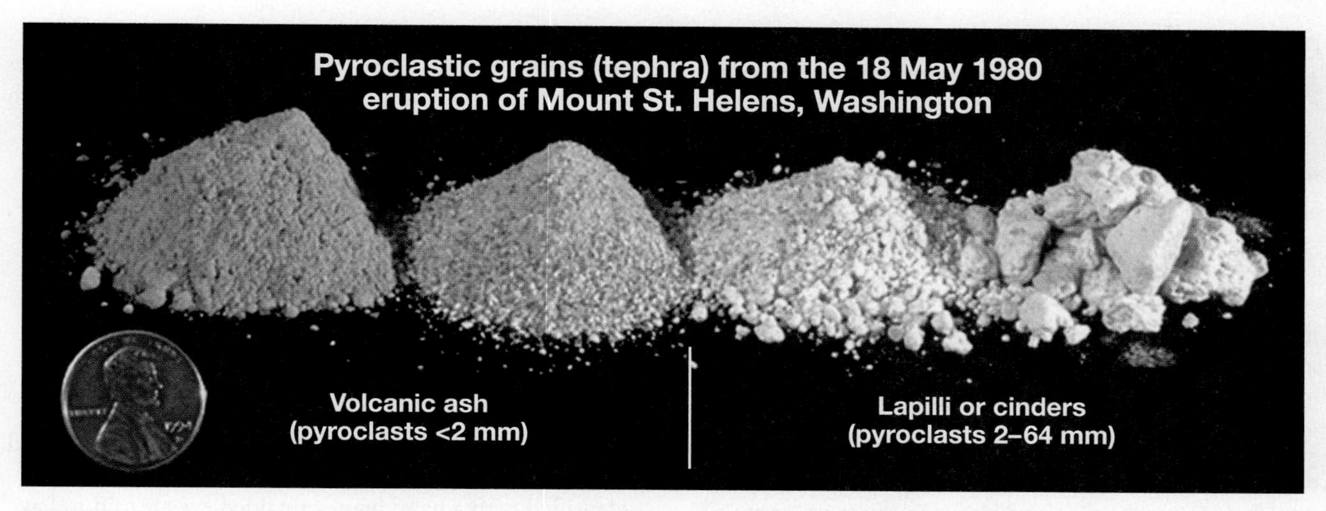

FIGURE 5.3 Pyroclastic grains (tephra). These samples of volcanic ash and lapilli (cinders) were ejected from Mount St. Helens, then collected and photographed by D. Wieprecht. (Image courtesy of U.S. Geological Survey. Scale x1.)

Using the Mafic Color Index (MCI)

The **mafic color index (MCI)** of an igneous rock is the percentage of its green, dark gray, and black mafic (ferromagnesian) mineral crystals. If the rock has no visible mineral crystals, then the overall color of the rock is used to estimate its mafic color index and corresponding compositional group. A white, pale gray, or pink rock has a felsic MCI (0–15%) and compositonal group. A moderately medium-gray rock has an intermediate MCI (16–45%) and compositional group. A very dark gray rock has a mafic MCI (46–85%) and compositional group. A black or dark green rock has a mafic MCI (above 85%) and compositional group.

If the rock has visible crystals, then you should use a Visual Estimation of Percent chart to estimate the mafic color index as closely as possible.

The mafic color index of an igneous rock is only an approximation of the rock's mineral composition, because there are some exceptions to the generalization that "light-colored equals felsic" and "dark-colored equals mafic." For example, labradorite feldspar (felsic) can be dark gray to black. Luckily, it can be identified by its characteristic play of iridescent colors that flash on and off as the mineral is rotated and reflects light. Olivine (mafic) is sometimes a pale yellow-green color (instead of medium to dark green). Volcanic glass (obsidian) is also an exception to the mafic color index rules. Its dark color suggests that it is mafic when, in fact, most obsidian has a very high weight percentage of silica and less than 15% ferromagnesian constituents. (Ferromagnesian-rich obsidian does occur, but only rarely.)

Using Point Counting

Point counting is counting the number of times that each kind of mineral crystal occurs in a specified area of the sample, or along a line randomly drawn across the sample, then calculating the relative percentage of each mineral.

ACTIVITY

5.4 Glassy and Vesicular Textures of Igneous Rocks

THINK About It **What are igneous rock textures? How is texture used to classify and interpret igneous rocks?**

OBJECTIVE Experiment with molten sugar to produce glassy and vesicular textures, then apply your knowledge to interpret rock samples.

PROCEDURES

1. **Before you begin**, read about Textures of Igneous Rocks below. Also, this is **what you will need:**
   - ____ Activity 5.4 Worksheet (p. 146) and pencil
   - ____ sugar
   - ____ materials provided in lab: hot plate, small metal sauce pan with handle or 500 mL Pyrex™ beaker and tongs, water (~50 mL), safety goggles, aluminum foil, hand lens, sugar (~50 mL, 1/8 cup), and hot plate
   - ____ collection of numbered igneous rock samples
2. **Then follow your instructor's directions** for completing the worksheet.

Textures of Igneous Rocks

Texture of an igneous rock is a description of its constituent parts and their sizes, shapes, and arrangement. You must be able to identify the common textures of igneous rocks described below and understand how they form. Notice the list of textures and their origins in **FIGURE 5.4**.

ACTIVITY

5.5 Crystalline Textures of Igneous Rocks

THINK About It **What are igneous rock textures? How is texture used to classify and interpret igneous rocks?**

OBJECTIVE Review a crystallization experiment, infer how rate of cooling affects crystal size, and then apply your knowledge to interpret a rock with porphyritic texture.

PROCEDURES

1. **Before you begin**, read about Textures of Igneous Rocks (p. 133). Also, this is **what you will need:**

   ____ Activity 5.5 Worksheet (p. 147) and pencil

2. **Then follow your instructor's directions** for completing the worksheet.

Igneous rocks are also classified into *two textural groups*: intrusive (plutonic) versus extrusive (volcanic).

Intrusive (plutonic) rocks form deep underground, where they are well insulated (take a long time to cool) and pressurized. The pressure prevents gases from expanding, just like carbonation in a sealed soft drink. The cap seals in the pressure—an intrusive process. If you remove the cap, then the carbon dioxide inside the bottle expands and bubbles—an extrusive process. Therefore, **extrusive (volcanic) rocks** form near and on Earth's surface, where the confining pressure is low and gases begin to bubble out of the magma. This can help cause explosive eruptions and textures related to fragmenting of rocks. Cooler surface temperatures also rob thermal energy from magma, so it cools quickly.

The size of mineral crystals in an igneous rock generally indicates the rate at which the lava or magma cooled to form a rock and the availability of the chemicals required to form the crystals. Large crystals require a long time to grow, so their presence generally means that a body of molten rock cooled slowly (an intrusive process) and contained ample atoms of the chemicals required to form the crystals. Tiny crystals generally indicate that the magma cooled more rapidly (an extrusive process). Volcanic glass (no crystals) can indicate that a magma was quenched (cooled immediately), but most volcanic glass is the result of poor nucleation as described below.

Nucleation and Rock Texture

The crystallization process depends on the ability of atoms in lava or magma to *nucleate*. *Nucleation* is the initial formation of a microscopic crystal, to which other atoms progressively bond. This is how a crystal grows. Atoms are mobile in a fluid magma, so they are free to nucleate. If such a fluid magma cools slowly, then crystals have time to grow—sometimes to many centimeters in length. However, if a magma is very viscous (thick and resistant to flow), then atoms cannot easily move to nucleation sites. Crystals may not form even by slow cooling. Rapid cooling of very viscous magma (with poor nucleation) can produce igneous rocks with a **glassy texture** (see FIGURE 5.4), which indicates an extrusive (volcanic) origin.

Textures Based on Crystal Size

Several common terms are used to describe igneous rock texture on the basis of crystal size (FIGURE 5.4). Igneous rocks made of crystals that are too small to identify with the naked eye or a hand lens (generally <1 mm) have a very fine-grained **aphanitic texture** (from the Greek word for *invisible*). Those made of visible crystals that can be identified with a hand lens or unaided eye are said to have a **phaneritic texture** (coarse-grained; crystals 1–10 mm) or **pegmatitic texture** (very coarse-grained; >1 cm).

Some igneous rocks have two distinct sizes of crystals. This is called **porphyritic texture** (see FIGURE 5.4). The large crystals are called *phenocrysts,* and the smaller, more numerous crystals that surround them form the *groundmass,* or *matrix* (FIGURE 5.4). Porphyritic textures may generally indicate that a body of magma cooled slowly at first (to form the large crystals) and more rapidly later (to form the small crystals). However, recall from above that crystal size can also be influenced by changes in magma composition or viscosity.

Combinations of igneous-rock textures also occur. For example, a *porphyritic-aphanitic* texture signifies that phenocrysts occur within an aphanitic matrix. A *porphyritic-phaneritic* texture signifies that phenocrysts occur within a phaneritic matrix.

Vesicular and Pyroclastic Textures

When gas bubbles get trapped in cooling lava they are called *vesicles,* and the rock is said to have a **vesicular texture**. Scoria is a textural name for a rock having so many vesicles that it resembles a sponge. Pumice has a glassy texture and so many tiny vesicles (like frothy meringue on a pie) that it floats in water.

Recall that *pyroclasts* (from Greek meaning *fire broken*) are rocky materials that have been fragmented and/or ejected by explosive volcanic eruptions (FIGURE 5.3). They include *volcanic ash* fragments (pyroclasts < 2 mm), *lapilli* or *cinders* (pyroclasts 2–64 mm), and *volcanic bombs* or *blocks* (pyroclasts > 64 mm). Igneous rocks composed mostly of pyroclasts have a **pyroclastic texture** (see FIGURE 5.4).

How to Identify Igneous Rocks

The identification and interpretation of an igneous rock is based on its composition and texture (FIGURES 5.4 and 5.5). ***Follow these steps to classify and identify an igneous rock:***

Steps 1 and 2: Identify the rock's mafic color index (MCI). Then, if possible, identify the minerals that make up the rock and estimate the percentage of each.

ACTIVITY

5.6 Rock Analysis, Classification, and Origin

THINK About It **How are rock composition and texture used to classify, name, and interpret igneous rocks?**

OBJECTIVE Analyze the composition and textures of rocks, identify them, and infer how they formed.

PROCEDURES

1. **Before you begin**, read about How to Identify Igneous Rocks (p. 134). Also, this is **what you will need:**

   ___ Activity 5.6 Worksheet (p. 148) and pencil

2. **Then follow your instructor's directions** for completing the worksheet.

- If the rock is very fine-grained (aphanitic or porphyritic-aphanitic), then you must estimate mineralogy based on the rock's mafic color index. *Felsic* fine-grained rocks tend to be pink, white, or pale gray/brown. *Intermediate* fine-grained rocks tend to be greenish gray to medium gray. *Mafic* and *ultramafic* fine-grained rocks tend to be green, dark gray, or black.
- If the rock is coarse-grained (phaneritic or pegmatitic), then estimate the mafic color index (MCI) and percentage abundance of each of the specific felsic and mafic minerals. With this information, you can also characterize the rock as felsic, intermediate, mafic, or ultramafic.

Step 3: Identify the rock's texture(s) using **FIGURE 5.4**.

Step 4: Determine the name of the rock using the flowchart in **FIGURE 5.4** or the expanded classification chart in **FIGURE 5.5**.

- Use textural terms, such as porphyritic or vesicular, as adjectives. For example, you might identify a pink, aphanitic (fine-grained), igneous rock as a rhyolite. If it contains scattered phenocrysts, then you would call it a *porphyritic rhyolite*. Similarly, you should call a basalt with vesicles a *vesicular basalt*.
- The textural information can also be used to infer the origin of a rock. For example, vesicles (vesicular textures) imply that the rock formed by cooling of a gas-rich lava (vesicular and aphanitic). Pyroclastic texture implies violent volcanic eruption. Aphanitic texture implies more rapid cooling than phaneritic texture.

ACTIVITY

5.7 Thin Section Analysis and Bowen's Reaction Series

THINK About It **How are rock composition and texture used to classify, name, and interpret igneous rocks?**

OBJECTIVE Analyze composition and texture of two thin sections of igneous rock, and then interpret their origin relative to Bowen's Reaction Series.

PROCEDURES

1. **Before you begin**, read about Bowen's Reaction Series and Plate Tectonics and Igneous Rocks below. Also, this is **what you will need:**

   ___ Activity 5.7 Worksheet (p. 149) and pencil

2. **Then follow your instructor's directions** for completing the worksheet.

Bowen's Reaction Series

When magma intrudes Earth's crust, it cools into a mass of mineral crystals and/or glass. Yet when geologists observe and analyze the igneous rocks in a single dike, sill, or batholith, they often find that it contains more than just one kind of igneous rock. Apparently, more than one kind of igneous rock can form from a single homogeneous body of magma as it cools. American geologist, Norman L. Bowen made such observations in the early 1900s. He then devised and carried out laboratory experiments to study how magmas might evolve in ways that could explain how more than one kind of igneous rock could form from a single body of magma. His work is commonly summarized in a diagram (**FIGURE 5.6**) called **Bowen's Reaction Series**, which shows how different kinds of igneous rocks can evolve from a single body of magma as it cools.

Bowen's Experiment

Other geologic investigations had already suggested that the top of Earth's mantle is made of peridotite. So Bowen placed pieces of peridotite into *bombs*, strong pressurized ovens used to melt the rocks at high temperatures (1200–1400°C). Once the rocks had melted to form peridotite magma, Bowen would allow the magma to cool to a given temperature and remain at that temperature for a while in hopes of having it begin to crystallize. The rock was then quickly removed from the bomb and quenched (cooled by dunking it in water) to make any remaining molten rock form glass. Bowen then identified the mineral crystals that had formed at each temperature. His experiments showed that as magma cools in an otherwise unchanging environment, two series of silicate minerals crystallize in a predictable order.

Discontinuous Crystallization of Mafic Minerals (Left Branch). The left branch of Bowen's Reaction Series (**FIGURE 5.6**) shows the predictable series of mafic minerals that crystallize from a peridotite magma that is allowed to cool slowly. This series is discontinuous because one mafic mineral replaces another as the magma cools. For example, olivine is first to crystallize at very high temperature. But if the magma

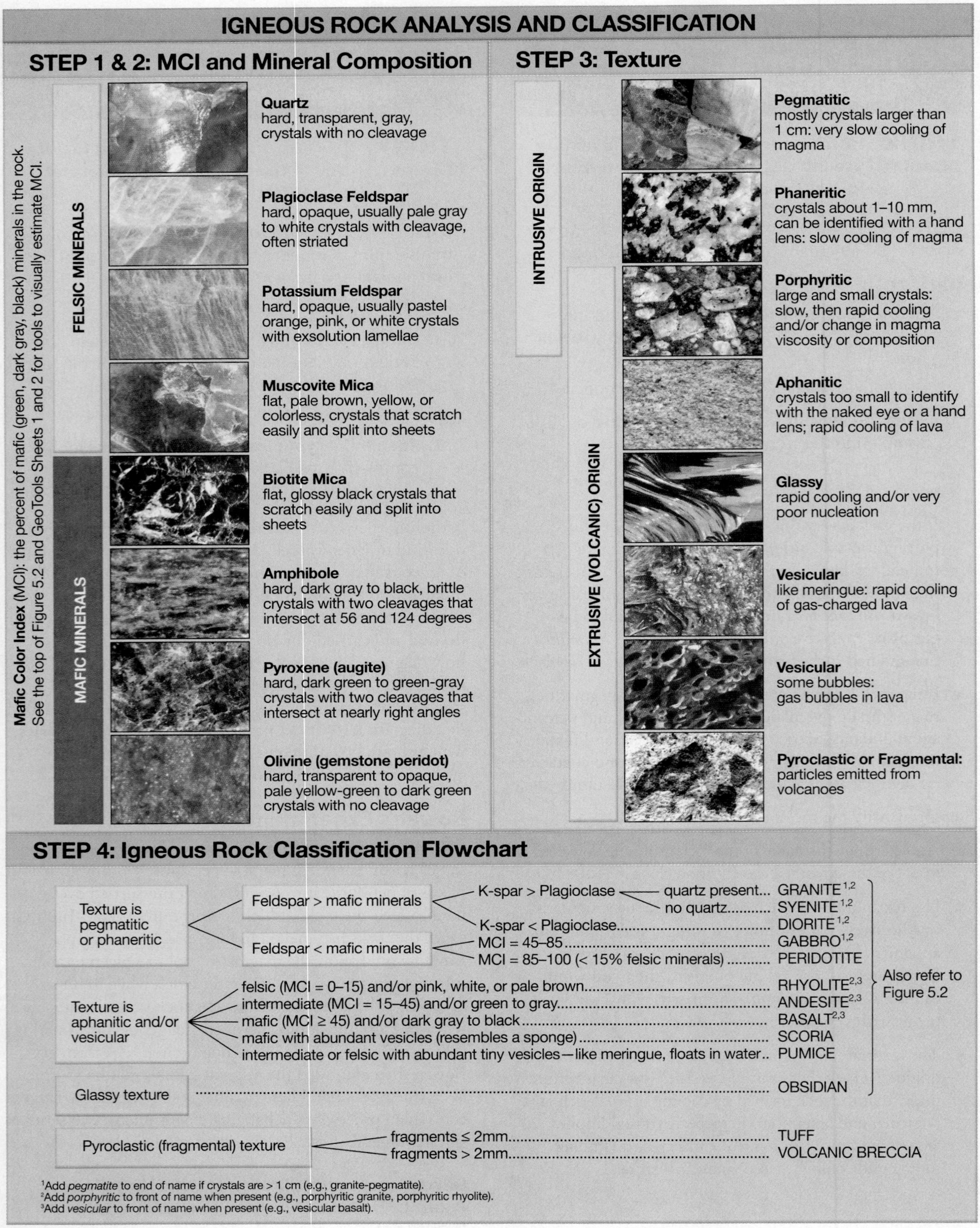

IGNEOUS ROCK ANALYSIS AND CLASSIFICATION

STEP 1 & 2: MCI and Mineral Composition

Mafic Color Index (MCI): the percent of mafic (green, dark gray, black) minerals in the rock.
See the top of Figure 5.2 and GeoTools Sheets 1 and 2 for tools to visually estimate MCI.

FELSIC MINERALS

Quartz
hard, transparent, gray, crystals with no cleavage

Plagioclase Feldspar
hard, opaque, usually pale gray to white crystals with cleavage, often striated

Potassium Feldspar
hard, opaque, usually pastel orange, pink, or white crystals with exsolution lamellae

Muscovite Mica
flat, pale brown, yellow, or colorless, crystals that scratch easily and split into sheets

MAFIC MINERALS

Biotite Mica
flat, glossy black crystals that scratch easily and split into sheets

Amphibole
hard, dark gray to black, brittle crystals with two cleavages that intersect at 56 and 124 degrees

Pyroxene (augite)
hard, dark green to green-gray crystals with two cleavages that intersect at nearly right angles

Olivine (gemstone peridot)
hard, transparent to opaque, pale yellow-green to dark green crystals with no cleavage

STEP 3: Texture

INTRUSIVE ORIGIN

Pegmatitic
mostly crystals larger than 1 cm: very slow cooling of magma

Phaneritic
crystals about 1–10 mm, can be identified with a hand lens: slow cooling of magma

Porphyritic
large and small crystals: slow, then rapid cooling and/or change in magma viscosity or composition

EXTRUSIVE (VOLCANIC) ORIGIN

Aphanitic
crystals too small to identify with the naked eye or a hand lens; rapid cooling of lava

Glassy
rapid cooling and/or very poor nucleation

Vesicular
like meringue: rapid cooling of gas-charged lava

Vesicular
some bubbles: gas bubbles in lava

Pyroclastic or Fragmental:
particles emitted from volcanoes

STEP 4: Igneous Rock Classification Flowchart

Texture is pegmatitic or phaneritic
- Feldspar > mafic minerals
  - K-spar > Plagioclase
    - quartz present... GRANITE[1,2]
    - no quartz........... SYENITE[1,2]
  - K-spar < Plagioclase....... DIORITE[1,2]
- Feldspar < mafic minerals
  - MCI = 45–85....... GABBRO[1,2]
  - MCI = 85–100 (< 15% felsic minerals)....... PERIDOTITE

Texture is aphanitic and/or vesicular
- felsic (MCI = 0–15) and/or pink, white, or pale brown....... RHYOLITE[2,3]
- intermediate (MCI = 15–45) and/or green to gray....... ANDESITE[2,3]
- mafic (MCI ≥ 45) and/or dark gray to black....... BASALT[2,3]
- mafic with abundant vesicles (resembles a sponge)....... SCORIA
- intermediate or felsic with abundant tiny vesicles—like meringue, floats in water.. PUMICE

Glassy texture .... OBSIDIAN

(GRANITE through OBSIDIAN: Also refer to Figure 5.2)

Pyroclastic (fragmental) texture
- fragments ≤ 2mm....... TUFF
- fragments > 2mm....... VOLCANIC BRECCIA

[1]Add *pegmatite* to end of name if crystals are > 1 cm (e.g., granite-pegmatite).
[2]Add *porphyritic* to front of name when present (e.g., porphyritic granite, porphyritic rhyolite).
[3]Add *vesicular* to front of name when present (e.g., vesicular basalt).

FIGURE 5.4 Igneous rock analysis and classification. Step 1—Estimate the rock's mafic color index (MCI). Step 2—Identify the main rock-forming minerals if the mineral crystals are large enough to do so, and estimate the relative abundance of each mineral (using a Visual Estimation of Percent chart from GeoTools Sheet 1 or 2). Step 3—Identify the texture(s) of the rock. Step 4—Use the Igneous Rock Classification Flowchart to name the rock. Start on the left side of the flowchart, and work toward the right side to the rock name.

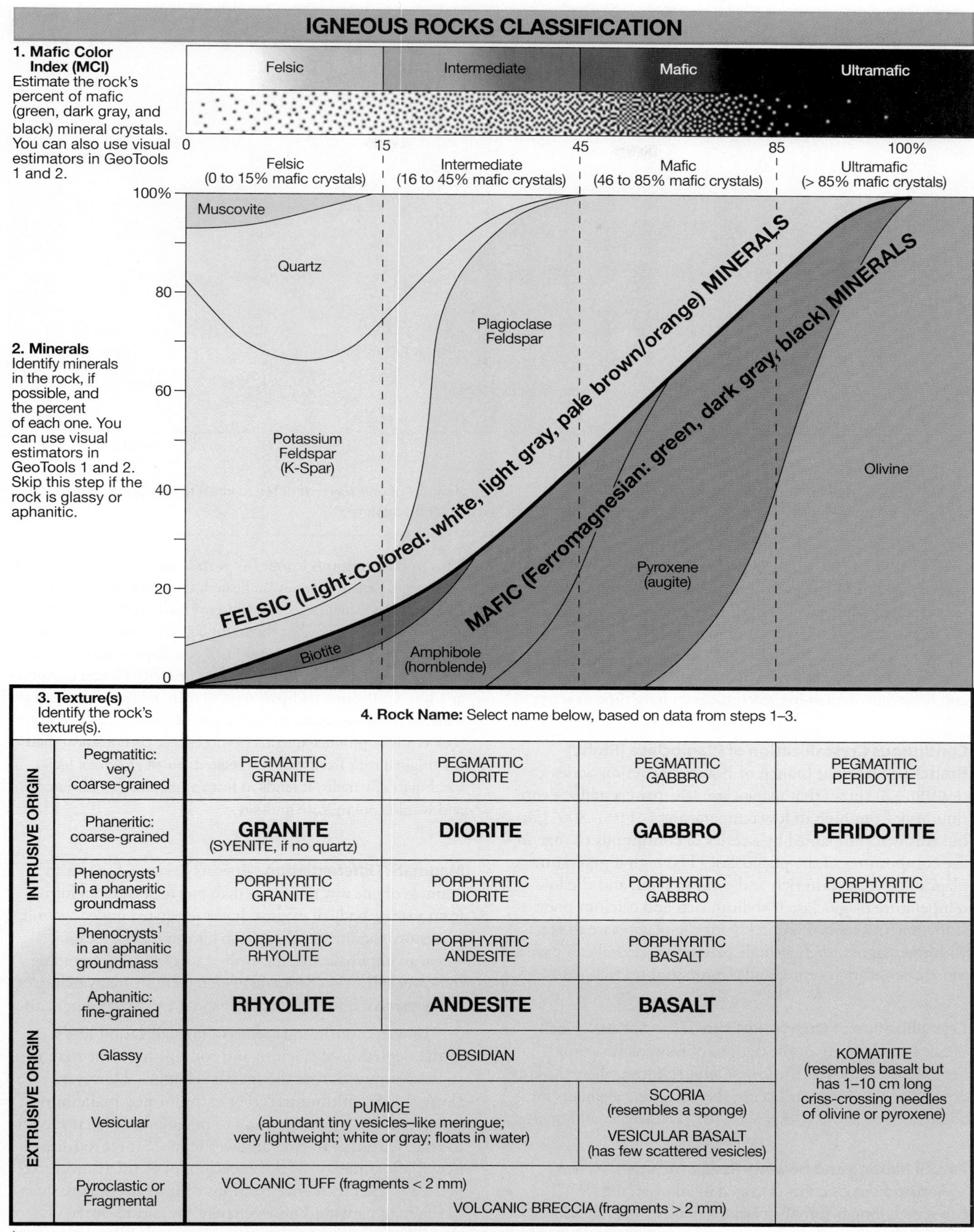

FIGURE 5.5 Igneous Rock Classification Chart. Obtain data about the rock in Steps 1–3, then use that data to select the name of the rock (Step 4). Also refer to **FIGURE 5.4** and the examples of classified igneous rocks in **FIGURES 5.8–5.14**.

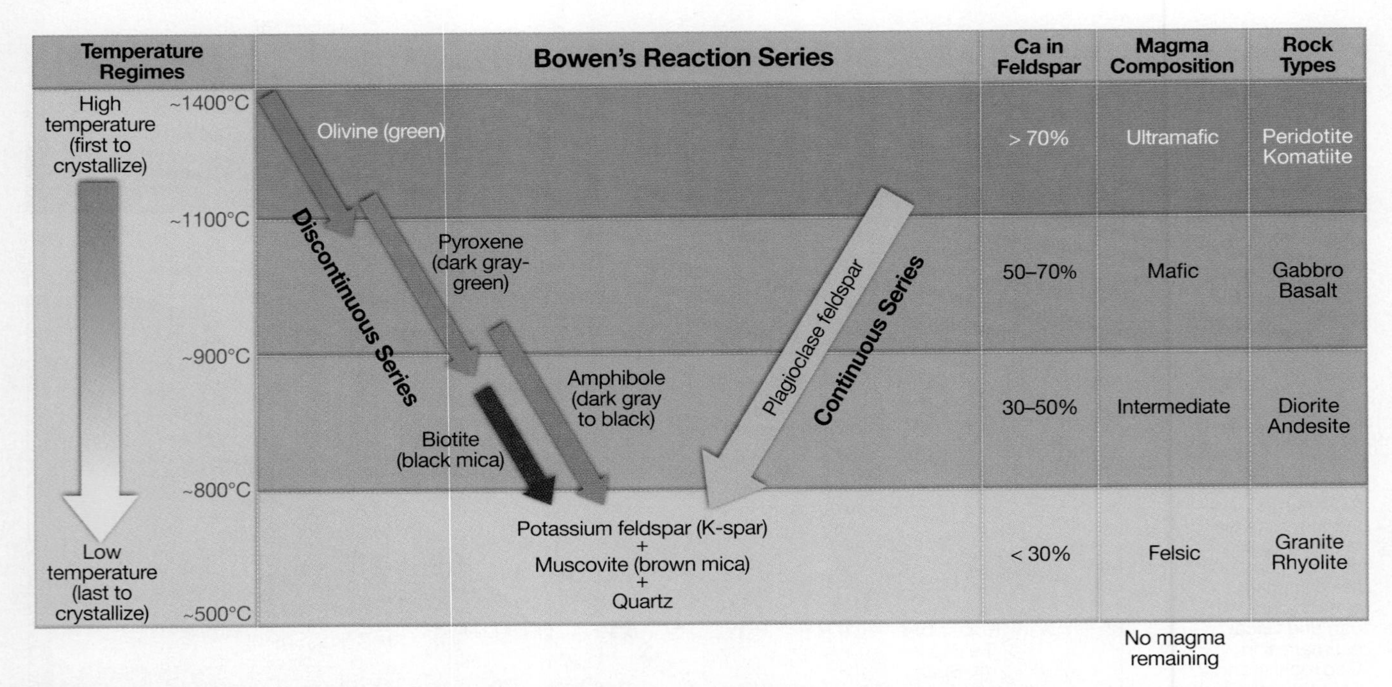

FIGURE 5.6 Bowen's Reaction Series—A laboratory-based conceptual model of one way that different kinds of igneous rocks can differentiate from a single, homogeneous body of magma as it cools. See text for discussion.

cools to about 1100° C, then the olivine starts to react with it and dissolve as pyroxene (next mineral in the series) starts to crystallize. More cooling of the magma causes pyroxene to react with the magma as amphibole (next mineral in the series) starts to crystallize, and so on. If the magma cools too quickly, then rock can form while one reaction is in progress and before any remaining reactions even have time to start.

Continuous Crystallization of Plagioclase (Right Branch). The right branch of Bowen's Reaction Series (FIGURE 5.6) shows that plagioclase feldspar crystallizes continuously from high to low temperatures (~1100–800° C), but this is accompanied by a series of continuous change in the composition of the plagioclase. The high temperature plagioclase is calcium rich and sodium poor, and the low temperature plagioclase is sodium rich and calcium poor. If the magma cools too quickly for the plagioclase to react with the magma, then a single plagioclase crystal can have a more calcium rich center and a more sodium rich rim.

Crystallization of Quartz (Bottom of the Series). Finally, notice what happens at the bottom of Bowen's Reaction Series (FIGURE 5.6). At the lowest temperatures, where the last crystallization of magma occurs, the remaining elements form abundant potassium feldspar (K-spar), muscovite, and quartz.

Partial Melting and Bowen's Reaction Series. When a plastic tray of ice cubes is heated in an oven, the ice cubes melt long before the plastic tray melts (i.e., the ice cubes melt at a much lower temperature). As rocks are heated, their different mineral crystals also melt at different temperatures. Therefore, at a given temperature, it is possible to have rocks that are partly molten and partly solid. This phenomenon is known as *partial melting*. When minerals of Bowen's Reaction Series are heated, they melt at different temperatures. The plagioclase feldspars melt continuously from about 1100–1500° C, but the ferromagnesian minerals, quartz, and K-feldspar melt discontinuously. K-feldspar melts at about 1250° C, pyroxene at 1400° C, quartz at 1650° C, and olivine at 1800° C. Because feldspars tend to melt at lower temperatures than the ferromagnesian minerals, partial melting of an igneous rock tends to produce magma of more felsic composition than the original rock from which it melted. So when a rock like basalt partially melts, it tends to form a magma that is more felsic and would cool to form andesite.

Magmatic Differentiation. Bowen's Reaction Series is an example of one way that more than one rock type can form from a single body of magma. It was generated under controlled laboratory conditions. There is no known natural location where an ultramafic magma evolved to a felsic one according to Bowen's Reaction Series. However, there are many examples where parts of Bowen's Reaction Series have occurred in nature.

Bowen's continuous series of crystallization leads to the depletion of calcium and sodium from the magma, so the composition of the magma changes. However, along the discontinuous series, early-formed mafic mineral crystals in a cooling body of magma have been shown to react with the magma at lower temperatures to form new mafic minerals. If this recycling of elements occurred perfectly, then the concentrations of iron and magnesium in the magma would never change. In nature, some of the early-formed crystals either settle out of the magma or are encrusted with different minerals before they can react, so they can no longer react with the original magma. This is called *fractional crystallization*. On the other hand,

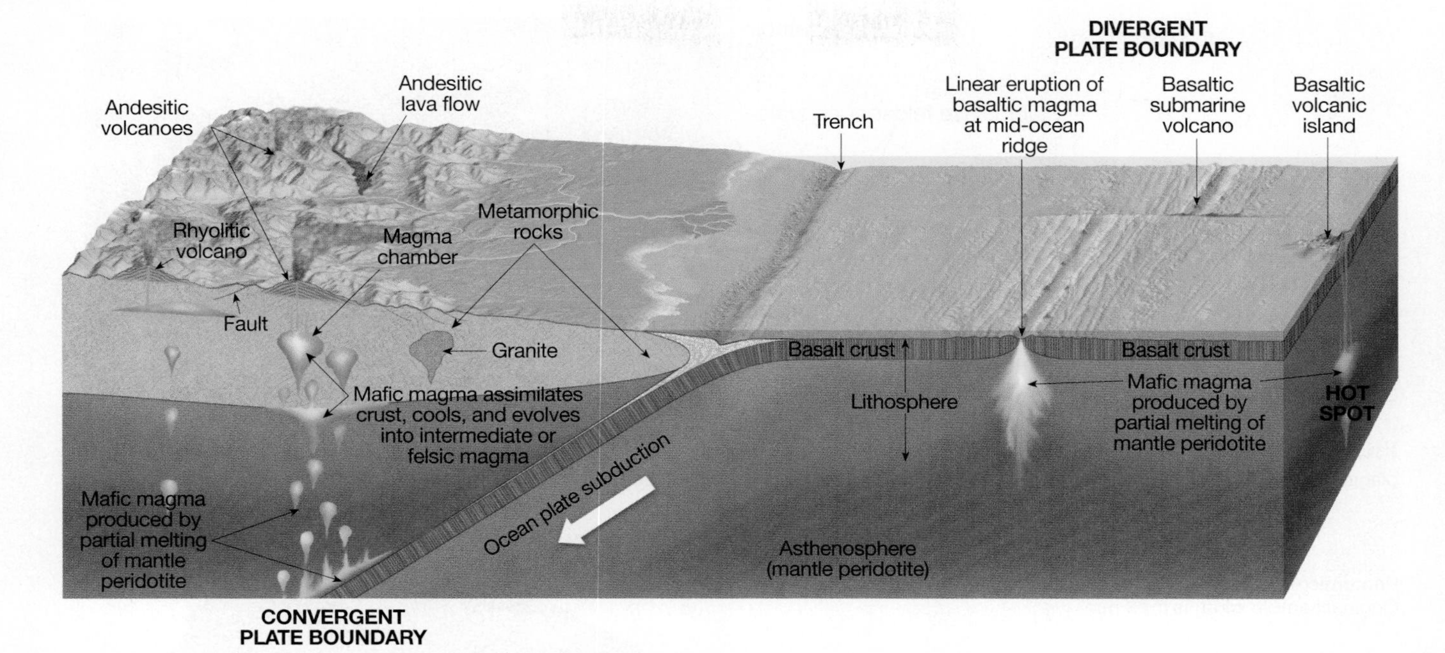

FIGURE 5.7 Tectonic settings where igneous rocks form. Different types of igneous rocks are formed in different geologic setiings: a hot spot (such as the Hawaiian Islands), divergent plate boundary (mid-ocean ridge), convergent plate boundary (subduction zone), and Earth's mantle. See text for discussion.

a magma may melt some of the wall rocks surrounding it and assimilate its elements. This is called *assimilation*. *Magma mixing* may also occur. Bowen's continuous series of crystallization, fractional crystallization, assimilation, and magma mixing are all factors that can contribute to **magmatic differentiation** (any process that causes magma composition to change). Magmatic differentiation produces more than one rock type from a single body of magma.

Plate Tectonics and Igneous Rocks

The four compositional groups of igneous rocks occur in specific tectonic settings (FIGURE 5.7).

Ultramafic Rocks Occur in the Mantle

Ultramafic igneous rocks, like peridotite, are associated with Earth's mantle. They are denser than rocks of the crust, so they are not normally found at Earth's surface. Billions of years ago, when the body of Earth was much hotter and the crust was thinner, ultramafic magmas occasionally erupted to the surface. However, no such eruptions have occurred for more than 60 million years. Xenoliths of peridotite found in some volcanic rocks are thought to have originated in the mantle (FIGURE 5.8).

Mafic Rock at Divergent Plate Boundaries and Ocean Hot Spots

Partial melting of mantle peridotite beneath ocean hot spots and mid-ocean ridges produces mafic magma rather than ultramafic magma (FIGURE 5.7). When the mafic magma cools along the mid-ocean ridges and ocean hot spots (e.g., Hawaiian Islands), it forms gabbro (FIGURE 5.9) and seafloor basalt (FIGURE 5.10).

FIGURE 5.8 Peridotite (ultramafic, intrusive, phaneritic). Peridotite—an intrusive, phaneritic igneous rock having a very high MCI (>85%) and mostly made of ferromagnesian mineral crystals. This sample is a peridotite xenolith extracted from a basaltic volcanic rock. It is made mostly of olivine mineral crystals.

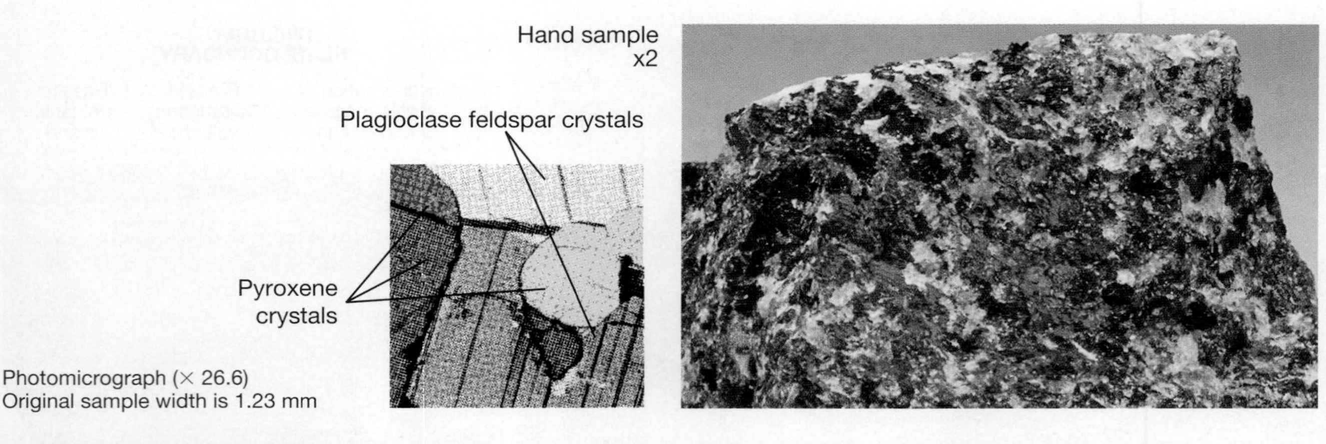

FIGURE 5.9 Gabbro (mafic, intrusive, phaneritic). Gabbro—a mafic, phaneritic igneous rock made up chiefly of ferromagnesian and plagioclase mineral crystals. The ferromagnesian mineral crystals usually are pyroxene (augite). Quartz is absent.

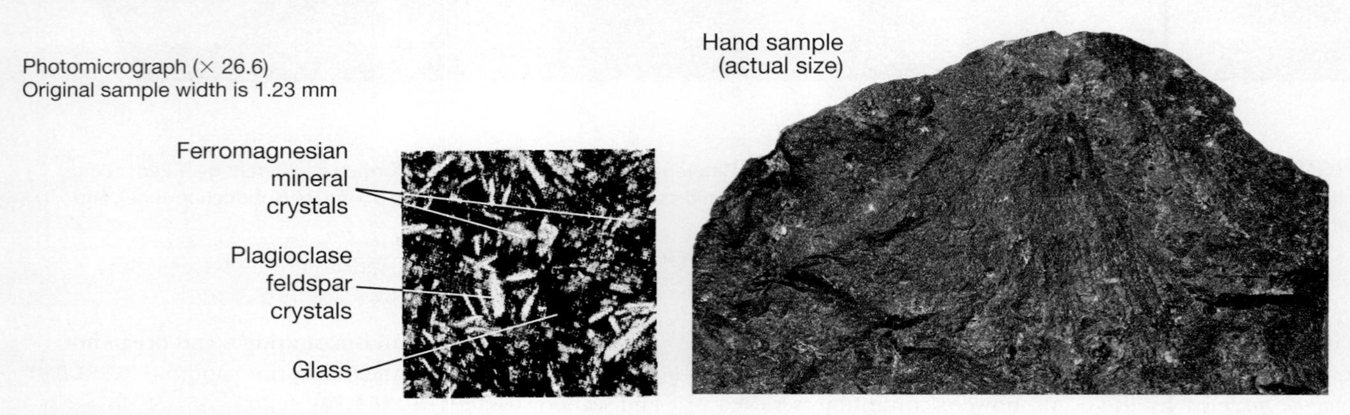

FIGURE 5.10 Basalt (mafic, extrusive, aphanitic). Basalt is a mafic, aphanitic igneous rock that is the extrusive equivalent of gabbro, so it is dark gray to black. Microscopic examination of basalts reveals that they are made up chiefly of plagioclase, ferromagnesian mineral crystals, and glass. The ferromagnesian mineral crystals generally are pyroxene, but they also may include olivine or magnetite. Basalt forms the floors of all modern oceans (beneath the mud and sand) and is the most abundant aphanitic igneous rock on Earth.

FIGURE 5.11 Diorite (intermediate, intrusive, phaneritic). Diorite—an intrusive, phaneritic igneous rock that has an intermediate MCI and is made up chiefly of plagioclase feldspar and ferromagnesian mineral crystals. The ferromagnesian mineral crystals are usually amphibole (hornblende). Quartz is only rarely present and only in small amounts (<5%).

FIGURE 5.12 Andesite (intermediate, extrusive, aphanitic). Andesite—an intermediate, aphanitic igneous rock that is the extrusive equivalent of diorite. It is usually medium gray. This sample has a porphyritic-aphanitic texture, because it contains phenocrysts of black amphibole (hornblende) set in the aphanitic groundmass.

Intermediate Rocks at Convergent Plate Boundaries

Factors that contribute to the formation of intermediate igneous rocks are plate subduction (**FIGURE 5.7**), water, and magmatic differentiation. Subduction carries water-rich seafloor rocks (basalt) back down into the mantle. Water, which can lower the melting point of rocks, causes partial melting of mantle peridotite above the subducting plate and development of mafic magma. Although some of this mafic magma may erupt to the surface as basaltic volcanoes, the mafic magma tends to differentiate by one process or many and take on an intermediate composition. When it cools, it forms diorite (**FIGURE 5.11**) and andesite (**FIGURE 5.12**). Thus, arcs of volcanoes, called *magmatic arcs*, develop above subduction zones and are often of intermediate composition.

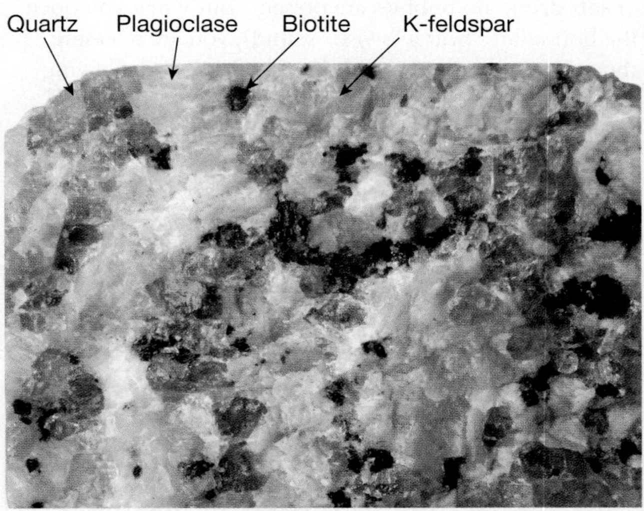

FIGURE 5.13 Granite (felsic, intrusive, phaneritic). Granite—an intrusive, phaneritic igneous rock that has a low MCI (light color) and is made up chiefly of quartz and feldspar mineral crystals. Mafic (ferromagnesian) mineral crystals in granites generally include biotite and amphibole (hornblende). Granites rich in pink potassium feldspar appear pink like this one, whereas those with white K-spar appear gray or white. Felsic rocks that resemble granite, but contain no quartz, are called *syenites*.

FIGURE 5.14 Rhyolite (felsic, extrusive, aphanitic). Rhyolite—a felsic, aphanitic igneous rock that is the extrusive equivalent of a granite. It is usually light gray or pink.

Felsic Rocks of the Continents

Felsic rocks like granite (**FIGURE 5.13**) and rhyolite (**FIGURE 5.14**) form within the crust of the continents. Felsic magmas are generated where there is extreme differentiation of mafic magmas or partial melting of intermediate rocks associated with continental magmatic arcs (**FIGURE 5.7**). Felsic magmas are also normally generated above continental hot spots, like the Yellowstone Park hot spot.

ACTIVITY

5.8 Analysis and Interpretation of Igneous Rocks

THINK About It | **How are rock composition and texture used to classify, name, and interpret igneous rocks?**

OBJECTIVE Analyze composition and texture of igneous rock samples, and then infer how they formed.

PROCEDURES

1. **Before you begin**, read about Analysis and Interpretation of Igneous Rock Samples below. Also, this is **what you will need:**
   ___ Activity 5.8 Worksheets (pp. 150–151) and pencil
   ___ set of igneous rocks (obtained as directed by your instructor)
2. **Then follow your instructor's directions** for completing the worksheets.

Analysis and Interpretation of Igneous Rock Samples

Before you begin this activity, compare the named rock types in **FIGURES 5.8–5.14** with the igneous rock classification charts in **FIGURES 5.4** and **5.5**. Also consider the origin of each rock type relative to Bowen's Reaction Series (**FIGURE 5.6**) and plate tectonic setting (**FIGURE 5.7**). Read about Bowen's Reaction Series and Plate Tectonics and Igneous Rocks (starting on page 134) if you have not already done so.

Intrusion, Eruption, and Volcanic Landforms

Magma is under great pressure (like a bottled soft drink that has been shaken) and is less dense than the rocks that confine it. Like the blobs of heated "lava" in a lava lamp, the magma tends to rise and squeeze into Earth's cooler crust along any fractures or zones of weakness that it encounters. A body of magma that pushes its way through Earth's crust is called an **intrusion,** and it will eventually cool to form a body of igneous rock.

ACTIVITY

5.9 Geologic History of Southeastern Pennsylvania

THINK About It **How can the shapes of bodies of igneous rock be used to classify them and infer their origin?**

OBJECTIVE Analyze bodies of igneous rock in southeastern Pennsylvania, using a geologic map, and infer their origin.

PROCEDURES

1. **Before you begin**, read about Intrusion, Eruption, and Volcanic Landforms. Also, this is **what you will need:**

   ___ Activity 5.9 Worksheet (p. 152) and pencil

2. **Then follow your instructor's directions** for completing the worksheet.

Intrusions have different sizes and shapes. *Batholiths* (**FIGURE 5.15**) are massive intrusions (often covering regions of 100 km^2 or more in map view) that have no visible bottom. They form when small bodies of lava amalgamate (mix together) into one large body. To observe one model of this amalgamation process, watch the blobs of "lava" in a lighted lava lamp as they rise and merge into one large body (batholith) at the top of the lamp.

Smaller intrusions (see **FIGURE 5.15**) include *sills* (sheet-like intrusions that force their way between layers of bedrock), *laccoliths* (blister-like sills), *pipes* (vertical tubes or pipe-like intrusions that feed volcanoes), and *dikes* (sheet-like intrusions that cut across layers of bedrock). The dikes can occur as *sheet dikes* (nearly planar dikes that often occur in parallel pairs or groups), *ring dikes* (curved dikes that form circular patterns when viewed from above; they typically form under volcanoes), or *radial dikes* (dikes that develop from the pipe feeding a volcano; when viewed from above, they radiate away from the pipe).

When magma is extruded onto Earth's surface it is called **lava**. The lava may erupt gradually and cause a blister-like *lava dome* to form in the neck of a volcano or a *lava flow* to run from a volcano. The lava may also erupt explosively to form *pyroclastic deposits* (accumulations of rocky materials that have been fragmented and ejected by explosive volcanic eruptions). All of these extrusive (volcanic) igneous processes present geologic hazards that place humans at risk.

When you examine an unopened pressurized bottle of soft drink, no bubbles are present. But when you open the bottle (and hear a "swish" sound), you are releasing the pressure that was containing the drink and allowing bubbles of carbon dioxide gas to escape from the liquid. Recall that magma behaves similarly. When its pressure is released near Earth's surface, it's dissolved gases expand and make bubbly lava that may erupt from a volcano. In fact, early stages of volcanic eruptions are eruptions of steam and other gases separated from magma just beneath Earth's surface. If the hot, bubbly lava cannot escape normally from the volcano, then the volcano may explode (like the top blowing off of a champagne bottle).

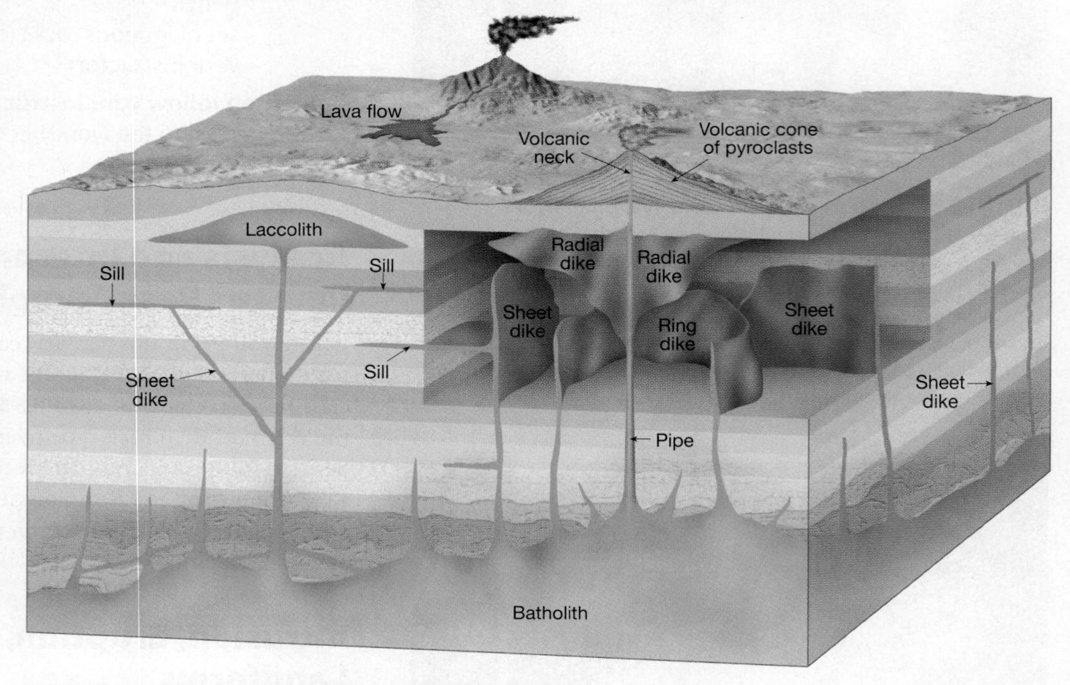

FIGURE 5.15 **Intrusive and extrusive igneous rock bodies.**

ACTIVITY 5.1 Igneous Rock Inquiry

Name: ______________________ **Course/Section:** ______________________ **Date:** ____________

A. Analyze the igneous rocks below (and actual rock samples of them if available). They are all x1. Beneath each picture, describe the rock's **color**, **composition** (what it is made of), and **texture** (the size, shape, and arrangement of its parts). Use your current knowledge, and complete the worksheet with your current level of ability. Do not look up terms or other information.

1

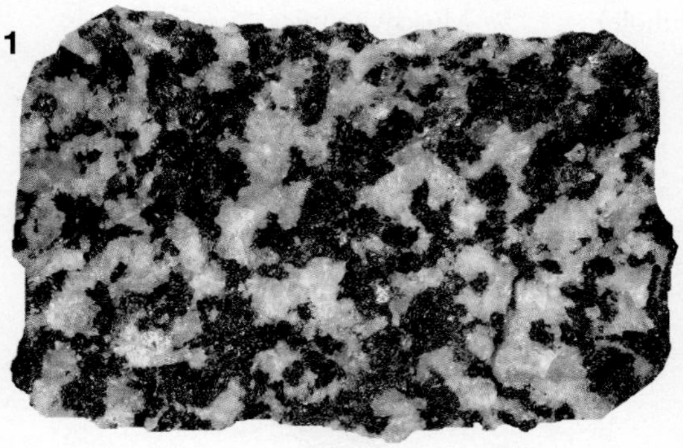

2

3

4

5

6

B. REFLECT & DISCUSS Describe how you would classify the rocks above into groups. Be prepared to discuss your classification with other geologists.

ACTIVITY 5.2 Minerals That Form Igneous Rocks

Name: ______________________ **Course/Section:** ______________________ **Date:** ____________

A. Using the Minerals Database (pages 93–97), look up each mineral name listed below and identfify its picture below. Below each picture, write the **name of the mineral**, its **chemical composition** (chemical formula or chemical name), and **words that describe it** in a way that you will be able to recognize it. (All samples below are shown at their actual size, x1.)

Augite (pyroxene)	Biotite mica	Hornblende (amphibole)	Muscovite mica
Olivine	Plagioclase feldspar	Potassium feldspar	Quartz

1

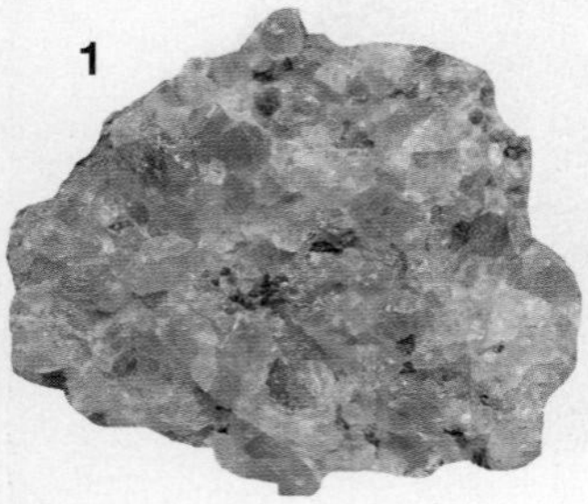

2

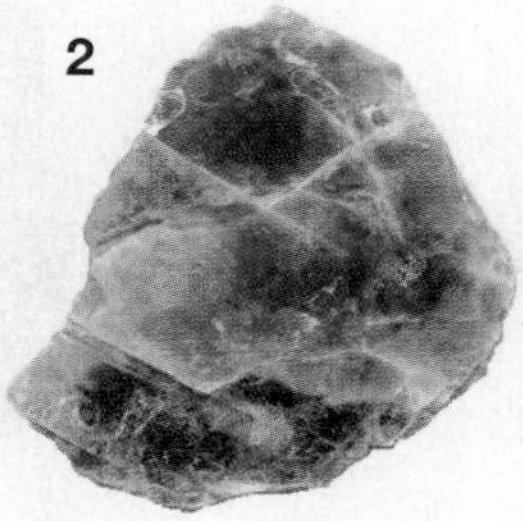

3

4

5

6

7

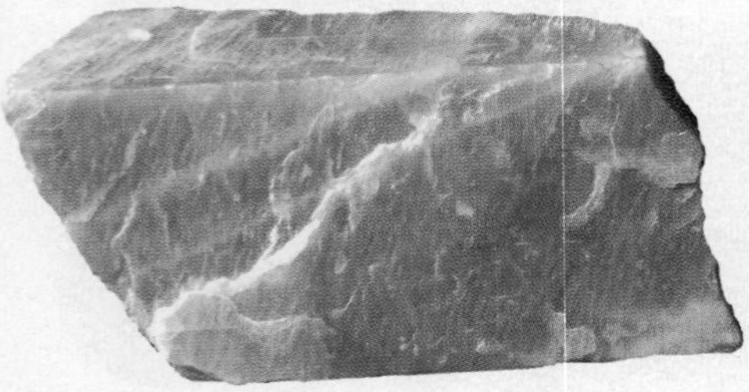

8

B. **REFLECT & DISCUSS** Which specific minerals are mafic and which ones are felsic? Why?

ACTIVITY 5.3 Estimate Rock Composition

Name: ____________________ **Course/Section:** ____________________ **Date:** ____________

A. Estimate mafic color index (MCI) for each rock below, using a Visual Estimation of Percent Chart (cut from GeoTools 1 or 2 at the back of the manual). For the rocks with visible crystals, determine the percent of each mineral by point counting *along the line provided* and by using the Visual Estimation of Percent Chart. Based on all methods, determine the rock's compositional group. All of these rocks are shown at their actual size (x1).

Mafic Color Index (MCI): ____________________

Compositional Group: ____________________

Mafic Color Index (MCI): ____________________

Compositional Group: ____________________

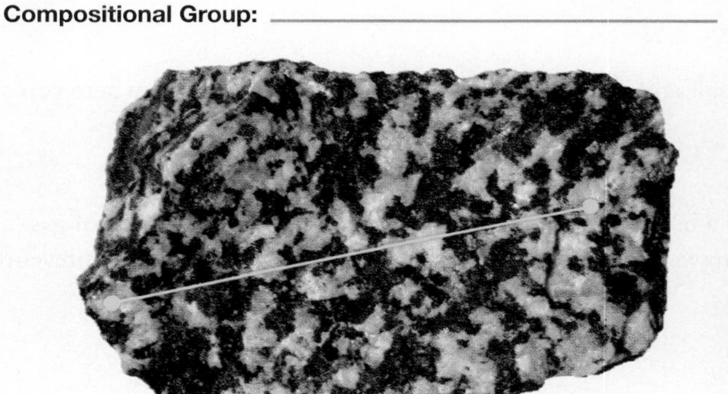

Mafic Color Index (MCI): ____________________

Mineral	Percent Based on Visual Estimation	Percent Based on Point Counting
Plagioclase feldspar		
Hornblende		

Compositional Group: ____________________

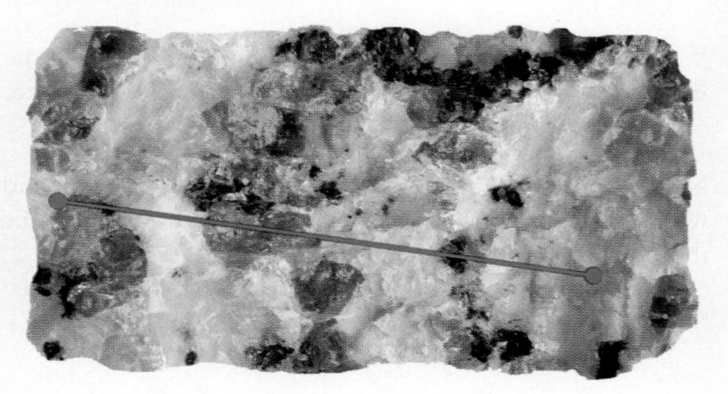

Mafic Color Index (MCI): ____________________

Mineral	Percent Based on Visual Estimation	Percent Based on Point Counting
Potassium feldspar		
Plagioclase feldspar		
Quartz		
Biotite		

Compositional Group: ____________________

B. REFLECT & DISCUSS

1. Which method do you think is the best one to use for determining the rock's compositional group? Why?

2. Is there any benefit to using two or all three methods to determine the rock's compositional group? Why?

ACTIVITY 5.4 Glassy and Vesicular Textures of Igneous Rocks

Name: ______________________ **Course/Section:** ______________________ **Date:** ____________

Place equal parts of sugar (sucrose, $C_{12}H_{22}O_{11}$) and water in the pan/beaker and heat on medium high. Do not touch the hot plate, beaker/pan, or boiling sugar, because it is very hot! Notice that steam is given off after the sugar dissolves and the solution boils. After a few minutes there will be no more steam, and the remaining molten sugar will be have a very thick (viscous) consistency. At this point (before the sugar begins to burn), pour the thick molten sugar onto a piece of aluminum foil on a flat table. DO NOT TOUCH the molten sugar, but lift a corner of the foil to observe how it flows and behaves until it hardens (2–3 minutes).

A. Viscosity is a measure of how much a fluid resists flow. Water has low viscosity. Honey is more viscous than water. How did the viscosity of the sugar solution change as the water boiled off?

B. What happened to the viscosity of the molten sugar as it cooled on the aluminum foil?

C. When the molten sugar has cooled to a solid state, break it in half and observe its texture. Look about the room where you are now seated and name two objects that have this same texture.

D. Now observe the texture of the cooled solid mass of sugar with a hand lens. Notice that there are some tiny bubbles of gas within it. Geoscientists call these "vesicles," and rocks containing vesicles are said to have a "vesicular texture." What prevented the gas bubbles from escaping to the atmosphere?

E. **REFLECT & DISCUSS** When a sugar solution is permitted to slowly evaporate, sugar crystals form. The process of crystallization depends on the ability of atoms to move about in the solution and bond together in an orderly array. What two things may have prevented crystals from forming in the molten sugar as it cooled on the aluminum foil in this experiment?

F. In your collection of numbered igneous rock samples, do any of the samples have the texture that you just observed in Part C? If yes, which one(s)?

G. In your collection of numbered igneous rock samples, do any of the samples have the texture that you just observed in Part D? If yes, which one(s)?

ACTIVITY 5.5 Crystalline Textures of Igneous Rocks

Name: ______________________ **Course/Section:** ______________________ **Date:** ____________

John and Sarah are doing an experiment to find out if crystal size in igneous rocks can be related to the speed of cooling a magma/lava. They did not have equipment to melt rock, so they used thymol to model pieces of rock. Thymol melts easily at low temperature on a hot plate, and it cools and recrystallizes quickly. Thymol is a transparent, crystalline organic substance derived from the herb, thyme, and is used in antiseptics and disinfectants. It is not toxic, but does give off a very strong pungent odor that can irritate skin and eyes and cause headaches. Therefore, John and Sarah used a spoon to handle the thymol and did all of their work under a fume hood with supervision from their teacher. Sarah placed some thymol in a small Pyrex beaker and melted it completely under a fume hood to model the formation of magma. John poured one half of the molten thymol into a cold Petri dish and the other half into a hot Petri dish of the same size.

A. The results of John and Sarah's experiment are shown below. Notice that the images are enlarged. Beside each image below, measure and record the actual size range of the crystals (in mm) that formed.

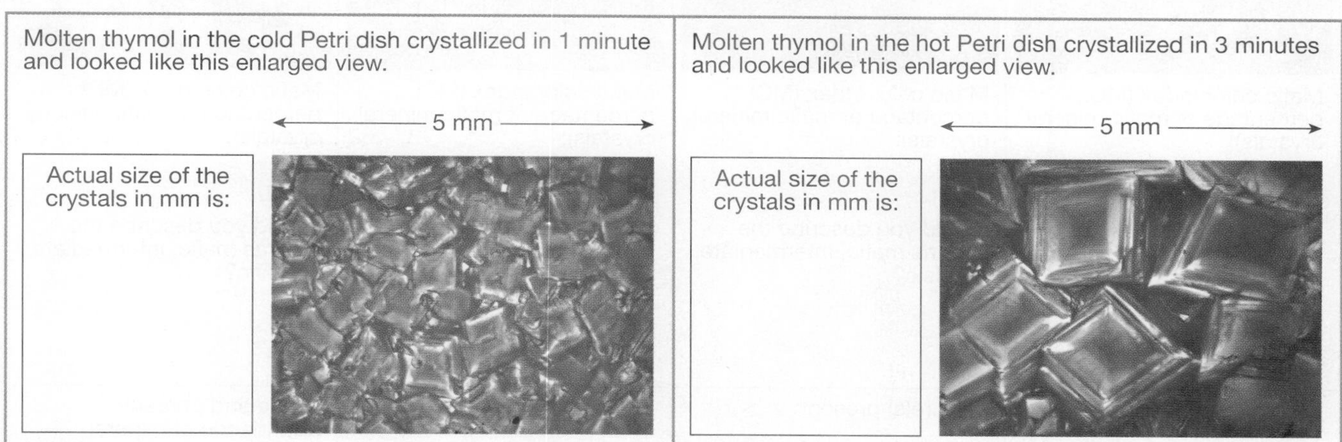

B. Igneous rocks that are made of crystals too small to see with your naked eyes or hand lens are said to have an **aphanitic texture** (from the Greek word for invisible). Those made of visible crystals are said to have a **phaneritic texture** (crystals ~1–10 mm) or **pegmatitic texture** (crystals greater than 1 cm). Which of these three igneous rock textures probably represents the most rapid cooling of magma/lava?

C. REFLECT & DISCUSS This rock has a "**porphyritic texture,**" which means that it contains two sizes of crystals. The large white plagioclase crystals are called phenocrysts and sit in a green-gray "groundmass" of more abundant, smaller (aphanitic) crystals. Based on your work above, explain how this texture may have formed (more than one answer is possible).

D. In your collection of numbered igneous rock samples, record the sample numbers with these textures:

Sample(s) with porphyritic texture:	Sample(s) with phaneritic texture:
Sample(s) with pegmatitic texture:	Sample(s) with aphanitic texture:

ACTIVITY 5.6 Rock Analysis, Classification, and Origin

Name: ______________________ **Course/Section:** ______________________ **Date:** __________

A. Analyze and classify each igneous rock below, then infer the origin of each rock based on its texture. Refer to **FIGURE 5.4** and **5.5** as needed. The rocks are shown ×1 (actual size).

Mafic color index (MCI, percentage of mafic mineral crystals):	Mafic color index (MCI, percentage of mafic mineral crystals):	Mafic color index (MCI, percentage of mafic mineral crystals):	Mafic color index (MCI, percentage of mafic mineral crystals):
Would you describe the rock as mafic, intermediate, or felsic?	Would you describe the rock as mafic, intermediate, or felsic?	Would you describe the rock as mafic, intermediate, or felsic?	Would you describe the rock as mafic, intermediate, or felsic?
Texture(s) present:	Texture(s) present:	Texture(s) present:	Name and percent abundance of mineral crystals:
			Texture present:
The name of this rock is:	The name of this rock is:	The name of this rock is:	The name of this rock is:
Based on its texture, how did this rock form?	Based on its texture, how did this rock form?	Based on its texture, how did this rock form?	Based on its texture, how did this rock form?

ACTIVITY 5.7 Thin Section Analysis and Bowen's Reaction Series

Name: ______________________ **Course/Section:** ______________________ **Date:** ____________

A thin section of a rock can be made by grinding one side of it flat, gluing the flat side to a glass slide, and then grinding the rock so thin that light passes through it. The thin section is then viewed with a polarizing microscope. The view in plane polarized light is the same as looking at the thin section through a pair of sunglasses. If you place a second pair of sunglasses behind the thin section and hold it perpendicular to the first pair, then you are viewing the thin section through cross polarized light. These images of thin sections were made by geologist LeeAnn Srogi, West Chester University of Pennsylvania.

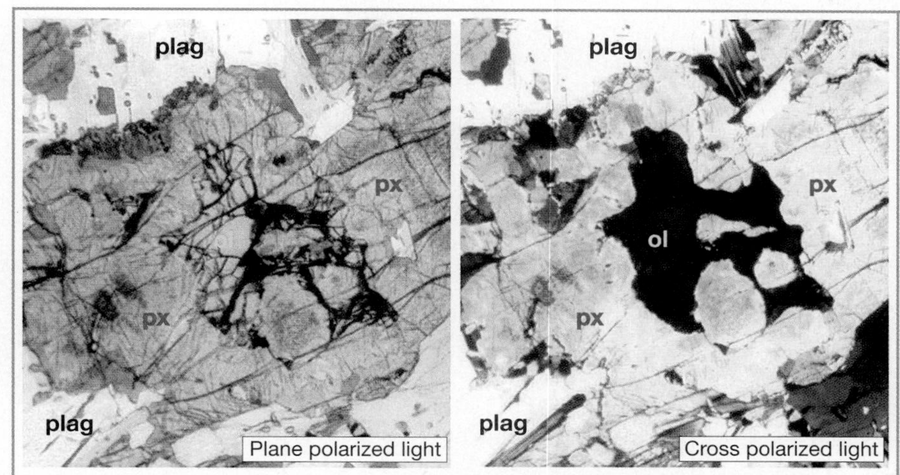

Thin section 1: This thin section has a crystal of olivine (ol) that is medium gray in plane polarized light and black in cross polarized light. The olivine crystal once had a rectangular outline, but is now surrounded and partly replaced by the mineral pyroxene (px), which has a different composition and crystal structure. Pyroxene is pale brown in plane polarized light, but yellow- to orange-brown in cross polarized light. The other white to light gray crystals in this rock are plagioclase (plag).

← 1 mm →

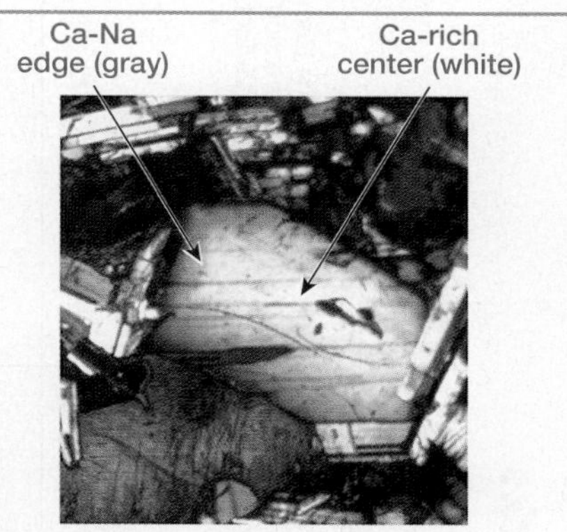

Thin section 2: This thin section is shown in cross polarized light. There are crystals of amphibole (brown to green color) and plagioclase (white to gray color). The large plagioclase crystal in the center of the image is "zoned." It has a calcium rich (sodium poor) center surrounded by zones that also have progressively more and more sodium. The zone at its outer edge is equally rich in both calcium and sodium.

1 mm ↔

A. Based on Bowen's Reaction Series (**FIGURE 5.3**), explain as exactly as you can what may have caused the relationship between olivine and pyroxene observed in thin section 1.

B. Based on Bowen's Reaction Series (**FIGURE 5.3**), explain as exactly as you can what may have caused the large plagioclase crystal in thin section 2 to be zoned as it is.

C. **REFLECT & DISCUSS** Based on your work above, circle and label the parts of this Bowen's Reaction Series diagram to indicate the exact path of crystallization and reaction represented in thin section 1, and then 2. Refer to **FIGURE 5.6** as needed.

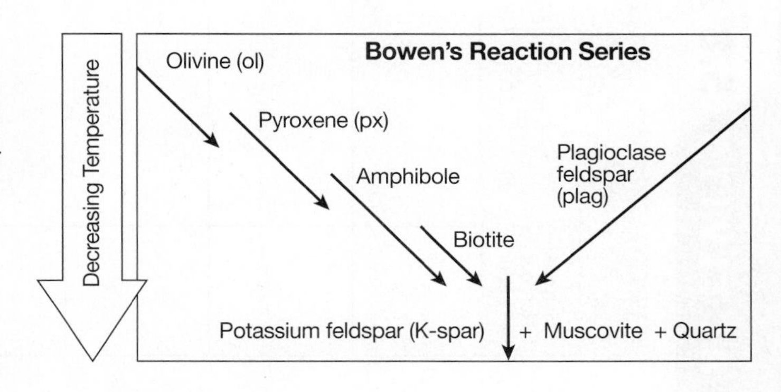

Name: ______________ Course/Section: ______________ Date: ______________

ACTIVITY 5.8 Analysis and Interpretation of Igneous Rocks

IGNEOUS ROCKS WORKSHEET

Sample Number or Letter	Texture(s) Present (Figure 5.4)	Minerals Present and Their % Abundance (Figure 5.4)	Mafic Color Index (Figure 5.5)	Rock Name from Figure 5.4 or 5.5	How Did the Rock Form Relative to Bowen's Reaction Series (Figure 5.6) and Intrusive/Extrusive Processes?

Name: ____________________ Course/Section: ____________________ Date: __________

IGNEOUS ROCKS WORKSHEET

Sample Number or Letter	Texture(s) Present (Figure 5.4)	Minerals Present and Their % Abundance (Figure 5.4)	Mafic Color Index (Figure 5.5)	Rock Name from Figure 5.4 or 5.5	How Did the Rock Form Relative to Bowen's Reaction Series (Figure 5.6) and Intrusive/Extrusive Processes?

ACTIVITY 5.9 Infer the Geologic History of Southeastern Pennsylvania

Name: ______________________ **Course/Section:** ______________________ **Date:** ____________

Review **FIGURE 5.15**. Then study the portion of a geologic map of Pennsylvania below. The green-colored areas are exposures of 200–220 million-year-old Mesozoic sand and mud that were deposited in lakes, streams, and fields of a long, narrow valley. The red-colored areas are bodies of basalt about 190 million years old. Paleozoic and Precambrian rocks are more than 252 million years old and colored pale brown.

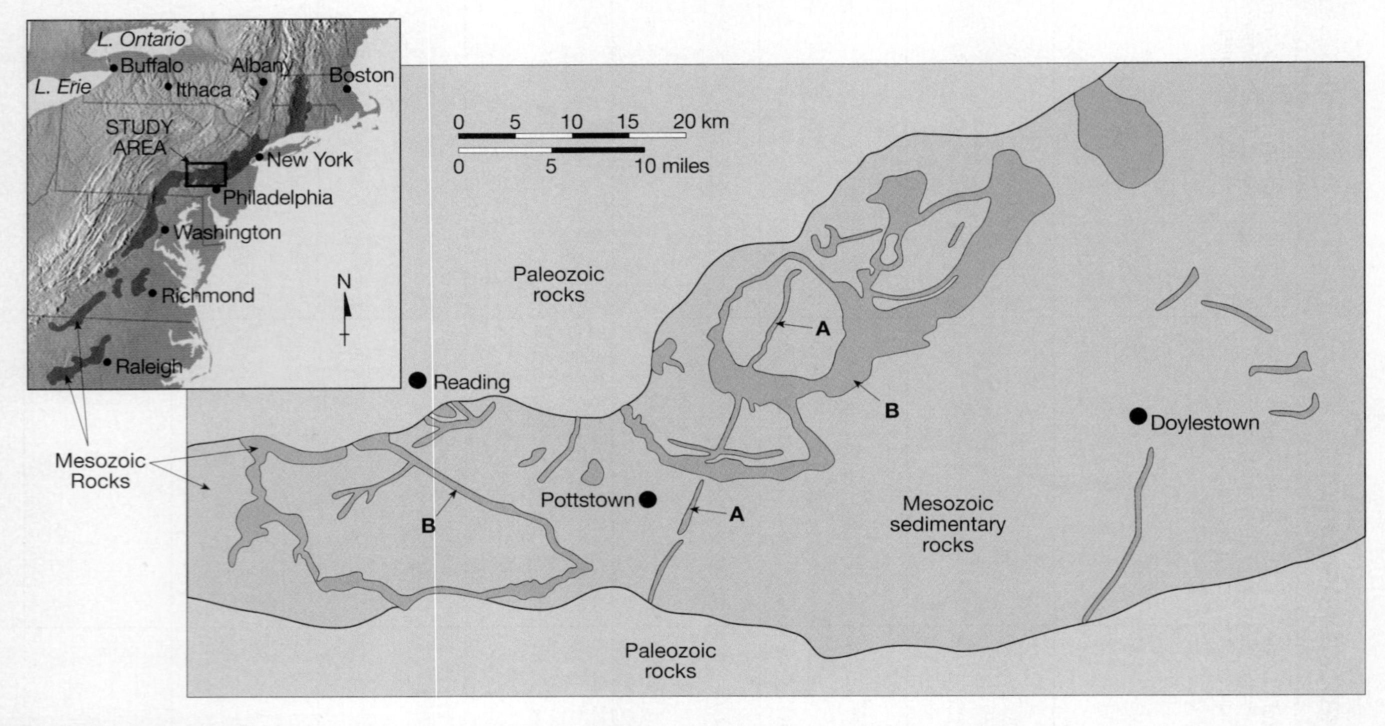

A. Based on their geometries (as viewed from above, in map view), what kind of igneous bodies on the map are labeled A?

B. Based on their geometries (as viewed from above, in map view), what kind of igneous bodies on the map are labeled B (more than one answer is possible)?

C. REFLECT & DISCUSS If you could have seen the landscape that existed in this part of Pennsylvania about 190 million years ago (when the bodies of igneous rock were lava), then what else would you have seen on the landscape besides valleys, streams, lakes, and fields? Explain your reasoning.

6

LABORATORY

Sedimentary Processes, Rocks, and Environments

CONTRIBUTING AUTHORS

Harold Andrews • *Wellesley College*

James R. Besancon • *Wellesley College*

Pamela J.W. Gore • *Georgia Perimeter College*

Margaret D. Thompson • *Wellesley College*

The grains in this detrital sedimentary rock are clasts (broken pieces) of older rocks (x1). Some of the clasts are freshly broken and angular. Others clasts had their corners worn down and are now rounded.

BIG IDEAS

Sediments are loose particles of Earth materials, including rock fragments, mineral grains weathered from rocks, animal shells, twigs, crystals precipitated from evaporating water, and chemical residues like rust. Sedimentary rocks form wherever the loose particles of sediment are compacted, cemented, or otherwise hardened to a solid mass. Layers of sediments and sedimentary rocks are like pages of a book. Their fossils and geologic structures tell us about Earth's history and past environments and ecosystems.

FOCUS YOUR INQUIRY

THINK About It | What do sedimentary rocks look like? How can they be classified into groups?

ACTIVITY 6.1 Sedimentary Rock Inquiry *(p. 154)*

THINK About It | What are sedimentary rocks made of, and how are they formed?

ACTIVITY 6.2 Mount Rainier Sediment Analysis *(p. 154)*

ACTIVITY 6.3 Clastic and Detrital Sediment *(p. 154)*

ACTIVITY 6.4 Biochemical and Chemical Sediment and Rock *(p. 155)*

ACTIVITY 6.5 Sediment Analysis, Classification, and Interpretation *(p. 155)*

THINK About It | How do geologists describe, classify, and identify sedimentary rocks?

ACTIVITY 6.6 Hand Sample Analysis and Interpretation *(p. 160)*

THINK About It | What can sedimentary rocks tell us about Earth's history and past environments and ecosystems?

ACTIVITY 6.7 Grand Canyon Outcrop Analysis and Interpretation *(p. 163)*

ACTIVITY 6.8 Using the Present to Imagine the Past—Dogs and Dinosaurs *(p. 163)*

ACTIVITY 6.9 Using the Present to Imagine the Past—Cape Cod to Kansas *(p. 166)*

ACTIVITY 6.10 "Reading" Earth History from a Sequence of Strata *(p. 167)*

ACTIVITY

6.1 Sedimentary Rock Inquiry

THINK About It | **What do sedimentary rocks look like? How can they be classified into groups?**

OBJECTIVE Analyze and describe samples of sedimentary rocks, then infer how they can be classified into groups.

PROCEDURES

1. **Before you begin**, do not look up definitions and information. Use your current knowledge, and complete the worksheet with your current level of ability. Also, this is **what you will need** to do the activity:
   ____ Activity 6.1 Worksheet (p. 171) and pencil
   ____ optional: a set of sedimentary rock samples (obtained as directed by your instructor)
2. **Analyze the rocks, and complete the worksheet in a way that makes sense to you.**
3. **After you complete the worksheet**, read the Introduction below, and be prepared to discuss your observations, interpretations, and inferences with others.

ACTIVITY

6.2 Mount Rainier Sediment Analysis

THINK About It | **What are sedimentary rocks made of, and how are they formed?**

OBJECTIVE Investigate sediment forming on and near Mount Rainier, WA.

PROCEDURES

1. **Before you begin**, read about Sedimentary Processes, Composition, and Textures of Sediments and Sedimentary Rocks below. Also, this is **what you will need:**
   ___ Activity 6.2 Worksheet (p. 172) and pencil
   ___ optional: computer with access to Google Earth™
2. **Then follow your instructor's directions** for completing the worksheets.

ACTIVITY

6.3 Clastic and Detrital Sediment

THINK About It | **What are sedimentary rocks made of, and how are they formed?**

OBJECTIVE Analyze clastic and detrital sediment and infer the environment in which sedimentary grains formed.

PROCEDURES

1. **Before you begin**, read about Sedimentary Processes and Composition and Textures of Sediments and Sedimentary Rocks below. Also, this is **what you will need:**
   ___ Activity 6.3 Worksheet (p. 174) and pencil
   ___ hand lens or stereo zoom microscope
   ___ grain size scale cut from GeoTools Sheet 1 or 2
   ___ small piece of shale
   ___ medium quartz sandpaper
   ___ 2 small pieces of granite or diorite
   ___ optional: computer with access to Google Earth™
2. **Then follow your instructor's directions** for completing the worksheets.

Introduction

Sedimentary rocks form when sediments are compressed, cemented, or otherwise hardened together. Some sedimentary rocks form by a process similar to mud hardening in the Sun to form *adobe*. Others form when masses of intergrown mineral crystals precipitate from aqueous (water-based) solutions and lock together to form crystalline rock, like rock salt that remains when ocean water is evaporated.

Sediments are loose grains and chemical residues of Earth materials, including rock fragments, mineral grains, parts of plants or animals like seashells and twigs, and chemical residues like rust (hydrated iron oxide residue). Grains of sediment are affected by chemical and physical weathering processes until they are buried in a sedimentary deposit or else disintegrate to invisible atoms and molecules dissolved in water (aqueous solutions), like groundwater (water beneath Earth's surface), lakes, streams, and the ocean. The salty taste of ocean water or salty lake water (e.g., Great Salt Lake or the Dead Sea) is a clue that many Earth materials are dissolved into it, but even fresh water has some materials dissolved in it (just not as many). Only distilled water has no materials dissolved in it.

ACTIVITY

6.4 Biochemical and Chemical Sediment and Rock

THINK About It | **What are sedimentary rocks made of, and how are they formed?**

OBJECTIVE Analyze characteristics of biochemical and chemical sediment and rock and infer how they form.

PROCEDURES

1. **Before you begin**, read about Sedimentary Processes and Composition and Textures of Sediments and Sedimentary Rocks below. Also, this is **what you will need:**
   - ___ Activity 6.4 Worksheet (p. 176) and pencil
   - ___ dilute HCl (hydrochloric acid) in dropper bottle
   - ___ seashells, charcoal briquette
   - ___ coal, dolomite
   - ___ hand lens
   - ___ plastic sandwich bags
   - ___ piece of chalk from the chalkboard
2. **Then follow your instructor's directions** for completing the worksheets.

ACTIVITY

6.5 Sediment Analysis, Classification, and Interpretation

THINK About It | **What are sedimentary rocks made of, and how are they formed?**

OBJECTIVE Describe and classify samples of sediment in terms of texture and composition, and then infer environments in which they formed.

PROCEDURES

1. **Before you begin**, read about Sedimentary Processes, Composition, and Textures of Sediments and Sedimentary Rocks below. Also, this is **what you will need:**
   - ___ Activity 6.5 Worksheet (p. 178) and pencil
   - ___ Visual Estimation of Percent chart from GeoTools 1 or 2
2. **Then follow your instructor's directions** for completing the worksheets.

Sedimentary Processes

Sedimentary processes (**FIGURE 6.1**) include everything from the time and place that sediment forms to the time and place where it is *lithified* (hardened into sedimentary rock).

Formation of Chemical Sediment

Water is a *solvent* (a liquid capable of dissolving and dispersing solid materials), so all natural bodies of water are aqueous solutions. This means that they are filled with chemicals that are "in solution," dissolved and dispersed from the materials over and through which the water has flowed. When water full of dissolved chemicals (an aqueous solution) evaporates, the chemicals in the water combine and precipitate (form solids from the solution) as mineral crystals and chemical residues called **chemical sediment**. Chemical sediment is generally *in situ,* meaning that it formed where it is found. For example, think of the intergrown halite crystals in rock salt that formed in an evaporating sea. The crystals are intergrown and locked together as sedimentary rock as they form. Oxide residues, like rust, are often deposited *in situ* (in place, where the rust formed) as coatings on surfaces of rocks, but they can also form as powdery residues in the water and be carried by the water to new locations.

Chemical sediment is the end product of *chemical weathering*—the decomposition or dissolution of Earth materials. For example, feldspars are a group of the most common minerals in Earth's crust. When potassium feldspar decomposes in acidic groundwater, it chemically decays to clay minerals (kaolinite) plus chemicals (potassium and silica) in solution. This is the main way that clay forms to make soil. Olivine decomposes to iron and magnesium in solution, and then they combine with oxygen to make oxide residues, like rust. Chemical residues commonly coat the surfaces of visible grains of sediment and either discolor them or serve as a cement to "glue" them together and form sedimentary rock.

Formation of Clastic (Detrital) and Biochemical Sediment

Physical (mechanical) weathering is the cracking, crushing, and wearing away (scratching, abrasion, transportation) of Earth materials. Cracking and crushing processes cause big rocks to be fragmented into *clasts* (broken pieces: from the Greek *klastós*, meaning broken in pieces) or *clastic sediment*, including *rock fragments* and *mineral grains* (whole crystals or fragments of crystals). Continental bedrock, rich in silicate minerals, is fragmented into **siliciclastic sediment** made of quartz grains, feldspar grains, and rock fragments. Sediment worn and transported from the land, generally siliciclastic, is also called **detrital sediment** (from the Latin *detritus*, participle of *detero*, meaning to weaken, wear away, rub off). Rock fragments and mineral crystals

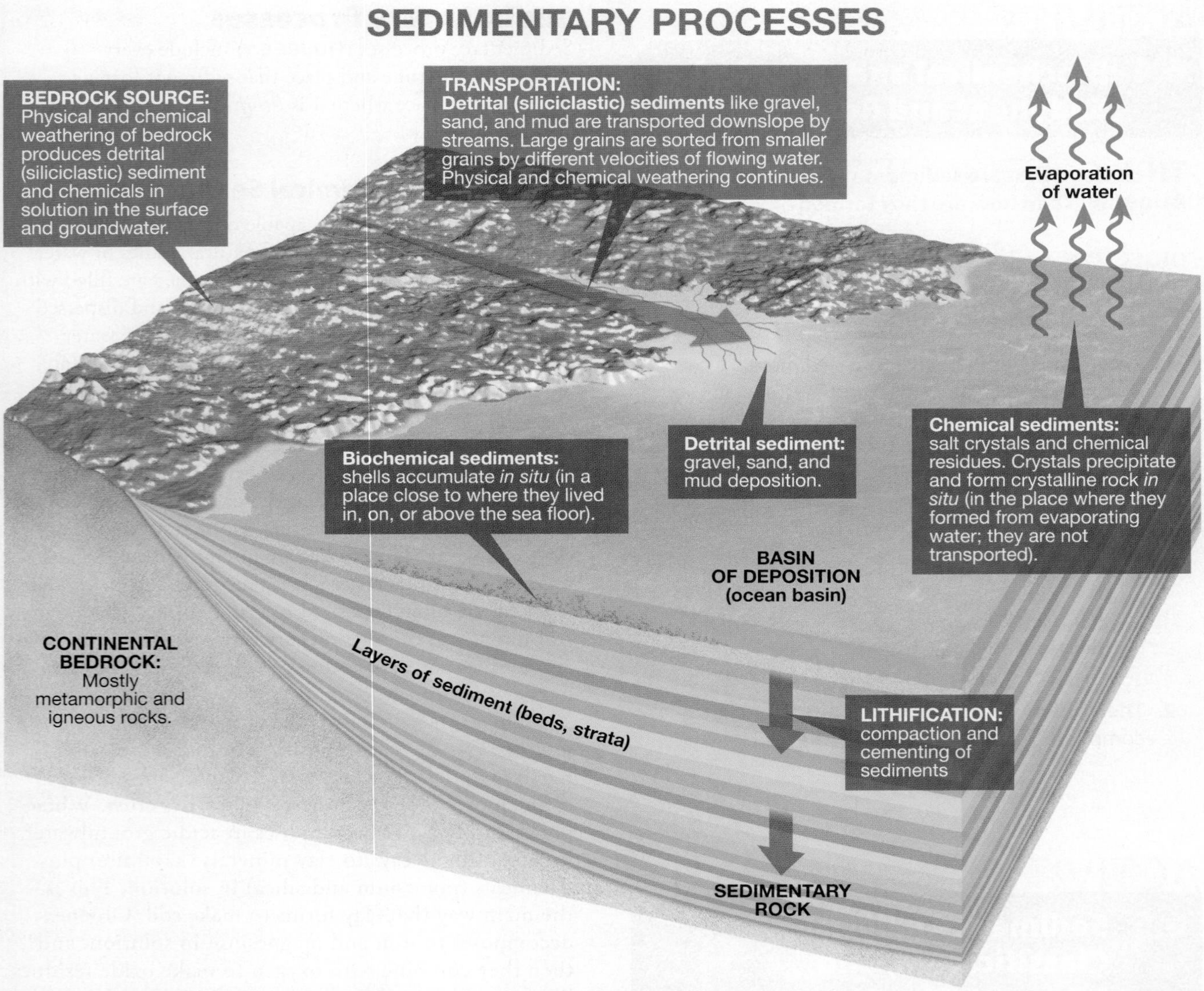

SEDIMENTARY PROCESSES

BEDROCK SOURCE: Physical and chemical weathering of bedrock produces detrital (siliciclastic) sediment and chemicals in solution in the surface and groundwater.

TRANSPORTATION: **Detrital (siliciclastic) sediments** like gravel, sand, and mud are transported downslope by streams. Large grains are sorted from smaller grains by different velocities of flowing water. Physical and chemical weathering continues.

Evaporation of water

Biochemical sediments: shells accumulate *in situ* (in a place close to where they lived in, on, or above the sea floor).

Detrital sediment: gravel, sand, and mud deposition.

Chemical sediments: salt crystals and chemical residues. Crystals precipitate and form crystalline rock *in situ* (in the place where they formed from evaporating water; they are not transported).

BASIN OF DEPOSITION (ocean basin)

CONTINENTAL BEDROCK: Mostly metamorphic and igneous rocks.

Layers of sediment (beds, strata)

LITHIFICATION: compaction and cementing of sediments

SEDIMENTARY ROCK

FIGURE 6.1 Sedimentary processes. Sedimentary processes include everything from the formation of detrital (siliciclastic), biochemical (bioclstic), and chemical sediments to the lithification (hardening) of sediments that results in sedimentary rock.

broken and transported away from bedrock surfaces (cliffs, valley walls, other outcrops) are detrital grains comprising detrital sediment. Detrital sediment is not *in situ*; it is transported away from its source. Plants and animals are fragmented into bioclastic **biochemical sediment** made of things like shells, fragmented shells, twigs, and leaves. This kind of sediment is easily broken, worn, and chemically decayed, so it is generally *in situ*. If you find a **fossil** (any evidence of ancient life), then the organism probably lived where it was fossilized.

Erosion, Transportation, and Deposition of Sediment

The place where sediment originates or forms is called its *source*. Although most biochemical and chemical sediment remains close to where it formed (is *in situ*), detrital sediment is **eroded** (loosened, removed) from its *source* and **transported** (moved, carried) over great distances. Agents of erosion and transportation include wind, water, ice, organisms, and gravity. For example, gravity forces water to flow downhill, and water is a physical agent that picks up and carries sediment. Eventually, the water flows into a *basin* (depression where water and sediment accumulate), becomes part of a lake or ocean, and sediment deposition occurs. **Deposition** is what happens when transportation stops and sediment accumulates by settling out of the water (or air or melting ice) that carried it. (In contrast, chemical and biochemical sediment is usually not transported, so it is deposited *in situ*—where it forms.)

Layering of Sediment

The result of deposition is a **deposit** of sediment. So erosion, transportation, and deposition are a sequence of related events. The events are also episodic (happen infrequently, not continuously). Erosion happens when it rains, transportation happens when it floods, and deposition happens when flood waters accumulate in a lake or ocean and stop moving (and sediment settles out or precipitates out of the water). The net result is, therefore, a layered deposit. Each time a new episode of flood water washes into the lake or ocean, a new layer of sediment is deposited on top of the last (older) one. In between the depositional events, there is *nondeposition* (a time during which no deposition occurs). The times of nondeposition become surfaces, called **bedding planes**, between the layers of sediment (called **beds**, **bedding,** or **strata**).

Lithification of Sediment

Lithification is the process of changing loose particles of sediment (unconsolidated sediment) to solid rock (consolidated sediment). This happens most often when sediment is *compacted* (squeezed together) or *cemented* (glued together by tiny crystals or chemical residues).

Composition and Textures of Sediments and Sedimentary Rocks

Sediment and sedimentary rocks are described, classified, named, and interpreted on the basis of their composition and textures.

Composition of Sediment and Sedimentary Rocks

The **composition** of a sediment or sedimentary rock is a description of the kinds and abundances of grains that compose it (**FIGURE 6.2**). Sediments and sedimentary rocks are classified as biochemical (bioclastic), chemical, or detrital (siliciclastic) based on their composition. **Biochemical** sediments and rocks consist of whole and broken **(bioclastic)** parts of organisms, such as shells and plant fragments. **Chemical** sediments and rocks consist of chemical residues and intergrown mineral crystals precipitated from aqueous solutions. The precipitated minerals commonly include gypsum, halite, hematite, limonite, calcite, dolomite, and chert (microcrystalline variety of quartz). **Detrital** sediments and rocks consist of **siliciclastic** grains (rock fragments, quartz, feldspar, clay minerals) that are also *detrital* grains—rock fragments and mineral grains that were worn and transported away from the landscape.

Textures of Sediment and Sedimentary Rocks

Processes of weathering, transportation, precipitation, and deposition that contribute to the formation of a sediment or sedimentary rock also contribute to forming its texture. The **texture** of a sediment or sedimentary rock is a description of its parts and their sizes, shapes, and arrangement (**FIGURE 6.3**).

Grain Size. The particles that make up sedimentary rocks are called **grains**. Size of the grains is commonly expressed in these *Wentworth classes,* named after C. K. Wentworth, an American geologist who devised the scale in 1922:

- **Gravel** includes grains larger than 2 mm in diameter (granules, pebbles, cobbles, and boulders).
- **Sand** includes grains from 1/16 mm to 2 mm in diameter (in decimal form, 0.0625 mm to 2.000 mm). This is the size range of grains in a sandbox. The grains are visible and feel very gritty when rubbed between your fingers.
- **Silt** includes grains from 1/256 mm to 1/16 mm in diameter (in decimal form, 0.0039 mm to 0.0625 mm). Grains of silt are usually too small to see, but you can still feel them as very tiny gritty grains when you rub them between your fingers or teeth.
- **Clay** includes grains less than 1/256 mm diameter (in decimal form, 0.0039 mm). Clay-sized grains are too small to see, and they feel smooth (like chalk dust) when rubbed between your fingers or teeth. Note that the word *clay* is used not only to denote a grain size, but also a clay mineral. However, clay mineral crystals are usually clay-sized.

Rounding of Sediment. All sediment has a *source* (place of origin; **FIGURE 6.1**). Sediments deposited quickly at or near their source tend to lack abrasion. Sediments that have been moved about locally (as in waves on a beach) or transported away from their source are abraded (worn). **Roundness** is a description of the degree to which the sharp corners and points of a fragmented grain have been worn away and its profile has become round (**FIGURE 6.3**). A newly formed clast is *very angular*. As it is transported and worn it will become *subangular,* then *subround,* and then *well rounded.* A freshly broken rock fragment, mineral grain, or seashell has sharp edges and is described as *angular.* The more rounded a grain becomes, the smaller it generally becomes. Gravel gets broken and abraded down into sand, and sand gets broken and abraded into silt and clay-sized grains. When combined, the silt plus clay mixture is called *mud.*

Sorting of Sediment. Different velocities of wind and water currents are capable of transporting and naturally separating different densities and sizes of sediments from one another. **Sorting** is a description of the degree to which one size class of sediment has been separated from the others (**FIGURE 6.3**). *Poorly sorted* sediments consist of a mixture of many different sizes of grains. *Well-sorted* sediments consist of grains that are of similar size and/or density.

COMPOSITIONAL CLASSIFICATION OF SEDIMENT AND SEDIMENTARY ROCKS

A. DETRITAL (SILICICLASTIC) SEDIMENT AND SEDIMENTARY ROCK IS MOSTLY ONE OR MORE OF THESE:

Rock fragments: may be angular or rounded; can include detrital chert grains (see "chert" below)

Quartz grains: angular grains freshly broken from their source and pebbles rounded during transportation

Feldspar grains: large angular grains freshly broken from their source and small subangular grains

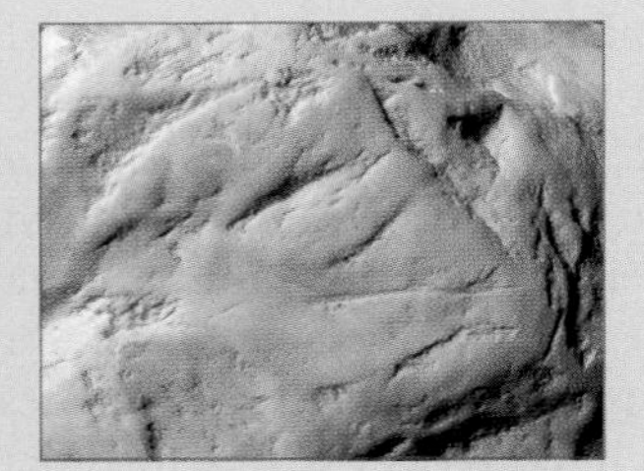

Clay: commonly forms from chemical decay of feldspars and micas

B. BIOCHEMICAL SEDIMENT AND SEDIMENTARY ROCK IS MOSTLY EITHER OR BOTH OF THESE:

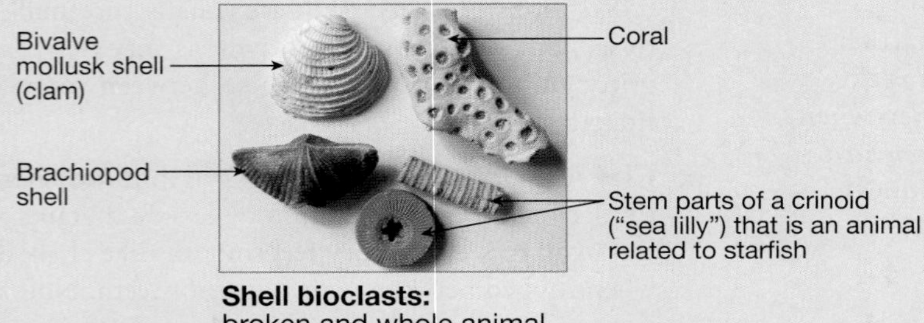

Shell bioclasts: broken and whole animal shells

Plant fragments: are brown in peat and black in coal

C. CHEMICAL SEDIMENT AND SEDIMENTARY ROCK IS MOSTLY MADE OF ONE OR MORE OF THESE:

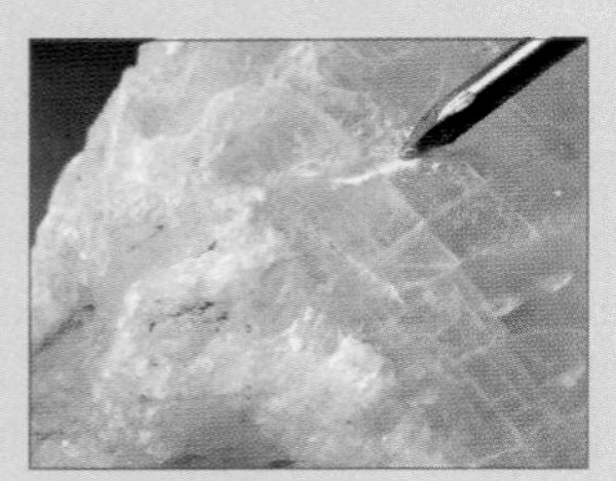

Gypsum: white or gray, easily scratched with your fingernail

Calcite spar (crystals): reacts with dilute HCl, breaks into rhombohedral shapes

Dolomite: usually cryptocrystalline; reacts with dilute HCl only if it is powdered

Halite: gray to red cubic crystals (often intergrown as rock salt); salty taste

Ooids: tiny (< 2 mm) spheres of calcite or aragonite that resemble miniature pearls; reacts to dilute HCl

Limonite: opaque brown to yellow rusty-looking crusts, layers; cements sediment, making it look yellow to brown

Hematite: opaque brick red to silver gray layers; cements sediment, making it look red

Chert: a gray, red, brown or black cryptocrtstalline variety of quartz (may contain fossils, including silica microfossils

FIGURE 6.2 Composition of sedimentary rocks. Scale for all images is × 1 unless noted otherwise.

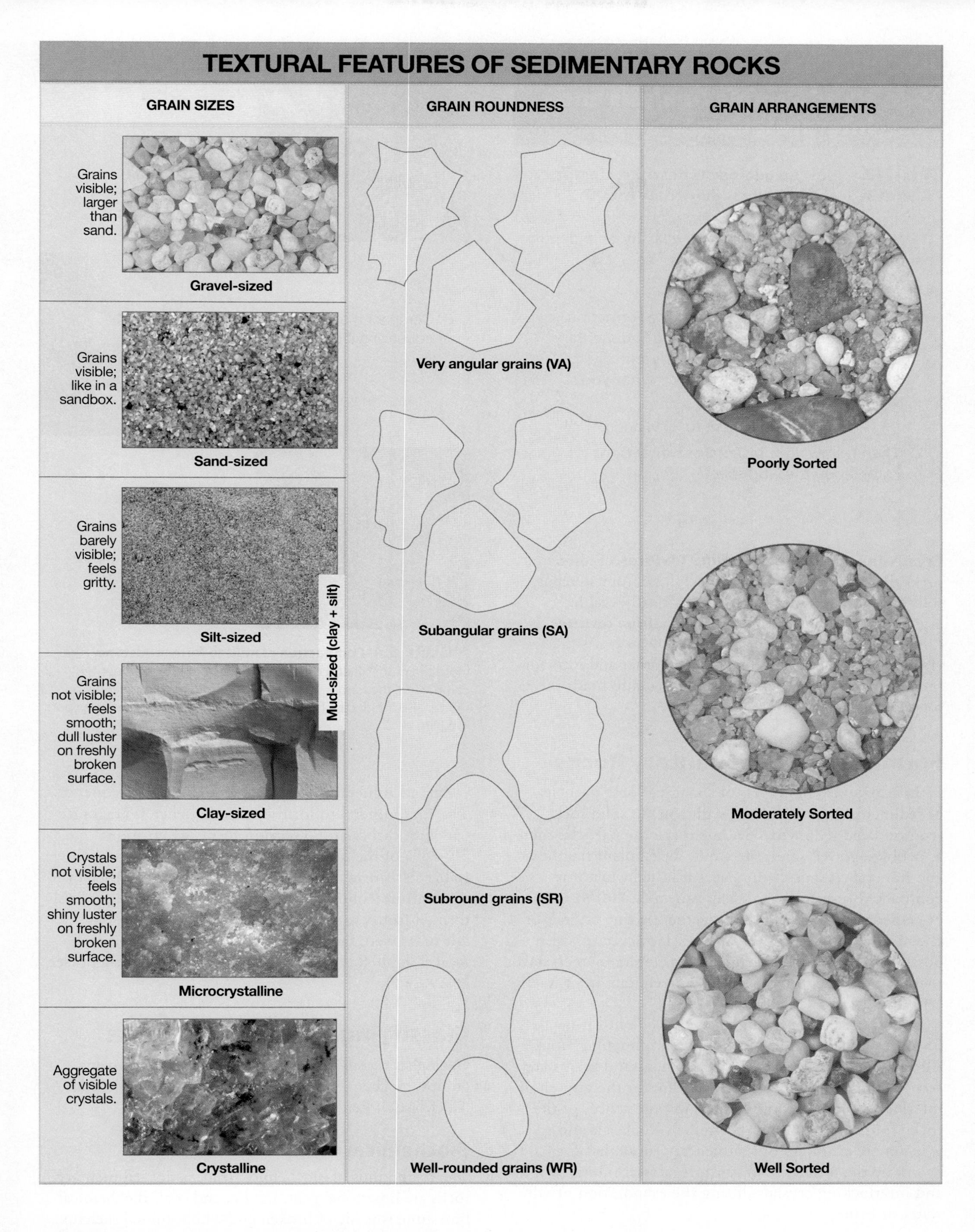

FIGURE 6.3 Textures of sedimentary rocks. Scale for all images is × 1.

ACTIVITY

6.6 Hand Sample Analysis and Interpretation

THINK About It | **How do geologists describe, classify, and identify sedimentary rocks?**

OBJECTIVE Be able to describe, classify, and identify hand samples of sedimentary rocks.

PROCEDURES

1. **Before you begin**, read about the Formation of Sedimentary Rocks, Classifying Sedimentary Rocks, and Hand Sample Analysis and Interpretation below. Also, this is **what you will need:**

   ___ Activity 6.6 Worksheet (p. 179) and pencil

2. **Then follow your instructor's directions** for completing the worksheets.

Crystalline and Microcrystalline Textures. Sedimentary rocks that form when crystals precipitate from aqueous solutions have a **crystalline texture** (clearly visible crystals; see **FIGURE 6.2**) or **microcrystalline texture** (crystals too small to identify; see **FIGURE 6.2**). As the crystals grow, they interfere with each other and form an intergrown and interlocking texture that also holds the rock together.

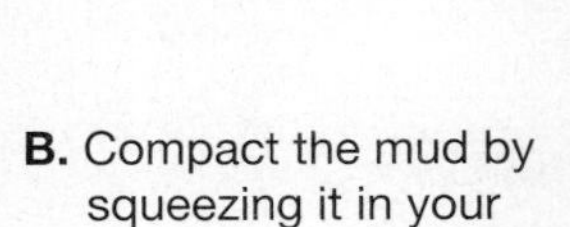

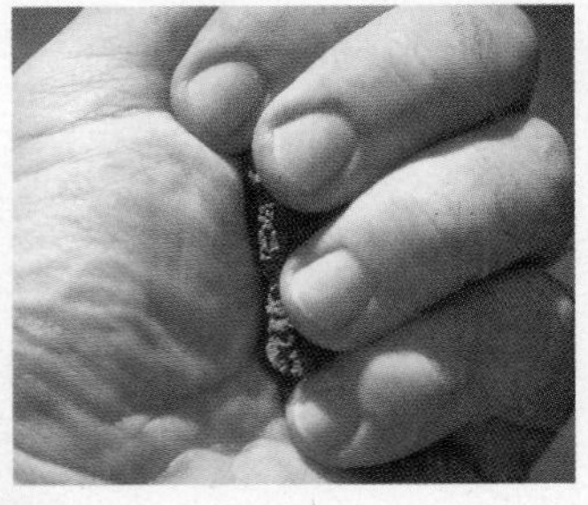

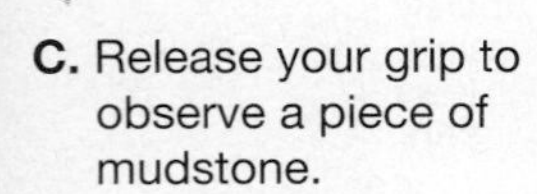

FIGURE 6.4 Compaction of mud to form mudstone. The more the mud (silt and clay sized grains of detrital sediment) is compacted, the harder (more lithified) it will become. Deeply buried mud is also lithified by heat as it is compacted, like baking clay pots in a kiln.

Formation of Sedimentary Rocks

Lithification is the process of changing loose particles of sediment (unconsolidated sediment) to solid rock (consolidated sediment). Sediment is loose particles such as pebbles, gravel, sand, silt, mud, shells, plant fragments, and mineral crystals. Sediment is lithified when it is **compacted** (pressure-hardened, squeezed: **FIGURE 6.4**) or **cemented** together (glued together by tiny crystals or chemical residues, **FIGURES 6.5, 6.6**). However, it is also possible to form a dense hard mass of intergrown crystals that lock together directly, as they precipitate from water (**FIGURES 6.7** and **6.8**).

Sand (a sediment) can be *compacted* until it is pressure-hardened into sandstone (a sedimentary rock). Alternatively, sandstone can form when sand grains are *cemented* together by chemical residues or the growth of interlocking microscopic crystals in pore spaces of the rock (void spaces among the grains). Rock salt and rock gypsum are examples of sedimentary rocks that form *in situ* by the *precipitation* of aggregates of intergrown and interlocking crystals during the evaporation of salt water or brine.

Ocean water is the most common aqueous solution and variety of salt water on Earth. As it evaporates, a variety of minerals precipitate in a particular sequence. The first mineral to form in this sequence is aragonite (calcium carbonate). Gypsum forms when about 50–75% of the ocean water has evaporated, and halite (table salt) forms when 90% has evaporated. Ancient rock salt units buried under modern Lake Erie probably formed from evaporation of an ancient ocean. The salt units were then buried under layers of mud and sand, long before Lake Erie formed on top of them (see **FIGURE 6.7**).

Classifying Sedimentary Rocks

Geologists classify sedimentary rocks into three main groups: biochemical, chemical (inorganic), and detrital (siliciclastic). Refer to **FIGURES 6.2, 6.9.** and **6.10.**

Biochemical Rocks

The main kinds of biochemical (bioclastic) sedimentary rocks are limestone, peat, lignite, and coal. Biochemical limestone is made of broken and whole animal skeletons (usually seashells, coral, or microscopic shells), as in **FIGURE 6.6**. Differences in the density and size of the

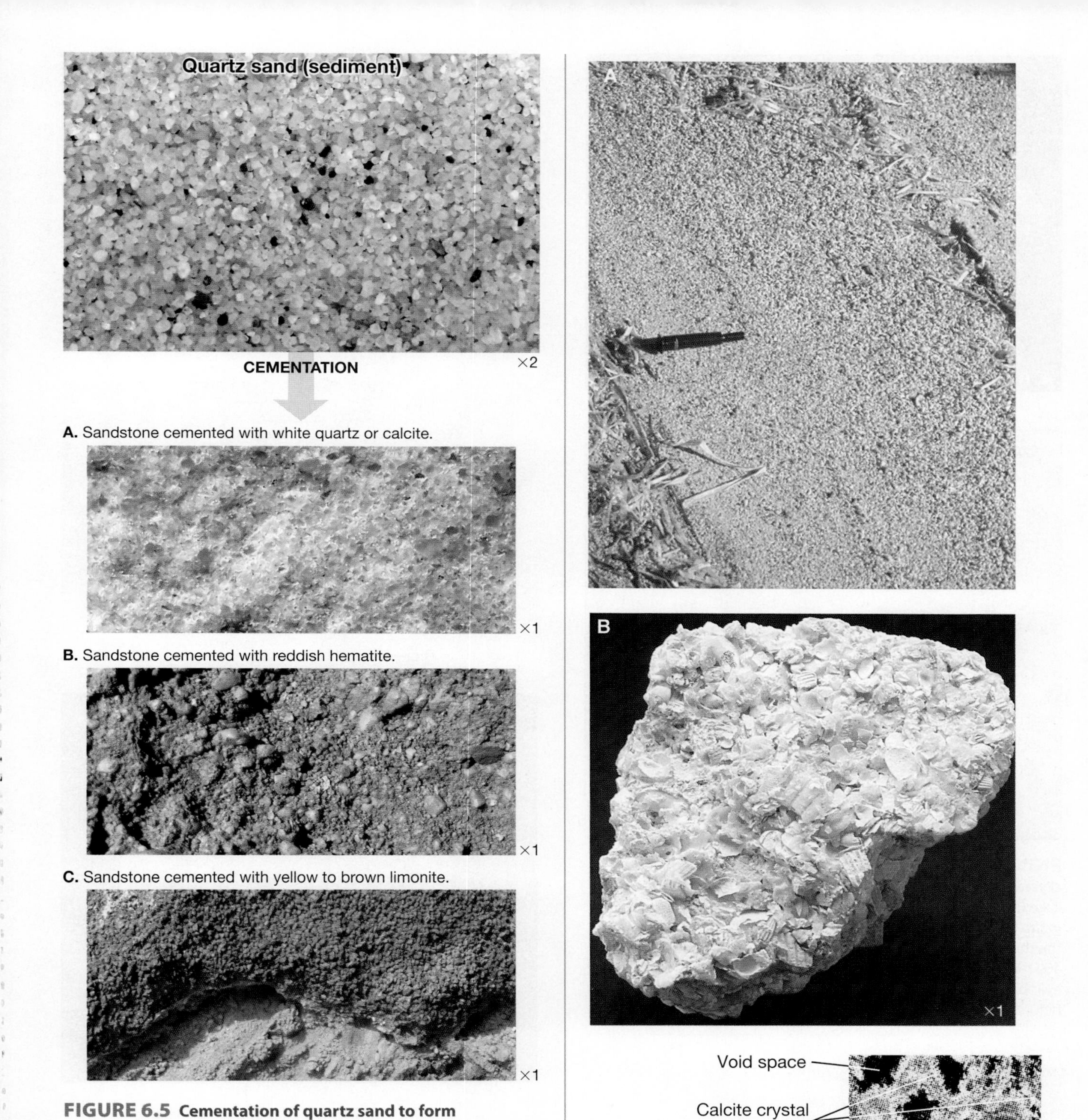

FIGURE 6.5 Cementation of quartz sand to form sandstone. Quartz and iron oxides (limonite, hematite) are the most common cements that help hold together quartz sandstone. Calcite (FIGURE 6.6) can also cement together sandstones. Compaction (FIGURE 6.4) and fusion of quartz sand grains (like pushing together two balls of clay) may accompany cementation in deeply buried layers of sandstone.

FIGURE 6.6 Formation of the biochemical (bioclastic) limestone. A. Shell gravel and blades of the sea grass *Thalassia* have accumulated on a modern beach of Crane Key, Florida. Note pen (12 cm long) for scale. B. Sample of gravel like that shown in part A, but it is somewhat older and has been cemented together with calcite to form limestone (coquina). C. Photomicrograph of a thin section of the sample shown in B. Note that the rock is very porous and that it is cemented with microscopic calcite crystals that have essentially glued the shells together.

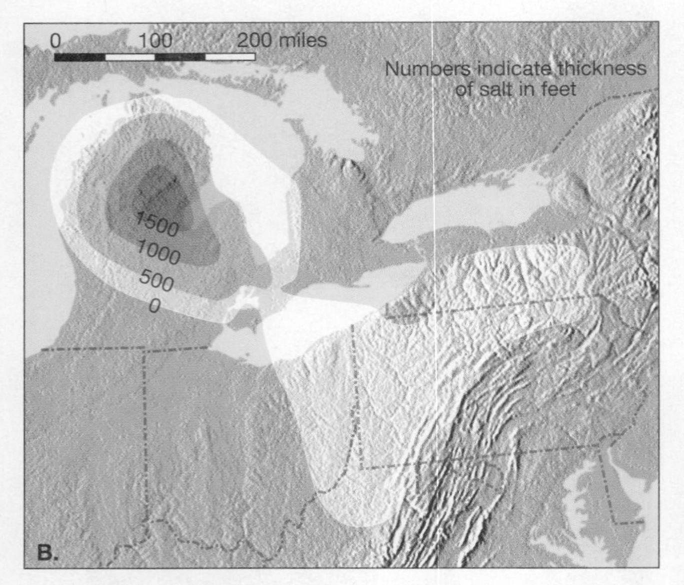

FIGURE 6.7 Rock salt, a chemical sedimentary rock with crystalline texture. A. Hand sample from mines deep below Lake Erie shows how crystals grew together to make the rock salt *in situ* (in place, where the crystals precipitated). B. Map showing the thickness and distribution of rock salt deposits formed about 400 million years ago, when a portion of the ocean was trapped and evaporated in what is now the Great Lakes region, millions of years before any lakes existed.

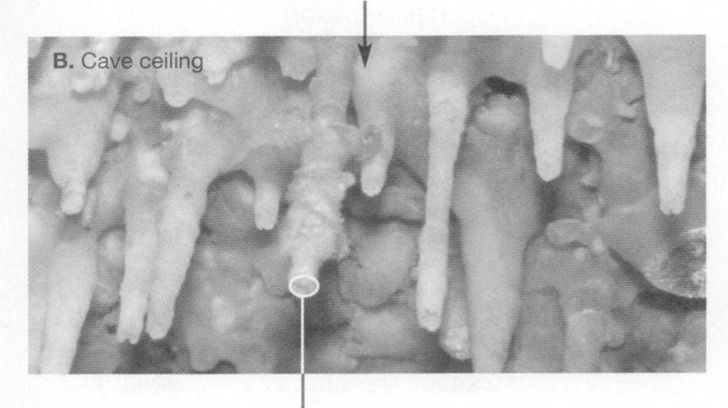

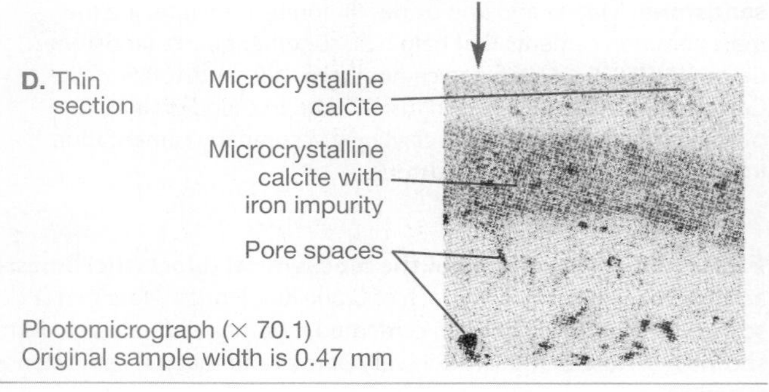

FIGURE 6.8 Formation of the chemical sedimentary rock, travertine. A. Limestone bedrock is dissolved by acidic rain near the Earth's surface. B. The resulting aqueous solution of water, calcium ions, and bicarbonate ions seeps into caves. As the solution drips from the roof of a cave, it forms icicle-shaped stalactites. C. Broken end of a stalactite reveals that it is actually an aggregate of *in situ* (in the place where they formed), chemically precipitated calcite crystals. D. Thin section photomicrograph reveals that the concentric laminations of the stalactite are caused by variations in iron impurity and porosity of the calcite layers.

constituent grains of a biochemical (bioclastic) limestone can also be used to call it a **coquina, calcarenite (fossiliferous limestone), micrite,** or **chalk** (FIGURE 6.9). **Peat** is a very porous brown rock with visible plant fragments that can easily be pulled apart from the rock. **Lignite** is brown but denser than peat. Its plant fragments cannot be pulled apart from the rock. **Bituminous coal** is a black rock made of sooty charcoal-like or else shiny brittle layers of carbon and plant fragments.

Chemical Rocks

There are seven main kinds of chemical (inorganic) sedimentary rocks in the classification in FIGURE 6.9. **Chemical limestone** refers to any mass of crystalline limestone that has no color banding or visible internal structures. **Travertine** is a mass of intergrown calcite crystals that may have light and dark color banding, cavities, or pores (FIGURE 6.8C). **Oolitic limestone** is composed mostly of tiny spherical grains (ooids, FIGURE 6.2) that resemble beads or miniature pearls and are made of concentric layers of microcrystalline aragonite or calcite. They form in intertidal zones of some marine regions (FIGURE 6.10) where the water is warm and detrital sediment is lacking. **Dolostone** (FIGURE 6.9) is an aggregate of dolomite mineral crystals that are usually microcrystalline. It forms in very salty lagoons and desert playa lakes (FIGURE 6.10). Because calcite and dolomite closely resemble one another, the best way to tell them apart is with the "acid test." Calcite will effervesce (fizz) in dilute HCl, but dolomite will effervesce *only* if it is powdered first. **Rock gypsum** is an aggregate of gypsum crystals, and **rock salt** is an aggregate of halite crystals (FIGURE 6.7). Two other chemical sedimentary rocks are **chert** (microcrystalline or even cryptocrystalline quartz) and **ironstone** (rock made mostly of hematite, limonite, or other iron-bearing minerals or chemical residues).

Detrital Rocks

The main kinds of detrital (siliciclastic) sedimentary rocks are mudstone, sandstone, breccia, and conglomerate (FIGURE 6.9). It is very difficult to tell the percentage of clay or silt in a sedimentary rock with the naked eye, so sedimentary rocks made of clay and/or silt are commonly called **mudstone.** Mudstone that is *fissile* (splits apart easily into layers) can be called **shale.** Mudstone can also be called siltstone or claystone, depending upon whether silt or clay is the most abundant grain size. Any detrital rock composed mostly of sand-sized grains is simply called **sandstone** (FIGURES 6.5 and 6.9); although you can distinguish among *quartz sandstone* (made mostly of quartz grains), *arkose* (made mostly of feldspar grains), *lithic sandstone* (made mostly of rock fragments), or *wacke* (made of a mixture of sand-sized and mud-sized grains). **Breccia** and **conglomerate** are both made of gravel-sized grains and are often poorly sorted or moderately sorted. The grains in breccia are very angular and/or subangular, and the grains in conglomerate are subrounded and/or well rounded.

ACTIVITY

6.7 Grand Canyon Outcrop Analysis and Interpretation

THINK About It **What can sedimentary rocks tell us about Earth's history and past environments and ecosystems?**

OBJECTIVE Analyze and interpret sedimentary rocks from the edge of the Grand Canyon.

PROCEDURES

1. **Before you begin**, read about Ancient Environments and Ecosystems and Indicators of Ancient Environments next. Also, this is **what you will need:**

   ___ Activity 6.7 Worksheet (p. 183) and pencil

2. **Then follow your instructor's directions** for completing the worksheets.

ACTIVITY

6.8 Using the Present to Imagine the Past—Dogs to Dinosaurs

THINK About It **What can sedimentary rocks tell us about Earth's history and past environments and ecosystems?**

OBJECTIVE Infer characteristics of an ancient environment by comparing modern dog tracks in mud with fossil dinosaur tracks in sedimentary rock.

PROCEDURES

1. **Before you begin**, read about Ancient Environments and Ecosystems and Indicators of Ancient Environments next. Also, this is **what you will need:**

   ___ Activity 6.8 Worksheet (p. 184) and pencil

2. **Then follow your instructor's directions** for completing the worksheets.

SEDIMENTARY ROCK ANALYSIS AND CLASSIFICATION

STEP 1: Composition. What materials comprise most of the rock?		STEP 2: What are the rock's texture and other distinctive properties?			STEP 3: Name the rock based on your analysis in steps 1 and 2.		
Detrital (Siliciclastic) sediment grains: fragmented rocks and/or silicate mineral crystals	Rock fragments and/or quartz grains and/or feldspar grains and/or clay minerals (e.g., kaolinite) Detrital sediment is derived from the mechanical and chemical weathering of continental (land) rocks, which consist mostly of silicate minerals. Detrital sediment is also called terrigenous (land derived) sediment.		Mostly angular and/or subangular gravel (grains larger than 2 mm)		BRECCIA*		Detrital (Siliciclastic) sedimentary rocks (clastic rocks)
			Mostly subround and/or well rounded gravel (grains larger than 2 mm)		CONGLOMERATE*		
		Mostly sand (1/16–2 mm grains). May contain fossils		Mostly quartz sand	QUARTZ SANDSTONE	SANDSTONE	
				Mostly feldspar sand	ARKOSE	SANDSTONE	
				Mostly rock fragment sand	LITHIC SANDSTONE	SANDSTONE	
				Sand is mixed with much mud	WACKE (GRAYWACKE)	SANDSTONE	
		Mud (< 1/16 mm)	Mostly silt. May contain fossils	Breaks into blocks or layers	SILTSTONE	MUDSTONE	
		No visible grains; Mud (< 1/16 mm)	Mostly clay. May contain fossils	Fissile (splits easily into layers)	SHALE	MUDSTONE	
				Crumbles into blocks	CLAYSTONE	MUDSTONE	
Biochemical (Bioclastic) sediment grains: fragments/shells of organisms	Plant fragments and/or charcoal	Brown porous rock with visible plant fragments that are easily broken apart from one another			PEAT		Biochemical (Bioclastic) sedimentary rocks (clastic rocks)
		Dull, dark brown, brittle rock; fossil plant fragments may be visible			LIGNITE		
		Black, layered, brittle rock; may be sooty or bright			BITUMINOUS COAL		
	Shells and shell/coral fragments, and/or calcareous microfossils (rocks that effervesce in dilute HCl)		Mostly gravel-sized shells and shell or coral fragments; (Figure 6.6)		COQUINA	LIMESTONE (carbonate/calcareous rocks)	
			Mostly sand-sized shell fragments; often contains a few larger whole fossil shells		CALCARENITE (FOSSILIFEROUS LIMESTONE)	LIMESTONE	
			Silty, earthy rock comprised of the microscopic shells of calcareous phytoplankton (microfossils); may contain a few visible fossils		CHALK	LIMESTONE	
		No visible grains	No visible grains in most of the rock. May break with conchoidal fracture. May contain a few visible fossils in the micrite		MICRITE	LIMESTONE	
Mineral crystals (inorganic) or chemical residues (e.g., rust)	Calcite crystals and/or calcite spheres and/or microcrystalline calcite/aragonite (rocks that effervesce in dilute HCl)	Mostly spherical grains that resemble miniature pearls (< 2 mm), called ooliths or ooids			OOLITIC LIMESTONE	LIMESTONE (carbonate/calcareous rocks)	Chemical sedimentary rocks
		Masses of visible crystals and/or microcrystalline; may have cavities, pores, or color banding (Figure 6.8); usually light colored			TRAVERTINE	LIMESTONE (carbonate/calcareous rocks; evaporite rocks)	
	Microcrystalline dolomite	Effervesces in dilute HCl only if powdered. Usually light colored. (Commonly forms from alteration of limestone)			DOLOSTONE	(carbonate/calcareous rocks; evaporite rocks)	
	Halite mineral crystals	Visible cubic crystals, translucent, salty taste (Figure 6.7)			ROCK SALT	(evaporite rocks)	
	Gypsum mineral crystals	Gray, white, or colorless. Visible crystals or microcrystalline. Can be scratched with your fingernail			ROCK GYPSUM	(evaporite rocks)	
	Iron-bearing minerals crystals or residues	Dark-colored, dense, amorphous masses (e.g. limonite), microcrystalline nodules or inter-layered with quartz or red chert (banded iron formation)			IRONSTONE		
	Microcrystalline varieties of quartz (flint, chalcedony, chert, jasper)	Microcrystalline, may break with a conchoidal fracture. Hard (scratches glass). Usually gray, brown, black, or mottled mixture of those colors. Chert can be regarded as biochemical if its silica came from dissolution of siliceous plankton (diatoms, radiolaria).			CHERT (a siliceous rock)		

*Modify name as quartz breccia/conglomerate, arkose breccia/conglomerate, lithic breccia/conglomerate, or wacke breccia/conglomerate as done for sandstones.

FIGURE 6.9 Sedimentary rock analysis and classification. See page 166 for steps to analyze and name a sedimentary rock.

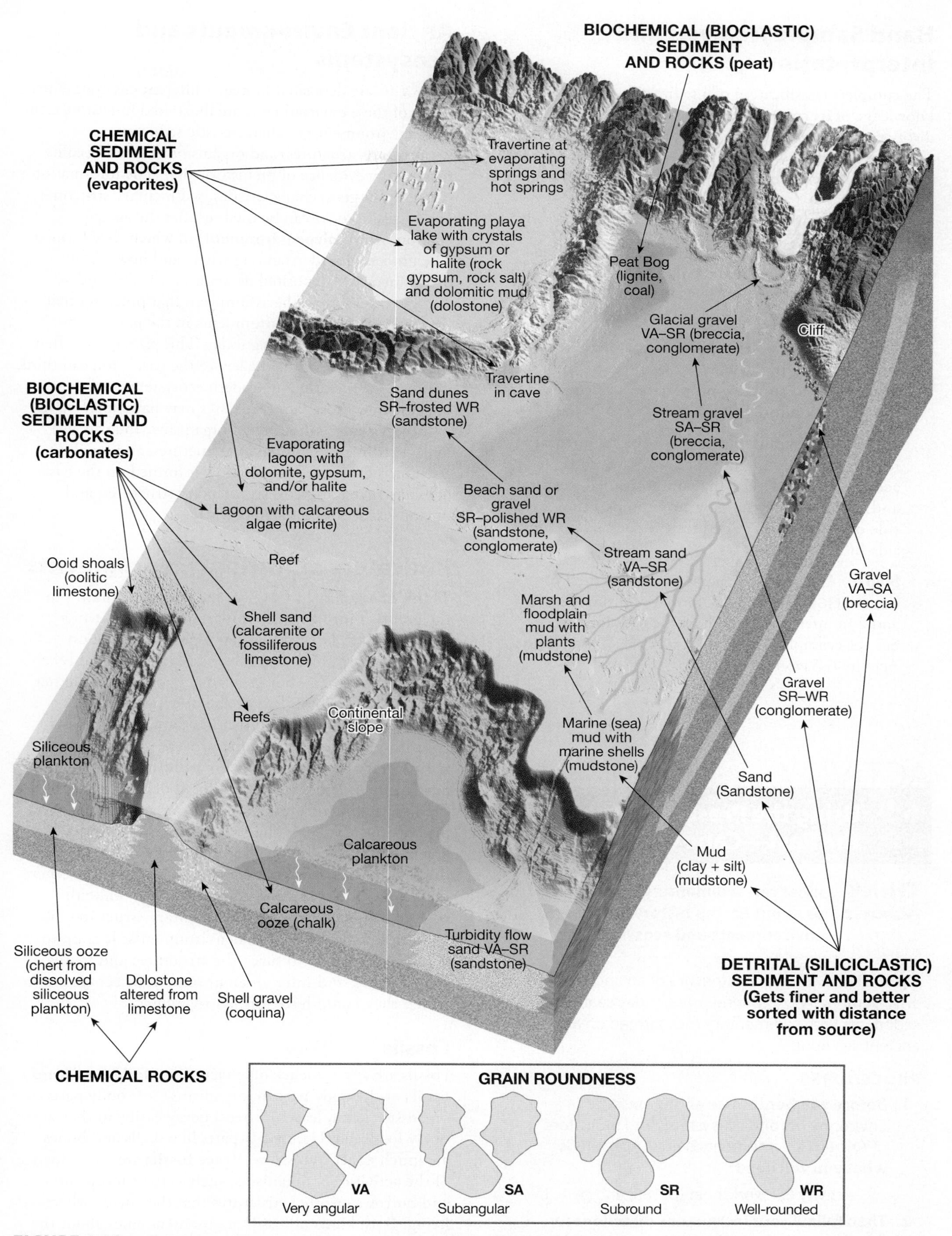

FIGURE 6.10 Sedimentary environments. Some named modern environments where specific kinds of sediments and sedimentary rocks are forming.

Hand Sample Analysis and Interpretation

The complete classification of a sedimentary rock requires knowledge of its composition, texture(s), and other distinctive properties. The same information can be used to infer where and how it formed (FIGURE 6.10). ***Follow these steps to analyze and interpret a sedimentary rock:***

Step 1: Determine and record the rock's general composition as *biochemical (bioclastic), chemical,* or *detrital (siliciclastic)* with reference to FIGURES 6.2 and 6.9, and record a description of the specific kinds and abundances of grains that make up the rock. Refer to the categories for composition in the left-hand column of FIGURE 6.9.

Step 2: Record a description of the rock's texture(s) with reference to FIGURE 6.3. Also record any other of the rock's distinctive properties as categorized in the center columns of FIGURE 6.9.

Step 3: Determine the name of the sedimentary rock by categorizing the rock from left to right across FIGURE 6.9. Use the compositional, textural, and special properties data from Steps 1 and 2 (left side of FIGURE 6.9) to deduce the rock name (right side of FIGURE 6.9).

Step 4: After you have named the rock, then you can use FIGURE 6.10 and information from Steps 1 and 2 to infer where and how the rock formed. See the example for sample X (FIGURE 6.11 and the Activity 6.7 worksheet).

ACTIVITY 6.9 Using the Present to Imagine the Past—Cape Cod to Kansas

THINK About It **What can sedimentary rocks tell us about Earth's history and past environments and ecosystems?**

OBJECTIVE Infer characteristics of an ancient environment by comparing present-day seafloor sediments with sedimentary rock formed on an ancient sea floor.

PROCEDURES

1. **Before you begin**, read about Ancient Environments and Ecosystems and Indicators of Ancient Environments below. Also, this is **what you will need:**

   ___ Activity 6.9 Worksheet (p. 185) and pencil

2. **Then follow your instructor's directions** for completing the worksheets.

Ancient Environments and Ecosystems

Sediments are deposited in many different environments. Some of these environments are illustrated in FIGURE 6.10. Each environment has characteristic sediments, sedimentary structures, and organisms that can become **fossils** (any evidence of prehistoric life). The information gained from grain characteristics, sedimentary structures, and fossils in rocks can be used to infer the ancient environment (**paleoenvironment**) in which they formed. The process of understanding where and how a body of sediment was deposited depends on the *Principle of Uniformitarianism*—the assumption that processes that shaped Earth and its environments in the past are the same as processes operating today. This principle is often stated as, "the present is the key to the past." You can think of processes operating in modern ecosystems and then imagine how those same processes may have operated in past ecosystems with different organisms. You can also look at sediment, sedimentary structures, and fossils in a sedimentary rock and infer how it formed on the basis of where such sediment, sedimentary structures, and organisms are found together today.

Indicators of Ancient Environments

Think of a goldfish. Chances are that your brain put the goldfish into context, and you imagined it in a bowl of water. Now if you saw a goldfish bowl on your neighbor's kitchen table, you would probably think that the neighbor is getting a goldfish. Whether you think of the goldfish or the bowl, you cannot help but imagine the goldfish in a bowl of water—a goldfish ecosystem. The same process is used to analyze sedimentary rocks and infer how and where they may have formed. If the rock has a fossil of a freshwater fish, then the sediment must have accumulated under water, in a stream or lake. If the rock is made of rounded gravel with pieces of tree bark, then the sediment in the rock must have accumulated in an ecosystem where there were both trees and rounded gravel—like the edge of a river. Fossils and sedimentary structures are good indicators of the paleoenvironments. It is up to you, the geologist, to place the structures and fossils into context, and infer an environment or ecosystem in which they could have formed together.

Fossils

Fossils are any evidence of ancient life. **Body fossils** are fossils or the body parts of organisms. Soft body parts of organisms (skin, leaves of trees) decay easily, so they are rarely fossilized. Hard body parts like shells and bones are much easier to fossilize. **Trace fossils** are any evidence of the activities of organisms, such as their footprints and burrows or other structures that they made when living. Both kinds of fossils are useful as clues about the ancient environment of deposition. Trace fossils cannot

FIGURE 6.11 Photograph of hand sample X (actual size).
Refer to the first row of the Activity 6.7 worksheet to see the example of how this rock's composition, texture, and origin were described.

ACTIVITY 6.10 "Reading" Earth History from a Sequence of Strata

THINK About It **What can sedimentary rocks tell us about Earth's history and past environments and ecosystems?**

OBJECTIVE Infer Earth history by "reading" (interpreting) a sequence of strata, from bottom to top.

PROCEDURES

1. **Before you begin**, read about Stratigraphic Sequences below. Also, this is **what you will need:**

   ___ Activity 6.10 Worksheet (p. 186) and pencil

2. **Then follow your instructor's directions** for completing the worksheets.

be transported, so they are *in situ* (formed where they are found). Body fossils, even those of hard shells, are worn away quickly if transported, so they are generally *in situ* as well.

Sedimentary Structures

Sedimentary structures are things like layers of sediment and fossil burrows in the layers. They are structures made of the sediment as it accumulated or after it accumulated (FIGURE 6.12). Some are the result of physical processes, and others are the result of the activities of plants or animals.

Stratigraphic Sequences

As sediments accumulate, they cover up the sediments that were already deposited at an earlier (older) time. Environments also change through time, as layers of sediment accumulate. Therefore, at any particular location, bodies of sediment have accumulated in different times and environments. These bodies of sediment then changed into rock units, which have different textures, compositions, and sedimentary structures.

An undisturbed succession of beds of rock strata can be divided into units of different color, composition, and texture. The succession of such units, one on top of the other, is called a *stratigraphic sequence.* If you interpret each rock unit of the stratigraphic sequence in order, from oldest (at the base) to youngest (at the top), then you will know what happened over a given portion of geologic history for the site where the stratigraphic sequence is located. This order of oldest on the bottom and youngest on the top is the definition of the Law of Superposition, one of the geologic principles that will be discussed in detail in Chapter 8.

SEDIMENTARY STRUCTURES

ILLUSTRATIONS	DESCRIPTIONS	ENVIRONMENTS
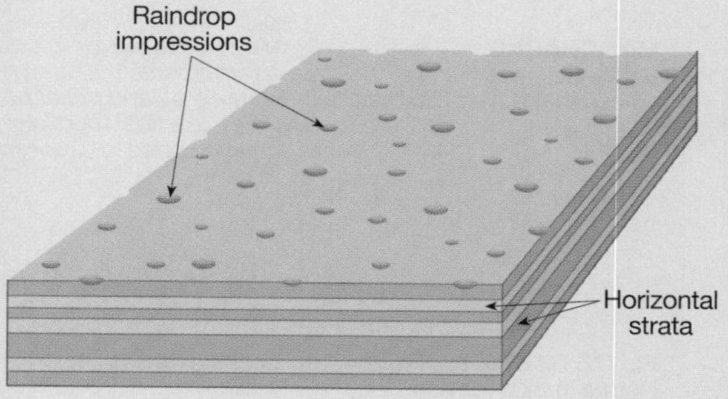 Raindrop impressions	**RAINDROP IMPRESSIONS:** Tiny craters formed by raindrops as they impact bedding plane surfaces.	Raindrop impressions *occur on muddy land surfaces.*
Horizontal strata	**HORIZONTAL STRATA:** Relatively flat *beds* (≥ 1cm thick) and *laminations* (< 1cm thick).	Horizontal strata *occur where sediments settle from a standing body of water or air; or where currents travel parallel to the surface on which sediments are accumulating.*
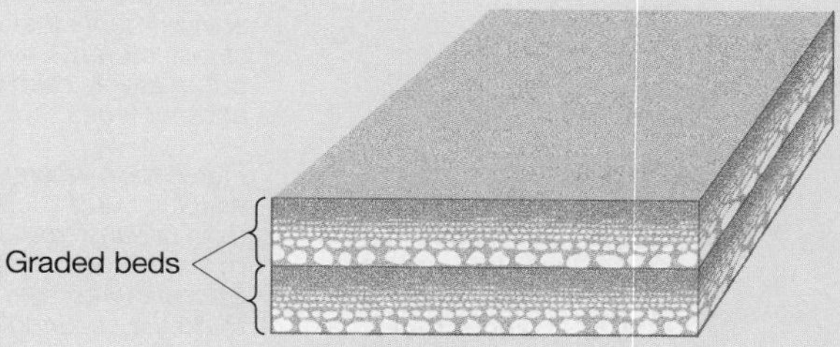 Graded beds	**GRADED BED:** Stratum that contains different sizes of sedimentary grains arranged from largest at the bottom of the bed to smallest at the top.	Graded beds *form when a turbulent body of water full of sediment (flood, wave, river) suddenly loses energy and calms down. Large particles settle out before small.*
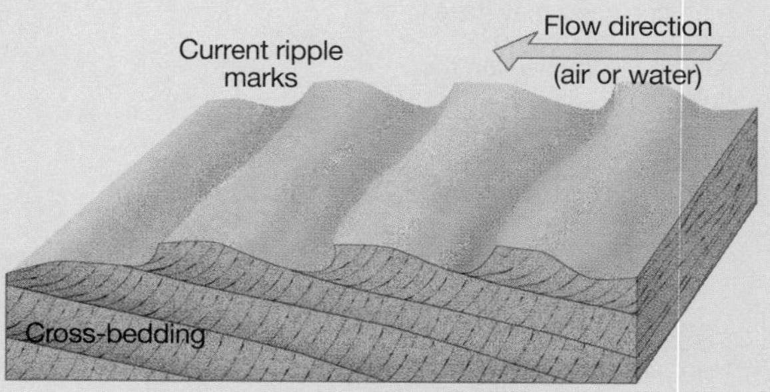 Current ripple marks; Flow direction (air or water); Cross-bedding	**CURRENT RIPPLE MARKS:** Asymmetrical ripple marks. The steep slope faces down current, and the gentle slope faces up current.	Current ripple marks *form in any environment where wind or water travels in one direction for some of the time: rivers, ocean currents, wind blowing sand dunes.*
	CROSS-BEDDING: Inclined beds or laminations.	Cross-bedding *forms wherever there are wind or water currents.*
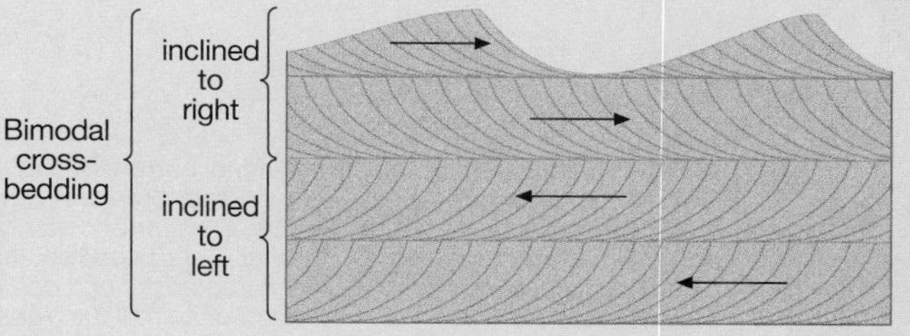 Bimodal cross-bedding: inclined to right; inclined to left	**BIMODAL CROSS-BEDDING:** Sequence of cross-bedding in which cross-bedding of current ripple marks is inclined in opposite directions.	Bimodal cross bedding *forms in environments where currents of wind or water flow back and forth in opposite directions. It is common in environments with tides.*
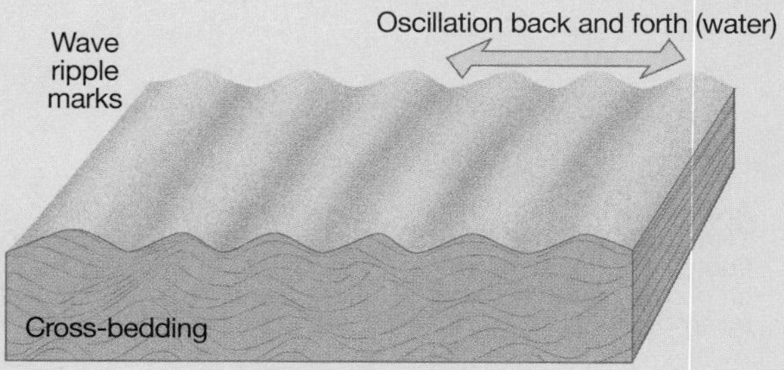 Wave ripple marks; Oscillation back and forth (water); Cross-bedding	**WAVE RIPPLE MARKS:** Symmetrical ripple marks.	Wave (symmetrical) ripple marks *form in any body of water where gentle waves barely touch bottom, or where weak currents move back and forth (oscillate) in shallow water.*

FIGURE 6.12 Sedimentary structures.

SEDIMENTARY STRUCTURES

ILLUSTRATIONS	DESCRIPTIONS	ENVIRONMENTS
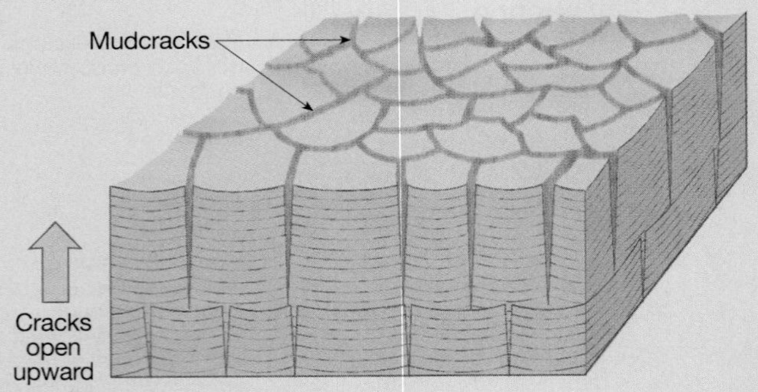 Mudcracks Cracks open upward	**MUDCRACKS:** Polygonal patterns of cracks that develop in mud as it dries.	Mudcracks *form in muddy environments that are wet sometimes and dry at other times, like tidal mudflats or land surfaces exposed to rain.*
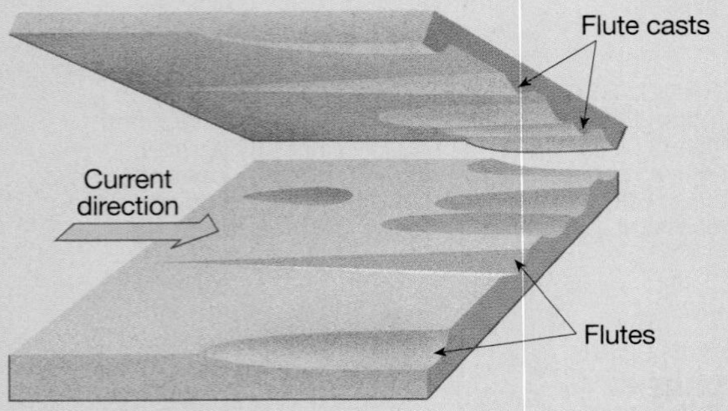 Flute casts	**FLUTE CASTS:** Natural molds formed when mud or sand fill up flutes.	Flute casts *form when sediment is deposited on current-scoured surfaces. Thus, flute casts develop in environments that have strong currents sometimes, but relatively calm conditions at other times.*
Current direction Flutes	**FLUTES:** U-shaped or V-shaped scrapes and gouges in mud or sand that were scoured out by currents. The opening of a V or U points in the downstream direction. The mud and sand may have turned to mudstone or sandstone, preserving the flutes.	Flutes *form wherever water or wind scours away mud or sand from land or submerged surfaces. Strong currents are required to do the scouring.*
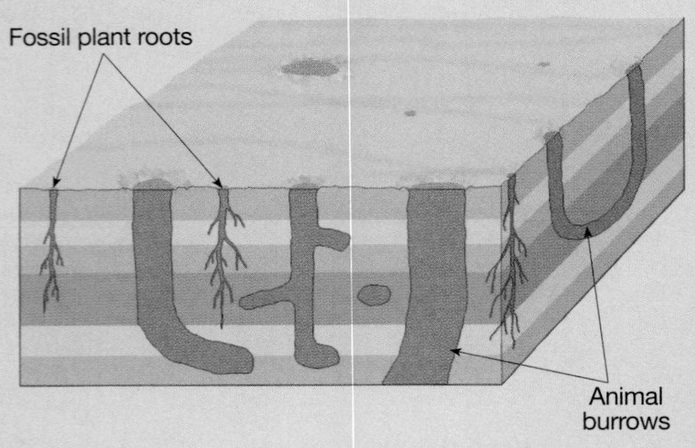 Fossil plant roots	**FOSSIL PLANT ROOTS:** Root-shaped fossils that narrow away from the main branch.	Fossil plant roots *indicate ancient soil zones where plants once grew.*
Animal burrows	**ANIMAL BURROWS:** All sizes of tunnels or tubes that cut into or across strata and maintain constant diameters with circular cross sections.	Animal burrows *occur wherever burrowing animals live, in water or on land. The shape of the burrow may be characteristic of a particular kind of animal that lives only in a specific environment.*
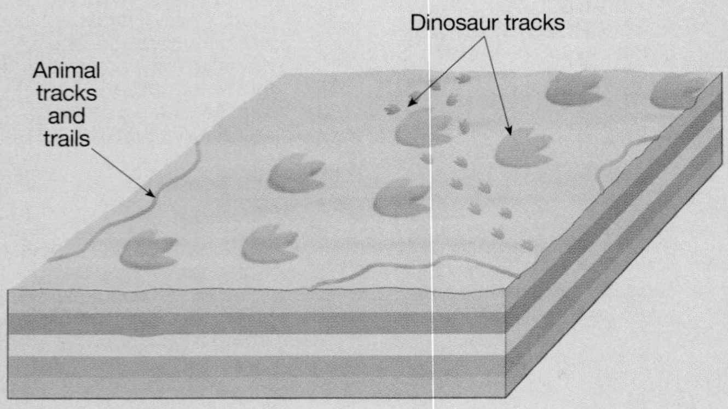 Dinosaur tracks Animal tracks and trails	**ANIMAL TRACKS, TRACKWAYS, AND TRAILS:** Footprints or grooves left on bedding plane surfaces by animals.	Animal tracks and trails *occur wherever animals live. Some are diagnostic of specific kinds of animals that live in specific environments.*

FIGURE 6.12 (continued)

ACTIVITY 6.1 Sedimentary Rock Inquiry

Name: ______________________ **Course/Section:** ______________________ **Date:** ____________

A. Analyze the sedimentary rocks below (and actual rock samples of them if available). Beside each picture, write words and phrases to describe the rock's **composition** (what it is made of) and **texture** (the size, shape, and arrangement of its parts). Use your current knowledge, and complete the worksheet with your current level of ability. Do not look up terms or other information.

B. **REFLECT & DISCUSS** Reflect on your observations and descriptions of sedimentary rocks in part A. Then describe how you would classify the rocks into groups. Be prepared to discuss your classification with other geologists.

ACTIVITY 6.2 Mount Rainier Sediment Analysis

Name: ______________________ **Course/Section:** ______________________ **Date:** ____________

A. These are images of rocks on or near Mount Rainier, WA, an andesitic volcano. Image A was taken at an outcrop of the andesite near the top of the volcano, and Image B was taken near the middle of the volcano's slope. Image C was taken in the Nisqually River that drains away from the base of the volcano. Image C was taken 30 km downstream, at a delta where the river enters Alder Lake. All images are 1/3 of actual size. Note how the sediment changes from A to D.

B. Mount Rainier slope, 5600 feet above sea level. 45 47 23.5N, 121 44 23.4W in Google Earth™ 2 kilometers from Location A

A. Mount Rainier outcrop, 6600 feet above sea level. 46 48 15.2N, 121 43 57.8W in Google Earth™

C. Nisqually River, near Longmire, southwest of Mount Rainier, 2600 feet above sea level. 46 44 26.4N, 121 49 27W in Google Earth™ 9 kilometers downhill from Location B

D. Alder Lake delta, at Elbe, southwest of Mount Rainier, 1200 feet above sea level. 46 45 52N, 122 11 45W in Google Earth™ 30 kilometers downstream from Location C

1. What is the grain size of the sediment at each location, expressed as one or more Wentworth size classes?

A.

B.

C.

D.

2. What is the grain roundness at each of the following locations?

A.

B.

C.

3. In general, would you describe the sediment in these images as detrital (siliciclastc), biochemical, or chemical? Why?

4. Name the kind of rock that the sediment in each image would form if it became lithified (**FIGURE 6.9**, Step 3).

A.

B.

C.

D.

5. Notice the yellow-orange color of the sedimentary grains at Location B. What is the yellow-orange material and where did it come from?

6. Each image is a photograph of materials that are the product of chemical and physical sedimentary processes. For each image, list the processes that must have occurred to form the sediment.

A.

B.

C.

D.

B. REFLECT & DISCUSS Based on your work, write a sentence that describes what happens to detrital (siliciclastic) sediment with distance from its source. Then describe how you could use your statement to interpret detrital (siliciclastic) rocks.

ACTIVITY 6.3 Clastic and Detrital Sediment

Name: ____________________ **Course/Section:** ____________________ **Date:** ____________

A. Obtain two pieces of granite or diorite. Hold one in each hand and tap them together over a piece of paper. As you do this you should notice that you are breaking tiny sedimentary grains from the larger rock samples. These broken pieces of rocks and minerals are called **clasts** (from the Greek *klastós*, meaning "broken in pieces").

1. Using a hand lens or microscope, observe the tiny clasts that you just broke from the larger rock samples. Describe what minerals make up the clasts and whether or not the clasts are fragments of mineral crystals, rock fragments, or a mixture of both.

2. Geologists commonly refer to several different kinds of clastic sediment. Circle the one that you just made.

- **pyroclastic sediment**—volcanic bombs and/or volcanic rocks fragmented by volcanic eruption
- **bioclastic sediment**—broken pieces of shells, plants, and/or other parts of organisms
- **siliciclastic sediment**—broken pieces of silicate mineral crystals and/or rocks containing them

3. Roundness is a measure of how much the profile of a grain of sediment resembles a circle. It is most often visually estimated using a chart like this one. Re-examine your clasts from Part A1 and sketch the outline of several of them. Compared to the chart, what is the roundness of the clasts that you sketched?

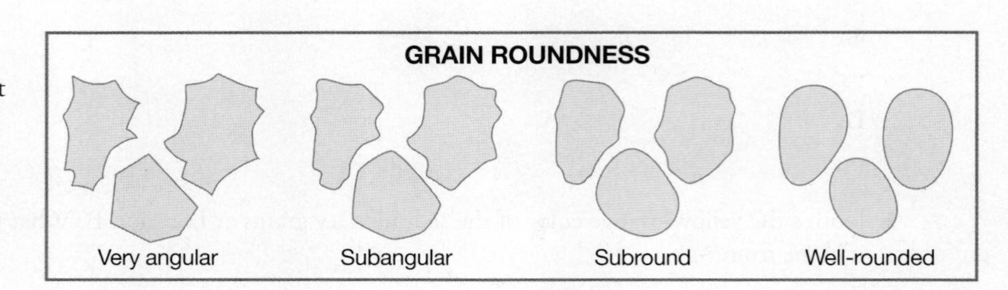

4. Using a grain size scale (from GeoTools 1 or 2 at the back of your manual), circle the Wentworth size class(es) of the clastic sediment that you made above.

gravel	sand	silt	clay
(grains > 2 mm)	(grains 1/16 to 2 mm)	(grains too small to see but you can feel them)	(grains too small to see or feel; like chalk dust)

5. Obtain a piece of quartz sandpaper and lay it flat on the table. Find a sharp corner on one of the granite/diorite samples that you used above and sketch its outline in the "before abrasion" box below. Next, rub that corner against the quartz sandpaper for about 10 seconds. Sketch its profile in the "after abrasion" box. What did this abrasion process do to the sharp corner?

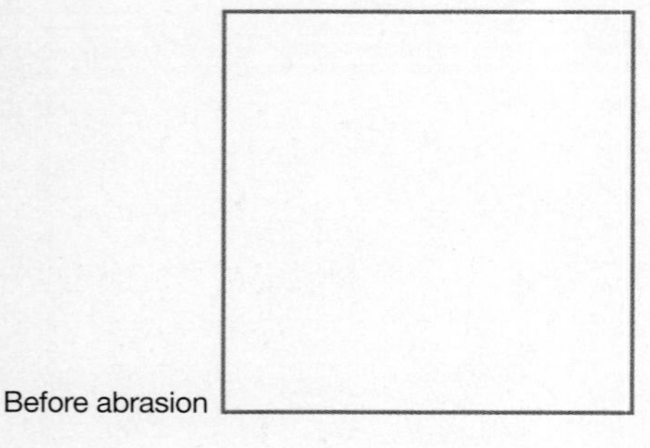

After abrasion

6. The sediment that you just made by wearing down the corner of a rock clast is called **detrital sediment** (from the Latin *detritus*, participle of *detero*, meaning "to weaken, wear away, rub off"). The term is also used to refer to all sediment that is terrigenous (from the land)—worn and transported away from landscapes (rock fragments, mineral grains, and rock material that has been weakened and decomposed by chemical weathering).

The Mississippi River carries detrital sediment that has been weathered from bedrock and worn away from the landscape of much of the United States. The river flows downhill under the influence of gravity and eventually flows into the Gulf of Mexico, where its load of detrital sediment temporarily accumulates at the mouth of the river on the edge of the Mississippi Delta. On this NASA satellite image of the Mississippi Delta, write a "D" to indicate where the main load of terrigenous detrital sediment is being deposited at the edge of the delta. How do you think the roundness of sediment in the river will change from a place upstream where it was broken from bedrock to the location where you placed your "D" on the image?

29 39 45N, 90 33 48W in Google Earth™ (© Google Earth)

B. Sediment falls and slides (rockslides) downhill under the influence of gravity and is transported by flowing agents like water, wind, and ice (glaciers). As grains are transported, they scrape, chip, brake, and generally increase in roundness.

1. Glacial ice holds detrital grains of sediment in its firm grip while the weight of the glacier exerts tremendous downward force and gravity pulls the glacier downhill. You can model this process and see what it does to grains of sediment. Place a piece of sandpaper flat on the table. Next, firmly grip (like glacial ice) a piece of shale with a somewhat flat side pointing down. In one motion press the shale firmly against the sandpaper and push it forward one time. Then use a hand lens to observe the shale surface that you just scraped over the sandpaper. To the right of this paragraph, draw the pattern of scratches that you observe. What would happen to the shale surface if you kept grinding it straight ahead on a 10-meter-long strip of sandpaper?

2. Grains of sediment carried by water and wind move generally in one main direction but are free to quickly change direction and roll about so that all of their sides scrape and impact other grains often. Imagine that the piece of shale above has been dropped from a melting glacier and is being transported by a melt water stream. To model what might happen to the shale grain, place it onto the sandpaper, grip it lightly, and move it about against the sandpaper in multiple directions. Turn the shale to a different side and repeat. Now observe the newly scraped surfaces with a hand lens. To the right of this paragraph, draw the pattern of scratches that you observe.

C. REFLECT & DISCUSS Based on your work above, how could you tell a grain of sediment that was abraded and shaped in a glacial environment from one that was abraded and shaped while being transported by water or wind?

ACTIVITY 6.4 Biochemical and Chemical Sediment and Rock

Name: ______________________ **Course/Section:** ______________________ **Date:** __________

A. Seashells are grains of sediment made by the biochemical processes of organisms, so they are grains of biochemical sediment. When you find a rock with a fossil seashell, then you have found evidence that the rock contains sediment deposited where the sea animal lived (i.e., in the ocean, in a marine environment). Some limestone is entirely made of the seashells or broken pieces of seashells. Obtain a seashell (e.g., hard clam shell) and draw it to the right of this paragraph. It may be easiest to trace it, then fill in the outline with details of what the shell looks like inside or out. Next, place the shell into a plastic sandwich bag and take the bag to the hammering station in your lab. Lightly tap the bag with the hammer to break up the shell into pieces. Return to your table and view the broken pieces of shell with a hand lens.

1. The shell fragments that you just made are called **clasts** (from the Greek *klastós*, meaning "broken in pieces"). Geologists commonly refer to several different kinds of clastic sediment. Circle the one that you just made.

- **pyroclastic sediment**—volcanic bombs and/or volcanic rocks fragmented by volcanic eruption
- **bioclastic sediment**—broken pieces of shells, plants, and/or other parts of organisms
- **siliciclastic sediment**—broken pieces of silicate mineral crystals and/or rocks containing them

2. Compared to **FIGURE 6.3**, what is the roundness of your clasts? ______________________

3. What is the roundness of the clasts in this picture (× 1 scale)? ______________________
Explain how and in what environment the shell clasts could have attained their roundness.

4. Some limestone is made of shells that are calcareous (calcite or aragonite), like visible seashells, but they are microscopic and cannot even be seen with a hand lens. Chalk is such a limestone. Some chalk used with modern blackboards is clay or plaster-of-Paris, rather than real chalk. Obtain a piece of chalk from your lab room or instructor. Explain how dilute HCl (hydrochloric acid) can be used to help you test your chalk and find out if it is real chalk or not. Then conduct your test and report the results of your test.

5. Based on **FIGURE 6.10** (page 165), how and where does chalk form?

B. Place a charcoal briquette into a plastic sandwich bag and take it to the hammering station in your lab. Lightly hammer the bag enough to break apart the briquette. Return to your table with the bag of charcoal.

1. View the broken pieces of charcoal with a hand lens. Describe what kinds of grains you see and their texture.

2. Charcoal is made by allowing wood to smolder just enough that an impure mass of carbon remains. In the presence of oxygen, the charcoal briquette will naturally combine with oxygen to make carbon dioxide. Over a period of many years, it will all react with oxygen and chemically weather to carbon dioxide. When you burn charcoal in your grill, you are simply speeding up the process. However, if plant fragments are buried beneath layers of sediment that keep oxygen away from them, then they can slowly convert to a charcoal-like rock (peat, lignite, or coal) and remain so for millions of years. Obtain a piece of coal and compare it to your charcoal. How is it different? Why?

C. **REFLECT & DISCUSS** Based on your observations in this activity, write a definition of biochemical sedimentary rock in your own words.

D. Bedrock can remain buried underground for millions to billions of years. However, when it is exposed to water and air at Earth's surface it weathers chemically and physically. For example, acidic water reacts with potassium and plagioclase feldspars to make clay minerals plus water containing dissolved silica (hydrosilicic acid) and metallic ions (K, Na, Ca). This is one of the main sources of clay found in soil and worn away into rivers and the ocean. The metals in many minerals oxidize (combine with oxygen) to form metal oxides like limonite ("rusty" iron) and hematite. Obtain and observe samples of both.

1. What is the color and chemical formula for hematite? (Refer to Minerals Database, page 95)

sandstone ×1

2. What is the color and chemical formula for limonite? (Refer to Minerals Database, page 96)

3. As iron oxides form, they act like glue to cement together grains of sediment, like the "sandstone" above. Which iron oxide mineral has cemented together this sandstone? How can you tell?

4. Powder some limonite in a mortar and pestle, and note its true streak color (yellow-brown). Put on safety goggles. In a fume hood or behind a glass shield, heat some of the powder in the Pyrex test tube over the Bunsen burner. Be sure to point the test tube at an angle, away from people. After about a minute of heating, pour the hot limonite powder onto the foil on the table. What happened to the yellow-brown limonite? Why?

5. **REFLECT & DISCUSS** The *rapid* chemical change that you observed above can occur quickly only at temperatures like those above the Bunsen burner. However, some modern desert soils do contain hematite and appear red. How can that be?

ACTIVITY 6.5 Sediment Analysis, Classification, and Interpretation

Name: ______________________ **Course/Section:** ______________________ **Date:** __________

Complete parts 1 through 6 for each sample below. Refer to FIGURES 6.2 and 6.3 as needed.

SAMPLE A

0 1 mm

Ooids

1. Grain size range in mm: ______________________
2. Percent of each Wentworth size class:
   clay _______ silt _______ sand _______ gravel _______
3. Grain sorting (circle):
   Poor Moderate Well
4. Grain roundness (circle):
   Angular Subround Well-rounded
5. Sediment composition (circle):
   Detrital Biochemical (Siliciclastic) Chemical (Bioclastic)
6. Describe how and in what environment (**FIGURE 6.10**) this sediment may have formed.

SAMPLE B

0 1 2 mm

1. Grain size range in mm: ______________________
2. Percent of each Wentworth size class:
   clay _____ silt _____ sand _____ gravel _____
3. Grain sorting (circle):
   Poor Moderate Well
4. Grain roundness (circle):
   Angular Subround Well-rounded
5. Sediment composition (circle):
   Detrital (Siliciclastic) Biochemical (Bioclastic) Chemical
6. Describe how and in what environment (**FIGURE 6.10**) this sediment may have formed.

SAMPLE C

0 10 mm

1. Grain size range in mm: ______________________
2. Percent of each Wentworth size class:
   clay _______ silt _____ sand _____ gravel _____
3. Grain sorting (circle):
   Poor Moderate Well
4. Grain roundness (circle):
   Angular Subround Well-rounded
5. Sediment composition (circle):
   Detrital (Siliciclastic) Biochemical (Bioclastic) Chemical
6. Describe how and in what environment (**FIGURE 6.10**) this sediment may have formed.

D. REFLECT & DISCUSS Imagine that these sediments are rocks. Which of the samples do you think would be the least diagnostic of a specific ancient environment? Why?

ACTIVITY 6.6 Hand Sample Analysis and Interpretation

Name: ________________ **Course/Section:** ________________ **Date:** ____________

SEDIMENTARY ROCKS WORKSHEET

Sample Number or Letter	Composition (Figures 6.2 and 6.9)	Textural and Other Distinctive Properties (Figures 6.3 and 6.9)	Rock Name (Figure 6.9)	How Did the Rock Form? (See Figure 6.10)
Fig. 6.11	Detrital (Siliciclastic): • Mostly orange feldspar grains (~85%) • Some quartz (~10%) • Green silty matrix (~5%)	• Mostly (~95%) angular to subangular gravel-sized grains • Poorly sorted (The gravel is mixed with some sand and green silt)	Breccia (Arkose breccia)	Preexisting rock exposed on land (probably granite) was weathered. Grains were not rounded or sorted much, so they were not transported very far from their source. Grains were mixed with some green silt, deposited, and hardened (compaction?) into rock.
201	Clastic	• Orange feldspar • Some crystals • some rocks	Arkose	
203	Chemical	• Fizzes when scratched. • Clay looking but isn't, orangish.	Dolostone	
204	Clastic	• light green sandish • Breaks into blocks.	Siltstone	
205	Clastic	• Angular • gravel • rocky.	Breccia	

Name: ______________________ Course/Section: ______________________ Date: ______________

SEDIMENTARY ROCKS WORKSHEET

Sample Number or Letter	Composition (Figures 6.2 and 6.9)	Textural and Other Distinctive Properties (Figures 6.3 and 6.9)	Rock Name (Figure 6.9)	How Did the Rock Form? (See Figure 6.10)
206	Biochemical	• gravel shells • coral fragments	Coquina	
208	Biochemical	Sand-sized shell fragments.	Calcarenite	
209	Clastic	• rounded gravel	Conglomerate	
210	Biochemical	• No visible grains	Micrite	
211	Chemical	• gray, white. • easily scratch.	Rock Gypsum	

Name: ________________ Course/Section: ________________ Date: ________

SEDIMENTARY ROCKS WORKSHEET

Sample Number or Letter	Composition (Figures 6.2 and 6.9)	Textural and Other Distinctive Properties (Figures 6.3 and 6.9)	Rock Name (Figure 6.9)	How Did the Rock Form? (See Figure 6.10)
212	Chemical	Visible crystals Salty-taste.	Rock Salt	
213	Clastic	quartz sand	Quartz Sandstone	
214	Clastic	Fissile easily layered	Shale	

Name: ____________________ **Course/Section:** ____________________ **Date:** ____________

SEDIMENTARY ROCKS WORKSHEET

Sample Number or Letter	Composition (Figures 6.2 and 6.9)	Textural and Other Distinctive Properties (Figures 6.3 and 6.9)	Rock Name (Figure 6.9)	How Did the Rock Form? (See Figure 6.10)

ACTIVITY 6.7 Grand Canyon Outcrop Analysis and Interpretation

Name: ______________________ **Course/Section:** ______________________ **Date:** ____________

A. Analyze the images above, from the South Rim of the Grand Canyon, near Grand Canyon Village. The edge of the canyon here is made of a Permian calcarenite (sand–sized fossilifereous limestone) called the Kiabab Limestone. It is about 270 million years old.

1. Notice that some of the beds in the outcrop are cross-bedded. Draw an arrow on the picture to show the direction that the water moved here to make this cross bedding. Refer to **FIGURE 6.12** as needed.

2. Which kind of cross bedding is this? (**FIGURE 6.12**)? ______________________________

3. **REFLECT & DISCUSS** Describe (as well as you can) what the environment was like here about 270 million years ago and the evidence and logic that you used to reach your conclusion.

ACTIVITY 6.8 Using the Present to Imagine the Past—Dogs and Dinosaurs

Name: ______________________ **Course/Section:** ______________________ **Date:** ____________

A. Analyze photographs X and Y below.

X. Modern dog tracks in mud with mudcracks on a tidal flat, St Catherines Island, Georgia (x1)

Y. Triassic rock (about 215 m.y. old) from southeast Pennsylvania with the track of a three-toed *Coelophysis* dinosaur (x1)

1. How are the modern environment (Photograph X) and Triassic rock (Photograph Y) the same?

2. How are the modern environment (Photograph X) and Triassic rock (Photograph Y) different?

3. Describe what the Pennsylvania ecosystem (environment + organisms) was like when *Coelophysis* walked there about 215 million years ago.

B. REFLECT & DISCUSS Use what you learned about sediment and sedimentary rocks. Develop a hypothesis about how the dinosaur footprint in Photograph Y was preserved.

ACTIVITY 6.9 Using the Present to Imagine the Past—Cape Cod to Kansas

Name: ______________________ **Course/Section:** ______________________ **Date:** __________

A. Analyze photographs A and B below of a Kansas rock and the modern-day seafloor near Cape Cod.

A. Pennsylvanian-age rock from Kansas (290 m.y. old)

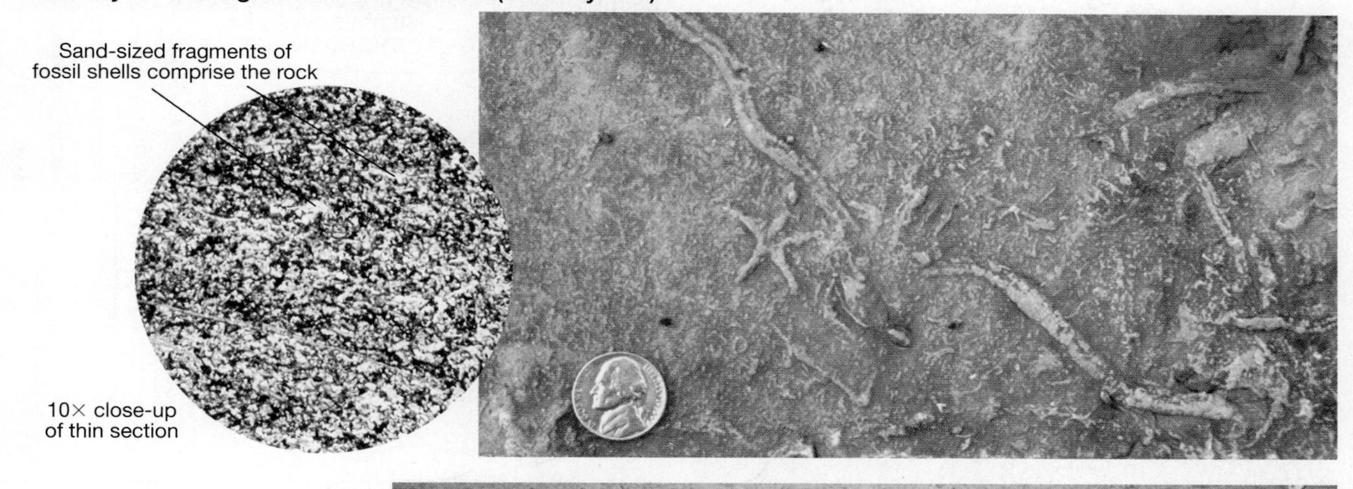

B. Modern sea-floor environment, 40 m (130 ft) deep, near Massachusetts (10 miles north of Cape Cod). Detrital (siliciclastic) sediment:
- **1% gravel**
- **90% sand**
- **9% mud**

1. How are the modern environment (Photograph B) and Kansas rock (Photograph A) the same?

2. How are the modern environment (Photograph B) and Kansas rock (Photograph A) different?

3. Today, this part of Kansas is rolling hills and farm fields. Describe what the Kansas ecosystem (environment + organisms) was like when the sediment in this rock sample (Photograph A) was deposited there about 290 million years ago.

B. REFLECT & DISCUSS What would have to happen to the sediment in Photograph B to turn it into sedimentary rock?

ACTIVITY 6.10 "Reading" Earth History from a Sequence of Strata

Name: ______________________ **Course/Section:** ______________________ **Date:** ____________

A. Permian strata (about 270 million years old) exposed along Interstate Route 70 in northeastern Kansas. Describe the paleoenvironment (pink column), then apply it to infer the record of change (purple column).

OUTCROP	HAND SAMPLE Bedding plane surface	DESCRIPTION OF ROCK UNIT	DESCRIPTION OF PALEOENVIRONMENT REPRESENTED BY THE ROCK UNIT	RECORD OF CHANGE				
				ocean (marine)	muddy bay/estuary	evaporating bay	peat bog or swamp	land
		7. Tan skeletal limestone with shells of many kinds of marine organisms, bimodal cross-bedding, oscillation ripple marks, animal burrows, flutes, flute casts, and chert.						
		6. Gray silty mudstone (shale) with animal burrows, fossil clams, fossil plant fragments, and current ripple marks.						
		5. Red and gray silty mudstone with raindrop impressions, fossil roots, and mudcracks.						
		4. Gray silty mudstone with abundant gypsum layers and crystals.						
		3. Tan skeletal limestone with bimodal cross-bedding.						
		2. Coal.	peat bog or swamp				(shaded)	
1 METER		1. Gray silty mudstone with mudcracks and fossil ferns.	Probably moist muddy land where ferns grew; mudcracks formed in dry periods.					(shaded)

B. REFLECT & DISCUSS What could have caused the sea level to rise and fall in this way about 270 million years ago?

PRE-LAB VIDEO

LABORATORY 7

Metamorphic Rocks, Processes, and Resources

CONTRIBUTING AUTHORS

Harold E. Andrews • *Wellesley College*
James R. Besancon • *Wellesley College*
Margaret D. Thompson • *Wellesley College*

Marble is metamorphosed limestone. It is being quarried here for use in construction, table tops, decorative tiles, and sculptures. (Fotografiche/Shutterstock)

BIG IDEAS

Metamorphic rocks are rocks that have changed to a new and different form as a result of intense heat, intense pressure, and/or the action of watery hot fluids. The mineralogy and texture of a metamorphic rock can be used to deduce its original form (parent rock) and infer the geologic history of how and why it changed. Metamorphic rocks are widely used in the arts and construction industries and are sources of industrial minerals and energy.

FOCUS YOUR INQUIRY

THINK About It | What do metamorphic rocks look like? How can they be classified into groups?

ACTIVITY 7.1 Metamorphic Rock Inquiry *(p. 188)*

THINK About It | What are the characteristics of metamorphic rocks, and how are they formed?

ACTIVITY 7.2 Metamorphic Rock Analysis and Interpretation *(p. 189)*

THINK About it | How are rock composition and texture used to classify, name, and interpret metamorphic rocks?

ACTIVITY 7.3 Hand Sample Analysis, Classification, and Origin *(p. 196)*

THINK About it | What can metamorphic rocks tell us about Earth's history and the environments in which the rocks formed?

ACTIVITY 7.4 Metamorphic Grades and Facies *(p. 198)*

Introduction

The word *metamorphic* is derived from Greek and means "of changed form." **Metamorphic rocks** are rocks changed from one form to another (metamorphosed) by intense heat, intense pressure, or the action of watery hot fluids. Think of metamorphism as it occurs in your home. *Heat* can be used to metamorphose bread into toast, *pressure* can be used to compact an aluminum can into a flatter and more compact form, and the chemical action of *watery hot fluids* (boiling water, steam) can be used to change raw vegetables into cooked forms. Inside Earth, all of these metamorphic processes are more intense and capable of changing a rock from one form to another. Thus metamorphism can change a rock's size, shape, texture, color, and/or mineralogy.

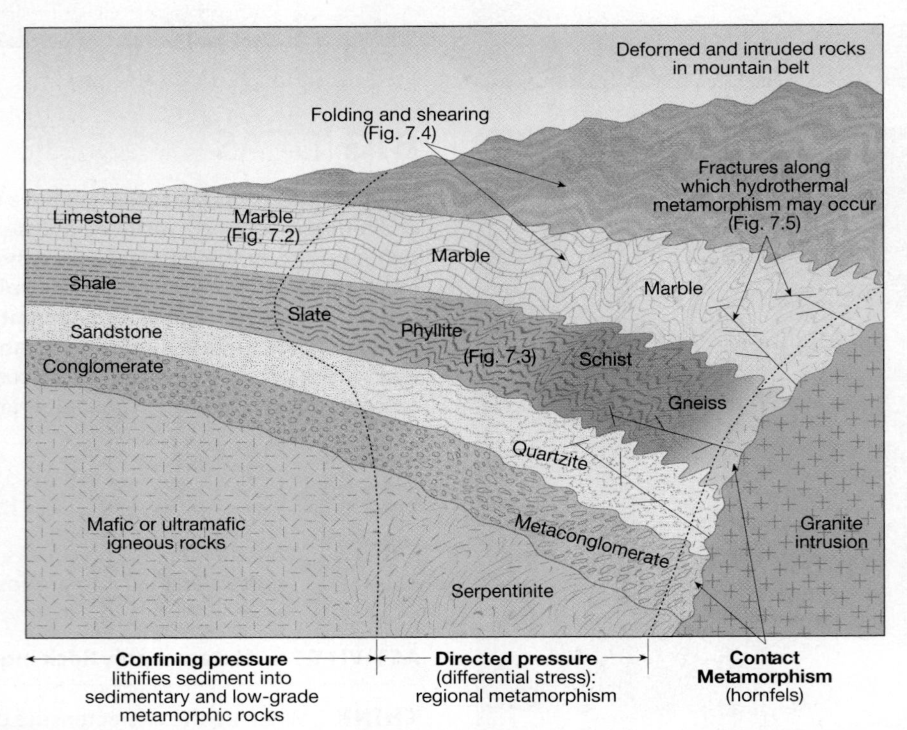

FIGURE 7.1 Generalized diagram of metamorphism. This hypothetical diagram shows how heat (from a body of granitic magma), directed pressure (as in a mountain belt at a convergent plate boundary), and the chemical action of watery hot (hydrothermal) fluids drive the process of metamorphism. *Parent rocks* far from the intrusion and directed pressure remain unchanged. In the region of folding and igneous intrusion, mafic igneous rocks were metamorphosed to serpentinite. Sedimentary conglomerate, sandstone, and limestone parent rocks were metamorphosed to *metaconglomerate, quartzite,* and *marble*. Shale was metamorphosed to *slate, phyllite, schist,* and *gneiss* depending on the grade (intensity) of metamorphism from low-grade (slate) to medium-grade (phyllite, schist), to high-grade (gneiss). *Contact metamorphism* occurred in narrow zones next to the contact between parent rock and intrusive magma. Hydrothermal metamorphism occurred along fracture systems along which the fluids migrated through the rocks.

Every metamorphic rock has a **parent rock** (or *protolith*), the rock type that was metamorphosed. Parent rocks can be any of the three main rock types: igneous rock, sedimentary rock, or even metamorphic rock (i.e., metamorphic rock can be metamorphosed again), and the degree that a parent rock is metamorphosed can vary. As temperature and pressure increases, so does the metamorphic grade. **Metamorphic grade** refers to the intensity of metamorphism, from low grade (least intense metamorphism) to high grade (most intense metamorphism).

FIGURE 7.1 is a highly generalized illustration of metamorphism at part of a convergent plate boundary, where rocks were highly compressed at great depths within a mountain belt. A body of granitic magma also intruded part of the region. Note how the rocks were folded and changed. Mafic and ultramafic igneous rocks were metamorphosed to serpentinite. Sedimentary conglomerate, sandstone, and limestone parent rocks were metamorphosed to *metaconglomerate*, *quartzite*, and *marble*. Shale was metamorphosed to *slate*, *phyllite*, *schist*, and *gneiss*, depending on the grade of metamorphism from low-grade (slate) to medium-grade (phyllite, schist), to high-grade (gneiss). *Hornfels* formed only in a narrow zone of "contact" metamorphism next to the intrusion of magma. Watery hot fluids, called **hydrothermal fluids**, traveled along faults and fractures, where they leached chemicals from the rocks while hot and deposited mineral crystals as they cooled.

ACTIVITY 7.1 Metamorphic Rock Inquiry

THINK About it | **What do metamorphic rocks look like, and how can they be classified into groups?**

OBJECTIVE Analyze and describe samples of metamorphic rock, then infer how they can be classified into groups.

PROCEDURES

1. **Before you begin**, do not look up definitions and information. Use your current knowledge, and complete the worksheet with your current level of ability. Also, this is **what you will need** to do the activity:
   ____ Activity 7.1 Worksheet (p. 199) and pencil
   ____ optional: a set of metamorphic rock samples (obtained as directed by your instructor)
2. **Analyze the rocks, and complete the worksheet in a way that makes sense to you.**
3. **After you complete the worksheet**, be prepared to discuss your observations, interpretations, and inferences with others.

ACTIVITY

7.2 Metamorphic Rock Analysis and Interpretation

THINK About it — **What are metamorphic rocks composed of? How is composition used to classify and interpret igneous rocks?**

OBJECTIVE Be able to describe and interpret textural and compositional features of metamorphic rocks.

PROCEDURES

1. **Before you begin**, read about Metamorphic Processes and Rocks below. Also, this is **what you will need:**
   ___ Activity 7.2 Worksheets (pp. 200–201) and pencil
   ___ optional: a set of mineral samples (obtained as directed by your instructor)
   ___ optional: a set of mineral analysis tools (obtained as directed by your instructor)
2. **Then follow your instructor's directions** for completing the worksheets.

Agents of Metamorphism

Temperature, pressure, and hydrothermal fluids (watery hot fluids) are known as agents of metamorphism. Wherever metamorphism is occurring, one or more of these agents is involved in the metamorphic process.

Pressure Effects on Rocks

Confining Pressure is pressure (stress) applied equally in all directions (**FIGURE 7.2**). When you jump into a swimming pool you feel the confining pressure of the water pushing on every part of your body with equal force. If you dive down deep under the water, the pressure increases all around you. The same thing happens with rocks. Confining pressure increases with depth below Earth's surface and is equal in all directions. The deeper the rocks, the greater the confining pressure. This is what compacts rocks from sediment into sedimentary rock. The rock gets more dense because it is squeezed into less space. Unequal-sized

FIGURE 7.2 Effects of confining pressure (equal stress). As rocks get buried, they experience confining pressure that is equal in all directions (equal stress). The rocks become more dense as pore space is squeezed, may recrystallize to crystals of equal size, and remain nonfoliated.

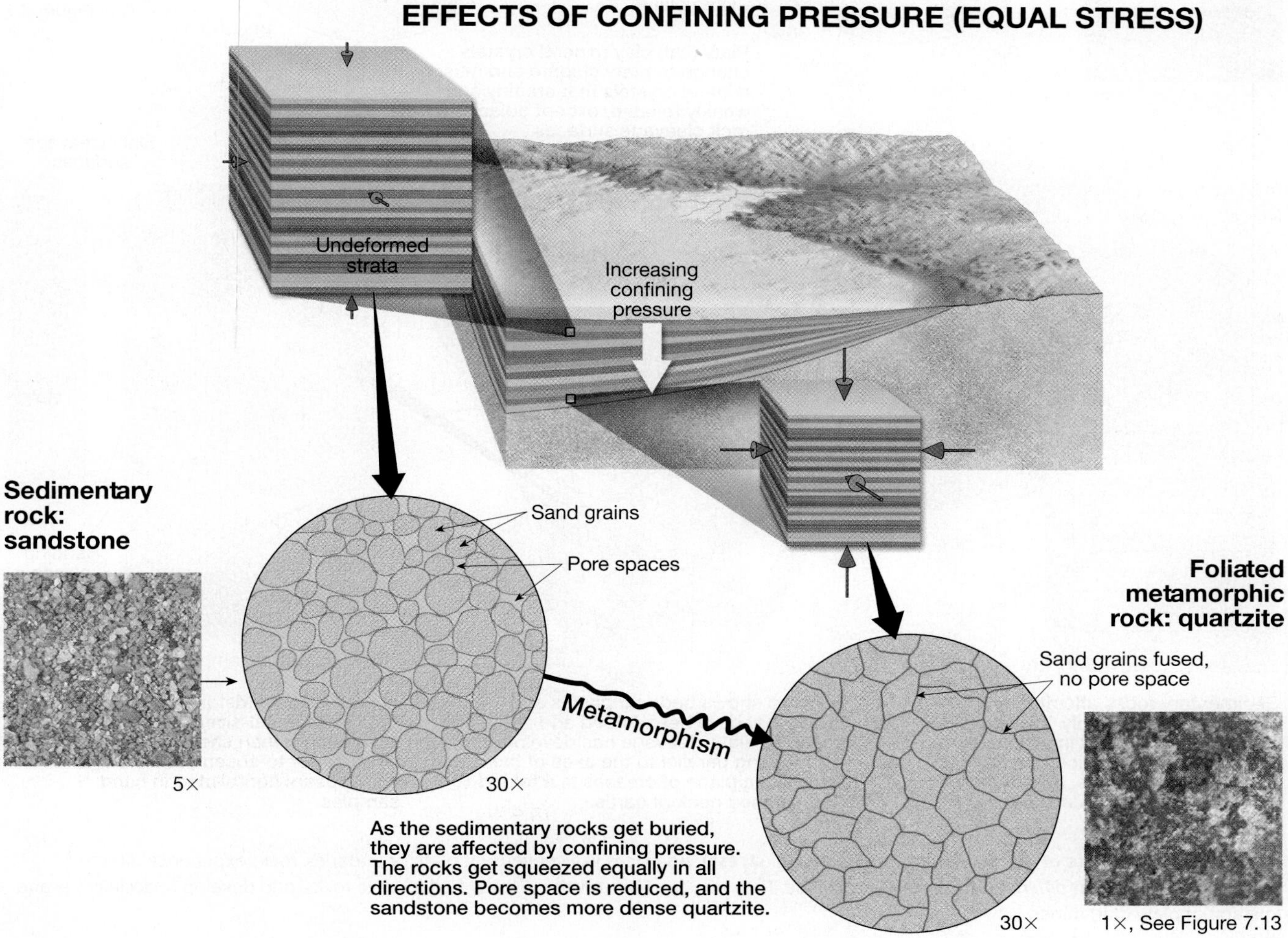

EFFECTS OF CONFINING PRESSURE (EQUAL STRESS)
Undeformed strata
Increasing confining pressure
Sedimentary rock: sandstone
Sand grains
Pore spaces
5×
30×
Metamorphism
Foliated metamorphic rock: quartzite
Sand grains fused, no pore space
As the sedimentary rocks get buried, they are affected by confining pressure. The rocks get squeezed equally in all directions. Pore space is reduced, and the sandstone becomes more dense quartzite.
30×
1×, See Figure 7.13

aragonite seashells or calcite mineral crystals will both recrystallize to a mass of small equal-sized crystals in the metamorphic rock called marble. Quartz sandstone becomes quartzite.

Directed pressure (differential stress) is pressure that is not equal in all directions. This causes the rock to get more compressed in one direction than any other (FIGURE 7.3). If you roll a lump of dough into a ball, then you are rolling and squeezing it equally in all directions to make the ball. But if you place the dough on a table and press on it with your hand, it gets squashed and shortened in the direction of the directed pressure. This causes flat minerals to get **foliated**—flatten out parallel to one another and perpendicular to the stress.

Directed stress occurs on a large scale at convergent plate boundaries, where the edges of two plates push together.

Temperature Effects on Rocks

Temperature is a measure of thermal energy. The greater the thermal energy, the higher the temperature and more energized the atoms and molecules are in the rock. When temperature exceeds 200°C (twice the boiling point of water), the molecules get highly energized. If the rock is under directed pressure, then it may fold in a ductile (like plastic) manner and become foliated. (FIGURE 7.4). Some bonds in the minerals begin to break and reform in more stable configurations. This may cause recrystallization or neomorphism.

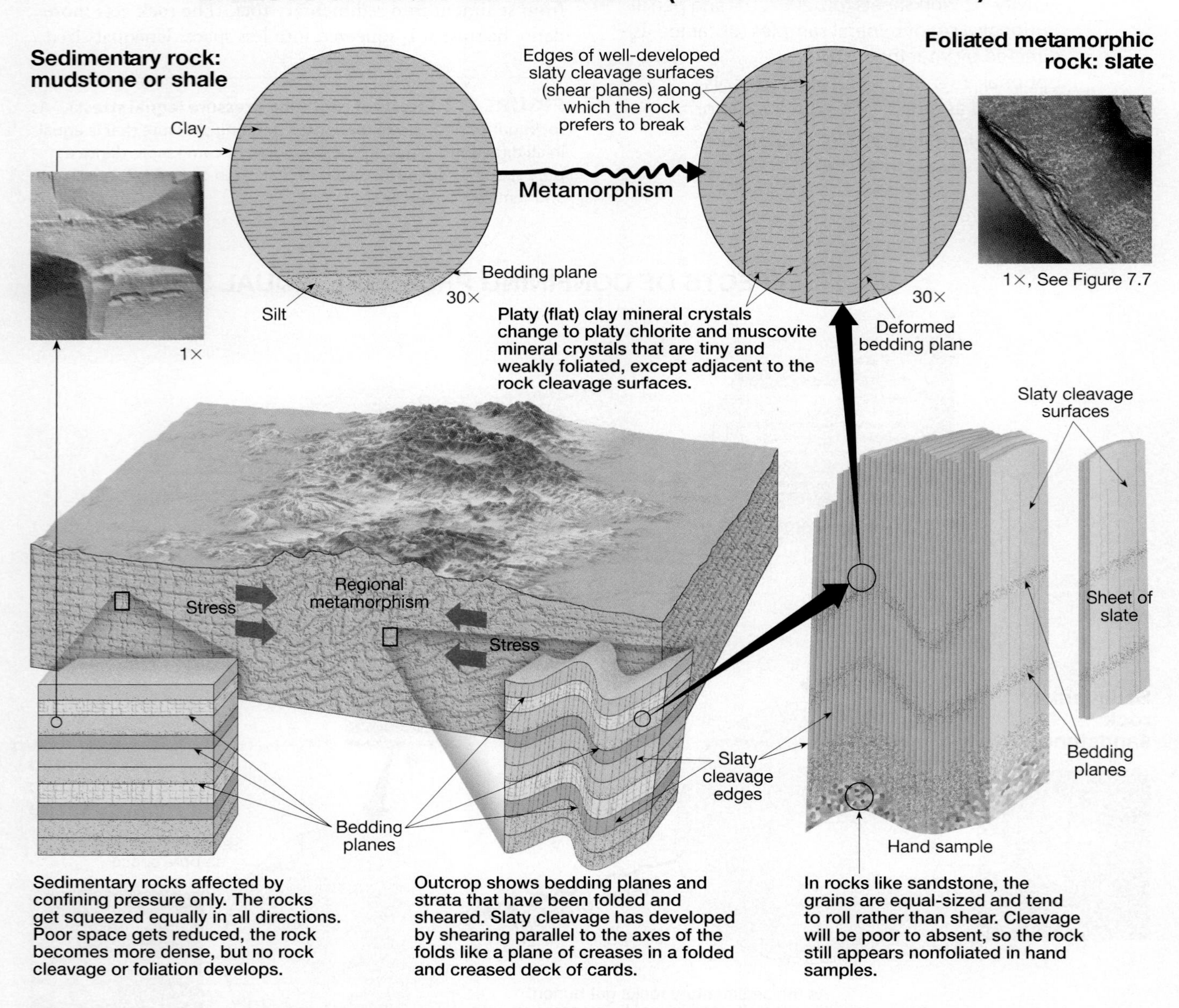

FIGURE 7.3 Effects of directed pressure (unequal stress). In places like convergent plate boundaries, rocks experience directed pressure, also known as *differential stress*, as they collide. They may fracture (brittle rocks) or fold (ductile rocks) and develop rock cleavage and foliation of platy (flat) minerals.

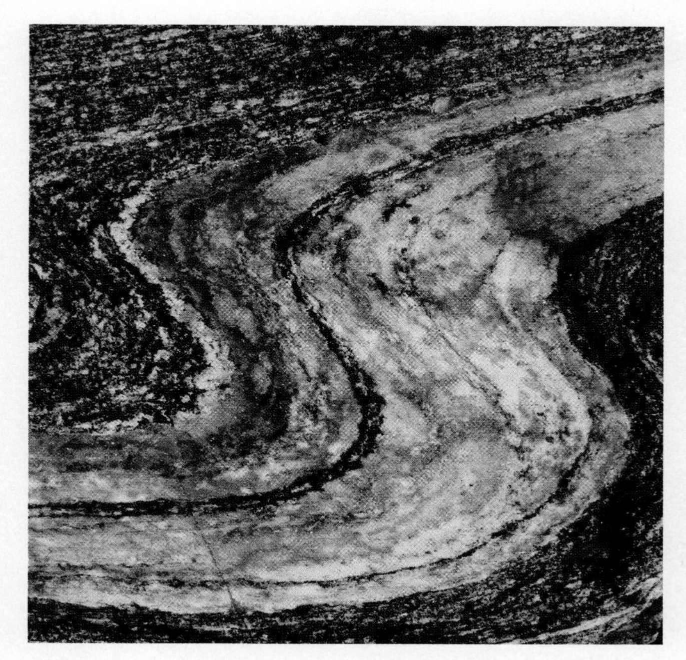

FIGURE 7.4 Folded and foliated (layered) gneiss. The dark minerals are muscovite, and the white minerals are quartz. Some of the quartz has been stained brown by iron. Regional metamorphism caused this normally rigid and brittle rock to be bent into *folds* without breaking. Pressure applied to the flat mica mineral grains has caused them to shear (slide parallel to and past one another) into layers called *foliations*. Metamorphic rocks with a layered appearance or texture are *foliated* metamorphic rocks. FIGURE 7.5 is a *nonfoliated* metamorphic rock because it lacks layering.

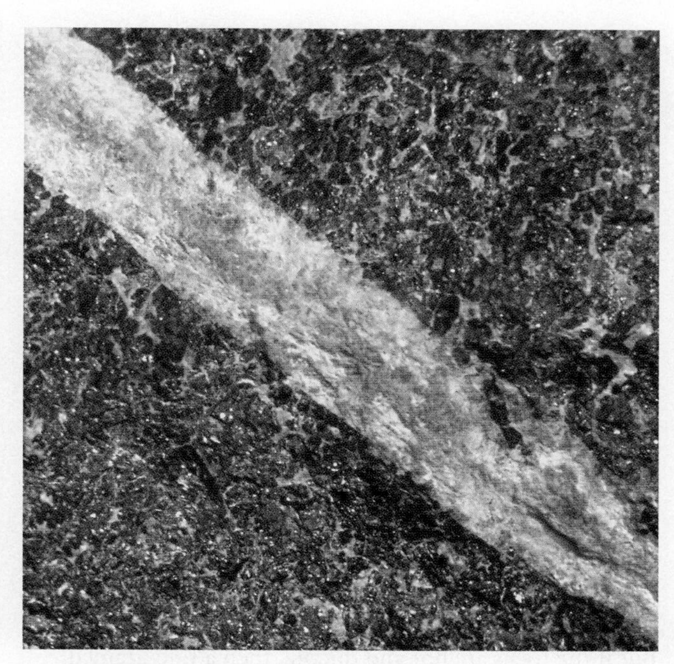

FIGURE 7.5 Hydrothermal mineral deposits. The dark part of this rock is chromite (chromium ore) that was precipitated from *hydrothermal fluids* (watery hot fluids). The light-colored minerals form a *vein of* zeolites (a group of light-colored hydrous aluminum silicates formed by low-grade metamorphism). The vein formed when directed pressure fractured the chromite deposit, hydrothermal fluids intruded the fracture, and the zeolites precipitated from the hydrothermal fluids as they cooled (making a *healed* fracture and a *vein* of zeolites).

Recrystallization is a process whereby unequal-sized crystals of one mineral slowly convert to equal-sized crystals of the same mineral, without melting of the rock. The longer the process continues, the larger the crystals become. For example, microscopic calcite crystals in chemical limestone (travertine, as in a cave stalactite) can recrystallize to form a mass of visible calcite crystals in metamorphic marble. Mineral composition of the rock stays the same, but texture of the rock changes.

Neomorphism is a process whereby mineral crystals not only recrystallize but also form different minerals from the same chemical elements. This happens when bonds of the original minerals break, and the chemical elements rearrange themselves into different crystalline structures and/or different molecules. For example, shales consisting mainly of clay minerals, quartz grains, and feldspar grains may change to a metamorphic rock consisting mainly of muscovite and garnet.

Hydrothermal Fluid Effects on Rocks

Just as hot water can cook vegetables and remove their color by breaking down molecules within them, it can also change the composition and form of rocks. Thus, water is an important agent of **metasomatism**, the loss or addition of new chemicals during metamorphism. Hornfels sometimes has a spotted appearance caused by the partial decomposition of just some of its minerals. In still other cases, one mineral may decompose (leaving only cavities or molds where its crystals formerly existed) and be simultaneously replaced by a new mineral of slightly or wholly different composition. When the hydrothermal fluids cool, minerals precipitate in the fractures and "heal" them (FIGURE 7.5).

Types of Metamorphism

Metamorphism can occur at different scales and in different types of environments.

Burial metamorphism is the most common type of metamorphism and occurs on a regional scale as rocks form and get buried. The metamorphism is caused by confining pressure (FIGURE 7.2).

Regional metamorphism, as the name implies, occurs on a regional scale, but the term now refers specifically to large-scale metamorphism at convergent plate boundaries, where there is directed pressure (differential stress) and high temperature that causes folding and foliation of the rocks. It is also called dynamothermal (pressure-temperature) metamorphism.

Contact metamorphism occurs locally, adjacent to igneous intrusions. It involves conditions of low to moderate pressure and intense heating. The intensity of contact metamorphism is greatest at the contact between parent rock and intrusive magma. The intensity then decreases rapidly over a short distance from the magma or hydrothermal fluids. Thus, zones of contact metamorphism are usually narrow, on the order of millimeters to tens-of-meters thick but some are kilometers wide.

The intruding magma thermally metamorphoses the rock in a narrow zone adjacent to the heat source (magma).

Hydrothermal metamorphism occurs along fractures that are in contact with the watery hot (hydrothermal) fluids. Like contact metamorphism, there is high heat and low pressure.

Dynamic metamorphism occurs along fault zones where there is local-to-regional shearing and crushing of rocks. If the rocks are brittle, then shearing produces fault breccia. But if the rocks are hot and ductile, then a fine-grained metamorphic rock called mylonite may result. Mylonite is a hard, dense, fine-grained rock that lacks cleavage but may have a banded coloration.

Minerals of Metamorphic Rocks

The **mineralogical composition** of a metamorphic rock is a description of the kinds and *relative* abundances of mineral crystals that make up the rock. Information about the relative abundances of the minerals is important for constructing a complete name for the rock and understanding metamorphic changes that formed the mineralogy of the rock.

Mineralogical composition of a parent rock may change during metamorphism as a result of changing pressure, changing temperature, and/or the chemical action of hydrothermal fluids, and processes like neomorphism and metasomatism. In general, as temperature and pressure increase, so does the **metamorphic grade**—the intensity of metamorphism, from low grade (least intense metamorphism) to high grade (most intense metamorphism). One group of minerals that was stable at a low temperature and/or pressure will eventually neomorphose to different minerals at a higher temperature and/or pressure. An **index mineral** is a mineral that is stable under a specific range of temperature and pressure and thus characterizes a grade of metamorphism (FIGURE 7.6).

Textures of Metamorphic Rocks

Texture of a metamorphic rock is a description of its constituent parts and their sizes, shapes, and arrangements. Two main groups of metamorphic rocks are distinguished on the basis of their characteristic textures, *foliated* and *nonfoliated.*

Foliated Metamorphic Rocks

Foliated metamorphic rocks (foliated textures) exhibit **foliations**—*layering* and parallel alignment of platy (flat) mineral crystals, such as micas. The foliations form when directed pressure causes the platy (flat) mineral crystals to slide parallel to and past one another (shear). This can happen as they recrystallize. Crystals of minerals such as tourmaline, hornblende, and kyanite can also be foliated because their crystalline growth occurred during metamorphism and had a preferred orientation in relation to the directed pressure. Specific kinds of foliated textures are described below:

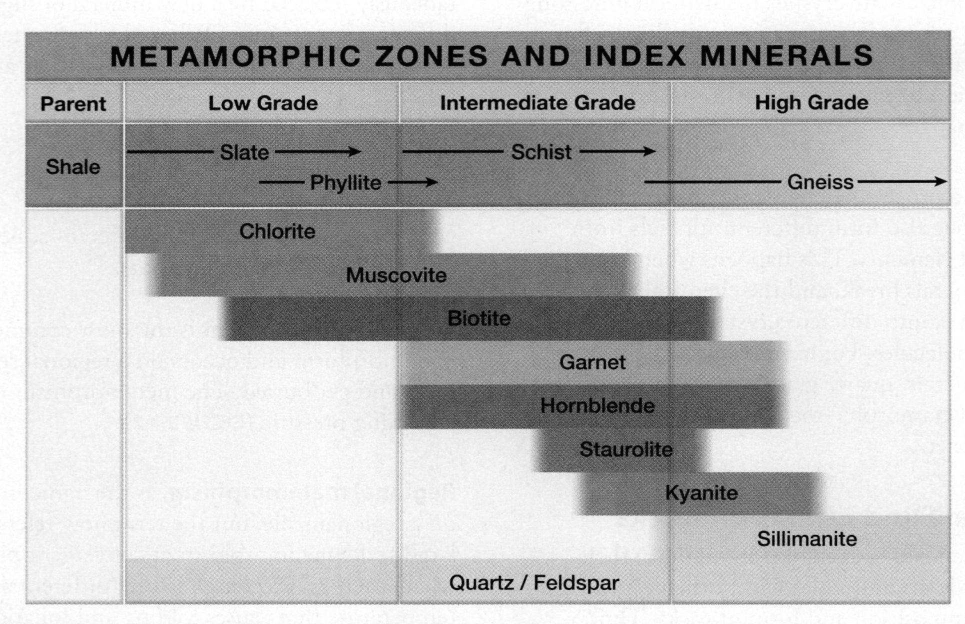

FIGURE 7.6 Index minerals of regionally-metamorphosed clay and mica-rich rocks. Sedimentary rocks rich in clay minerals neomorphose at low grades to larger foliated crystals of platy (flat) minerals like chlorite, muscovite, and biotite. These minerals neomorphose to garnet and staurolite at an intermediate grade, and then to sillimanite at a high grade of metamorphism.

- **Slaty rock cleavage**—*a very flat foliation* (resembling mineral cleavage) developed along flat, parallel, closely spaced shear planes (microscopic faults) in tightly folded clay- or mica-rich rocks (**FIGURE 7.3**). Rocks with excellent slaty cleavage are called *slate* (**FIGURE 7.7**), which is used to make roofing shingles and classroom blackboards. The flat surface of a blackboard or sheet of roofing slate is a slaty cleavage surface.
- **Phyllitic texture**—*a wavy and/or wrinkled foliation* of fine-grained *platy minerals* (mainly muscovite or chlorite crystals) that gives the rock a satiny or metallic luster. Rocks with phyllite texture are called *phyllite* (**FIGURE 7.8**). The phyllite texture is normally developed oblique or perpendicular to a weak slaty cleavage, and it is a product of intermediate-grade metamorphism.

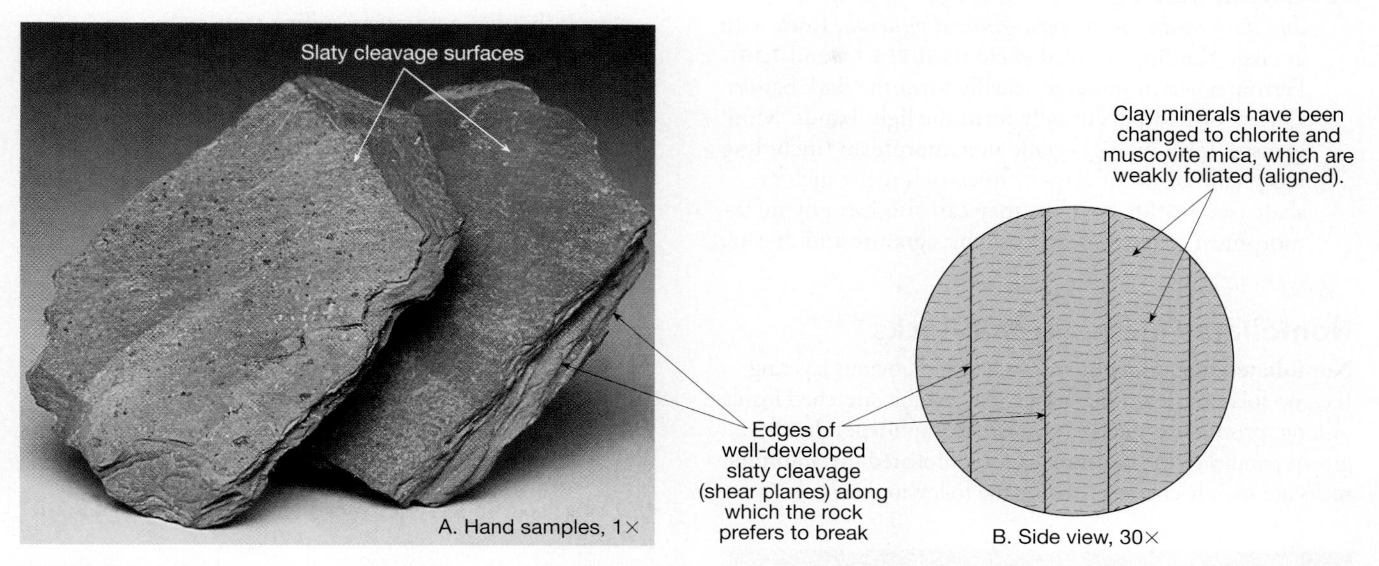

FIGURE 7.7 Slate. Slate is a foliated metamorphic rock with dull luster, excellent slaty cleavage, and no visible grains. Slate forms from low-grade metamorphism of mudstone (shale, claystone). Clay minerals of the mudstone parent rock change to foliated chlorite and muscovite mineral crystals. Slate splits into hard, flat sheets (usually less than 1 cm thick) along its well-developed *slaty cleavage* (**FIGURE 7.3**). It is used to make roofing shingles and classroom blackboards.

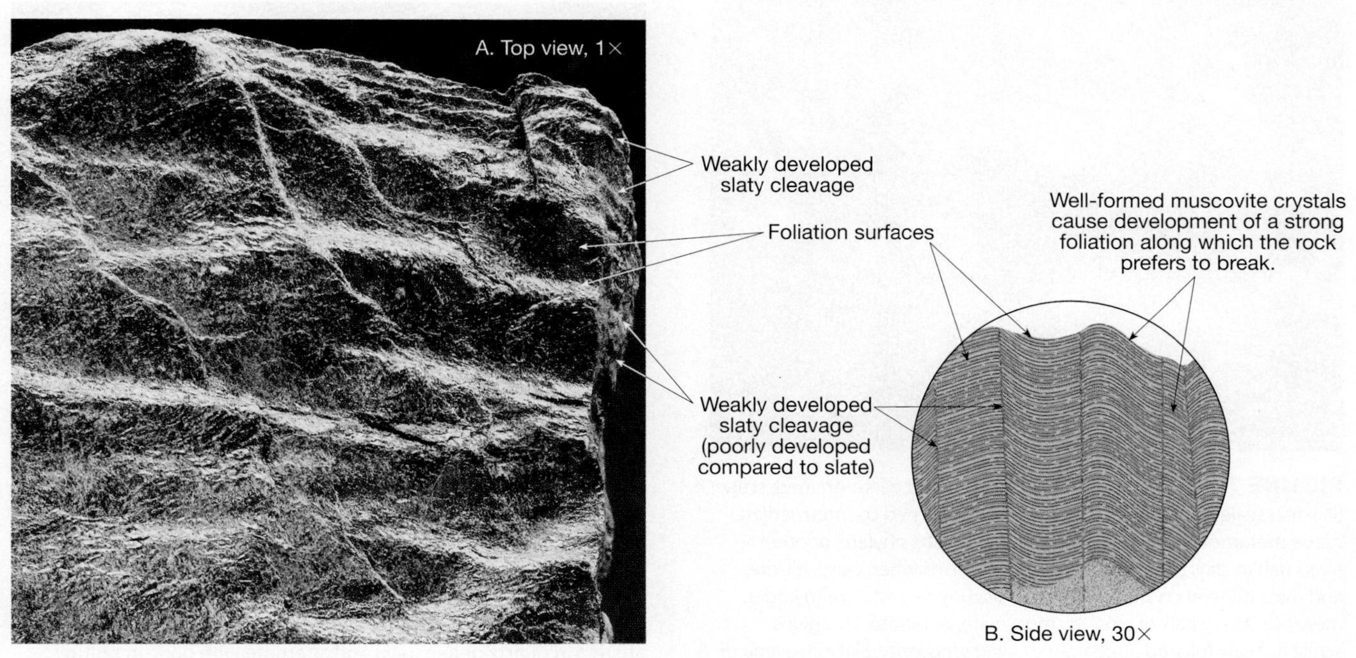

FIGURE 7.8 Phyllite. Phyllite is a foliated, fine-grained metamorphic rock, with a satiny, green, silver, or brassy metallic luster and a wavy foliation with a wrinkled appearance (*phyllite texture*). Phyllite forms from low-grade metamorphism of mudstone (shale, claystone), slate, or other rocks rich in clay, chlorite, or mica. When the very fine-grained mineral crystals of clay, chlorite, or muscovite in dull mudstone or slate are metamorphosed to form the phyllite, they become recrystallized to larger sizes and are aligned into a wavy and/or wrinkled foliation (*phyllite texture*) that is satiny or metallic. This is the wavy foliation along which phyllite breaks. Slaty cleavage may be poorly developed. It is not as obvious as the wavy and/or wrinkled foliation surfaces. The phyllite grade of metamorphism is between the low grade that produces slate (**FIGURE 7.7**) and the intermediate grade that produces schist (**FIGURE 7.9**).

- **Schistosity**—*a scaly glittery layering* of visible (medium- to coarse-grained) *platy minerals* (mainly micas and chlorite) *and/or linear alignment of long prismatic crystals* (tourmaline, hornblende, kyanite). Rocks with schistosity break along scaly, glittery foliations and are called *schist* (**FIGURE 7.9**). Schists are a product of intermediate-to-high grades of metamorphism.
- **Gneissic banding**—*alternating layers or lenses of light and dark medium- to coarse-grained minerals.* Rock with gneissic banding is called *gneiss* (**FIGURES 7.4** and **7.10**). Ferromagnesian minerals usually form the dark bands. Quartz or feldspars usually form the light bands. Most gneisses form by high-grade metamorphism (including recrystallization) of clay- or mica-rich rocks such as shale (see **FIGURE 7.1**), but they can also form by metamorphism of igneous rocks such as granite and diorite.

Nonfoliated Metamorphic Rocks

Nonfoliated metamorphic rocks have no obvious layering (i.e., no foliations), although they may exhibit stretched fossils or long, prismatic crystals (tourmaline, amphibole) that have grown parallel to the pressure field. Nonfoliated metamorphic rocks are mainly characterized by the following textures:

FIGURE 7.9 Schist. Schist is a medium- to coarse-grained, scaly (like fish scales), foliated metamorphic rock formed by intermediate-grade metamorphism of mudstone, shale, slate, phyllite, or other rocks rich in clay, chlorite, or mica. Schist forms when clay, chlorite, and mica mineral crystals are foliated as they recrystallize to larger, more visible crystals of chlorite, muscovite, or biotite. This gives schist its scaly foliated appearance called *schistosity*. Slaty cleavage or *crenulations* (sets of tiny folds) may be present, but schist breaks along its scaly, glittery schistosity. It often contains porphyroblasts of garnet, kyanite, sillimanite, or tourmaline mineral crystals. The schist grade of metamorphism is intermediate between the lower grade that produces phyllite (**FIGURE 7.8**) and the higher grade that produces gneiss (**FIGURES 7.10**). Also see chlorite schist in **FIGURE 7.15**.

- **Crystalline texture (nonfoliated)**—a medium- to coarse-grained aggregate of intergrown, usually equal-sized (equigranular), visible crystals. *Marble* is a nonfoliated metamorphic rock that typically exhibits an equigranular crystalline texture (**FIGURE 7.11**).
- **Microcrystalline texture**—a fine-grained aggregate of intergrown microscopic crystals (as in a sugar cube). *Hornfels* (**FIGURE 7.12**) is a nonfoliated metamorphic rock that has a microcrystalline texture.
- **Sandy texture**—a medium- to coarse-grained aggregate of fused, sand-sized grains that resembles sandstone. *Quartzite* is a nonfoliated metamorphic rock with a sandy texture (**FIGURE 7.13**) remaining from its sandstone parent rock, but the sand grains cannot be rubbed free of the rock because they are fused together.
- **Glassy texture**—a homogeneous texture with no visible grains or other structures and breaks along glossy surfaces; said of materials that resemble glass, such as *anthracite coal* (**FIGURE 7.14**).

Besides the main features that distinguish foliated and nonfoliated metamorphic rocks, there are some features that can occur in any metamorphic rock. They include the following:

- **Stretched or sheared grains**—deformed pebbles, fossils, or mineral crystals that have been stretched out, shortened, or sheared.

FIGURE 7.10 Gneiss. Gneiss is a medium- to coarse-grained metamorphic rock with *gneissic banding* (alternating layers or lenses of light and dark minerals). Generally, light-colored layers are rich in quartz or feldspars and alternate with dark in biotite mica, hornblende, or tourmaline. Most gneisses form by high-grade metamorphism (including recrystallization) of clay or mica-rich rocks such as shale (**FIGURE 7.1**), mudstone, slate, phyllite, or schist. However, they can also form by metamorphism of igneous rocks such as granite and diorite. The compositional name of the rock in this picture is biotite quartz gneiss.

FIGURE 7.11 Marble. Marble is a fine- to coarse-grained, nonfoliated metamorphic rock with a crystalline texture formed by tightly interlocking grains of calcite or dolomite. Marble forms by intermediate- to high-grade metamorphism of limestone or dolostone. Marble is a dense aggregate of nearly equal-sized crystals (see photograph), in contrast to the porous texture and/or odd-sized grains of its parent rock.

FIGURE 7.12 Hornfels. Hornfels is a fine-grained, nonfoliated metamorphic rock having a dull luster and a microcrystalline texture (that may appear smooth or sugary). It is usually very hard and dark in color, but it sometimes has a spotted appearance caused by patchy chemical reactions with the metamorphosing magma or hydrothermal fluid. Hornfels forms by contact metamorphism of any rock type.

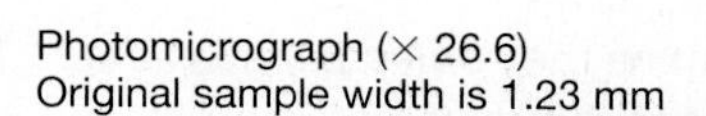

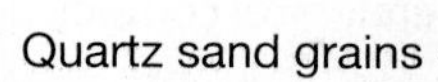

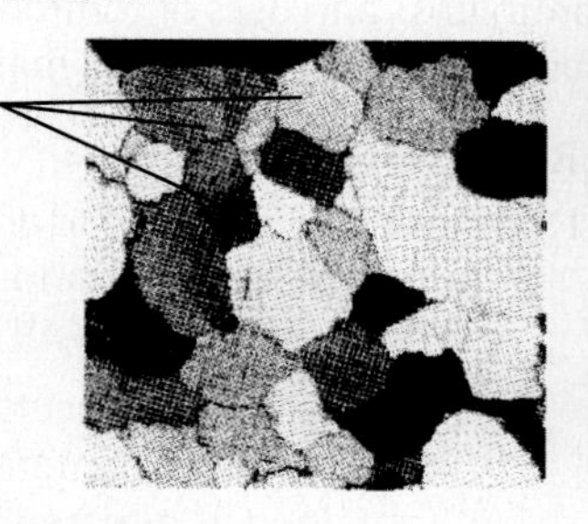

FIGURE 7.13 Quartzite. Quartzite is a medium- to coarse-grained, nonfoliated metamorphic rock consisting chiefly of fused quartz grains that give the rock its *sandy texture*. Compare the fused quartz grains of this quartzite sample (see photomicrograph) with the porous sedimentary fabric of quartz sandstone in **FIGURE 7.2**. Sand grains can often be rubbed from the edges of a sandstone sample, but never from quartzite (because the grains are fused together).

FIGURE 7.14 Anthracite coal. Anthracite is a fine-grained, nonfoliated metamorphic rock, also known as *hard coal* (because it cannot easily be broken apart like its parent rock, bituminous or soft coal). Anthracite has a smooth, homogeneous, glassy texture and breaks along glossy, curved (conchoidal) fractures. It is formed by low- to intermediate-grade metamorphism of bituminous coal, lignite, or peat.

FIGURE 7.15 Porphyroblastic texture. This texture is characterized by large, visible crystals of one mineral occur in a fine-grained groundmass of one or more other minerals. This medium-grained chlorite schist contains porphyroblasts of pyrite (brassy metallic cubes) in a groundmass of chlorite. The rock can be called porphyroblastic chlorite schist or pyrite chlorite schist or pyrite greenschist.

- **Porphyroblastic texture**—an arrangement of large crystals, called *porphyroblasts*, set in a finer-grained groundmass (FIGURE 7.15). It is analogous to porphyritic texture in igneous rocks.
- **Hydrothermal veins**—fractures "healed" (filled) by minerals that precipitated from hydrothermal fluids (see FIGURE 7.5).
- **Folds**—bends in rock layers that were initially flat, like a folded stack of paper (see FIGURE 7.4).
- **Lineations**—lines on rocks at the edges of foliations, shear planes, slaty cleavage, folds, or aligned crystals.

Classification of Metamorphic Rocks

Metamorphic rocks are mainly classified according to their texture and mineralogical composition. This information is valuable for naming the rock and determining how it formed from a parent rock (protolith). It is also useful for inferring how the metamorphic rock could be used as a commodity for domestic or industrial purposes. You can analyze and classify metamorphic rocks with the aid of FIGURE 7.16, which also provides information about parent rocks and how the metamorphic rocks are commonly used.

ACTIVITY

7.3 Hand Sample Analysis, Classification, and Origin

THINK About It | **How are rock composition and texture used to classify, name, and interpret metamorphic rocks?**

OBJECTIVE Determine the names, parent rocks (protoliths), and uses of common metamorphic rocks, based on their textures and mineralogical compositions.

PROCEDURES

1. **Before you begin**, read about Description and Interpretation of Metamorphic Rock Samples. Also, this is **what you will need:**
   ___ Activity 7.3 Worksheets (pp. 202–204) and pencil
   ___ optional: a set of metamorphic rock samples (obtained as directed by your instructor)
2. **Then follow your instructor's directions** for completing the worksheets.

METAMORPHIC ROCK ANALYSIS AND CLASSIFICATION

		STEP 1: What are the rock's textural features?	STEP 2: What are the rock's mineralogical composition and/or other distinctive features?	STEP 3: Metamorphic rock name	STEP 4: What was the parent rock?	STEP 5: What is the rock used for?
FOLIATED	Fine-grained or no visible grains	Flat slaty cleavage is well developed	Dull luster; breaks into hard flat sheets along the slaty cleavage	SLATE[1] (INCREASING METAMORPHIC GRADE ↓)	Mudstone or shale	Roofing slate, table tops, floor tile, and blackboards
FOLIATED	Fine-grained or no visible grains	Phyllite texture well developed more than slaty cleavage	Breaks along wrinkled or wavy foliation surfaces with shiny metallic luster	PHYLLITE[1]	Mudstone, shale, or slate	Construction stone, decorative stone, sources of gemstones
FOLIATED	Medium- to coarse-grained	Schistosity: foliation formed by alignment of visible crystals; rock breaks along scaly foliation surfaces; crystalline texture	Mostly blue or violet needle-like crystals (blue amphibole)	Blueschist (SCHIST[1])	Mudstone, shale, slate, or phyllite	
			Mostly visible sparkling crystals of chlorite +/– actinolite (green amphibole)	Greenschist (SCHIST[1])		
			Mostly visible sparkling crystals of muscovite	Muscovite schist (SCHIST[1])		
			Mostly visible sparkling crystals of biotite	Biotite schist (SCHIST[1])		
FOLIATED	Medium- to coarse-grained	Gneissic banding: minerals segregated into alternating layers gives the rock a banded texture in side view; crystalline texture	Visible crystals of two or more minerals in alternating light and dark foliated layers	GNEISS[1]	Mudstone, shale, slate, phyllite, schist, granite, or diorite	Construction stone, decorative stone, sources of gemstones
FOLIATED OR NONFOLIATED		Medium- to coarse-grained crystalline texture	Mostly visible glossy black amphibole (hornblende) in blade-like crystals	AMPHIBOLITE	Basalt, gabbro, or ultramafic igneous rocks	Construction stone
FOLIATED OR NONFOLIATED		Crystalline texture	Green pyroxene + red garnet	ECLOGITE	Basalt, gabbro	Titanium ore
NONFOLIATED	Fine-grained or no visible grains	Glassy texture; slaty cleavage may barely be visible	Black glossy rock that breaks along uneven or conchoidal fractures (Figure 7.12)	ANTHRACITE COAL	Peat, lignite, bituminous coal	Highest grade coal for clean burning fossil fuel
NONFOLIATED	Fine-grained or no visible grains	Microcrystalline texture	Usually a dull dark color; very hard	HORNFELS	Any rock type	Decorative stone
NONFOLIATED	Fine-grained or no visible grains	Microcrystalline texture or no visible grains. May have fibrous asbestos form	Serpentine; dull or glossy; color usually shades of green	SERPENTINITE	Basalt, gabbro, or ultramafic igneous rocks	Decorative stone
NONFOLIATED	Fine-grained or no visible grains	Microcrystalline or no visible grains	Talc; can be scratched with your fingernail; shades of green, gray, brown, white	SOAPSTONE	Basalt, gabbro, or ultramafic igneous rocks	Art carvings, electrical insulators, talcum powder
NONFOLIATED	Fine- to coarse-grained	Sandy texture	Quartz sand grains fused together; grains will not rub off like sandstone; usually light colored	QUARTZITE[1]	Sandstone	Construction stone, decorative stone
NONFOLIATED	Fine- to coarse-grained	Microcrystalline (resembling a sugar cube) or medium to coarse crystalline texture	Calcite (or dolomite) crystals of nearly equal size and tightly fused together; calcite effervesces in dilute HCl; dolomite effervesces only if powdered	MARBLE[1]	Limestone	Art carvings, construction stone, decorative stone, source of lime for agriculture
NONFOLIATED	Fine- to coarse-grained	Conglomeratic texture, but breaks across grains	Pebbles may be stretched or cut by rock cleavage	META-CONGLOMERATE	Conglomerate	Construction stone, decorative stone

[1]Modify rock name by adding names of minerals in order of increasing abundance. For example, garnet muscovite schist is a muscovite schist with a small amount of garnet.

FIGURE 7.16 Five-step chart for metamorphic rock analysis and classification. See text for description of steps (page 198).

Description and Interpretation of Metamorphic Rock Samples

The complete classification of a metamorphic rock requires knowledge of its composition, texture(s), and other distinctive properties. ***Follow these steps to analyze and classify a metamorphic rock:***

Step 1: *Determine and record the rock's textural features.* Determine and record if the rock is foliated or nonfoliated, and what other specific kinds of textural features are present. Use this information to work from left to right across the three columns of Step 1 in **FIGURE 7.16**, and match the rock texture to one of the specific categories there.

Step 2: *Determine and record the rock's mineralogical composition and/or other distinctive features.* List the minerals in order of increasing abundance, and distinguish between porphyroblasts and mineralogy of the groundmass making up most of the rock. Use this information and any other distinctive features to match the rock to one of the categories in Step 2 of **FIGURE 7.16**.

Step 3: Recall how you categorized the rock in Steps 1 and 2. Use this information to work from left to right across **FIGURE 7.16** and determine the name of the rock. You can also modify the rock name by adding the names of minerals present in the rock in order of their increasing abundance. If the rock is porphyroblastic, then you can add this to the name as well (e.g., **FIGURE 7.15**).

Step 4: After you have determined the metamorphic rock name in Step 3, look to the right along the same row of **FIGURE 7.16** and find the name of a parent rock (protolith) for that kind of metamophic rock.

Step 5: After you have determined the parent rock in Step 4, look to the right along the same row of **FIGURE 7.16** and find out what the rock is commonly used for.

Example:
Study the metamorphic rock in **FIGURE 7.4**. Step 1—The rock has obvious layering, so it is foliated. Step 2—The rock has alternating layers of light and dark colored minerals. Step 3—By moving from left to right across the chart in **FIGURE 7.16**, notice that the rock name is gneiss. Step 4—The parent rock for gneiss is mudstone, shale, slate, phyllite, schist, granite, or diorite. Step 5—Gneiss is commonly used as a construction stone, decorative stone, or as the source of some gemstones.

ACTIVITY

7.4 Metamorphic Grades and Facies

THINK About It **What can metamorphic rocks tell us about Earth's history and the environments in which the rocks formed?**

OBJECTIVE Infer regional geologic history and the relationship of metamorphic facies to plate tectonics using index minerals, pressure-temperature diagrams, and geologic maps.

PROCEDURES

1. **Before you begin**, read about Using Index Minerals to Interpret Metamorphic Grades and Facies. Also, this is **what you will need:**

   ___ Activity 7.4 Worksheets (pp. 205–206) and pencil

2. **Then follow your instructor's directions** for completing the worksheets.

Using Index Minerals to Interpret Metamorphic Grades and Facies

The changes in metamorphic grade that are indicated by textural changes, like schist to gneiss, are also accompanied by mineralogical changes. Minerals that indicate specific grades of metamorphism are called *index minerals* (**FIGURE 7.6**). Assemblages of index minerals make up metamorphic facies, which can be interpreted using pressure-temperature diagrams.

ACTIVITY 7.1 Metamorphic Rock Inquiry

Name: ______________________ **Course/Section:** ______________________ **Date:** ____________

A. Analyze the metamorphic rocks below (and actual rock samples of them if available). Beneath or beside each picture, write words that describe the rock's **composition** (what it is made of), and **texture** (the size, shape, and arrangement of its parts). Use your current knowledge, and complete the worksheet with your current level of ability. Do not look up terms or other information. All samples x1.

1

2

3

4

5

6

B. REFLECT & DISCUSS Reflect on your observations and descriptions of sedimentary rocks in part A. Then describe how you would classify the rocks into groups. Be prepared to discuss your classification with other geologists.

ACTIVITY 7.2 Metamorphic Rock Analysis and Interpretation

Name: ____________________ **Course/Section:** ____________________ **Date:** ____________

A. Analyze these samples of sedimentary limestone and metamorphic marble.

Limestone ×1

Marble ×1

1. These rocks are both composed of the same mineral. What is it? ____________________
What test could you perform on the rocks to be sure?

2. How is the texture of these two rocks different?

B. The sequence of increasing grades of metamorphism of a mudstone (shale, claystone, or siltstone) parent rock (protolith) is: slate (lowest grade), phyllite, schist, gneiss (highest grade).

Slate ×1

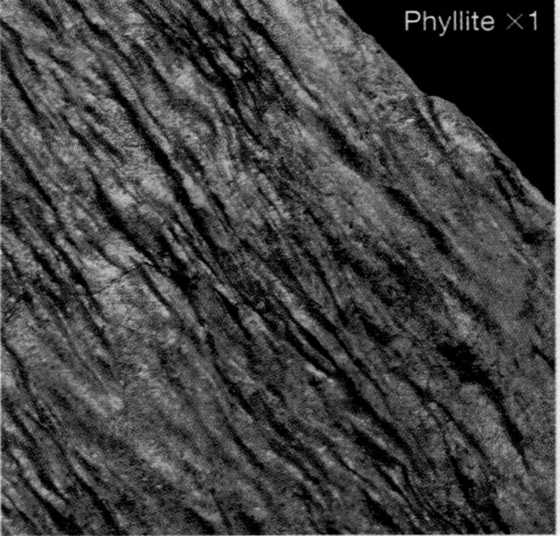

Phyllite ×1

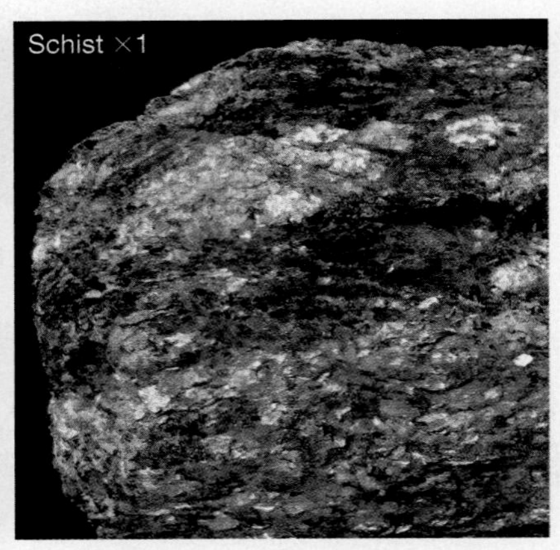

Schist ×1

1. Describe the change in grain size from slate to schist.

2. How does the texture of phyllite differ from that of schist?

3. Why do you think that the micas (flat minerals) in these rocks are all parallel, or nearly so, to one another?

C. Analyze the rock sample in **FIGURE 7.3**. The parent rock for this metamorphic rock had flat layers that were folded during metamorphism. Describe a process that could account for how this rigid gneiss was folded without breaking during regional metamorphism. (*Hint*: How could you bend a brittle candlestick without breaking it?)

D. Analyze this metamorphic rock sample using the classification chart in **FIGURE 7.16**.

1. Is this rock foliated or nonfoliated?

2. Notice that the rock consists mostly of muscovite but also contains scattered garnet crystals. What is the name for this kind of texture?

3. What is the name of this metamorphic rock?

4. What was the likely parent rock (protolith)?

E. Analyze this metamorphic rock sample using the classification chart in **FIGURE 7.16**.

1. Is this rock foliated or nonfoliated and how can you tell?

2. What is the name of this metamorphic rock?

3. What was the likely parent rock (protolith)?

F. **REFLECT & DISCUSS** Which one of the rocks in this activity do you think has the highest metamorphic grade? Explain your answer.

Name: ______ Course/Section: ______ Date: ______

ACTIVITY 7.3 Hand Sample Analysis, Classification, and Origin

METAMORPHIC ROCKS WORKSHEET

Sample Letter or Number	Texture(s) (Figure 7.16—Step 1)	Mineral Composition and Other Distinctive Properties (Figure 7.16, Step 2)	Rock Name (Figure 7.16, Step 3)	Parent Rock (Figure 7.16, Step 4)	Uses (Figure 7.16, Step 5)
301	☐ foliated ☒ nonfoliated Sugary Sandy	light colored quartz sandgrains fuzed together	Quartzite	Sandstone	
Black 302	☒ foliated ☐ nonfoliated Schositiosity	mostly visible sparkling crystal of biotite	Biotite Schist	mudstone Shale Slate Phyllite	
Gold 303	☒ foliated ☐ nonfoliated Schositiosity	mostly visible sparking crystalls of muslonite	Muscorite Schist	mudstone Shale Slate Phyllite	
Grey 304	☒ foliated ☐ nonfoliated Schositiosity	Small glass grains mostly visible crystals	garnet Schist	mudstone Shale Slate phyllite	
Clay 305	☒ foliated ☐ nonfoliated flat slaty clevage	Dull luster hard flat sheets	Slate	mudstone shale	

Name: ______________________ Course/Section: ______________________ Date: ____________

METAMORPHIC ROCKS WORKSHEET

Sample Letter or Number	Texture(s) (Figure 7.16—Step 1)	Mineral Composition and Other Distinctive Properties (Figure 7.16, Step 2)	Rock Name (Figure 7.16, Step 3)	Parent Rock (Figure 7.16, Step 4)	Uses (Figure 7.16, Step 5)
306 (really packed)	☒ foliated ☐ nonfoliated gneissic	visible crystals alternating dark & light foliated layers	gneiss	Shale/Schist mudstone Phyllite Slate granite diorite	
307 (white rock)	☐ foliated ☒ nonfoliated Microcrystalline texture	- reacts to acid - calcite crystals	Marble	Limestone	
308 (fossil rock)	☒ foliated ☐ nonfoliated Phyllite texture more than slaty cleavage	Shiny metallic luster Breaks along wrinkled or wavy foliated surface	Phyllite	mudstone Shale Slate	
	☐ foliated ☐ nonfoliated				
	☐ foliated ☐ nonfoliated				

Name: ______________________ Course/Section: ______________ Date: __________

METAMORPHIC ROCKS WORKSHEET

Sample Letter or Number	Texture(s) (Figure 7.16—Step 1)	Mineral Composition and Other Distinctive Properties (Figure 7.16, Step 2)	Rock Name (Figure 7.16, Step 3)	Parent Rock (Figure 7.16, Step 4)	Uses (Figure 7.16, Step 5)
	☐ foliated ☐ nonfoliated				
	☐ foliated ☐ nonfoliated				
	☐ foliated ☐ nonfoliated				
	☐ foliated ☐ nonfoliated				
	☐ foliated ☐ nonfoliated				

ACTIVITY 7.4 Metamorphic Grades and Facies

Name: ______________________ **Course/Section:** ______________ **Date:** __________

A. How much a parent rock (protolith) is metamorphosed is called its metamorphic **grade** and varies from low grade (low temperature and pressure) to high grade (high temperature and pressure). British geologist George Barrow mapped rocks in the Scottish Highlands that were metamorphosed by granitic igneous intrusions. He discovered that as he walked away from the granitic intrusive igneous rock, there was a sequence of mineral zones from the high grade to the low grade of metamorphism. He defined the following sequence of **index minerals**, which represent degrees of metamorphism along a gradient from low grade to high grade:

Chlorite (lowest grade), biotite, garnet, staurolite, kyanite, sillimanite (highest grade)

1. Boundaries between the index mineral zones of metamorphism are called isograds and represent lines/surfaces of equal metamorphic grade. In the geologic map below, color in the zone of maximum metamorphic grade.

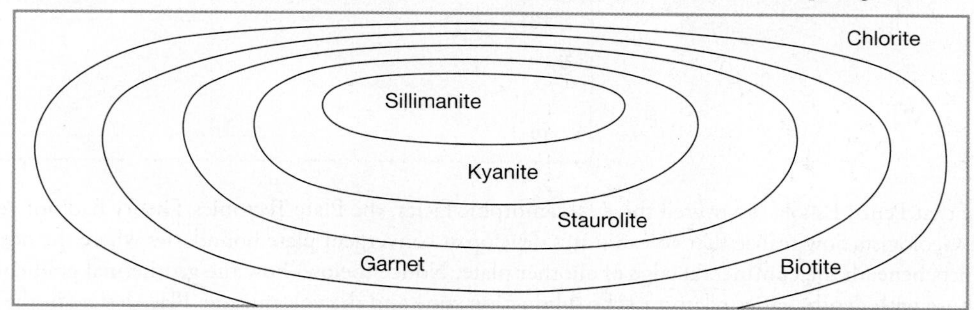

B. Most metamorphism is caused by increases of both temperature and pressure. Geologists represent these relationships on pressure-temperature (P-T) diagrams showing the stability of different index minerals. On this diagram andalusite, kyanite, and sillimanite are *polymorphs*, minerals that have the same chemical composition but different crystalline structure and physical properties that can be used to distinguish them. Note that any two of these minerals can occur together only along lines, and that the three minerals can only occur together at one specific point in temperature and pressure, 500˚C and 4 kilobars, which normally occurs about 15 km below Earth's surface.

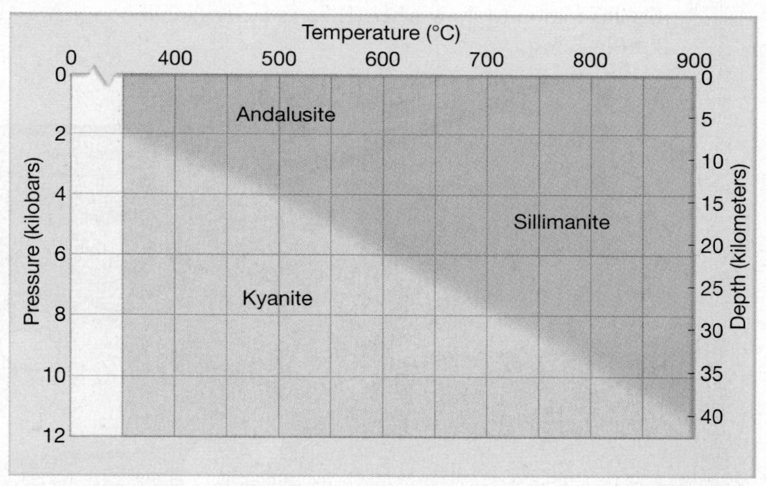

1. Study the mineral zones and isograds on the two maps below. Which region was metamorphosed at higher pressure, and how can you tell?

2. What was the minimum temperature at which the rocks in Map B were metamorphosed? ______________

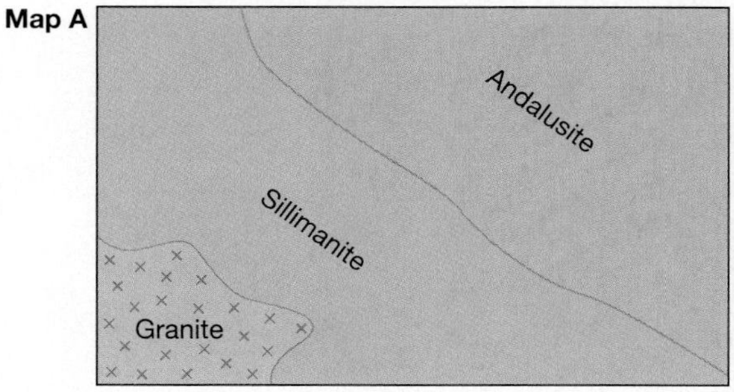

Map A

Andalusite

Sillimanite

Granite

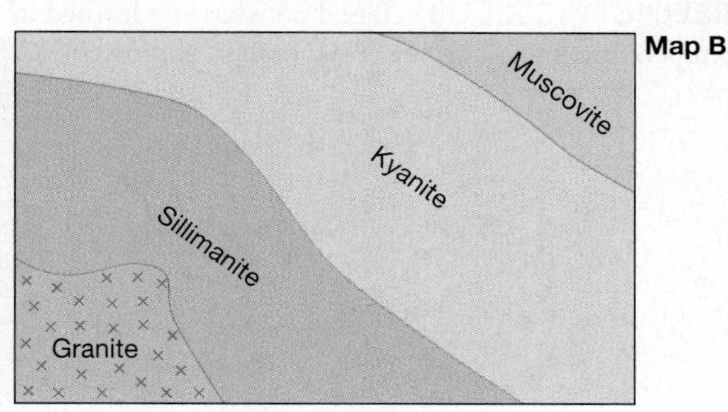

C. Finnish geologist, Pentti Eskola, first recognized in 1921 that basalt volcanic rock could be metamorphosed to distinctly different **metamorphic facies** (unique assemblages of several minerals) under changing conditions of temperature and pressure (depth).

- Amphibolite facies (low pressure, high temperature): black hornblende amphibole, sillimanite
- Greenschist facies (low pressure, low temperature): green actinolite amphibole and chlorite
- Eclogite facies (high pressure, high temperature): red garnet, green pyroxene
- Blueschist facies (high pressure, low temperature): blue amphibole (glaucophane, riebeckite)

1. Write the names of these metamorphic facies where they would occur in this pressure-temperature diagram.

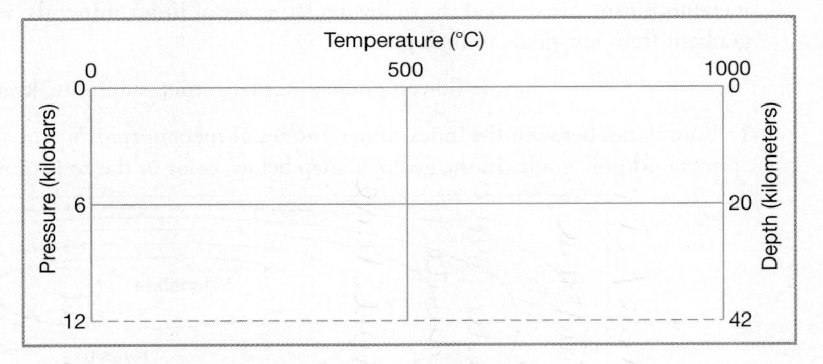

2. At the time that Pentti Eskola discovered these metamorphic facies, the Plate Tectonics Theory had not yet been developed. Geologists now realize that volcanic arcs develop at convergent plate boundaries where the oceanic edge of one plate subducts beneath the continental edge of another plate. Notice (below) how the geothermal gradient (rate of change in temperature with depth) varies relative to the subduction zone and the volcanic arc. Place letters in the white spaces on this illustration to show where Eskola's facies should occur: A = Amphibolite, G = Greenschist, E = Eclogite, B = Blueschist.

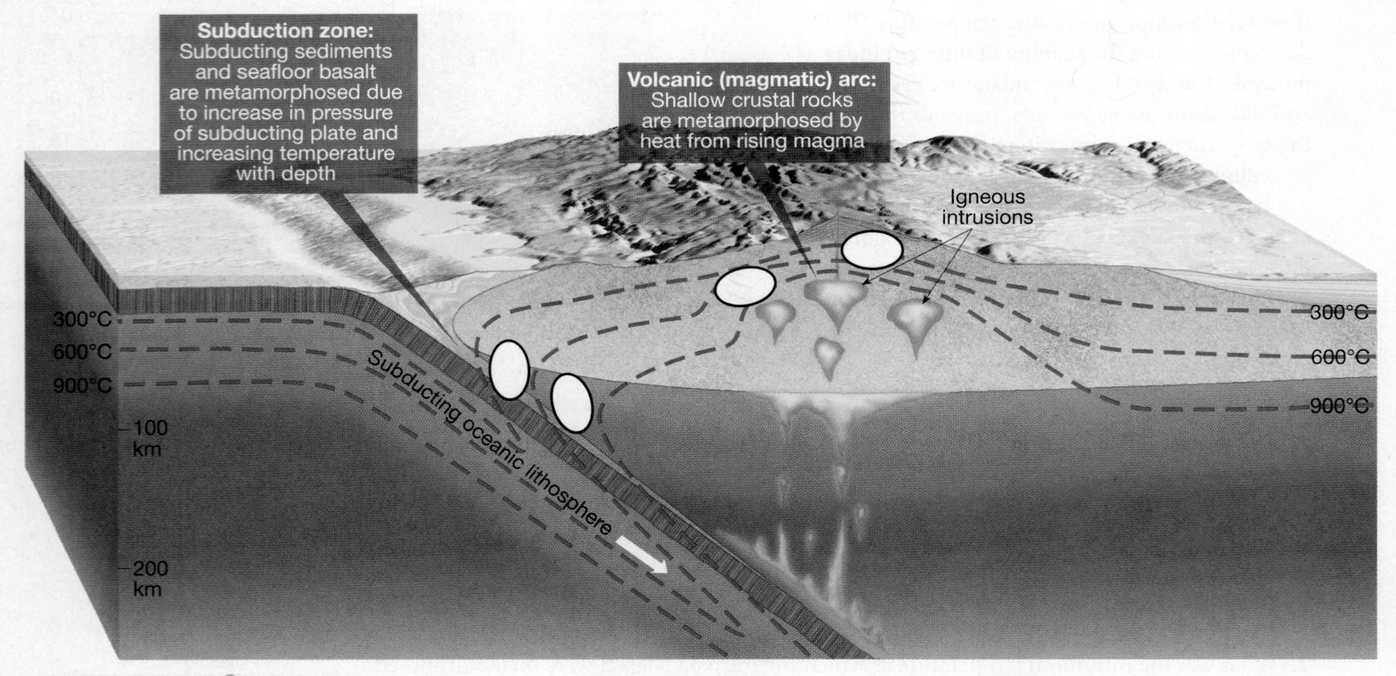

3. **REFLECT & DISCUSS** Based on what you learned in this activity, write a generalization that expresses how pressure, temperature, metamorphic environment, and metamorphic grade are related.

LABORATORY 8

Dating of Rocks, Fossils, and Geologic Events

CONTRIBUTING AUTHORS
Jonathan Bushee • *Northern Kentucky University*
John K. Osmond • *Florida State University*
Raman J. Singh • *Northern Kentucky University*

Fossil ferns, 310 million years old, from the Pennsylvanian (Carboniferous) System of rocks, Pottsville, Pennsylvania (x1). (Richard M. Busch)

BIG IDEAS

Geologists use relative and absolute dating techniques to infer the ages of geologic features and events in geologic history. Relative age dating is the process of determining what happened first, second, and so on, in relation to other geologic features and events. Absolute age dating is the process of determining when something formed or happened in exact units of time such as days, months, or years. The "geologic time scale" is a chart showing the chronological sequence (relative ages) of named rock units and cooresponding divisions of relative time arranged next to a scale of absolute age in years.

FOCUS YOUR INQUIRY

THINK About It How can you tell relative age relationships among the parts of geologic cross sections exposed in outcrops?

ACTIVITY 8.1 **Geologic Inquiry for Relative Age Dating** *(p. 208)*

THINK About It How can geologic cross sections be interpreted to establish the relative ages of rock units, contacts, and other geologic features?

ACTIVITY 8.2 **Determining Sequence of Events in Geologic Cross Sections** *(p. 208)*

THINK About It How are fossils used to tell geologic time and infer Earth's history?

ACTIVITY 8.3 **Using Index Fossils to Date Rocks and Events** *(p. 212)*

THINK About It How do geologists determine the absolute age, in years, of Earth materials and events?

ACTIVITY 8.4 **Absolute Dating of Rocks and Fossils** *(p. 214)*

THINK About It How are relative and absolute dating techniques used to analyze outcrops and infer geologic history?

ACTIVITY 8.5 **Infer Geologic History from a New Mexico Outcrop** *(p. 216)*

ACTIVITY 8.6 **CSI (Canyon Scene Investigation) Arizona** *(p. 216)*

ACTIVITY 8.1 Geologic Inquiry for Relative Age Dating

THINK About It | **How can you tell relative age relationships among the parts of geologic cross sections exposed in outcrops?**

OBJECTIVE Identify features of geologic cross sections exposed in outcrops, infer their relative ages, and suggest rules for relative age dating.

PROCEDURES

1. **Before you begin**, do not look up definitions and information. Use your current knowledge, and complete the worksheet with your current level of ability. Also, this is **what you will need** to do the activity:
   ____ Black or blue pen
   ____ Activity 8.1 Worksheet (p. 217) and pencil
2. **Complete the worksheet in a way that makes sense to you.**
3. **After you complete the worksheet**, be prepared to discuss your observations, interpretations, and inferences with others.

ACTIVITY 8.2 Determining Sequence of Events in Geologic Cross Sections

THINK About It | **How can geologic cross sections be interpreted to establish the relative ages of rock units, contacts, and other geologic features?**

OBJECTIVE Apply principles of relative age dating to analyze and interpret sequences of events in geologic cross sections.

PROCEDURES

1. **Before you begin**, read the Introduction and Relative Age Dating Based on Physical Relationships. Also, this is **what you will need:**
   ____ dark (black or blue) pen
   ____ Activity 8.2 Worksheet (p. 219) and pencil
2. **Then follow your instructor's directions** for completing the worksheets.

Introduction

If you could dig a hole deep into Earth's crust, you would encounter the **geologic record,** layers of rock stacked one atop the other like pages in a book. As each new layer of sediment or rock forms today, it covers the older layers of the geologic record beneath it and becomes the youngest layer of the geologic record. Thus, rock layers form a *sequence* from oldest at the bottom to youngest at the top. They also have different colors, textures, chemical compositions, and **fossils** (any evidence of ancient life) depending on the environmental conditions under which they were formed. Geologists have studied sequences of rock layers wherever they are exposed in mines, quarries, river beds, road cuts, wells, and mountain sides throughout the world. They have also *correlated* the layers (traced them from one place to another) across regions and continents. Thus, the geologic record of rock layers is essentially a stack of stone pages in a giant natural book of Earth's history. And like the pages in any old book, the rock layers have been folded, fractured (cracked), torn (faulted), and even removed by geologic events.

Geologists tell time based on relative and absolute dating techniques. **Relative age dating** is the process of determining when something formed or happened in relation to other events. For example, if you have a younger brother and an older sister, then you could describe your relative age by saying that you are younger than your sister and older than your brother. **Absolute age dating** is the process of determining when something formed or happened in exact units of time such as days, months, or years. Using the example above, you could describe your absolute age just by saying how old you are in years.

Geologists "read" and infer Earth's history from rocky outcrops and geologic cross sections by observing rock layers, recognizing geologic structures, and evaluating age relationships among the layers and structures. The so-called *geologic time scale* is a chart of named intervals of the geologic record and their ages in both relative and absolute time. It has taken thousands of geoscientists, from all parts of the world, more than a century to construct the present form of the geologic time scale.

Relative Age Dating Based on Physical Relationships

A geologist's initial challenge in the field is to subdivide the local sequence of sediments and bodies of rock into mappable units that can be correlated from one site to the next. Subdivision is based on color, texture, rock type, or other physical features of the rocks, and the mappable units are called **formations.** Formations can be subdivided into *members*, or even individual strata. Surfaces between any of these kinds of units are **contacts.**

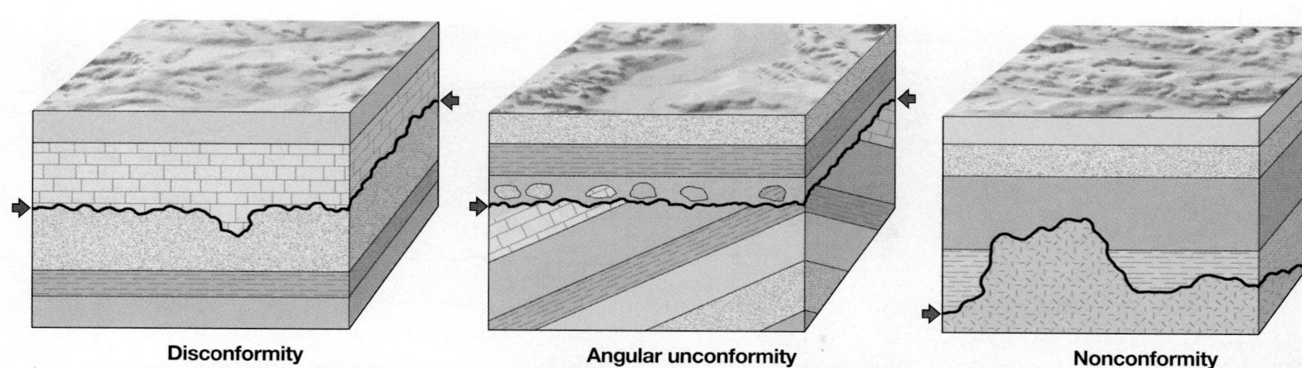

Disconformity
In a succession of rock layers (sedimentary strata or lava flows) parallel to one another, the disconformity surface is a gap in the layering. The gap may be a non-depositional surface where some layers never formed for a while, or the gap may be an erosional surface where some layers were removed before younger layers covered up the surface.

Angular unconformity
An angular unconformity is an erosional surface between two bodies of layered sedimentary strata or lava flows that are not parallel. The gap is because the older body of layered rock was tilted and partly eroded (rock was removed) before a younger body of horizontal rock layers covered the eroded surface.

Nonconformity
A nonconformity is an erosional surface between older igneous and/or metamorphic rocks and younger rock layers (sedimentary strata or lava flows). The gap is because some of the older igneous and/or metamorphic rocks were partly eroded (rock was removed) before the younger rock layers covered the eroded surface.

FIGURE 8.1 Three kinds of unconformities. Unconformities are surfaces that represent gaps (missing layers) in the geologic record; analogous to a gap (place where pages are missing) in a book. Red arrows point to the unconformity surface (bold black line) in each block diagram.

Laws for Determining Relative Age

Geologists use six basic laws for determining relative age relationships among bodies of rock based on their physical relationships. They are as follows:

- **Law of Original Horizontality**—*Sedimentary layers (**strata**) and lava flows were originally deposited as relatively horizontal sheets, like a layer cake.* If they are no longer horizontal or flat, it is because they have been displaced by subsequent movements of Earth's crust.
- **Law of Lateral Continuity**—*Lava flows and strata extend laterally in all directions until they thin to nothing (pinch out) or reach the edge of their basin of deposition.*
- **Law of Superposition**—*In an undisturbed sequence of strata or lava flows, the oldest layer is at the bottom of the sequence and the youngest is at the top.*
- **Law of Inclusions**—*Any piece of rock (clast) that has become included in another rock or body of sediment must be older than the rock or sediment into which it has been incorporated.* Such a clast (usually a rock fragment, crystal, or fossil) is called an **inclusion.** The surrounding body of rock is called the **matrix** (or groundmass). Thus, an inclusion is older than its surrounding matrix.
- **Law of Cross Cutting**—*Any feature that cuts across a rock or body of sediment must be younger than the rock or sediment that it cuts across.* Such cross cutting features include fractures (cracks in rock), faults (fractures along which movement has occurred), or masses of magma (*igneous intrusions*) that cut across preexisting rocks before they cooled. When a body of magma intrudes preexisting rocks, a narrow *zone of contact metamorphism* usually forms in the preexisting rocks adjacent to the intrusion.

Unconformities

Surfaces called **unconformities** represent gaps in the geologic record that formed wherever layers were not deposited for a time or else layers were removed by erosion. Most contacts between adjacent strata or formations are *conformities*, meaning that rocks on both sides of them formed at about the same time. An unconformity is a rock surface that represents a gap in the geologic record. It is like the place where pages are missing from a book. An unconformity can be a buried surface where there was a pause in sedimentation, a time between two lava flows, or a surface that was eroded before more sediment was deposited on top of it.

There are three kinds (**FIGURE 8.1**). A **disconformity** is an unconformity between *parallel* strata or lava flows. Most disconformities are very irregular surfaces, and pieces of the underlying rock are often included in the strata above them. An **angular unconformity** is an unconformity between two sets of strata that are not parallel to one another. It forms when new horizontal layers cover up older layers folded by mountain-building processes and eroded down to a nearly level surface. A **nonconformity** is an unconformity between younger sedimentary rocks and subjacent metamorphic or igneous rocks. It forms when stratified sedimentary rocks or lava flows are deposited on eroded igneous or metamorphic rocks.

Relative Age Dating Examples

Analyze and evaluate **FIGURES 8.2–8.9** to learn how the above laws of relative age dating are applied in cross sections of Earth's crust. These are the kinds of two-dimensional cross sections of Earth's crust that are exposed in road cuts, quarry walls, and mountain sides. *Be sure that you consider all of these examples before proceeding.*

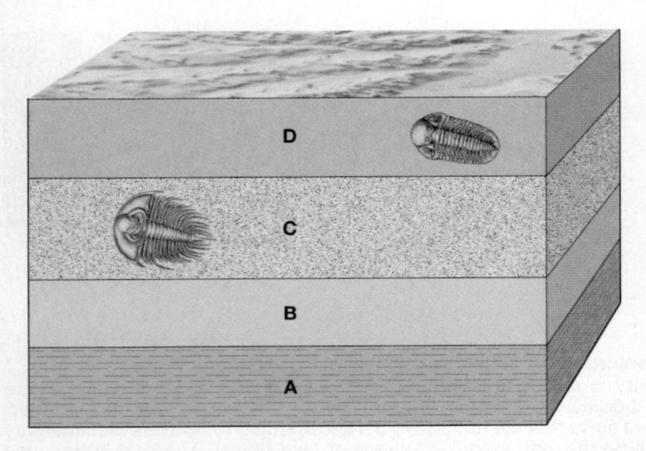

FIGURE 8.2 Law of superposition in horizontal strata. This is a sequence of strata that has maintained its original horizontality and does not seem to be disturbed. Therefore, Formation **A** is the oldest, because it is on the bottom of a sedimentary sequence of rocks. **D** is the youngest, because it is at the top of the sedimentary sequence. The sequence of events was deposition of **A, B, C,** and **D,** in that order and stacked one atop the other.

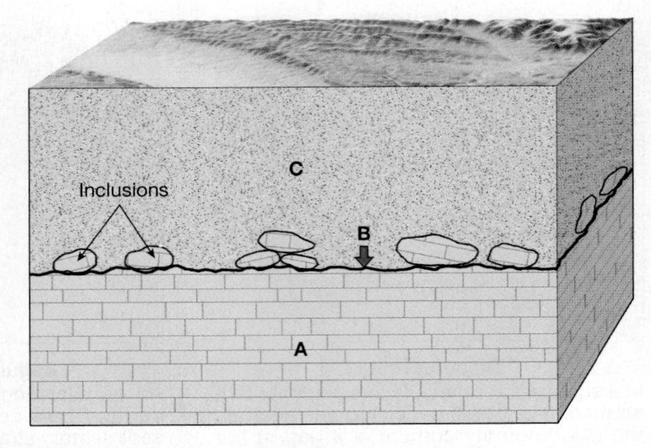

FIGURE 8.3 Inclusions on a disconformity. These strata are all horizontal. Limestone **A** is older, because it is on the bottom of a sequence of strata. Sandstone **C** is younger, because it is on top of the limestone and has inclusions (fragments of older rock) of the older limestone. Contact **C** is unconformable, because some of the limestone layers were eroded (making a gap in the rock layers) and became inclusions in the overlying sandstone. An unconformity between parallel strata, like this one, is a disconformity.

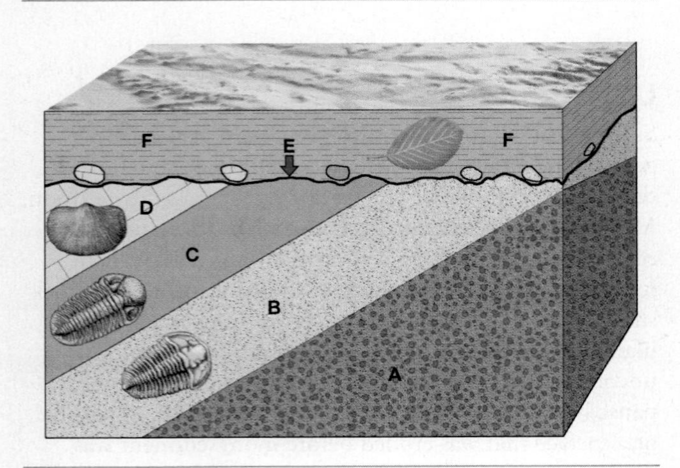

FIGURE 8.4 Angular unconformity. This is another sequence of strata, some of which do not have their original horizontality. Formation **A** is the oldest, because it is at the bottom of the sedimentary sequence. Formation **F** is youngest, because it forms the top of the sequence. Tilting and erosion of the sequence occurred after **D** but before deposition of Formation **F. E** is an angular unconformity.

The sequence of events began with deposition of **A, B, C,** and **D,** in that order and stacked one atop the other. The sequence of **A–D** was then tilted, and its top was eroded **(E).** Siltstone **F** was deposited horizontally on top of the erosional surface **(E),** which is now an angular unconformity.

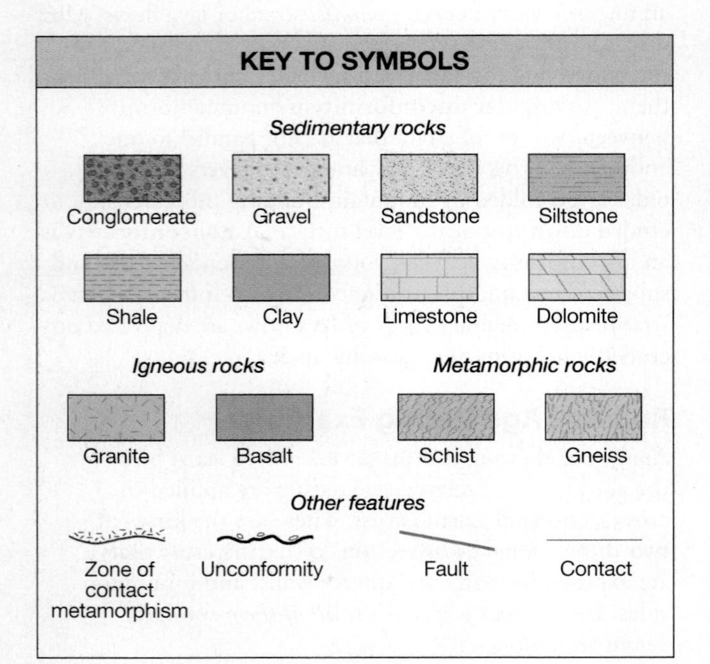

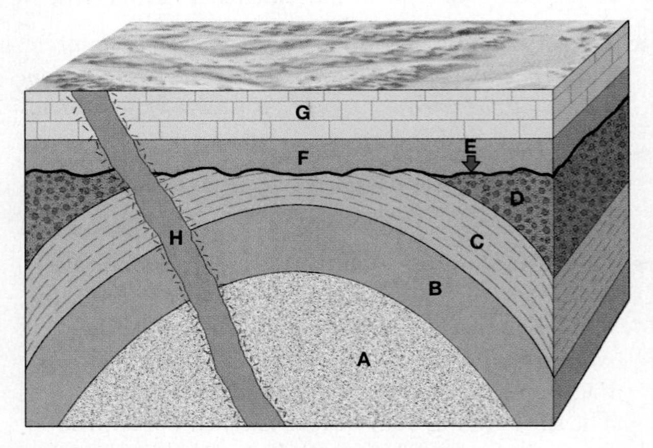

FIGURE 8.5 Law of cross-cutting. The body of igneous rock **H** is the youngest rock unit, because it cuts across all of the others. (When a narrow body of igneous rock cuts across strata in this way, it is called a **dike.**) **A** is the oldest formation because it is at the bottom of the sedimentary rock sequence that is cut by **H.** Folding and erosion occurred after **D** was deposited, but before **F** was deposited. **E** is an angular unconformity.

The sequence of events began with deposition of formations **A** through **D** in alphabetical order and one atop the other. That sequence was folded, and the top of the fold was eroded. Formation **F** was deposited horizontally atop the folded sequence and the erosional surface, which became angular unconformity **E. G** was deposited atop **F.** Lastly, a magma intruded across all of the strata and cooled to form basalt dike **H.**

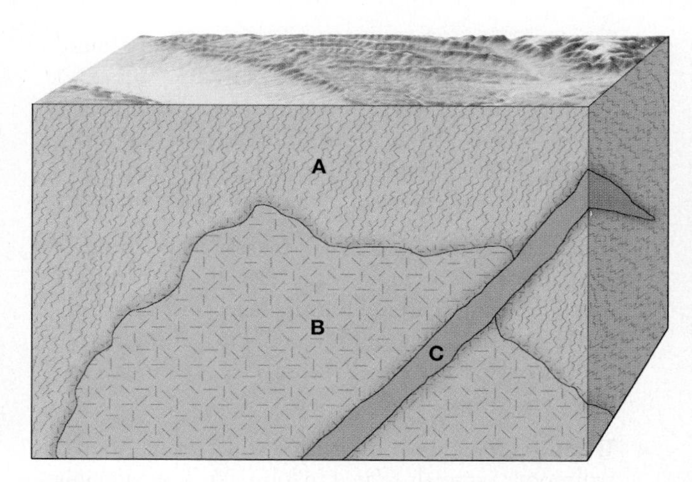

FIGURE 8.6 Igneous intrusions and cross-cutting. The body of granite **B** must have formed from the cooling of a body of magma that intruded the preexisting rock **A,** called **country rock.** The country rock is schist **A** containing a zone of contact metamorphism adjacent to the granite. Therefore, the sequence of events began with a body of country rock **A.** The country rock was intruded by a body of magma, which caused development of a zone of contact metamorphism and cooled to form granite **B.** Lastly, another body of magma intruded across both **A** and **B.** It caused development of a second zone of contact metamorphism and cooled to form basalt dike **C.**

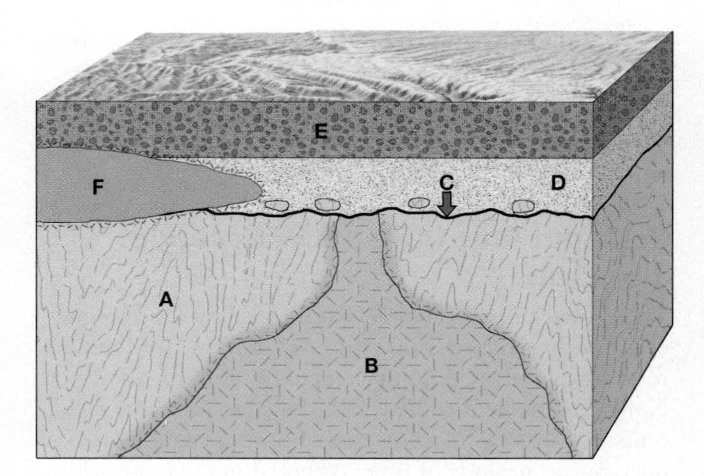

FIGURE 8.7 Nonconformity. At the base of this rock sequence there is gneiss **A,** which is separated from granite **B** by a zone of contact metamorphism. This suggests that a body of magma intruded **A,** then cooled to form the contact zone and granite **B.** There must have been erosion of both **A** and **B** *after* this intrusion (to form surface **C**), because there is no contact metamorphism between **B** and **D.** Formation **D** was deposited horizontally atop the eroded igneous and metamorphic rocks, forming nonconformity **C.** After **E** was deposited, a second body of magma **F** intruded across **A, C, D,** and **E**. Such an intrusive igneous body that is intruded along (parallel to) the strata is called a **sill (F).**

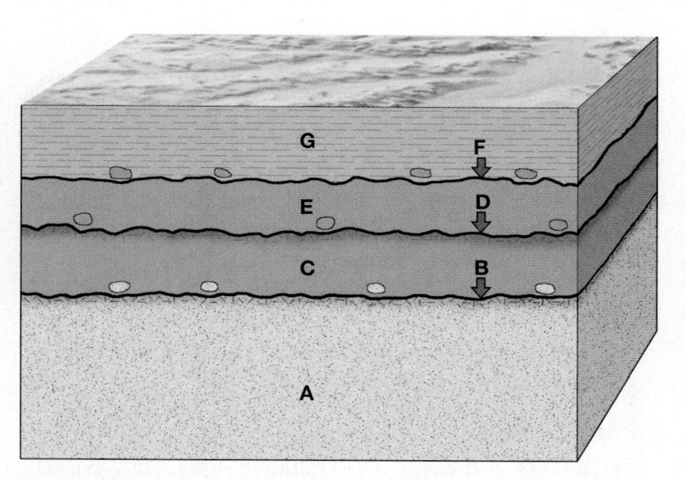

FIGURE 8.8 Disconformities. Notice that this is a sequence of strata and basalt lava flows (that have cooled to form the basalt). There are zones of contact metamorphism beneath both of the basalt lava flows (**C, E**). The sequence of events must have begun with deposition of sandstone **A,** because it is on the bottom. A lava flow was deposited atop **A** and cooled to form basalt **C.** This first lava flow caused development of the zone of contact metamorphism in **A** and the development of disconformity **B.** A second lava flow was deposited atop **C** and cooled to form basalt **E.** This lava flow caused the development of a zone of contact metamorphism and a disconformity **D.** An erosional surface developed atop **E,** and the surface became a disconformity **F** when shale **G** was deposited on top of it.

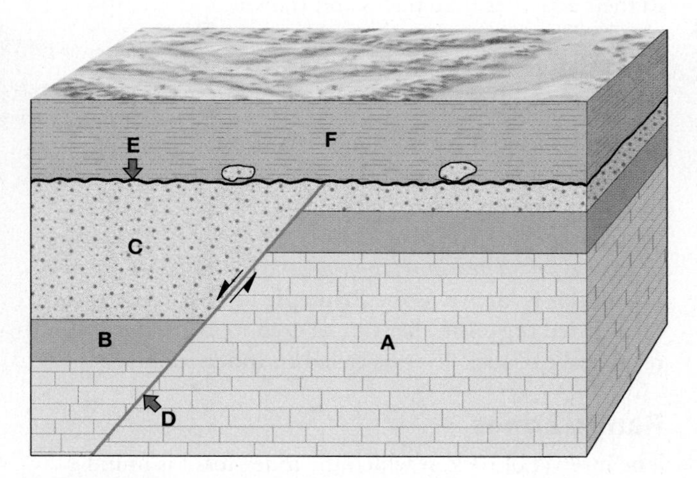

FIGURE 8.9 Cross-cutting by a fault. This is a sequence of relatively horizontal strata: **A, B, C,** and **F. A** must be the oldest of these formations because it is on the bottom. **F** is the youngest of these formations because it is on top. Formations **A, B,** and **C** are cut by a fault, which does not cut **F.** This means that the fault **D** must be younger than **C** and older than **F. E** is a disconformity. The sequence of events began with deposition of formations **A, B,** and **C,** in that order and one atop the other. This sequence was then cut by fault **D.** After faulting, the land surface was eroded. When siltstone **F** was deposited on the erosional surface, it became disconformity **E.**

ACTIVITY

8.3 Using Index Fossils to Date Rocks and Events

THINK About It | **How are fossils used to tell geologic time and infer Earth's history?**

OBJECTIVE Use index fossils to determine the relative ages (eras, periods) of rock bodies and infer some of Earth's history.

PROCEDURES

1. **Before you begin**, read Relative Age Dating Based on Fossils below. Also, this is **what you will need:**
   ____ calculator
   ____ Activity 8.3 Worksheet (p. 221) and pencil
2. **Then follow your instructor's directions** for completing the worksheets.

Relative Age Dating Based on Fossils

The sequence of strata that makes up the geologic record is a graveyard filled with the fossils of millions of kinds of organisms that are now extinct. Geologists know that they existed only because of their fossilized remains or the traces of their activities (like tracks and trails).

Principle of Fossil Succession and Index Fossils

Geologists have also determined that fossil organisms originate, co-exist, or disappear from the geologic record in a definite sequential order recognized throughout the world, so *any rock layer containing a group of fossils can be identified and dated in relation to other layers based on its fossils.* This is known as the **Principle of Fossil Succession.** A fossilized organism that can be used to identify the relative age of rock layers is called an **index fossil**.

Range Zones

The interval of rock in which the index fossil is found is called its **range zone** and corresponds to a particular interval of geologic time. The range zones of some well-known Phanerozoic index fossils are presented on the right side of FIGURE 8.10. Relative ages of the rocks containing these fossils are presented as *periods* and *eras* on the left side of FIGURE 8.10.

By noting the range zone of a fossil (vertical black line), you can determine the corresponding era(s) or period(s) of time in which it lived. For example, all of the different species of dinosaurs lived and died during the Mesozoic Era of time, from the middle of the Triassic Period to the end of the Cretaceous Period. Mammals have existed since late in the Triassic Period. If you found a rock layer with bones and tracks of both dinosaurs and mammals, then the age of the rock layer would be represented by the overlap of the dinosaur and mammal range zones (i.e., Middle Triassic to Late Cretaceous). Notice that FIGURE 8.10 also includes the following groups:

- **Brachiopods** (pink on chart): marine invertebrate animals with two symmetrical seashells of unequal size. They range throughout the Paleozoic, Mesozoic, and Cenozoic Eras, but they were most abundant in the Paleozoic Era. Only a few species exist today, so they are nearly extinct.
- **Trilobites** (orange on chart): an extinct group of marine arthropods (animals related to lobsters). They are only found in Paleozoic rocks, so they are a good index fossil for the Paleozoic Era and its named subdivisions.
- **Mollusks** (pink on chart): phylum of snails, cephalopods (squid, octopuses), and bivalves (oysters, clams; two asymmetrical shells of unequal size).
- **Plants** (dark green on chart).
- **Reptiles** (pale green on chart): the group of vertebrate animals that includes lizards, snakes, turtles, and dinosaurs. **Dinosaurs** are only found in Mesozoic rocks, so they are an index fossil for the Mesozoic and its subdivisions.
- **Mammals** (gray on chart): the group of vertebrate animals (including humans) that are warm blooded, nurse their young, and have hair.
- **Amphibians** (brown on chart): the group of vertebrate animals that includes modern frogs and salamanders.
- **Sharks** (blue on chart): a group of fish with teeth but no hard bones.

Rock and Time Units of the Geologic Time Scale

The geologic time scale in FIGURE 8.10 shows the ranges of index fossils in relation to named units of time and rock plus a scale of absolute ages in millions of years. Notice that there are two levels of named time and rock units in FIGURE 8.10. Long **eras** of time are subdivided into shorter **periods** of time. As noted in the table below, an era of time corresponds with an **erathem** of rock containing its characteristic index fossils. A period of time corresponds with a **system** of rock containing its characteristic index fossils.

Rock Units (Division of the Geologic Record)	Corresponding Geologic Time Units
Eonothem of rock	Eon of time (longest)
Erathem of rock	Era of time
System of rock	Period of time
Series of rock	Epoch of time

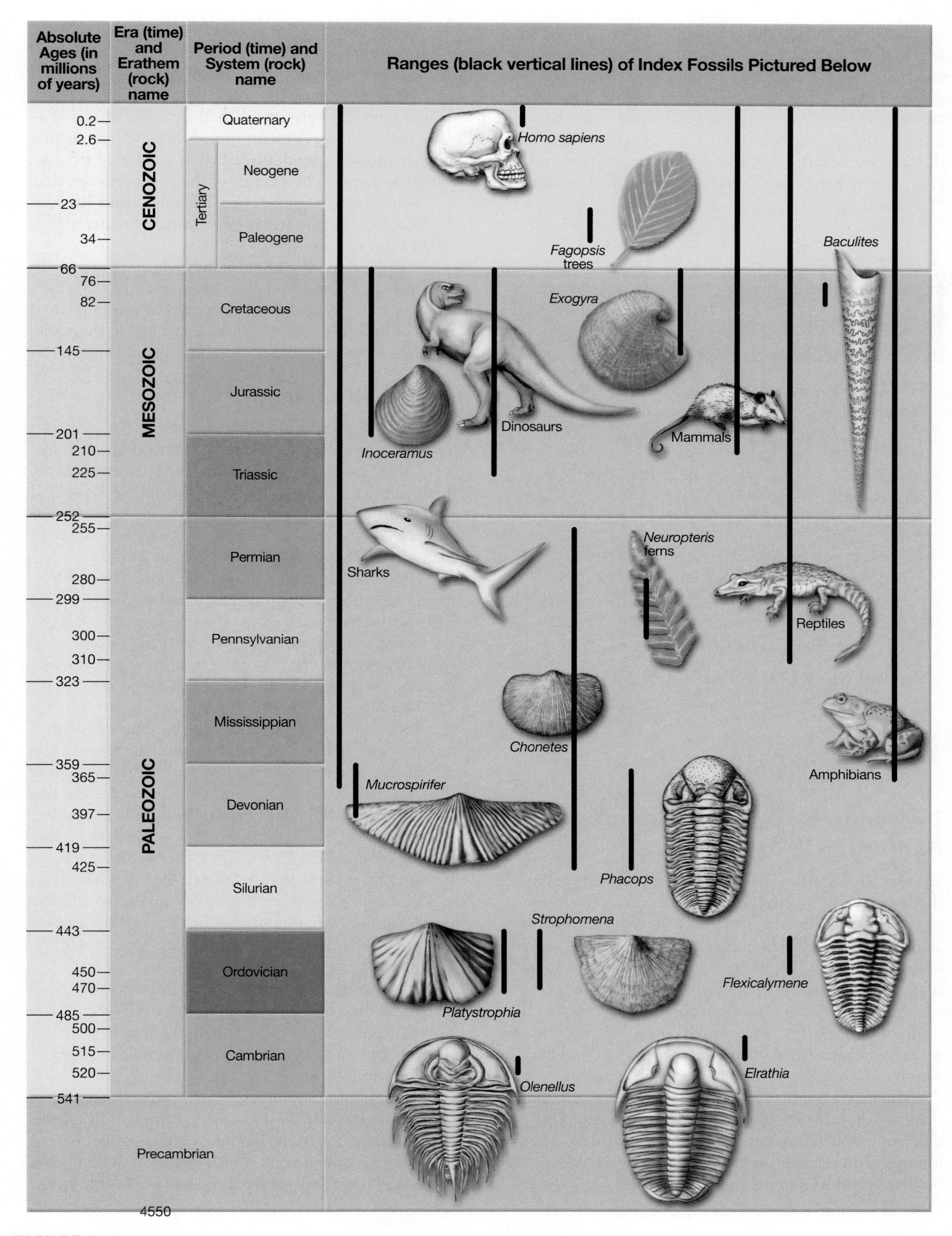

FIGURE 8.10 Range zones. Range zones (vertical bold black lines) of some well-known index fossils relative to named divisions of the geologic time scale.

There are times in this laboratory that you may be referring to a rock unit and other times when you may be referring to a time unit, so you will need to use the correct kind of unit for rock or time. For example, notice in FIGURE 8.10 that *Mucrospirifer* (a brachiopod) is an index fossil for the Devonian Period of time. When writing about this, you would write that *Mucrospirifer* is found in the Devonian System of rock, which represents the Devonian Period of time.

ACTIVITY

8.4 Absolute Dating of Rocks and Fossils

THINK About It **How do geologists determine the absolute age, in years, of Earth materials and events?**

OBJECTIVE Calculate absolute ages to date Earth materials and events.

PROCEDURES

1. **Before you begin**, read Determining Absolute Ages by Radiometric Dating below. Also, this is **what you will need:**
   ____ calculator
   ____ Activity 8.4 Worksheet (p. 222) and pencil
2. **Then follow your instructor's directions** for completing the worksheets.

Determining Absolute Ages by Radiometric Dating

You measure the passage of time based on the rates and rhythms at which regular changes occur around you. For example, you are aware of the rate at which hands move on a clock, the rhythm of day and night, and the regular sequence of the four seasons. These regular changes allow you to measure the passage of minutes, hours, days, and years.

Another way to measure the passage of time is by the regular rate of decay of radioactive isotopes. This technique is called **radiometric dating** and is one way that geologists determine absolute ages of some geologic materials.

You may recall that **isotopes** of an element are atoms that have the same number of protons and electrons but different numbers of neutrons. This means that the different isotopes of an element vary in atomic weight (mass number) but not in atomic number (number of protons).

About 350 different isotopes occur naturally. Some of these are *stable isotopes*, meaning that they are not radioactive and do not decay through time. The others are *radioactive isotopes* that decay spontaneously, at regular rates through time. When a mass of atoms of a radioactive isotope is incorporated into the structure of a newly formed crystal or seashell, it is referred to as a **parent isotope**. When atoms of the parent isotope decay to a stable form, they have become a **daughter isotope**. A parent isotope and its corresponding daughter are called a **decay pair**.

Atoms of a parent isotope always decay to atoms of their stable daughter isotope at an exponential rate that does not change. The rate of decay can be expressed in terms of **half-life**—the time it takes for half of the parent atoms in a sample to decay to stable daughter atoms.

Radiometric Dating of Geologic Materials

The decay parameters for all radioactive isotopes can be represented graphically as in FIGURE 8.11. Notice that the decay rate is exponential (not linear)—during the second half-life interval, only half of the remaining half of parent atoms will decay. All radioactive isotopes decay in this way, but each decay pair has its own value for half-life.

Half-lives for some isotopes used for radiometric dating have been experimentally determined by physicists and chemists, as noted in the top chart of FIGURE 8.11. For example, uranium-238 is a radioactive isotope (parent) found in crystals of the mineral zircon. It decays to lead-206 (daughter) and has a half-life of about 4500 million years (4.5 billion years).

To determine the age of an object, it must contain atoms of a radioactive decay pair that originated when the object formed. You must then measure the percent of those atoms that is parent atoms (**P**) and the percent that is daughter atoms (**D**). This is generally done in a chemistry laboratory with an instrument called a *mass spectrometer*. Based on **P** and **D** and the chart at the bottom of FIGURE 8.10, find the number of half-lives that have elapsed and the object's corresponding age in number of half-lives. Finally, multiply that number of half-lives by the known half-life for that decay pair (noted in the top chart of FIGURE 8.11).

For example, a sample of Precambrian granite contains biotite mineral crystals, so it can be dated using the potassium-40 to argon-40 decay pair. If there are three argon-40 atoms in the sample for every one potassium-40 atom, then the sample is 25.0% potassium-40 parent atoms (**P**) and 75.0% argon-40 daughter atoms (**D**). This means that two half-lives have elapsed, so the age of the biotite (and the granite) is 2.0 times 1.3 billion years, which equals 2.6 billion years. The useful dating ranges are also noted on FIGURE 8.11.

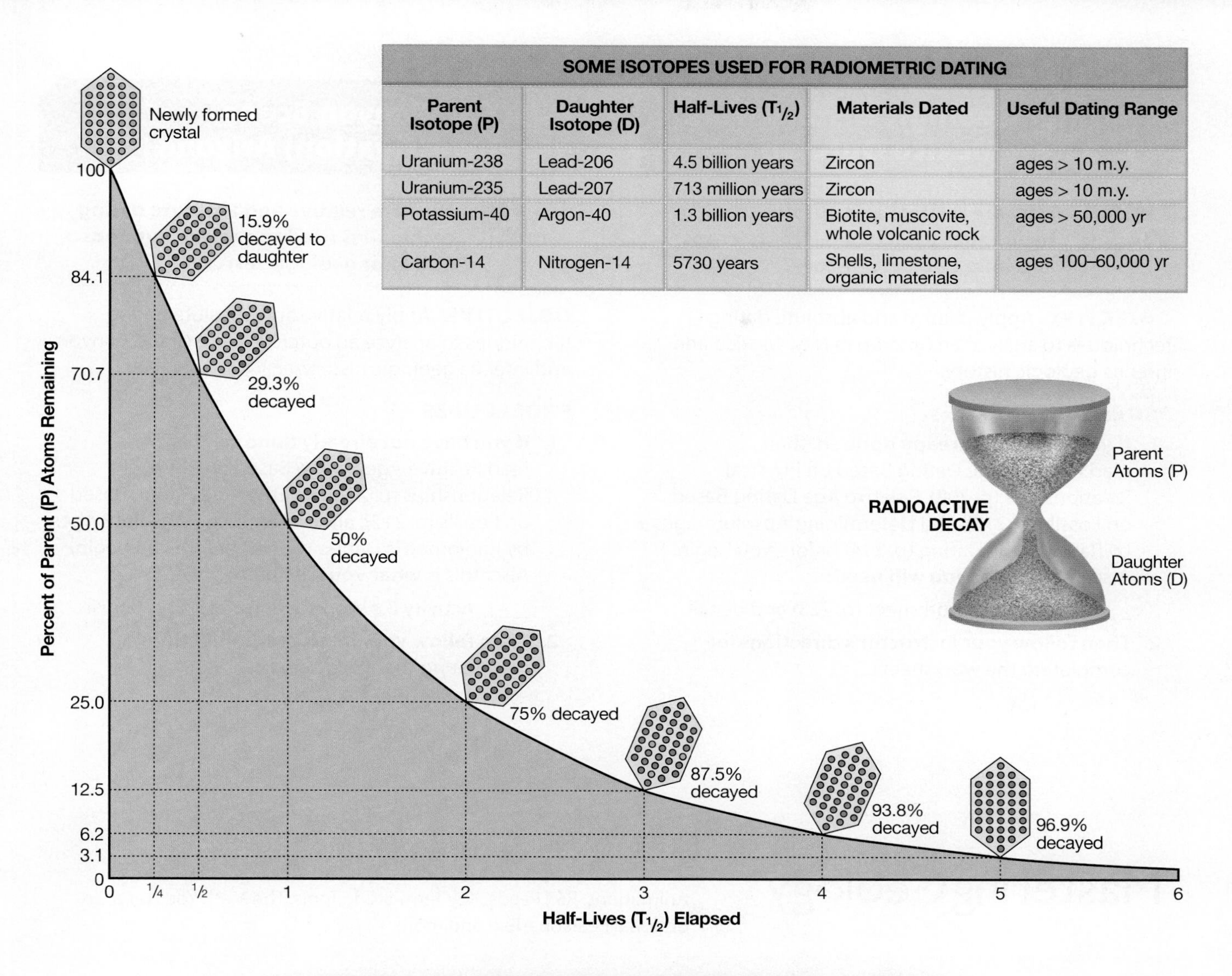

SOME ISOTOPES USED FOR RADIOMETRIC DATING				
Parent Isotope (P)	Daughter Isotope (D)	Half-Lives ($T_{1/2}$)	Materials Dated	Useful Dating Range
Uranium-238	Lead-206	4.5 billion years	Zircon	ages > 10 m.y.
Uranium-235	Lead-207	713 million years	Zircon	ages > 10 m.y.
Potassium-40	Argon-40	1.3 billion years	Biotite, muscovite, whole volcanic rock	ages > 50,000 yr
Carbon-14	Nitrogen-14	5730 years	Shells, limestone, organic materials	ages 100–60,000 yr

DECAY PARAMETERS FOR ALL RADIOACTIVE DECAY PAIRS			
Percent of Parent Atoms (P)	Percent of Daughter Atoms (D)	Half-Lives Elapsed	Age
100.0	0.0	0	0.000 x $T_{1/2}$
98.9	1.1	1/64	0.015 x $T_{1/2}$
97.9	2.1	1/32	0.031 x $T_{1/2}$
95.8	4.2	1/16	0.062 x $T_{1/2}$
91.7	8.3	1/8	0.125 x $T_{1/2}$
84.1	15.9	1/4	0.250 x $T_{1/2}$
70.7	29.3	1/2	0.500 x $T_{1/2}$
50.0	50.0	1	1.000 x $T_{1/2}$
35.4	64.6	1 1/2	1.500 x $T_{1/2}$
25.0	75.0	2	2.000 x $T_{1/2}$
12.5	87.5	3	3.000 x $T_{1/2}$
6.2	93.8	4	4.000 x $T_{1/2}$
3.1	96.9	5	5.000 x $T_{1/2}$

FIGURE 8.11 Radiometric dating. Some isotopes useful for radiometric dating, their decay parameters, and their useful ranges for dating. The half-life of each decay pair is different (top chart), but the graph and decay parameters (bottom charts) are the same for all decay pairs.

ACTIVITY

8.5 Infer Geologic History from a New Mexico Outcrop

THINK About It **How are relative and absolute dating techniques used to analyze outcrops and infer geologic history?**

OBJECTIVE Apply relative and absolute dating techniques to analyze an outcrop in New Mexico and infer its geologic history.

PROCEDURES

1. **If you have not already done so**, then read Relative Age Dating Based on Physical Relationships (p. 208), Relative Age Dating Based on Fossils (p. 212), and Determining Absolute Ages by Radiometric Dating (p. 214) before you begin. Also, this is **what you will need:**

   ____ Activity 8.5 Worksheet (p. 223) and pencil

2. **Then follow your instructor's directions** for completing the worksheets.

ACTIVITY

8.6 CSI (Canyon Scene Investigation) Arizona

THINK About It **How are relative and absolute dating techniques used to analyze outcrops and infer geologic history?**

OBJECTIVE Apply relative and absolute dating techniques to analyze an outcrop in the Grand Canyon and infer its geologic history.

PROCEDURES

1. **If you have not already done so**, then read Relative Age Dating Based on Physical Relationships (p. 208), Relative Age Dating Based on Fossils (p. 212), and Determining Absolute Ages by Radiometric Dating (p. 214) before you begin. Also, this is **what you will need:**

   ____ Activity 8.6 Worksheet (p. 225) and pencil

2. **Then follow your instructor's directions** for completing the worksheets.

ACTIVITY 8.1 Geologic Inquiry for Relative Age Dating

Name: ______________________ **Course/Section:** ______________________ **Date:** ____________

A. Analyze this block of layer cake. Each side of the block of cake is a vertical **cross section** of the layers. Also notice the surfaces between the layers, where two different layers touch each other. Geologists refer to surfaces between layers or other bodies of rock as **contacts**.

1. Think about the process used to construct the layer cake, from making and *depositing* (laying down) the first layer to making and depositing the last layer. On the left edge of the cake, number the layers to show the sequence of steps in which they were deposited to make the layer cake from 1 (first step) to n (the number of the last step).

2. Using a pen, draw lines on the layer cake to mark all of the contacts between layers. Then place arrows along the right edge of the cake that point to each contact. Label each arrow (contact) to show its relative age from 1 (the time when the first contact was created; the oldest contact) to "n" (the number corresponding to the last time a contact was created; the youngest contact).

B. The picture below is an outcrop about 5 meters thick near Sedona, Arizona. The red rock is an ancient body of soil. The brown layer in which grass is rooted is modern soil. The blocky brown-gray rock with wide fractures (cracks) is an ancient lava flow (basalt, a volcanic rock). This outcrop is a natural geologic cross section of rock layers, analogous to the cake.

1. Which layer is the oldest? How do you know?

2. Using a pen, draw a line on the picture that marks the exact position of:

a. the contact between the red ancient soil and the lava flow.

b. the exact contact between the top of the lava flow and the base of the darker brown modern soil in which grass is growing.

3. Notice the **fractures** (cracks) that cut across the lava flow layer. Are they older or younger than the lava flow? How do you know?

4. Notice that *clasts* (broken pieces) of the lava flow are included in the brown soil. Are they older or younger than the brown soil? How do you know?

C. Analyze this outcrop, photographed by geologist, Thomas McGuire. It is another natural geologic cross section with red sandstone layers on the bottom and a yellow conglomerate (gravel) rock layer on top. Notice that the red rock layers are not horizontal. They are bent up on the left and right, and down in the middle, as wave-like **folds** (like a crumpled rug).

1. Using a pen, trace two of the contacts between layers of the red sandstone as well as you can. Assuming that the red sandstone layers were originally horizontal, what may have caused them to be folded in this way?

2. On both sides of the picture, use an arrow to label the exact location of the contact between the red sandstone and the horizontal yellow conglomerate above it. This surface is an **unconformity**—a surface (contact) representing erosion of layers or a break in deposition of layers, like a place where pages are missing from a book. Something happened at the time represented by the surface, but no rock layer remains as a record of the event. What sequence of events may have happened to form the unconformity?

D. REFLECT & DISCUSS In all of your work above, you had to figure out the relative ages (from oldest to youngest) of rock layers, fractures, folds, and clasts included in soil. Based on your work, write down three rules that a geologist could follow to tell the relative ages of rock layers, fractures, clasts, and folds in geologic cross sections.

ACTIVITY 8.2 Determining Sequence of Events in Geologic Cross Sections

Name: ______________________ **Course/Section:** ______________________ **Date:** ____________

A. Review the legend of symbols at the bottom of the page. On the lines provided for each cross section, write letters to indicate the sequence of events from oldest (first in the sequence of events) to youngest (last in the sequence of events). Refer to **FIGURES 8.1–8.9** and the laws of relative age dating (page 209 as needed).

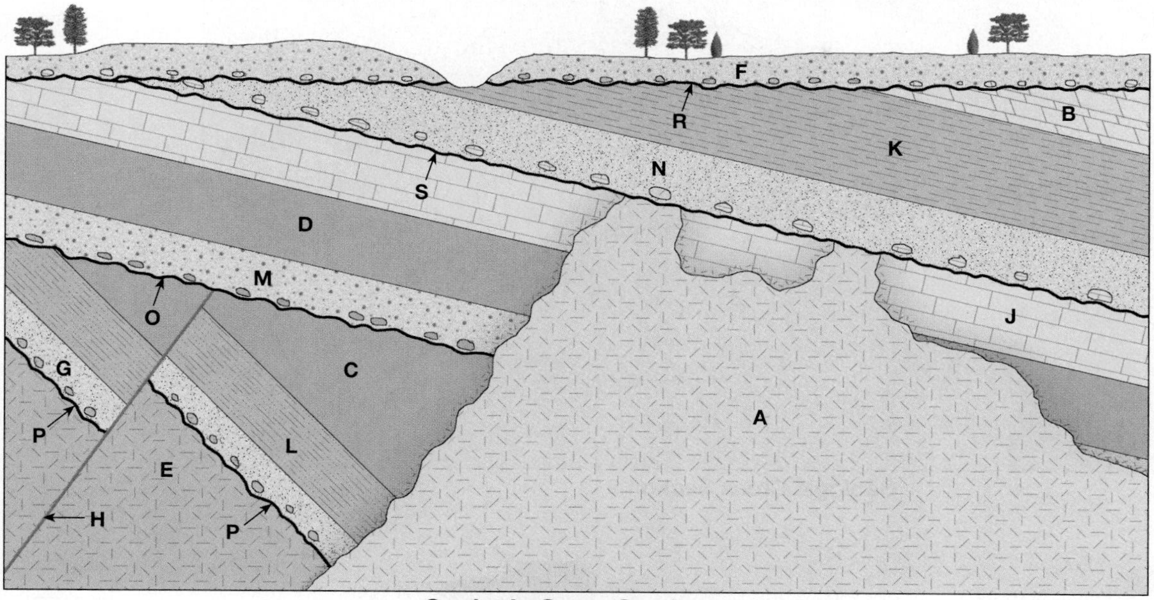

Youngest ______
______
______
______
______
______
______
______
______
______
______
______
______
______
______
______
Oldest ______

Geologic Cross Section 1

Youngest ______
______
______
______
______
______
______
______
______
______
______
______
______
______
______
Oldest ______

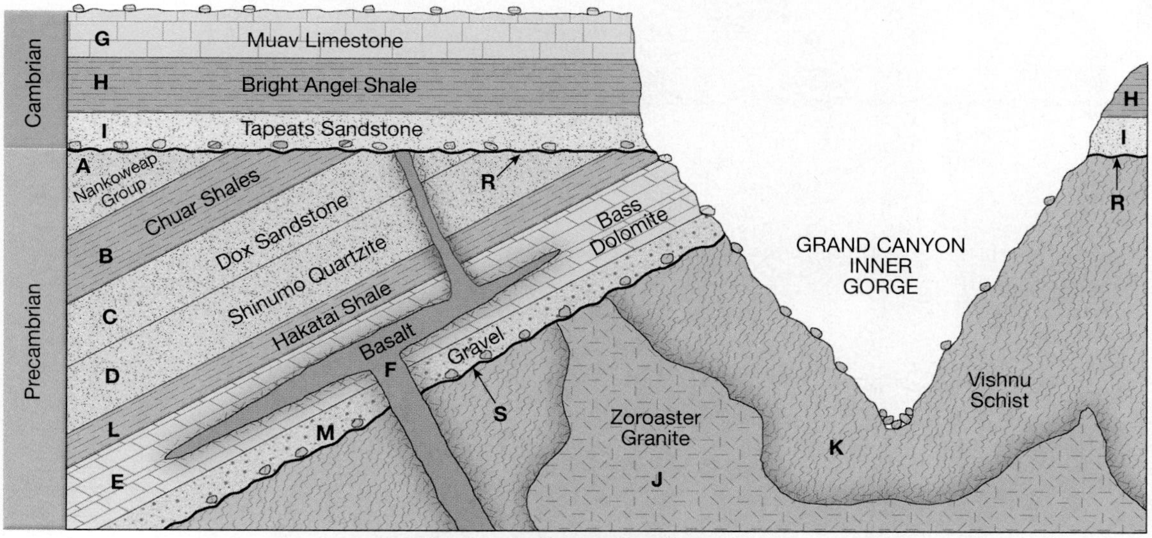

Geologic Cross Section 2

Cambrian: G Muav Limestone; H Bright Angel Shale; I Tapeats Sandstone

Precambrian: A Nankoweap Group; B Chuar Shales; C Dox Sandstone; D Shinumo Quartzite; L Hakatai Shale; E Bass Dolomite; F Basalt; M; R; S Gravel; J Zoroaster Granite; K Vishnu Schist

GRAND CANYON INNER GORGE

KEY TO SYMBOLS

Sedimentary rocks: Conglomerate, Gravel, Sandstone, Siltstone, Shale, Clay, Limestone, Dolomite

Igneous rocks: Granite, Basalt

Metamorphic rocks: Schist, Gneiss

Other features: Zone of contact metamorphism, Unconformity, Fault, Contact

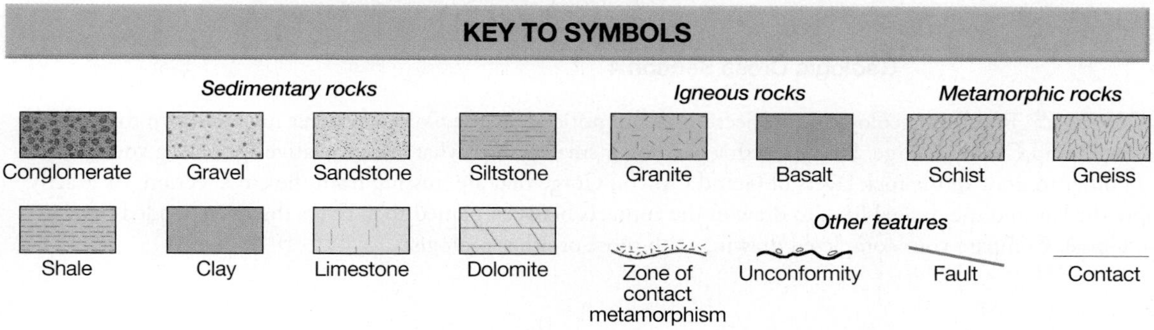

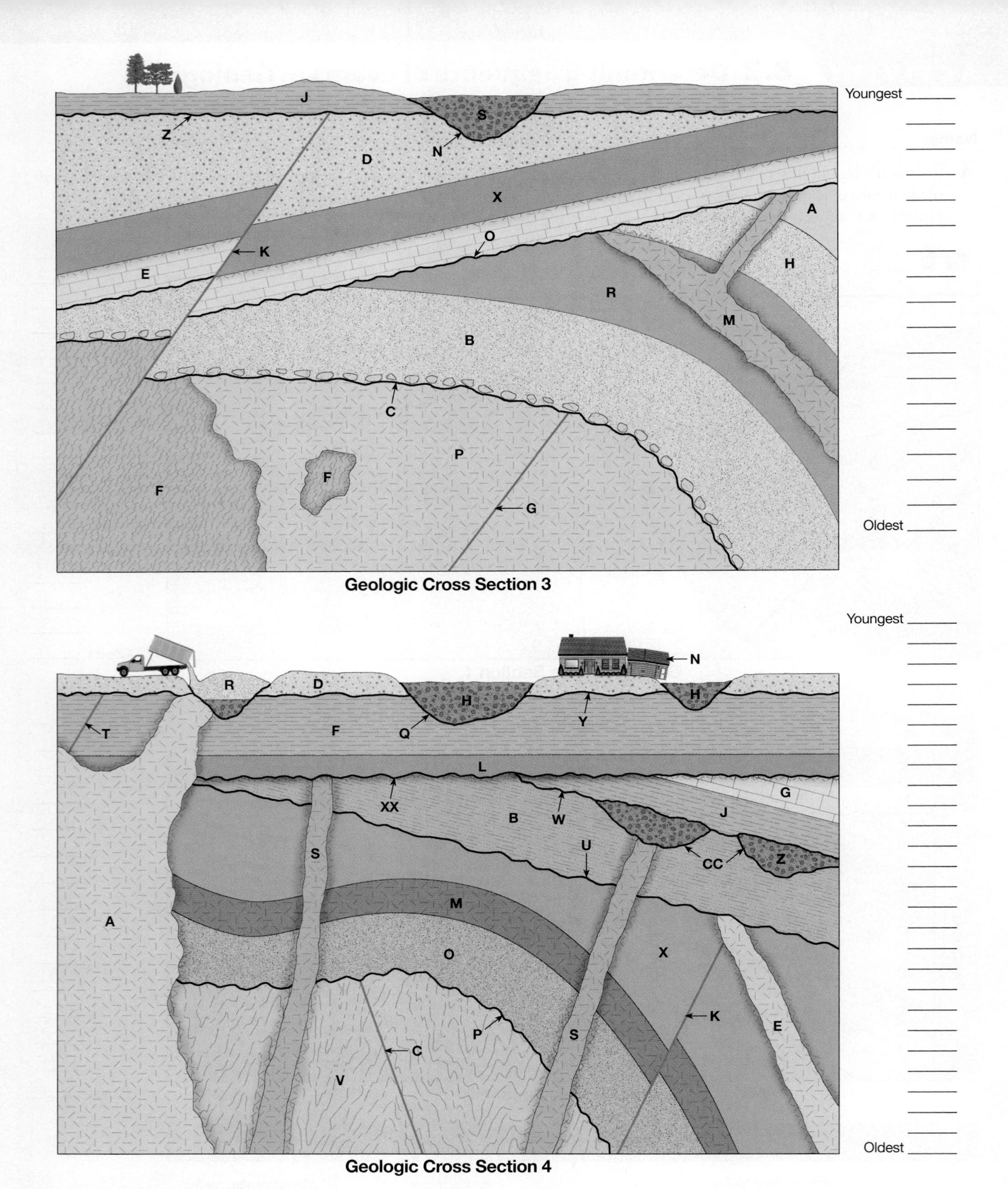

Geologic Cross Section 3

Geologic Cross Section 4

B. REFLECT & DISCUSS Return to geologic Cross Section 2, and notice how the Colorado River has cut down through the rocks to create the Grand Canyon Gorge. Discuss with a partner or small group, what law of relative age dating you would need to apply in order to draw in the rock layers of Grand Canyon Gorge that are missing from the cross section. As exactly as you can, apply the law and use dashed lines to draw in the contacts between named rock layers that were eroded away in Grand Canyon Gorge. Compare your completed drawing with those of other geologists.

ACTIVITY 8.3 Use Index Fossils to Date Rocks and Events

Name: ______________________________ **Course/Section:** ____________________ **Date:** __________

A. Analyze this fossiliferous rock from New York.

1. What index fossils from **FIGURE 8.10** are present?

2. Based on the overlap of range zones for these index fossils what is the relative age of the rock (expressed as the early, middle, or late part of one or more periods of time)?

3. Using **FIGURE 8.10**, what is the absolute age of the rock in Ma (millions of years old/ago), as a range from oldest to youngest?

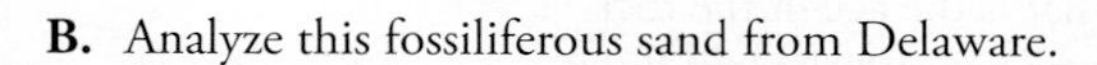

B. Analyze this fossiliferous sand from Delaware.

1. What index fossils from **FIGURE 8.10** are present?

2. Based on the overlap of range zones for these index fossils what is the relative age of the rock (expressed as the early, middle, or late part of one or more periods of time)?

3. Using **FIGURE 8.10**, what is the absolute age of the rock in Ma (millions of years old/ago), as a range from oldest to youngest?

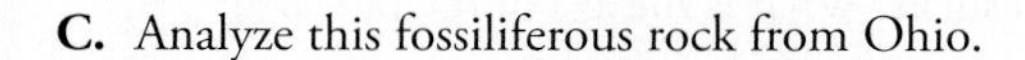

C. Analyze this fossiliferous rock from Ohio.

1. What index fossils from **FIGURE 8.10** are present?

2. Based on the overlap of range zones for these index fossils, what is the relative age of the rock (expressed as the early, middle, or late part of one or more periods of time)?

3. Using **FIGURE 8.10**, what is the absolute age of the rock in Ma (millions of years old/ago), as a range from oldest to youngest?

D. Using **FIGURE 8.10**, re-evaluate the geologic cross section in **FIGURE 8.2** based on its fossils.

1. Which one of the contacts (surfaces) between lettered layers is a disconformity?____________

2. A system is the rock/sediment deposited during a period of time. What system of rock is completely missing at the disconformity?

3. What amount of absolute time in m.y. (millions of years) is missing at the disconformity? ____________ m.y.

E. **REFLECT & DISCUSS** What geologic event occurred during the Mesozoic Era in the region where **FIGURE 8.4** is located? Explain.

ACTIVITY 8.4 Absolute Dating of Rocks and Fossils

Name: ______________________ **Course/Section:** ______________________ **Date:** __________

A. A solidified lava flow containing zircon mineral crystals is present in a sequence of rock layers that are exposed in a hillside. A mass spectrometer analysis was used to count the atoms of uranium-235 and lead-207 isotopes in zircon samples from the lava flow. The analysis revealed that 71% of the atoms were uranium-235, and 29% of the atoms were lead-207. Refer to **FIGURE 8.11** to help you answer the following questions.

1. About how many half-lives of the uranium-235 to lead-207 decay pair have elapsed in the zircon crystals? ______________

2. What is the absolute age of the lava flow based on its zircon crystals? Show your calculations.

3. What is the age of the rock layers above the lava flow? ______________

4. What is the age of the rock layers beneath the lava flow? ______________

B. Astronomers think that Earth probably formed at the same time as all of the other rocky materials in our solar system, including the oldest meteorites. The oldest meteorites ever found on Earth contain nearly equal amounts of both uranium-238 and lead-206. Based on **FIGURE 8.11**, what is Earth's age? Explain your reasoning.

C. If you assume that the global amount of radiocarbon (formed by cosmic-ray bombardment of atoms in the upper atmosphere and then dissolved in rain and seawater) is constant, then decaying carbon-14 is continuously replaced in organisms while they are alive. However, when an organism dies, the amount of its carbon-14 decreases as it decays to nitrogen-14.

1. The carbon in a buried peat bed has about 6% of the carbon-14 of modern shells. What is the age of the peat bed? Explain.

2. In sampling the peat bed, you must be careful to avoid any young plant roots or old limestone. Why?

D. Zircon ($ZrSiO_4$) forms in magma and lava as it cools into igneous rock. It is also useful for absolute age dating (**FIGURE 8.11**).

1. If you walk on a modern New Jersey beach, then you will walk on some zircon sand grains. Yet if you determine the absolute age of the zircons, it does not indicate a modern age (zero years) for the beach. Why?

2. Suggest a rule that geologists should follow when they date rocks based on the radiometric ages of crystals inside the rocks.

E. REFLECT & DISCUSS An "authentic dinosaur bone" is being offered for sale on the Internet. The seller claims that he had it analyzed by scientists who confirmed that it is a dinosaur bone and used carbon dating to determine that it is 400 million years old. Discuss the sellers claims with a partner or in a small group. Should you be suspicious of this bone's authenticity? Explain. (See **FIGURES 8.10, 8.11**).

ACTIVITY 8.5 Infer Geologic History from a New Mexico Outcrop

Name: ________________________ **Course/Section:** ________________ **Date:** __________

A. Refer to the image below, an outcrop in a surface mine (coal strip mine) in northern New Mexico. Note the sill, sedimentary rocks, fault, places where a fossil leaf was found, and isotope data for zircon crystals in the sill.

1. What is the relative age of the sedimentary rocks in this rock exposure? Explain your reasoning.

2. What is the absolute age of the sill? Show how you calculated the answer.

3. Locate the fault. How much displacement has occurred along this fault? ____________ meters

B. Make a numbered list of the geologic events of this region, starting with deposition of the sandstone (oldest event: 1) and ending with the time this picture was taken. Use names of relative ages of geologic time and absolute ages in your writing. *Your reasoning and number of events may differ from other geologists.*

1.

2.

3.

4.

5.

6.

7.

8.

C. **REFLECT & DISCUSS** Write a question that you have about the geologic history of this location. What geologic evidence would you need to answer the question?

ACTIVITY 8.6 CSI (Canyon Scene Investigation) Arizona

Name: ______________________ **Course/Section:** ______________________ **Date:** ____________

A. This is a photograph of part of the bottom of the Grand Canyon, which runs east to west across northern Arizona. You are standing west of Grand Canyon Villiage, on the south rim of the Canyon, and looking at the north side of the bottom of the canyon.

Carefully analyze the photograph for rock layering. The very bottom rock layers in the foreground are folded Precambrian metamorphic rock called the "Vishnu Schist," which contains narrow bodies of granite (colored white). The Vishnu Schist is overlain here by relatively horizontal layers of sedimentary rock.

1. Using a pen, draw a line exactly along the contact (boundary) between the Vishnu Schist and the relatively horizontal sedimentary rocks above it.

2. Based on **FIGURE 8.1**, what specific kind of unconformity did you trace above?

3. **REFLECT & DISCUSS** The Vishnu Schist has an absolute age of about 1700 million years. The Lower Cambrian Tapeats Sandstone sits on top of the unconformity that you drew in Part A. If you assume that the Tapeats Sandstone includes strata (sedimentary layers) that were deposited at the very start of the Cambrian Period, then how much of a gap in time exists at the unconformity (where you traced it with a pen)?

B. This is another photograph of part of the bottom of the Grand Canyon. You are standing east of Grand Canyon Villiage, on the south rim of the Canyon, and looking at the north side of the canyon. All of the rocks in this scene are sedimentary rocks.

1. Analyze this canyon scene. Then, using a pen, draw a line on the photograph to show the exact position of an unconformity.

2. What kind of unconformity did you identify?

C. **REFLECT & DISCUSS** What evidence did you apply to justify drawing an unconformity where you did on the above photograph?

LABORATORY 9

Topographic Maps and Orthoimages

CONTRIBUTING AUTHORS
Charles G. Higgins • *University of California*
John R. Wagner • *Clemson University*
James R. Wilson • *Weber State University*

Oblique Google Earth™ view of the Italian volcano, Vesuvius, which destroyed the Roman city of Pompeii in AD 79. Pliny the Younger described his eyewitness account of the eruption in two letters to Tacitus. (© Google Earth™)

BIG IDEAS

Topographic maps are two dimensional (flat) representations of three-dimensional landscapes, viewed from directly above. Horizontal (two-dimensional) positions of landscape features are represented with symbols, colors, and lines relative to geographic grid systems, specific scales, and directional data. The third dimension, elevation (height) of the landscape, is represented with contour lines marking certain elevations in feet or meters above sea level. The three-dimensional and quantitative aspect of topographic maps makes them valuable to geologists and other people who want to know the shapes and elevations of landscapes. They are often used in combination with orthoimages (aerial photographs that have been adjusted to the same scale as the map).

FOCUS YOUR INQUIRY

THINK About It How are specific places and quadrangles located using the latitude-longitude coordinate system, and how could geologists use Google Earth™ to study them?

ACTIVITY 9.1 Map and Google Earth™ Inquiry *(p. 228)*

THINK About It What are topographic quadrangle maps, and what geographic grid systems, scales, directional data, and symbols are represented on them?

ACTIVITY 9.2 Map Locations, Distances, Directions, and Symbols *(p. 228)*

THINK About It How are topographic maps constructed and interpreted?

ACTIVITY 9.3 Topographic Map Construction *(p. 239)*

ACTIVITY 9.4 Topographic Map and Orthoimage Interpretation *(p. 239)*

THINK About It How are topographic maps used to calculate the relief and gradients (slopes) of landscapes?

ACTIVITY 9.5 Relief and Gradient (Slope) Analysis *(p. 246)*

THINK About It How is a topographic profile constructed from a topographic map, and what is its vertical exaggeration?

ACTIVITY 9.6 Topographic Profile Construction *(p. 246)*

ACTIVITY

9.1 Map and Google Earth™ Inquiry

THINK About It **How are specific places and quadrangles located using the latitude-longitude coordinate system, and how could geologists use Google Earth™ to study them?**

OBJECTIVE Apply the latitude-longitude coordinate system to locate a country, quadrangle, and place of your choice, and then explore it in greater detail using Google Earth™ tools, layers, ruler, and historical imagery.

PROCEDURES

1. **Before you begin**, do not look up definitions and information. Use your current knowledge, and complete the worksheet with your current level of ability. Also, this is **what you will need** to do the activity:
   ____ Activity 9.1 Worksheet (p. 249) and pencil
   ____ computer with Internet access (on which Google Earth™ software is, or can be, loaded)
2. **Complete the worksheet in a way that makes sense to you.**
3. **After you complete the worksheet**, be prepared to discuss your observations and ideas with others.

ACTIVITY

9.2 Map Locations, Distances, Directions, and Symbols

THINK About It **What coordinate systems, scales, directional data, and symbols are used on maps?**

OBJECTIVE Identify and characterize features on topographic maps using printed information, compass bearings, scales, symbols, and three geographic systems: latitude and longitude, the U.S. Public Land Survey System (PLSS), and the Universal Transverse Mercator System (UTM).

PROCEDURES

1. **Before you begin**, read the following topics below: Introduction, Latitude-Longitude and Quadrangle Maps, Map Scales, Declination and Compass Bearings, Global Positioning System (GPS), UTM—Universal Transverse Mercator System, and Public Land Survey System. Also, this is **what you will need:**
   ___ Activity 9.2 Worksheet (p. 251) and pencil
   ___ calculator
2. **Then follow your instructor's directions** for completing the worksheet.

Introduction

Imagine that you are seated with a friend who asks you how to get to the nearest movie theater. To find the theater, your friend must know locations, distances, and directions. You must have a way of communicating your current location, the location of the movie theater, plus directions and distances from your current location to the movie theater. You may also include information about the topography of the route (whether it is uphill or downhill) and landmarks to watch for along the way. Your directions may be verbal or written, and they may include a map, satellite image, or aerial photograph (picture taken from an aircraft). Geologists are faced with similar circumstances in their field (outdoor) work. They must often characterize geologic features, the places where they occur, and their sizes, shapes, elevations, and locations in relation to other features. Satellite images and aerial photographs are used to view parts of Earth's surface from above (FIGURE 9.1A), and this information is summarized on maps.

A **map** is a flat representation of part of Earth's surface as viewed from above and reduced in size to fit a sheet of paper or computer screen. A **planimetric map** (FIGURE 9.1B) is a flat representation of Earth's surface that shows horizontal (two-dimensional) positions of features like streams, landmarks, roads, and political boundaries. A **topographic map** shows the same horizontal information as a planimetric map but also includes *contour lines* to represent elevations of hills and valleys. The contour lines are the distinguishing features of a topographic map and make it appear three dimensional. Thus topographic maps show the shape of the landscape in addition to horizontal directions, distances, and a system for describing exact locations.

Most United States topographic maps are published by the U.S. Geological Survey (USGS) and available at their US Topo website (**http://store.usgs.gov**). Canadian topographic maps are produced by the Centre for Topographic Information of Natural Resources Canada (NRCAN: **http://maps.nrcan.gc.ca**). State and provincial geological surveys, and the national geological surveys of other countries, also produce and/or distribute topographic maps.

Latitude-Longitude and Quadrangle Maps

Earth is a spherical body or globe, and specific points on the globe can be defined exactly using a geographic coordinate system in which points are defined by the

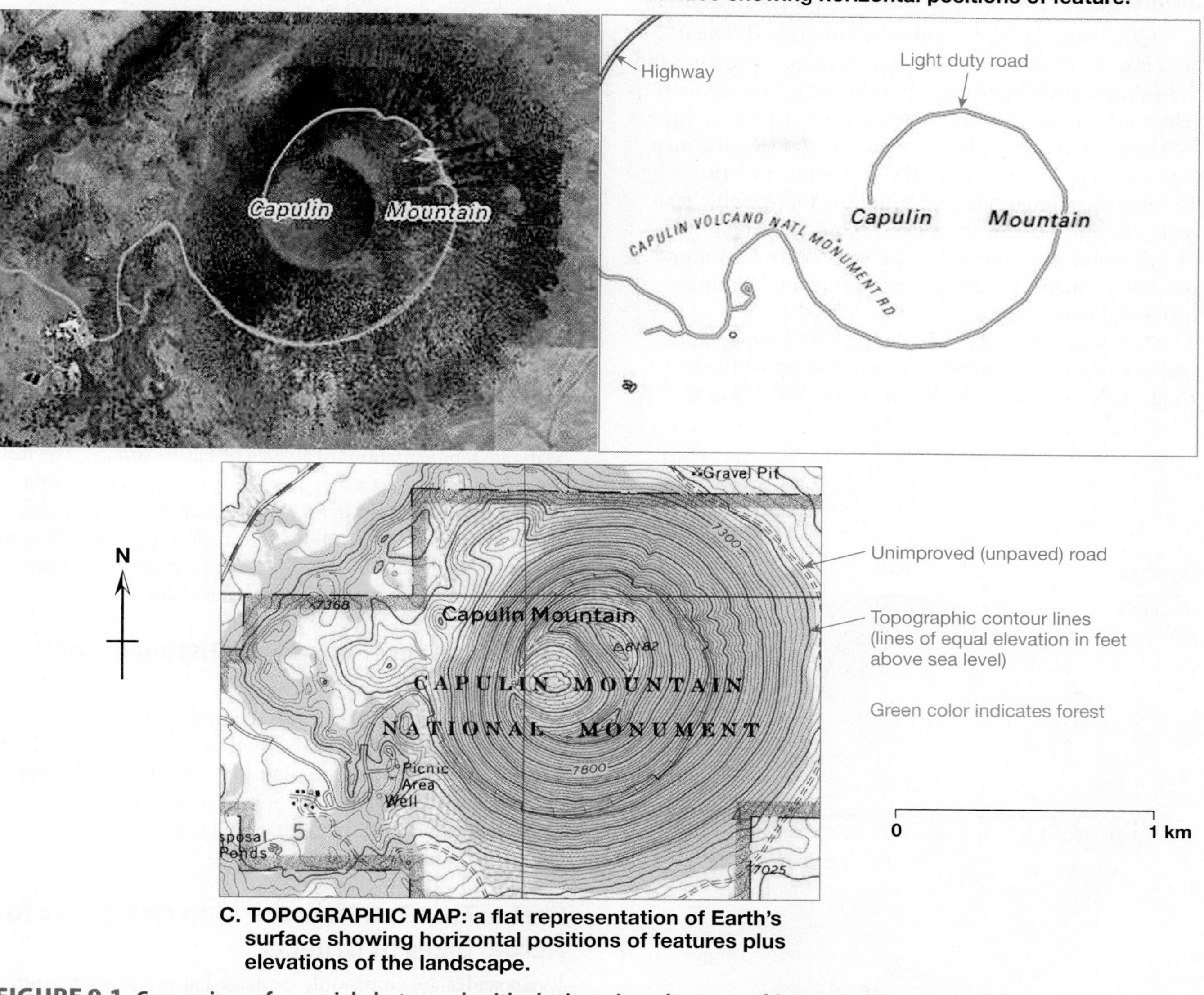

FIGURE 9.1 Comparison of an aerial photograph with planimetric and topographic maps. (Courtesy of USGS)

intersection of imaginary reference lines. The most traditional geographic coordinate system consists of reference lines of geographic latitude and longitude.

Latitude-Longitude Coordinate System

Earth's spherical surface is divided into lines of latitude (*parallels*) that go around the world parallel to the Equator, and lines of longitude (*meridians*) that go around the world from pole to pole (**FIGURE 9.2**). There are 360 degrees (360°) around the entire Earth, so the distance from the Equator to a pole (one-fourth of the way around Earth) is 90° of latitude. The Equator is assigned a value of zero degrees (0°) latitude, the North Pole is 90 degrees north latitude (90°N), and the South Pole is 90 degrees south latitude (90°S). The *prime meridian* is zero degrees of longitude and runs from pole to pole through Greenwich, England. Locations in Earth's Eastern Hemisphere are located in degrees east of the prime meridian, and points in the Western Hemisphere are located in degrees west of the prime meridian. Therefore, any point on Earth (or a map) can be located by its latitude-longitude coordinates. The latitude coordinate of the point is its position in degrees north or south of the Equator. The longitude coordinate of the point is its position in degrees east or west of the prime meridian. For example, point **A** in **FIGURE 9.2** is located at coordinates of: 20° north latitude, 120° west longitude. For greater detail, each degree of latitude and longitude can also be subdivided into 60 minutes 60′, and each minute can be divided into 60 seconds (60″).

Quadrangle Maps. Most depict rectangular sections of Earth's surface, called quadrangles. A **quadrangle** is a relatively rectangular area of Earth's surface, bounded by lines of latitude at the top (north) and bottom (south) and by lines of longitude on the left (west) and right (east)—see FIGURE 9.2. A *quadrangle map* is the map of a quadrangle.

Quadrangle maps are published in many different sizes but the most common USGS sizes are 15-minute and 7.5-minute quadrangle maps (FIGURE 9.2). The numbers refer to the amount of area that the maps depict, in degrees of latitude and longitude. A 15-minute topographic map represents an area that measures 15 minutes of latitude by 15 minutes of longitude. A 7.5-minute topographic map represents an area that measures 7.5 minutes of latitude by 7.5 minutes of longitude. Therefore, four 7.5-minute quadrangle maps (FIGURE 9.2) comprise one 15-minute quadrangle map.

A reduced copy of a 7.5-minute USGS topographic map is provided in FIGURE 9.3. Notice its name (Ritter Ridge, CA) and size (7.5 Minute Series, SW 1/4 of the Lancaster 15' Quadrangle) in the upper right and lower right corners of the map, respectively. Also notice that the map has colors, patterns, and symbols (FIGURE 9.4) that are used to depict water bodies, vegetation, roads, buildings, political boundaries, place names, and other natural and cultural features of the landscape. The lower right corner of the map indicates that the map was originally published in 1958, but it was photorevised in 1974. *Photorevised* means that aerial photographs (from airplanes) were used to discover changes on the landscape, and the changes are overprinted on the maps in a standout color like purple, red, or gray. The main new features shown on this 1974 photorevised map are the California Aqueduct (that carries water south, from the Sierra Nevada Mountains to the southern California desert) and several major highways.

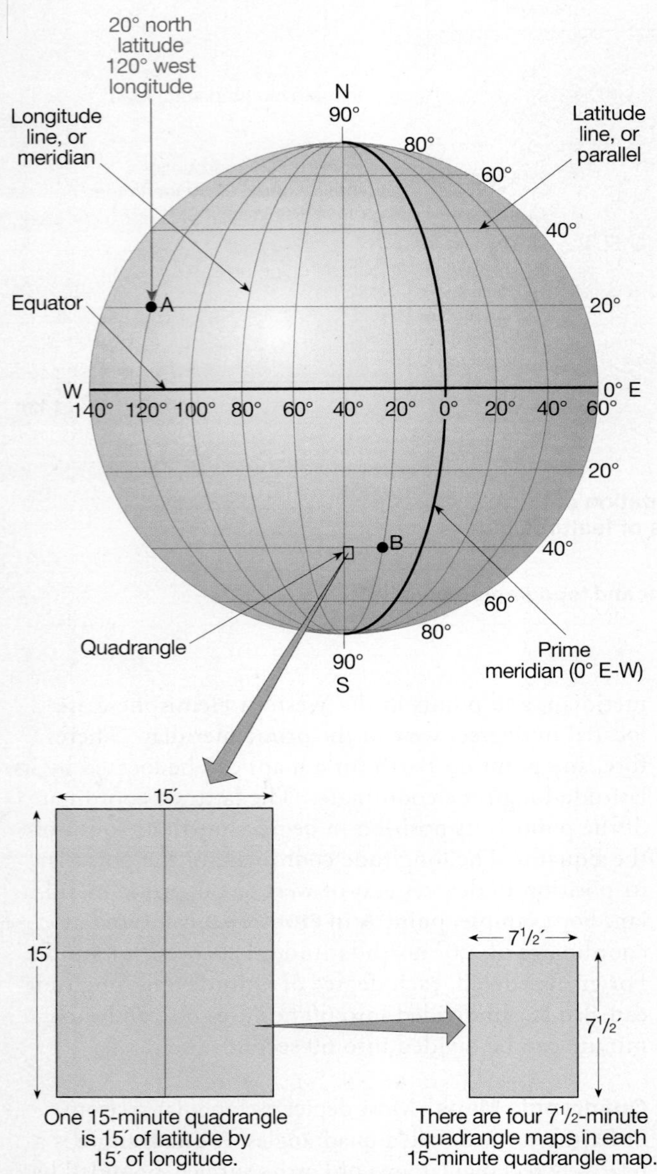

FIGURE 9.2 Latitude and longitude coordinate system and quadrangles. Point **A** is located at coordinates of 20° north latitude, 120° west longitude. Refer to text for discussion.

Map Scales

Maps are representations of an area of Earth's surface. The real sizes of everything on a map have been reduced so they fit a sheet of paper or computer screen. So maps are scale models. To understand how the real world is depicted by the map, you must refer to the map scales. Topographic maps commonly have any or all of the following kinds of scales.

Bar Scales for Measuring Distances on the Map

The most obvious scales on topographic maps are the **bar scales (graphic scales)** printed in their lower margins (FIGURE 9.3). Bar scales are rulers for measuring distances on the map. U.S. Geological Survey topographic maps generally have four different bar scales: miles, feet, kilometers, and meters.

Scales That Tell How the Map Compares to Actual Sizes of Objects

Ratio scales are commonly expressed above the bar scales in the bottom margins of topographic maps and express the ratio of a linear dimension on the map to the actual dimension of the same feature on the ground (in real life). For example, the ratio scale of the map in FIGURE 9.3 is written as "SCALE 1:24,000." This indicates that any unit (inch, centimeter, foot, etc.) on the map is actually 24,000 of the same units (inches, centimeters, feet) on the ground. So 1 cm on the map represents 24,000 cm on the ground, or your thumb width on the map represents 24,000 thumb widths on the ground. The ratio scale can also be interpreted as a **fractional scale**, which indicates how much smaller something is than its actual size on the ground. A map ratio scale of 1:24,000 equals a fractional scale of 1/24,000. This means that everything on the map is 1/24,000th of its actual size on the ground.

Verbal Scales Express Map Proportions in Common Terms

Verbal scales are sentences that help readers understand map proportions in relation to common units of measurement. For example, reconsider the map in FIGURE 9.3 with "SCALE

1:24,000." Knowing that 1 inch on the map equals 24,000 inches on the ground is not very convenient, because no one measures big distances in thousands of inches! However, if you divide the 24,000 inches by 12 to get 2000 ft, then the scale suddenly becomes useful: "1 in. on the map = 2000 ft on the ground." An American football field is 100 yards (300 ft) long, so: "1 in. on the map = $6\frac{2}{3}$ football fields."

On a map with a scale of 1:63,360, "1 inch equals 63,360 inches" is again not meaningful in daily use. But there are 63,360 inches in a mile. So, the verbal scale, "1 inch equals 1 mile" is very meaningful. A standard 1:62,500 map (15-minute quadrangle map commonly used in parts of Alaska) is very close to this scale, so "one inch equals approximately one mile" is often written on such a map. Note that verbal scales are often approximate because their sole purpose is to help the reader make general sense of how the map relates to sizes of real objects on the ground.

Declination and Compass Bearings

Directional information is summarized as a trident-shaped symbol like the one in the lower left corner of FIGURE 9.3 (Ritter Ridge Quadrangle). Because longitude lines form the left and right boundaries of a topographic map, north is always at the top of the quadrangle. This is called grid north (GN) and is usually very close to the same direction as *true north* on the actual Earth. Unfortunately, magnetic compasses are not attracted to grid north or true north (the geographic North Pole). Instead, they are attracted to the *magnetic north pole* (MN), currently located northwest of Hudson Bay in Northern Canada, about 700 km (450 mi) from the true North Pole.

What Is Declination?

The trident-shaped symbol on the bottom margin of topographic maps shows the **declination** (difference in degrees) between compass north (MN) and true north (usually a *star* symbol). Also shown is the declination between true north (*star* symbol) and grid north (GN). The magnetic pole migrates very slowly, so the declination is exact only for the year listed on the map. You can obtain the most recent magnetic data for your location from the NOAA National Geophysical Data Center (**http://www.ngdc.noaa.gov/geomag-web/#declination**).

What Is a Compass Bearing?

A **bearing** is the *compass direction* along a line from one point to another. If expressed in degrees east or west of true north or south, it is called a *quadrant bearing*. Or it may be expressed in degrees between 0 and 360, called an *azimuth bearing*, where north is 0° (or 360°), east is 90°, south is 180°, and west is 270°. Linear geologic features (faults, fractures, dikes), lines of sight and travel, and linear property boundaries are all defined on the basis of their bearings. But because a compass points to Earth's *magnetic north* (MN) pole rather than the true North Pole, one must correct for this difference. If the MN arrow is to the east of true north (star symbol), then subtract the degrees of declination from your compass reading (imagine that you are rotating your compass counter-clockwise to compensate for declination). If the MN arrow is to the west of true north, then add the degrees of declination to your compass reading (imagine that you rotated your compass clockwise). These adjustments will mean that your compass readings are synchronized with the map (so long as you used the latest declination values obtained from NOAA).

How to Set a Compass for Declination

Some compasses allow you to rotate their basal ring graduated in degrees to correct for the magnetic declination. If the MN arrow is 5° east (right) of true north, then you would rotate the graduated ring 5° east (clockwise, to subtract 5° from the reading). If the MN arrow is 5° west (left) of true north, then you would rotate the graduated ring 5° west (counter-clockwise, to add 5° to the reading).

How to Determine a Compass Bearing on a Map

To determine a compass bearing on a topographic map, follow the directions in FIGURE 9.5. Then imagine that you are buying a property for your dream home. The boundary of the property is marked by four metal rods driven into the ground, one at each corner of the property. The location of these rods is shown on the map in FIGURE 9.5 (left side) as points ***A*, *B*, *C*,** and ***D*.** The property deed notes the distances between the points *and* bearings between the points. This defines the shape of the property. Notice that the northwest edge of your property lies between two metal rods located at points ***A*** and ***B*.** You can measure the distance between the points using a tape measure. How can you measure the bearing?

First, draw a line (very lightly in pencil so that it can be erased) through the two points, ***A*** and ***B*.** Make sure the line also intersects an edge of the map. In both parts of FIGURE 9.5, a line was drawn through points ***A*** and ***B*** so that it also intersects the east edge of the map. Next, orient a protractor so that its 0° and 180° marks are on the edge of the map, with the 0° end toward geographic north. Place the origin of the protractor at the point where your line ***A–B*** intersects the edge of the map. You can now read a bearing of 43° east of north. We express this as a quadrant bearing of "North 43° East" (written N43°E) or as an azimuth bearing of 43°. If you were to determine the opposite bearing, from ***B*** to ***A*,** then the bearing would be pointing southwest and would be read as "South 43° West," or as an azimuth of 223°. Remember that a compass points to Earth's *magnetic north* pole (MN) rather than true north or grid north (GN). When comparing the bearing read directly from the map to a bearing read from a compass, you must adjust your compass reading to match true north or grid north (GN) of the map, as described above.

You also can use a compass to read bearings, as shown in FIGURE 9.5 (right). Ignore the compass needle and use the compass as if it were a circular protractor.

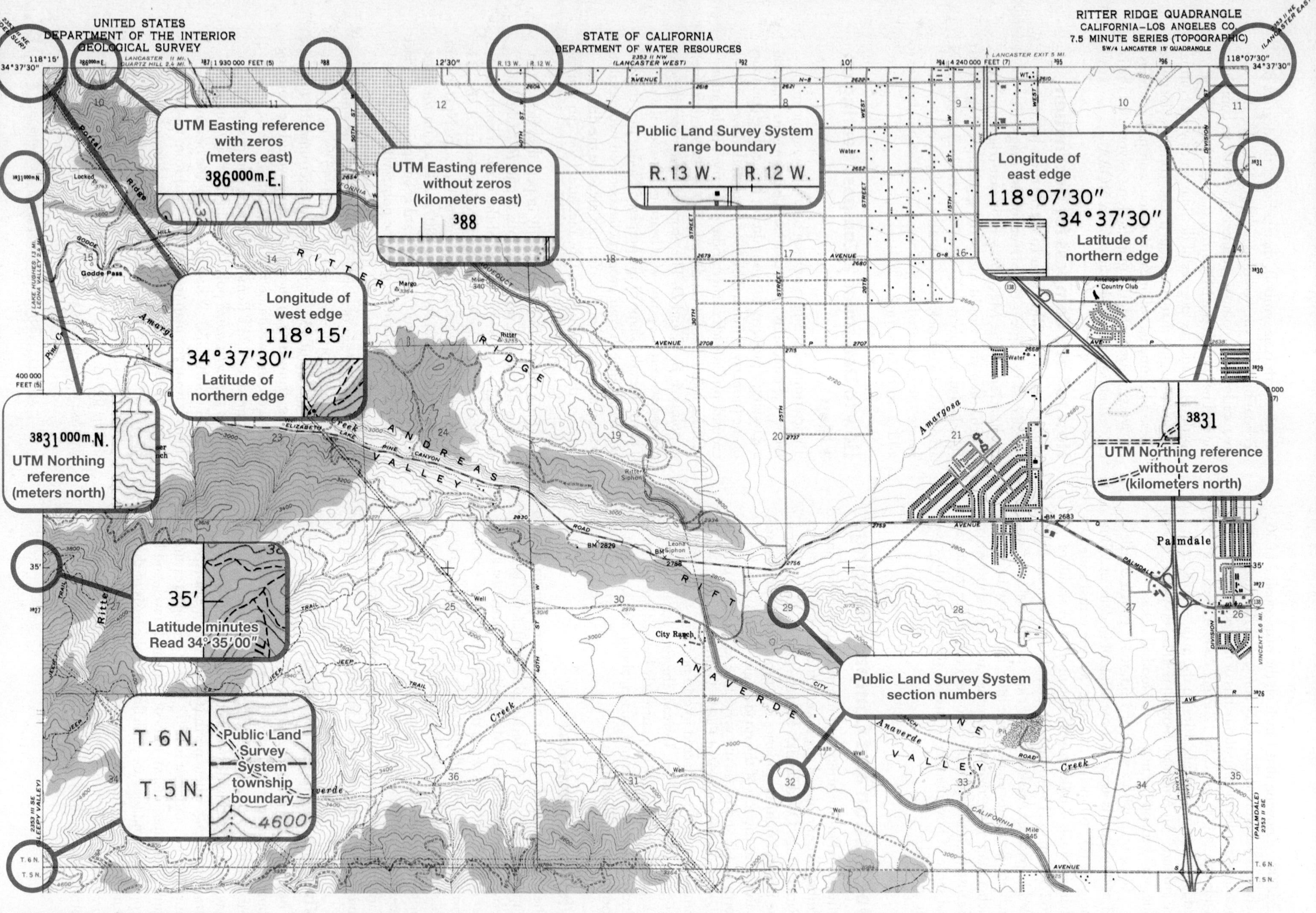

FIGURE 9.3 (NORTHERN HALF) USGS (US Topo) Ritter Ridge, CA 7.5–minute topographic quadrangle map. (Northern half of map reduced to about 55% of its actual size) (Courtesy of USGS)

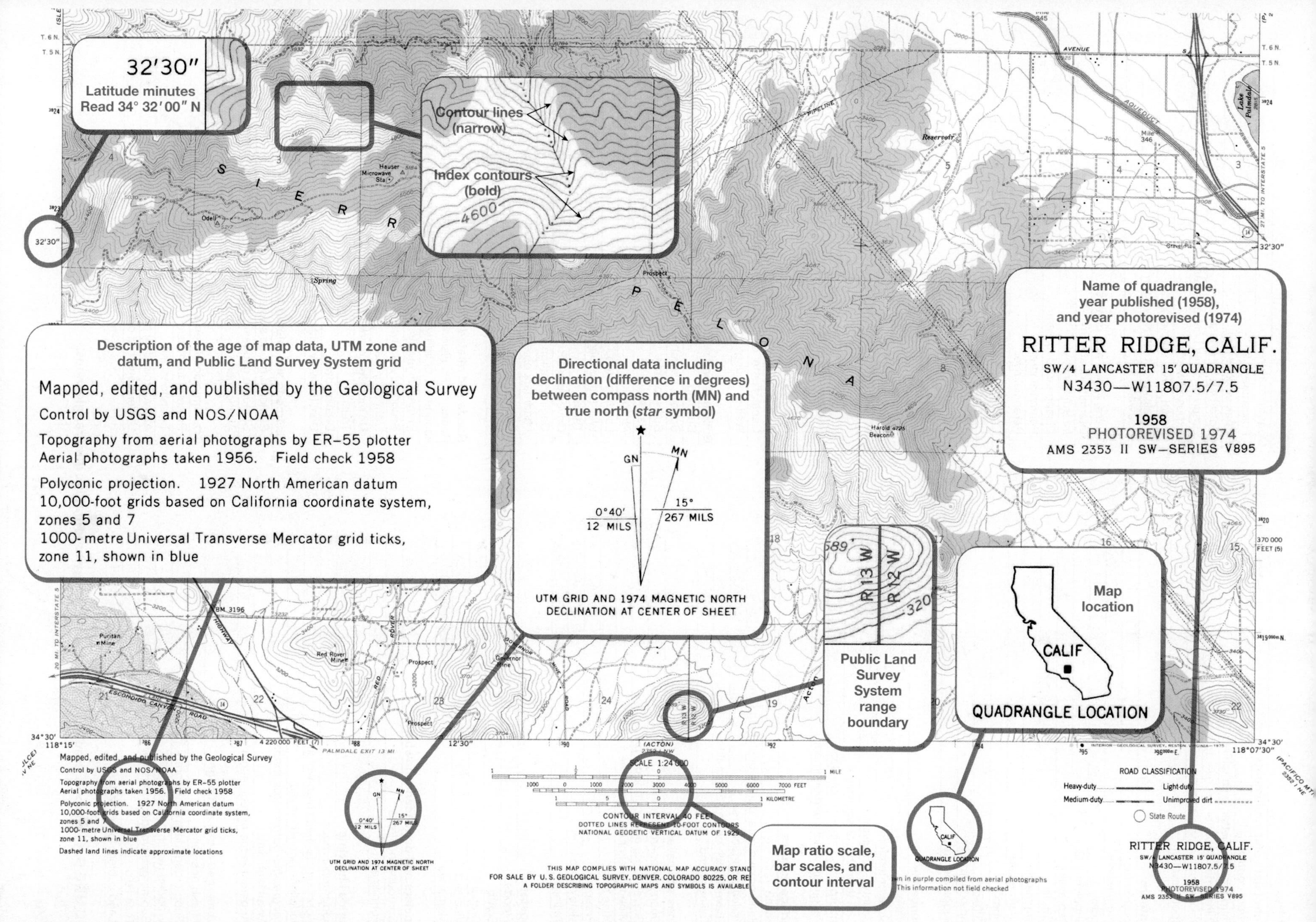

FIGURE 9.3 (SOUTHERN HALF) USGS Ritter Ridge, CA 7.5–minute topographic quadrangle map. (Southern half of map reduced to about 55% of its actual size) (Courtesy of USGS)

Control data and monuments

Vertical control	
Third order or better, with tablet	BM×16.3
Third order or better, recoverable mark	×120.0
Bench mark at found section corner	BM 18.6
Spot elevation	×5.3

Contours

Topographic	
Intermediate	
Index	
Supplementary	
Depression	
Cut; fill	
Bathymetric	
Intermediate	
Index	
Primary	
Index primary	
Supplementary	

Boundaries

National	
State or territorial	
County or equivalent	
Civil township or equivalent	
Incorporated city or equivalent	
Park, reservation, or monument	

Surface features

Levee	Levee
Sand or mud area, dunes, or shifting sand	Sand
Intricate surface area	Strip mine
Gravel beach or glacial moraine	Gravel
Tailings pond	Tailings pond

Mines and caves

Quarry or open pit mine	
Gravel, sand, clay, or borrow pit	
Mine tunnel or cave entrance	
Mine shaft	
Prospect	X
Mine dump	Mine dump
Tailings	Tailings

Vegetation

Woods	
Scrub	
Orchard	
Vineyard	
Mangrove	Mangrove

Glaciers and permanent snowfields

Contours and limits	
Form lines	

Marine shoreline

Topographic maps	
Approximate mean high water	
Indefinite or unsurveyed	
Topographic-bathymetric maps	
Mean high water	
Apparent (edge of vegetation)	

Submerged areas and bogs

Marsh or swamp	
Submerged marsh or swamp	
Wooded marsh or swamp	
Submerged wooded marsh or swamp	
Rice field	Rice
Land subject to inundation	Max pool 431

Coastal features

Foreshore flat	Mud
Rock or coral reef	Reef
Rock bare or awash	
Group of rocks bare or awash	
Exposed wreck	
Depth curve; sounding	3
Breakwater, pier, jetty, or wharf	
Seawall	

Rivers, lakes, and canals

Intermittent stream	
Intermittent river	
Disappearing stream	
Perennial stream	
Perennial river	
Small falls; small rapids	
Large falls; large rapids	
Masonry dam	
Dam with lock	
Dam carrying road	
Perennial lake; Intermittent lake or pond	
Dry lake	Dry lake
Narrow wash	
Wide wash	Wide wash
Canal, flume, or aquaduct with lock	
Well or spring; spring or seep	

Buildings and related features

Building	
School; church	
Built-up area	
Racetrack	
Airport	
Landing strip	
Well (other than water); windmill	
Tanks	
Covered reservoir	
Gaging station	
Landmark object (feature as labeled)	
Campground; picnic area	
Cemetery: small; large	Cem

Roads and related features

Roads on Provisional edition maps are not classified as primary, secondary, or light duty. They are all symbolized as light duty roads.

Primary highway	
Secondary highway	
Light duty road	
Unimproved road	
Trail	
Dual highway	
Dual highway with median strip	

Railroads and related features

Standard gauge single track; station	
Standard gauge multiple track	
Abandoned	

Transmission lines and pipelines

Power transmission line; pole; tower	
Telephone line	Telephone
Aboveground oil or gas pipeline	
Underground oil or gas pipeline	Pipeline

FIGURE 9.4 Symbols used on U.S. Geological Survey topographic quadrangle maps.

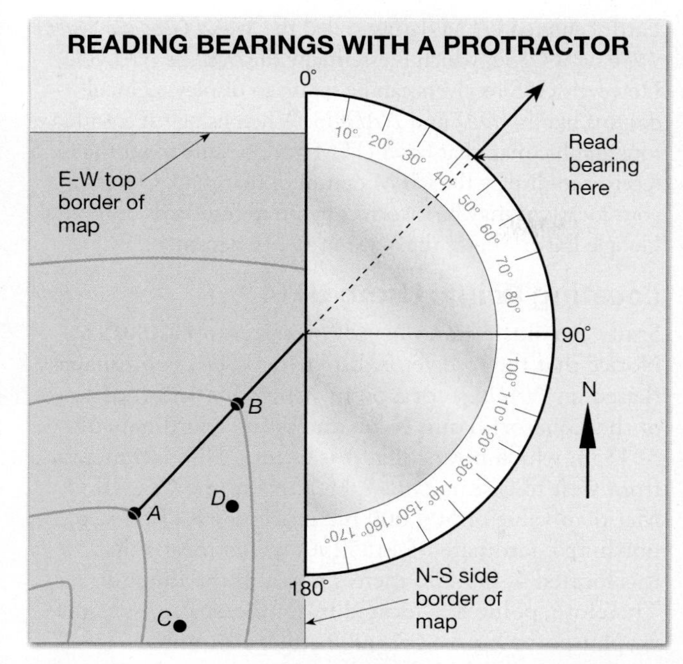

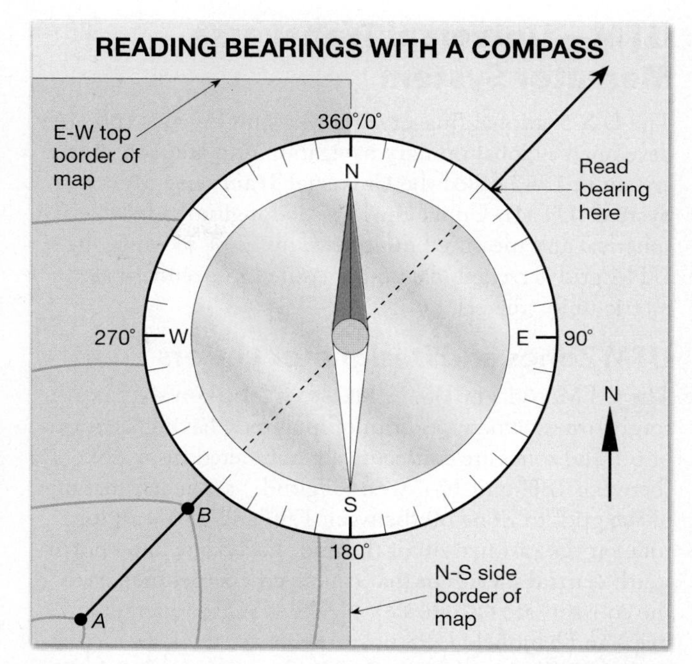

FIGURE 9.5 How to read a bearing (compass direction) on a map. A bearing is read or plotted on a map, from one point to another, using a protractor (left) or compass (right). To determine a bearing on a map, draw a straight line from the starting point to the destination point and also through any one of the map's borders. For example, to find the bearing from ***A*** to ***B***, a line was drawn through both points and the east edge of the map. Align a protractor (left drawing) or the N-S or E-W directional axis of a compass (right drawing) with the map's border and read the bearing in degrees toward the direction of the destination. In this example, notice that the *quadrant bearing* from point ***A*** to ***B*** is North 43° East (left map, using protractor) or an *azimuth bearing* of 43°. If you walked in the exact opposite direction, from ***B*** to ***A***, then you would walk along a quadrant bearing of South 43° West or an azimuth bearing of 223° (i.e., 43° + 180° = 223°). Remember that a compass points to Earth's magnetic north pole (MN) rather than true north (GN, grid north). When comparing the bearing read directly from the map to a bearing read from a compass, you must adjust your compass reading to match grid north (GN) of the map, as described in the text.

Some compasses are graduated in degrees, from 0–360, in which case you read an azimuth bearing from 0–360°. Square azimuth protractors for this purpose are provided in GeoTools Sheets 3 and 4 at the back of this manual.

GPS—Global Positioning System

The Global Positioning System (GPS) is a technology used to make *precise* (exact) and *accurate* (error free) measurements of the location of points on Earth. It is used for geodesy—the science of measuring changes in Earth's size and shape, and the position of objects, over time. GPS technology is based on a constellation of about 30 satellites that take just 12 hours to orbit Earth. They are organized among six circular orbits (20,200 km, or 12,625 mi above Earth) so that a minimum of six satellites will be in view to users anywhere in the world at any time. The GPS constellation is managed by the United States Air Force for operations of the Department of Defense, but they allow anyone to use it anywhere in the world.

How GPS Works

Each GPS satellite communicates simultaneously with fixed ground-based Earth stations and other GPS satellites, so it knows exactly where it is located relative to the center of Earth and Universal Time Coordinated (UTC, also called Greenwich Mean Time). Each GPS satellite also transmits its own radio signal on a different channel, which can be detected by a fixed or handheld GPS receiver. If you turn on a handheld GPS receiver in an unobstructed outdoor location, then the receiver immediately acquires (picks up) the radio channel of the strongest signal it can detect from a GPS satellite. It downloads the navigational information from that satellite channel, followed by a second, third, and so on. A receiver must acquire and process radio transmissions from at least four GPS satellites to triangulate a determination of its exact position and elevation—this is known as a **fix**. But a fix based on more than four satellites is more accurate. In North America and Hawaii, the accuracy of the GPS constellation is enhanced by WAAS (Wide Area Augmentation System) satellites operated by the Federal Aviation Administration. WAAS uses ground-based reference stations to measure small variations in GPS satellites signals and correct them. The corrections are transmitted up to geostationary WAAS satellites, which broadcast the corrections back to WAAS-enabled GPS receivers on Earth.

GPS Accuracy

The more channels a GPS receiver has, the faster and more accurately it can process data from the most satellites. The best GPS receivers have millimeter accuracy, but handheld WAAS-enabled GPS receivers and smartphones with GPS are accurate to within 3 meters. Receivers lacking WAAS are only accurate to within about 9 meters.

UTM—Universal Transverse Mercator System

The U.S. National Imagery and Mapping Agency (NIMA) developed a global military navigation grid and coordinate system in 1947 called the **Universal Transverse Mercator System (UTM).** Unlike the latitude-longitude grid that is spherical and measured in degrees, minutes, seconds, the UTM grid is rectangular and measured in decimal-based metric units (meters, km).

UTM Zones and Designator Letters

The UTM grid (top of FIGURE 9.6) is based on sixty north–south **zones,** which are strips of longitude having a width of 6°. The zones are consecutively numbered from Zone 01 (between 180° and 174° west longitude) at the left margin of the grid, to Zone 60 (between 174° and 180° east longitude) at the east margin of the grid. Each zone has a north-south **central meridian** that is perfectly perpendicular to the equator (see FIGURE 9.6). Newer USGS topographic maps and handheld GPS instruments use the U.S. Department of Defense, Military Grid Reference System (MGRS) that divides the zones into east-west (horizontal) rows identified by designator letters (FIGURE 9.6). These rows are 8° wide and lettered consecutively from C (between 80° and 72° south latitude) through X (between 72° and 84° north latitude). Letters I and O are not used, because they could be confused with numbers 0 and 1.

UTM Coordinates

The UTM coordinates of a point on Earth's surface include the zone, designator letter, easting, and northing (FIGURE 9.6). The **easting** coordinate is the west-to-east distance in meters within the zone, where the Central Meridian is assigned a false easting of 500,000 meters. The **northing** coordinate is the distance from the Equator measured in meters. In the Northern Hemisphere, northings are given in meters north of the Equator (from 0 m at the Equator to 9,300,000 m at 84° North latitude). To avoid negative numbers for northings in the Southern Hemisphere, NIMA assigned the Equator a reference northing of 10,000,000 meters. So northings in the southern hemisphere range from 1,100,000 m at 80° South latitude, to 10,000,000 m at the Equator. When recording UTM coordinates, it is commonplace to write the zone number and designator letter, then the easting, and then the northing (i.e., 18S 384333 4455250 in FIGURE 9.6). It is also wise to note what UTM datum was used to define the coordinates.

UTM Datums

Because satellites did not exist until 1957, and GPS navigational satellites did not exist until decades later, the UTM grid was applied for many years using regional ground-based surveys to determine locations of the grid boundaries. Each of these regional or continental surveys is called a **datum** and is identified by its location and the year it was surveyed. Examples include the *North American Datum 1927 (NAD27)* and the *North American Datum 1983 (NAD83)*, which appear on many Canadian and U.S. Geological Survey maps. The Global Positioning System (GPS) relies on an Earth-centered UTM datum called the *World Geodetic System 1984* or *WGS84*, which is essentially the same as *NAD83*. However, GPS receivers can be set up to display regional datums like *NAD27* and *NAD 83*. When using GPS with a topographic map that has a UTM grid, be sure to set the GPS receiver to display the UTM datum of that map. Otherwise, your locations may be incorrect by up to hundreds of meters. Google Earth™ uses the *WGS84* UTM datum.

Locating Points Using UTM

Study the illustration of a GPS receiver in FIGURE 9.6. Notice that the receiver is displaying UTM coordinates (based on *NAD27*) for a point **X** in Zone 18S (north of the Equator). Point **X** has an easting coordinate of 384333, which means that it is located 384,333 meters from west to east in Zone 18 relative to the Central Meridian value of 500,000 meters. Point **X** also has a northing coordinate of 4455250, which means that it is located 4,455,250 meters north of the Equator. Therefore, point **X** is located in southeast Pennsylvania. To plot point **X** on a 1:24,000 scale, $7\frac{1}{2}$-minute topographic quadrangle map, see FIGURE 9.7.

Point **X** is located within the Lititz, PA $7\frac{1}{2}$-minute (USGS, 1:24,000 scale) topographic quadrangle map (FIGURE 9.7). Information printed on the map margin indicates that the map has blue ticks spaced 1000 m apart along its edges that conform to *NAD27*, Zone 18. Notice how the ticks for northings (blue) and eastings (green) are represented on the northwest corner of the Lititz map—FIGURE 9.7B. One northing label is written out in full (4456000m N) and one easting label is written out in full (384000m E), but the other values are given in UTM shorthand for kilometers (i.e., they do not end in 000m). Because point **X** has an easting of 384333 within Zone 18S, it must be located 333 m east of the tick mark labeled 384000m E along the top margin of the map. Because Point **X** has a northing of 4455250, it must be located 250 m north of the tick mark labeled as 4455 in UTM shorthand (which stands for 4455 km or 4,455,000 m). Distances east and north can be measured using a ruler and the map's graphic bar scale as a reference (333 m = 0.333 km, 250 m = 0.250 km). However, you can also use the graphic bar scale to construct a UTM grid like the one in FIGURE 9.7C. If you construct such a grid and print it onto a transparency, then you can use it as a UTM *grid overlay*. To plot a point or determine its coordinates, place the grid overlay on top of the square kilometer in which the point is located. Then use the grid as a two-dimensional ruler for the northing and easting. Grid overlays for many different scales of UTM grids are provided in GeoTools Sheets 2–4 at the back of the manual for you to cut out and use.

Public Land Survey System

The **U.S. Public Land Survey System (PLSS)** was initiated in 1785 when the U.S. government required a way to divide public land of the Western Frontier (land west of the thirteen original colonies) into small parcels that could be transferred to ownership by private citizens. The U.S. Bureau of

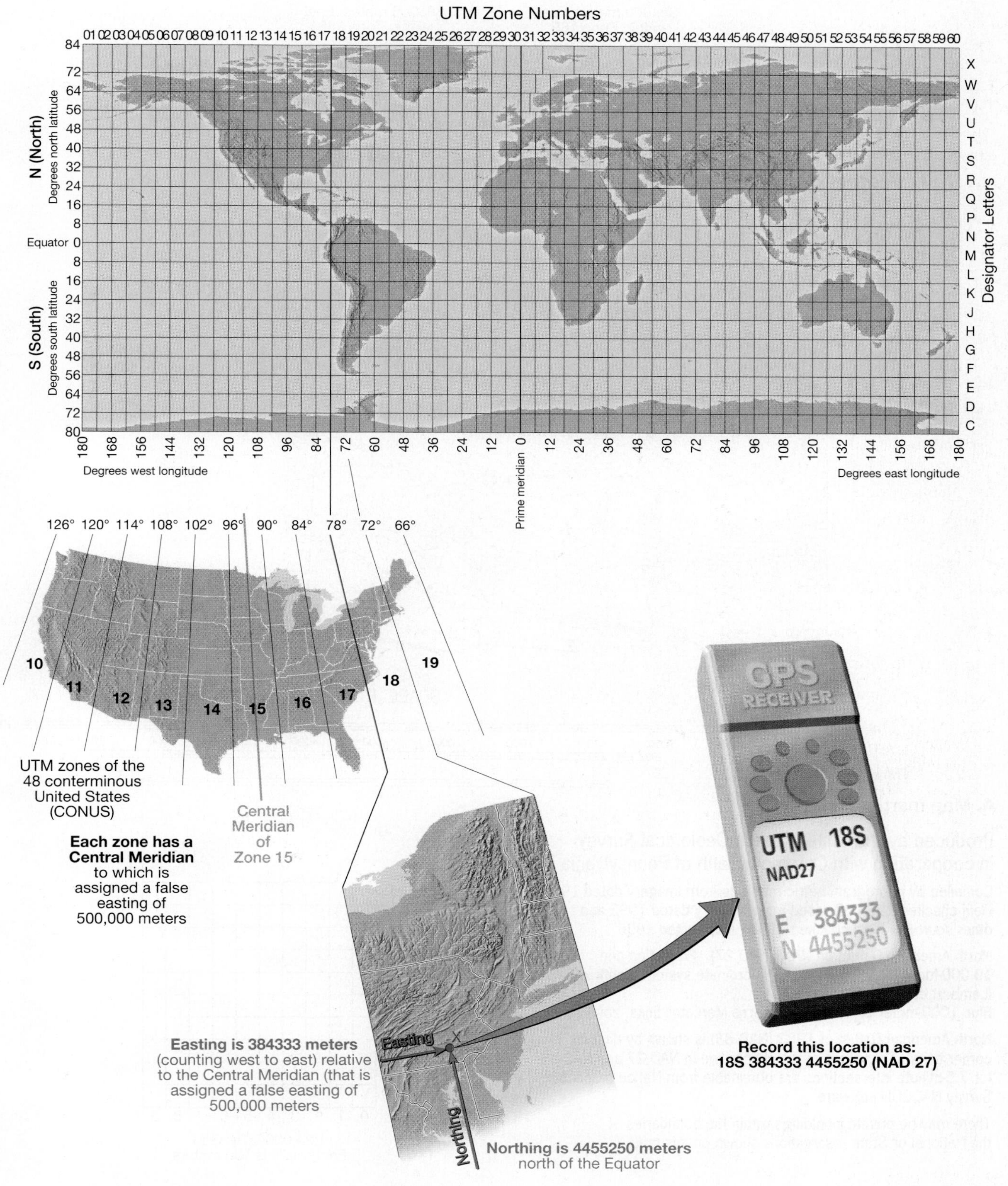

FIGURE 9.6 UTM with GPS. A handheld Global Positioning System (GPS) receiver is set to display the Universal Transverse Mercator (UTM) grid and coordinate system called *North American Datum 1927* (*NAD27*), same as the map grid in **FIGURE 9.8**. When operated at point **X**, it displays its exact location as a zone and designator letter (18S) plus an easting and northing. Most hand-held GPS devices use the *World Geodetic System of 1984* (*WGS84*) instead of *NAD27*. Be sure your grid system is the same as that of any map on which you may plot the GPS coordinates. Refer to the text for explanation.

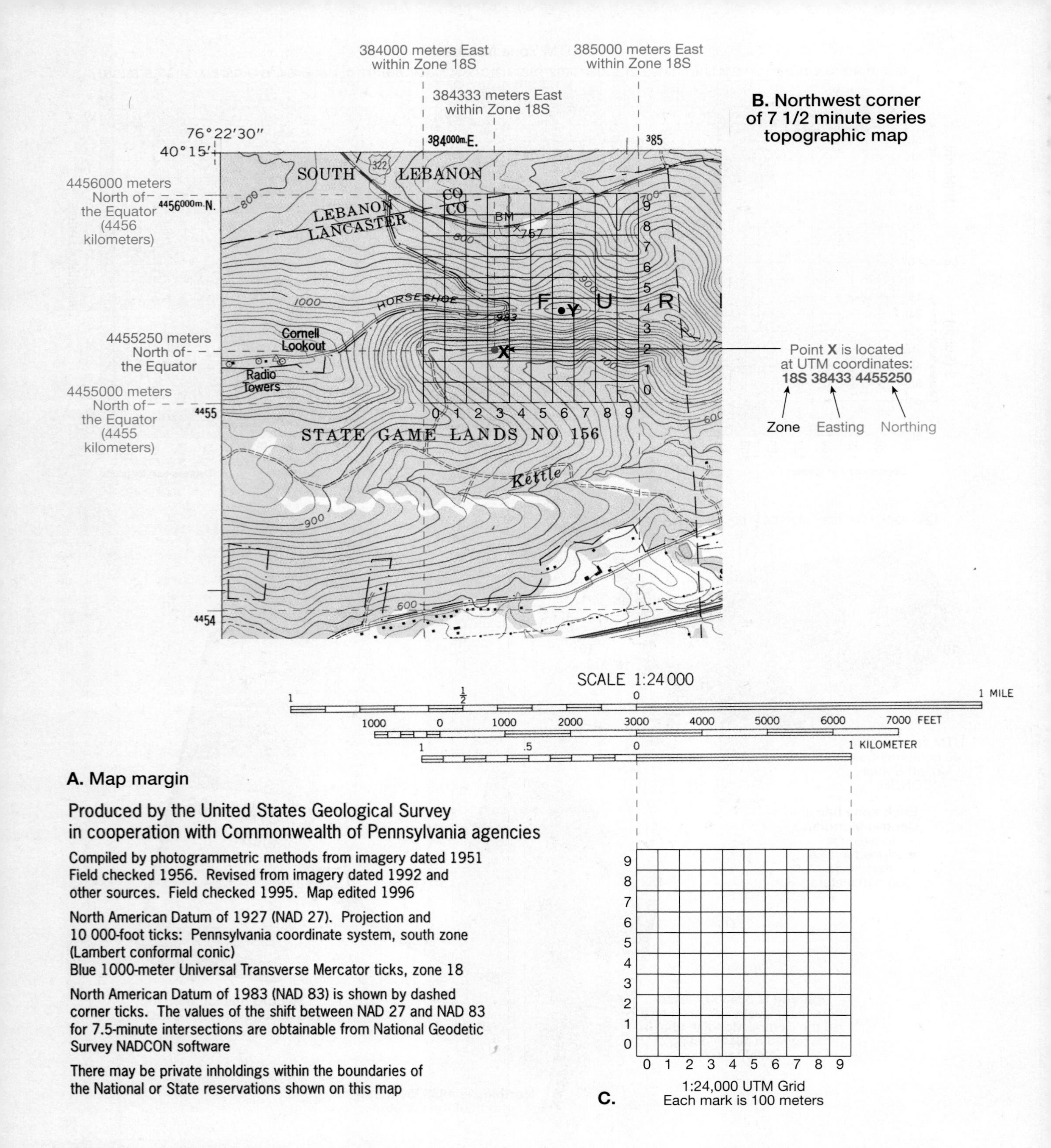

FIGURE 9.7 UTM with topographic maps. Refer to the text for discussion. Point **X** (from **FIGURE 9.7**) is located within the Lititz, PA $7\frac{1}{2}$-minute (USGS, 1:24,000 scale) topographic quadrangle map. A. Map margin indicates that the map includes UTM grid data based on North American Datum 1927 (NAD27, Zone 18) and represented by blue ticks spaced 1000 meters (1 km) apart along the map edges. B. Connect the blue 1000-m ticks to form a grid square, each representing 1 square kilometer. Northings (blue) are read along the N-S map edge, and eastings (green) are located along the E-W map edge. C. You can construct a 1-km grid (1:24,000 scale) from the map's bar scale, then make a transparency of it to form a grid overlay (see GeoTools Sheets 2 and 4 at back of manual). Place the grid overlay atop the 1-kilometer square on the map that includes point **X**, and determine the NAD27 coordinates of **X** as shown (red).

Land Management (BLM) regulates and maintains public land using the PLSS. It is also used as the basis for many legal surveys of private land that was once publicly owned.

PLSS Township-and-Range Grids

The PLSS is a square grid system centered on any one of dozens of **principal meridians** (lines running north and south) and **base lines** established among all but the thirteen original states and a few states derived from them. Once a principal meridian and base line was established, additional lines were surveyed parallel to them and 6 miles apart. This created a grid of 6 mi by 6 mi squares of land (FIGURE 9.8). The north–south squares of the grid are called **townships** and are numbered relative to the base line (Township 1 North, Township 2 North, etc.). The east–west squares of the grid are **ranges** and are numbered relative to the principal meridian (Range 1 West, Range 2 West, etc.). Each 6 mi by 6 mi square is, therefore, identified by its township and range position in the PLSS grid. For example, the township in FIGURE 9.8B is located at T1S (Township 1 South) and R2W (Range 2 West). Although each square like this is identified as both a township and a range within the PLSS grid, it is common practice to refer to the squares as townships rather than township-and-ranges.

Defining Land Areas Using PLSS

The PLSS is designed to define the location of square or rectangular subdivisions of land. The 6 mi by 6 mi townships are used as political subdivisions in some states and often have place names. Each township square is also divided into 36 small squares, each having an area of 1 square mile (640 acres). These square-mile subdivisions of land are called **sections.**

Sections are numbered from 1 to 36, beginning in the upper right corner of the township (FIGURE 9.8B). Sometimes these are shown on topographic quadrangle maps, like the red-brown grid of square numbered sections in FIGURE 9.3. Any tiny area or point can be located precisely within a section by dividing the section into quarters (labeled NW, NE, SW, SE). Each of these quarters can itself be subdivided into quarters and labeled (FIGURE 9.8C).

ACTIVITY 9.3 Topographic Map Construction

THINK About It | **How are topographic maps constructed and interpreted?**

OBJECTIVE Construct topographic maps by drawing contour lines based on maps showing elevations of specific points and a digital terrain model.

PROCEDURES

1. **Before you begin**, read the Introduction, What Are Topographic Maps, US Topo Maps and Orthoimages, and Rules for Contour Lines, (p. 228). Also, this is **what you will need:**

   ___ Activity 9.3 Worksheet (p. 253) and pencil
   ___ calculator

2. **Then follow your instructor's directions** for completing the worksheet.

ACTIVITY 9.4 Topographic Map and Orthoimage Interpretation

THINK About It | **How are topographic maps constructed and interpreted?**

OBJECTIVE Interpret ("read") topographic maps and determine their effectiveness in comparison to, and combination with, US Topo orthoimages.

PROCEDURES

1. **Before you begin**, read the Introduction, What Are Topographic Maps, US Topo Maps and Orthoimages, and Rules for Contour Lines (p. 228) if you have not already done so. Also, this is **what you will need:**

   ___ Activity 9.4 Worksheet (p. 254) and pencil

2. **Then follow your instructor's directions** for completing the worksheet.

What Are Topographic Maps?

Topographic maps are miniature models of Earth's three-dimensional landscape, printed on two-dimensional pieces of paper or displayed on a flat computer screen. Two of the dimensions are the lengths and widths of objects and landscape features, similar to a planimetric map (FIGURE 9.1B). But the third dimension, elevation (height), is shown using the *contour lines*, which are lines of equal elevation used to represent hills and valleys (FIGURES 9.1C, 9.3). But how are the contour lines determined, and how does one interpret them to "read" a topographic map?

Aerial Photographs and Stereograms

The production of a topographic map begins with overlapping pairs of aerial photographs, called *stereo pairs*. Each stereo pair is taken from an airplane making two closely spaced passes over a region at the same elevation. The passes are flown far enough apart to provide the stereo effect, yet close enough to be almost directly above the land that is to be mapped. Aerial photos commonly are overlapped to form a **stereogram** (FIGURE 9.9), which appears three-dimensional (stereo) when viewed through a stereoscope.

Topographic Map Construction

Stereo pairs of aerial photographs are used to build a digital file of terrain elevations that is converted into the first draft of contour lines for the topographic map. Angular

PUBLIC LAND SURVEY SYSTEM (PLSS)

A. The Township-and-Range Grid.
The grid is made of E-W *township* strips of land and N-S *range* strips of land (columns of land) surveyed relative to a *principal meridian* (N-S line) and its *base line* (E-W line). Township strips are 6 miles high and numbered T1N, T2N, and so on north of the base line and T1S, T2S, and so on south of the base line. Range strips (rows) of land are 6 miles wide and numbered R1E, R2E, and so on east of the principal meridian and R1W, R2W, and so on west of the principal meridian. Each intersection of a township strip of land with a range strip of land forms a square, called a *township*. Note the location of Township T1S, R2W.

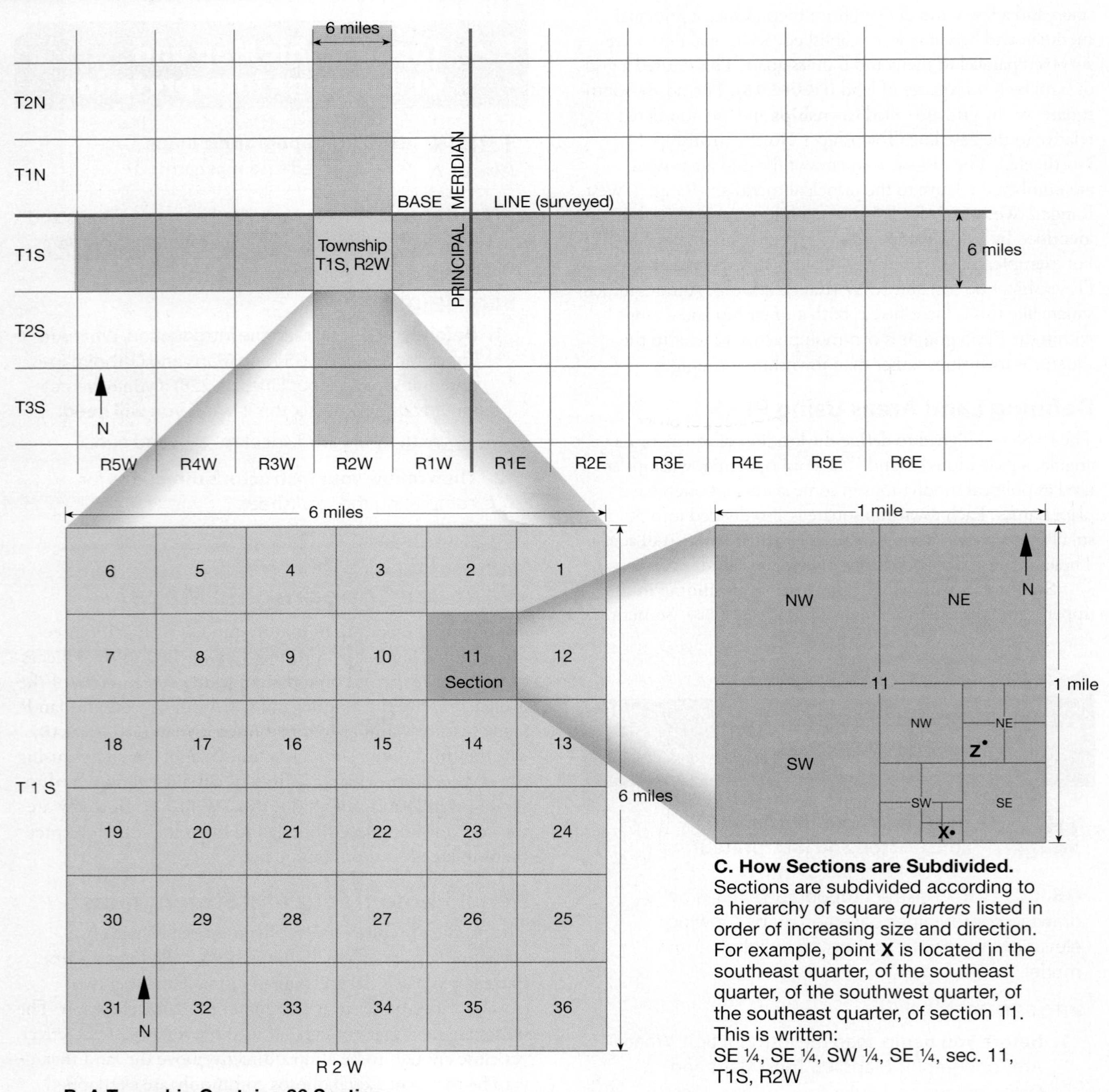

B. A Township Contains 36 Sections.
Each township is 6 miles wide by 6 miles long (36 square miles) and subdivided into 36 sections. Each section is 1 square mile (640 acres), called a *section*, and numbered as shown here.

C. How Sections are Subdivided.
Sections are subdivided according to a hierarchy of square *quarters* listed in order of increasing size and direction. For example, point **X** is located in the southeast quarter, of the southeast quarter, of the southwest quarter, of the southeast quarter, of section 11. This is written:
SE ¼, SE ¼, SW ¼, SE ¼, sec. 11, T1S, R2W

FIGURE 9.8 U.S. Public Land Survey System (PLSS). This survey system is based on grids of square townships, which are identified relative to *principal meridians* (N-S lines) of longitude and *base lines* (E-W lines, surveyed perpendicular to the principal meridian) that are unique to specific states or regions.

To view this stereogram without a stereoscope, cross your eyes until a third white dot appears between the first two. The image should then appear three-dimentional.

HOW TO USE A POCKET STEREOSCOPE

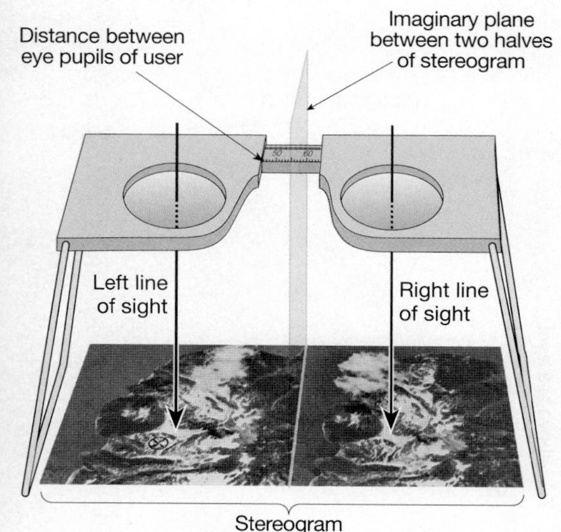

First have a partner measure the distance between the pupils of your eyes, in mm. Set that distance on the stereoscope as shown above. Then center the stereoscope over the stereogram, as above. Look through the stereoscope, and move it around slightly until the image becomes three-dimensional.

FIGURE 9.9 Stereogram (left) of Mount Meru region and how to view it. There are two methods used to view a stereogram so it appears to be three-dimensional: use a pocket stereoscope or cross your eyes. The tallest feature in this stereogram is Mount Meru, a 4,566-meter-high volcano located in east Africa (Tanzania), about 70 km (44 mi.) west of Mount Kilimanjaro. Both images in the stereogram are 20 km wide and 37 km tall (courtesy of NASA/PJL/NIMA). To view Mount Meru In Google Earth™, search: 03 14 36S, 36 45 41E.

distortion is then removed, and the exact elevations of the contour lines on the map are "ground truthed" (checked on the ground) using very precise altimeters and GPS. The final product is a topographic map like the one in **FIGURE 9.10**.

Notice how the contour lines in **FIGURE 9.10** occur where the landscape intersects horizontal planes of specific elevations: 0, 50, and 100 feet. Zero feet of elevation is sea level, so it is the coastline of the imaginary island. You can think of the contour lines for 50 and 100 feet above sea level as additional water levels above sea level. An "x" or triangle is often used to mark the highest point on a hilltop, with the exact elevation noted beside it. The highest point on the map in **FIGURE 9.10** is above the elevation of the highest contour line (100 feet) but below 150 feet (because there is no contour line for 150 feet). In this case, the exact elevation of the highest point on the island is marked by spot elevation ("x" labeled with the elevation of 108 feet).

US Topo Maps and Orthoimages

Historic USGS map series (**FIGURES 9.1C** and **9.3**), and those of most other countries, are one-page paper maps. However, the latest series of USGS topographic maps are layered digital maps called "US Topo" maps (**FIGURE 9.11**). The map layers can be turned on or off, including an aerial photograph layer called an *orthoimage*. The digital products can be downloaded free of charge (no registration required) or they can be ordered as printed paper maps.

Aerial photographs (taken from airplanes) are usually taken at angles oblique to the landscape, but topographic maps are representations of the landscape as viewed from directly above. **Orthoimages** are digitized aerial photographs or satellite images that have been orthorectified, corrected for distortions until they have the same geometry and uniform scale as a topographic map. Therefore, an orthoimage correlates exactly with its topographic map and reveals visual attributes of the landscape that are not visible on the topographic map. The topographic map, orthoimage, and other orthorectified "layers" of data can be added or removed to give the viewer extraordinary perspectives of the landscape. All of this can be done at US Topo, courtesy of the USGS and their partners. One can display features like hydrography (water bodies), roads, and UTM grid lines on a topographic base (**FIGURE 9.11A**), or display the topographic map layer on an orthoimage base (**FIGURE 9.11B**). All layers can be enlarged with outstanding resolution (**FIGURE 9.11C**). To learn more about obtaining and using US Topo products, watch a 6-minute USGS video (**http://gallery.usgs.gov/videos/663**).

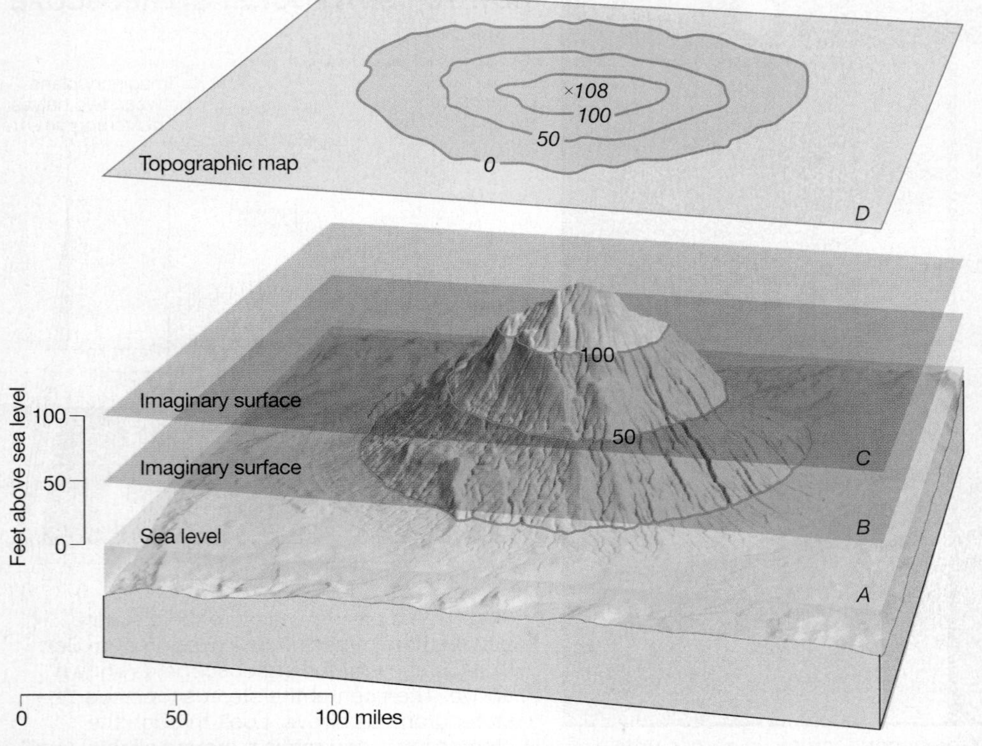

FIGURE 9.10 Topographic map construction. A *contour line* is drawn where a horizontal plane (**A**, **B**, or **C**) intersects the land surface. Where sea level (plane A) intersects the land, it forms the 0-ft contour line. Plane B is 50 ft above sea level, so its intersection with the land is the 50-ft contour line. Plane C is 100 ft above sea level, so its intersection with the land is the 100-ft contour line. D is the resulting topographic map of the island. It was constructed by looking down onto the island from above and tracing the 0, 50, and 100-ft contour lines. The elevation change between any two contour lines is 50 ft, so the map is said to have a 50-ft *contour interval*.

All contour lines on this map represent elevations in feet above sea level and are *topographic contour lines*. (Contours below sea level are called *bathymetric contour lines* and are generally shown in blue.)

Rules for Contour Lines

Each **contour line** connects all points on the map that have the same elevation above sea level (**FIGURE 9.12**, rule 1). Look at the topographic map in **FIGURE 9.3** and notice the light brown and heavy brown contour lines. The heavy brown contour lines are called **index contours**, because they have elevations printed on them (whereas the lighter contour lines do not; **FIGURE 9.12**, rule 6). Index contours are your starting point when reading elevations on a topographic map. For example, notice that every fifth contour line on **FIGURES 9.3** is an index contour. Also notice that the index contours are labeled with elevations in increments of 200 ft. This means that the map has five contours for every 200 ft of elevation, or a **contour interval** of 40 ft. This contour interval is specified at the center of the bottom margin of the map (**FIGURE 9.3**). All contour lines are multiples of the contour interval above a specific surface (almost always sea level). For example, if a map uses a 10-ft contour interval, then the contour lines represent elevations of 0 ft (sea level), 10 ft, 20 ft, 30 ft, 40 ft, and so on. Most maps use the smallest contour interval that will allow easy readability and provide as much detail as possible.

Additional rules for contour lines are also provided in **FIGURE 9.12** and the common kinds of landforms represented by contour lines on topographic maps (**FIGURE 9.13**). Your ability to use a topographic map is based on your ability to interpret what the contour lines mean (imagine the topography).

Reading Elevations

If a point on the map lies on an index contour, you simply read its elevation from that line. If the point lies on an unnumbered contour line, then its elevation can be determined by counting up or down from the nearest index contour. For example, if the nearest index contour is 300 ft, and your point of interest is on the fourth contour line *above* it, and the contour interval is 20 ft, then you simply count up by 20s from the index contour: 320, 340, 360, 380. The point is 380 ft above sea level. (Or, if the point is three contour lines *below* the index contour, you count down: 280, 260, 240; the point is 240 ft above sea level.)

If a point lies between two contour lines, then you must estimate its elevation by interpolation (**FIGURE 9.12**, rule 2). For example, on a map with a 20-ft contour interval, a point might lie between the 340 and 360-ft contours, so you know it is between 340 and 360 ft above sea level. If a point lies between a contour line and the margin of the map, then you must estimate its elevation by extrapolation (**FIGURE 9.12**, rule 3).

Depressions

FIGURE 9.14 shows how to read topographic contour lines in and adjacent to a depression. *Hachure marks* (short line segments pointing downhill) on some of the contour lines in these maps indicate the presence of a closed

COFFEYVILLE EAST QUADRANGLE
KANSAS
7.5-MINUTE SERIES

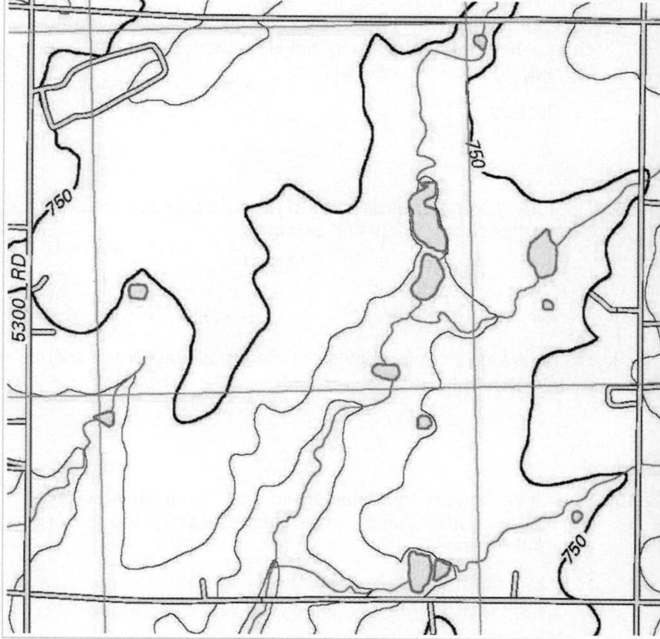

A. Topographic map base (contour lines), UTM grid lines (WGS84, Zone 15S), hydrography, and transportation features.

B. Orthoimage base with all other data layers: contour lines, UTM grid lines (WGS84, Zone 15S), geographic names and boundaries, hydrography, and transportation features.

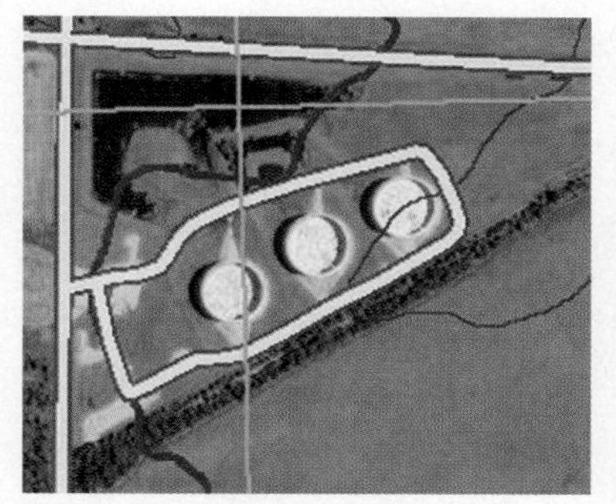

C. Enlarged portion of B.

FIGURE 9.11 US Topo (USGS) digital topographic maps and orthoimages of quadrangles. The latest series of USGS topographic maps are digital maps with data layers for planimetric features and an orthoimage. Choose layers displayed on a topographic map base **(A)**, or else combine the topographic base as one layer of data on an orthoimage base **(B)** that can be enlarged with high resolution **(C)**. Use Adobe Reader® (or Adobe Acrobat®) to print products in GeoPDF® format. To learn about US Topo GeoPDFs, watch a 6-minute USGS video (**http://gallery.usgs.gov/videos/663**). Using Adobe Acrobat®, you can save products as JPEG images for enhancement with any photo processing software. However, the location of all roads is ©TomTom.

depression (a depression from which water cannot drain) (**FIGURE 9.12**, rule 12). At the top of a hill, contour lines repeat on opposite sides of the rim of the depression. On the side of a hill, the contour lines repeat only on the downhill side of the depression.

Ridges and Valleys

FIGURE 9.15 shows how topographic contour lines represent linear ridge crests and valley bottoms. Ridges and valleys are roughly symmetrical, so individual contour lines repeat on each side (**FIGURE 9.12**, rule 14). To visualize this, picture yourself walking along an imaginary trail across the ridge or valley (dashed lines in **FIGURE 9.15**). Every time you walk up the side of a hill or valley, you cross contour lines. Then, when you walk down the other side of the hill or valley, you recross contour lines of the same elevations as those crossed walking uphill.

Spot Elevations and Benchmarks

Elevations of specific points on topographic maps (tops of peaks, bridges, survey points, etc.) sometimes are indicated directly on the maps as **spot elevations** beside a small triangle, black dot, or **x**-symbol at the exact spot of the elevation indicated. The elevations of prominent hilltops, peaks, or other features are often identified. For example, the highest point on the ridge in the west central part of **FIGURE 9.13B** has an elevation of 266 ft above sea level. The notation "BM" denotes a **benchmark,** a permanent marker (usually a metal plate) placed by the U.S. Geological Survey or Bureau of Land Management at the point indicated on the map (**FIGURE 9.7**).

RULES FOR CONTOUR LINES

1. Every point on a contour line is of the exact same elevation; that is, contour lines connect points of equal elevation. The contour lines are constructed by surveying the elevation of points, then connecting points of equal elevation.

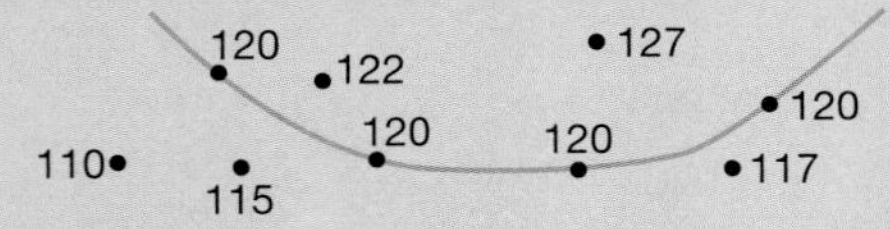

2. Interpolation is used to estimate the elevation of a point B located in line between points A and C of known elevation. To estimate the elevation of point B:

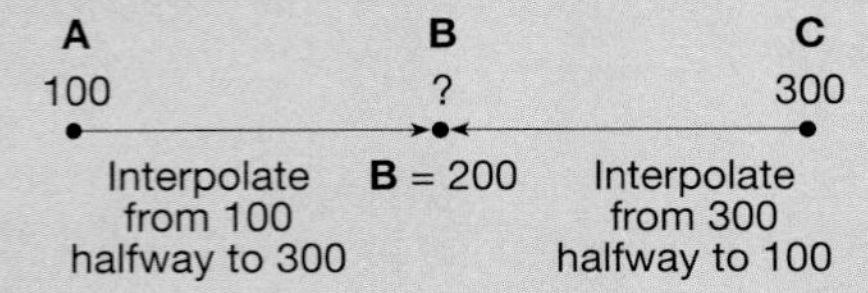

3. Extrapolation is used to estimate the elevations of a point C located in line beyond points A and B of known elevation. To estimate the elevation of point C, use the distance between A and B as a ruler or graphic bar scale to estimate in line to elevation C.

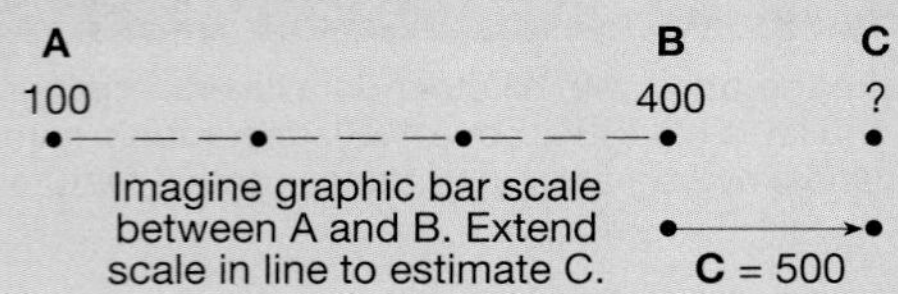

4. Contour lines always separate points of higher elevation (uphill) from points of lower elevation (downhill). You must determine which direction on the map is higher and which is lower, relative to the contour line in question, by checking adjacent elevations.

5. Contour lines always close to form an irregular circle. But sometimes part of a contour line extends beyond the mapped area so that you cannot see the entire circle formed.

6. The elevation between any two adjacent contour lines of different elevation on a topographic map is the *contour interval*. Often every fifth contour line is heavier so that you can count by five times the contour interval. These heavier contour lines are known as *index contours*, because they generally have elevations printed on them.

7. Contour lines never cross each other except for one rare case: where an overhanging cliff is present. In such a case, the hidden contours are dashed.

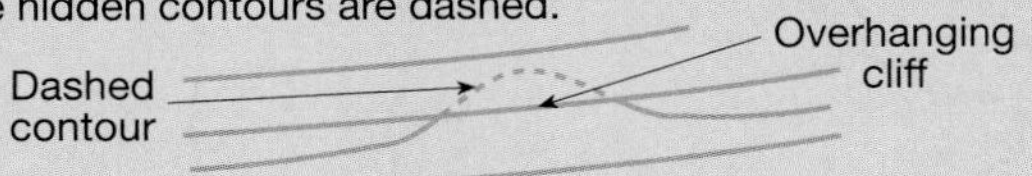

8. Contour lines can merge to form a single contour line only where there is a vertical cliff or wall.

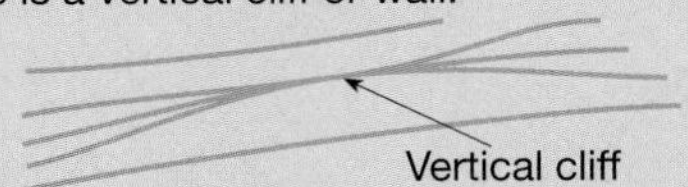

9. Evenly spaced contour lines of different elevation represent a uniform slope.

10. The closer the contour lines are to each other the steeper the slope. In other words, the steeper the slope the closer the contour lines.

11. A concentric series of closed contours represents a hill:

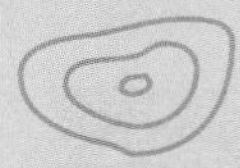

12. *Depression contours* have hachure marks on the downhill side and represent a closed depression:

See Figure 9.14

13. Contour lines form a V pattern when crossing streams. The apex of the V always points upstream (uphill):

14. Contour lines that occur on opposite sides of a valley or ridge always occur in pairs. See Figure 9.13.

FIGURE 9.12 Rules for constructing and interpreting contour lines on topographic maps.

Relief and Gradient (Slope)

Recall that **relief** is the difference in elevation between landforms, specific points, or other features on a landscape or map. *Regional relief* (total relief) is the difference in elevation between the highest and lowest points on a topographic map. The highest point is the top of the highest hill or mountain; the lowest point is generally where the major stream of the area leaves the map, or a coastline. **Gradient** is a measure of the steepness of a slope. One way to determine and express the gradient of a slope is by measuring its steepness as an angle of ascent or descent (expressed in degrees). On a topographic map, gradient is usually determined by dividing the relief (rise or fall) between two points on the map by the distance (run) between them (expressed as a fraction in feet per mile or meters per kilometer). For example, if points **A** and **B** on a map have elevations of 200 ft and 300 ft, and the points are located 2 miles apart, then:

$$\text{gradient} = \frac{\text{relief (amount of rise or fall between } \mathbf{A} \text{ and } \mathbf{B}\text{)}}{\text{distance between } \mathbf{A} \text{ and } \mathbf{B}}$$

$$= \frac{100 \text{ ft}}{2 \text{ mi}} = 50 \text{ ft/mi}$$

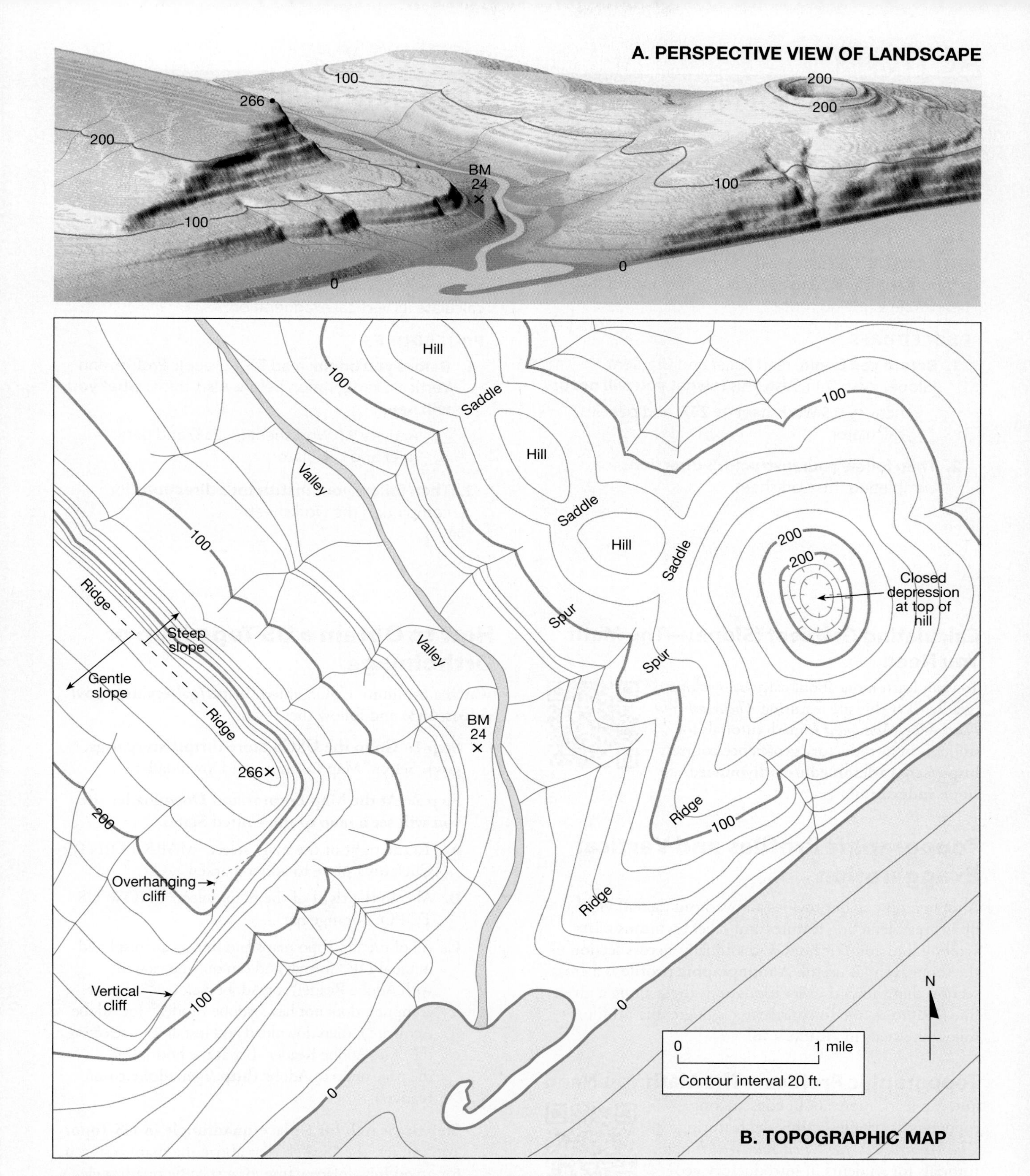

FIGURE 9.13 Names of landscape features observed on topographic maps. Note perspective view (A) and topographic map (B) features: **valley** (low-lying land bordered by higher ground), **hill** (rounded elevation of land; mound), **ridge** (linear or elongate elevation or crest of land), **spur** (short ridge or branch of a main ridge), **saddle** (low point in a ridge or line of hills; it resembles a horse saddle), **closed depression** (low point/area in a landscape from which surface water cannot drain; contour lines with hachure marks), **steep slope** (closely spaced contour lines), **gentle slope** (widely spaced contour lines), **vertical cliff** (merged contour lines), **overhanging cliff** (dashed contour line that crosses a solid one; the dashed line indicates what is under the overhanging cliff).

ACTIVITY

9.5 Relief and Gradient (Slope) Analysis

THINK About It **How are topographic maps used to calculate the relief and gradients (slopes) of landscapes?**

OBJECTIVE Calculate relief and gradients from a topographic map and apply the gradient data to determine a driving route.

PROCEDURES

1. **Before you begin**, read Relief and Gradient (Slope) on p. 244. Also, this is **what you will need:**
   ___ Activity 9.5 Worksheet (p. 257) and pencil
   ___ calculator
2. **Then follow your instructor's directions** for completing the worksheet.

ACTIVITY

9.6 Topographic Profile Construction

THINK About It **How is a topographic profile constructed from a topographic map, and what is its vertical exaggeration?**

OBJECTIVE Construct a topographic profile from a topographic map using the graph paper, and then calculate its vertical exaggeration.

PROCEDURES

1. **Before you begin**, read Topographic Profiles and Vertical Exaggeration below. Also, this is **what you will need:**
   ___ Activity 9.6 Worksheet (p. 258) and pencil
   ___ ruler and calculator
2. **Then follow your instructor's directions** for completing the worksheet.

Calculating Gradient (Slope)—The Math You Need

You can learn more about calculating slope (gradient) at this site featuring *The Math You Need, When You Need It* math tutorials for students in introductory geoscience courses: **http://serc.carleton.edu/mathyouneed/slope/index.html**

Topographic Profiles and Vertical Exaggeration

A topographic map provides an overhead (aerial) view of an area, depicting features and relief by means of its symbols and contour lines. Occasionally a cross section of the topography is useful. A **topographic profile** is a cross section that shows the elevations and slopes along a given line (FIGURE 9.16). To construct a topographic profile, follow the steps in FIGURE 9.16.

Topographic Profiles—The Math You Need

You can learn more about constructing topographic profiles at this site featuring *The Math You Need, When You Need It* math tutorials for students in introductory geoscience courses: **http://serc.carleton.edu/mathyouneed/slope/topoprofile.html**

How to Obtain a US Topo Map or Orthoimage

Watch a 6-minute USGS video (**http://gallery.usgs.gov/videos/663**) and follow these steps:

Step 1: Go to the USGS Store (http://store.usgs.gov). Select "Map Locator and Downloader."

Step 2: At the Map Locator and Downloader site, you will see a map of the United States.

A. To the right of the map, select "MARK POINTS: Click on a place to add a marker."

B. Also to the right of the map, select "SHOW US TOPO" (orange bar).

C. All of the US Topo maps and images are displayed in GeoPDF® format and can only be viewed with Adobe Reader® or Adobe Acrobat®. If your computer does not have Adobe Reader® (or Adobe Acrobat®), then download and install it by selecting the "Get Adobe Reader" bar at the bottom of the page or go to Adobe (**http://get.adobe.com/reader/**).

Step 3: Search for a place/quadrangle in US Topo. You can use the "Search" bar above the map to search for an address, place name, or a specific quadrangle (by name). You can also zoom in or out on the map and change it to a satellite or topographic view. As you zoom in, the outlines of quadrangles will appear

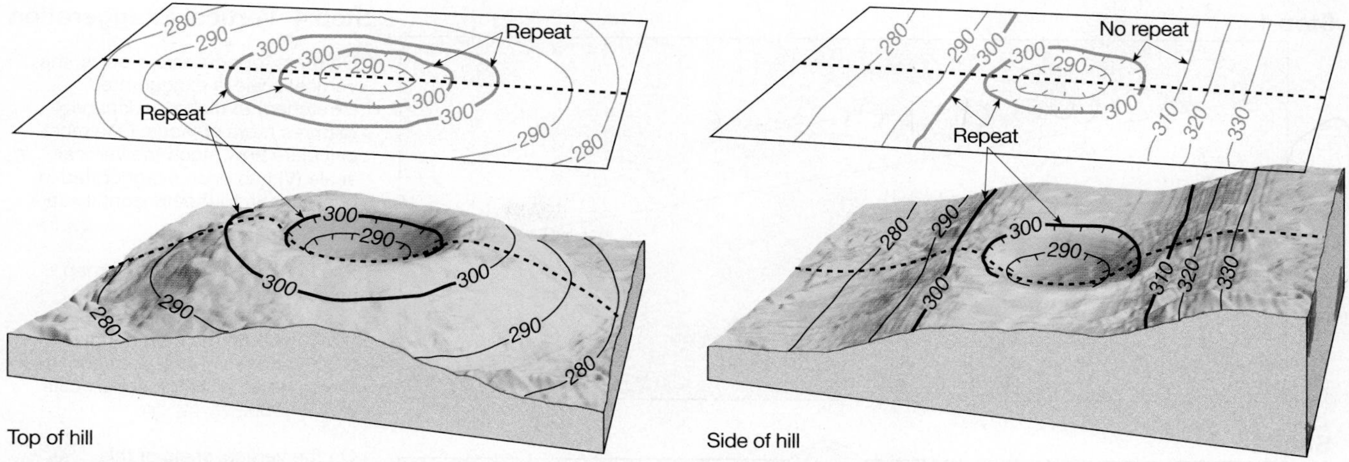

FIGURE 9.14 Contour lines for depressions. Contour lines repeat on opposite sides of a depression (left illustration), except when the depression occurs on a slope (right illustration).

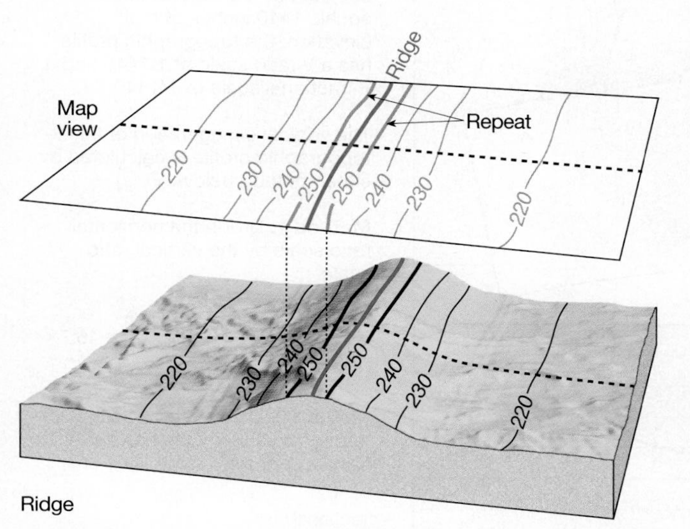

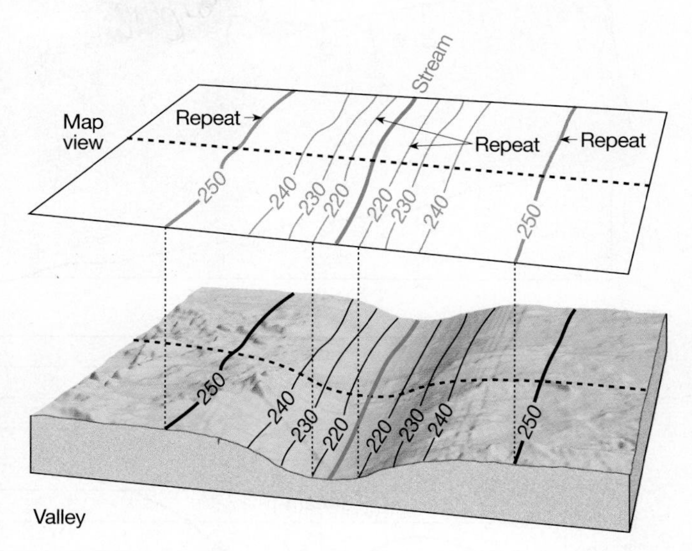

FIGURE 9.15 Contour lines for ridges and valleys. Contour lines repeat (occur in pairs) on opposite sides of linear ridges and valleys. For example, in the left illustration, if you walked the dashed line from left to right, you would cross the 220, 230, 240, and 250-ft contour lines, go over the crest of the ridge, and cross the 250, 240, 230, and 220-ft contour lines again as you walk down the other side. Note that the 250-ft contour lines on these maps are heavier than the other lines because they are *index contours*. On most maps, every fifth contour line above sea level is an index contour, so you can count by five times the contour interval. The *contour interval* (elevation between any two contour lines) of these maps is 10 ft, so the index contours are every 50 feet of elevation.

in black, with their names in black with yellow highlighting. Click on a quadrangle name to set a red balloon marker.

Step 4: Obtain digital products for the location of the balloon marker. Left click on the black dot in the center of the red balloon marker to obtain a list of products available for the Step-3 location. The latest map series is US Topo, but older maps are also listed. Left click on "download" to display or save your selection. The US Topo files are large (7–20 MB) and may take some time to download and open.

When you obtain and open a US Topo $7\frac{1}{2}$-minute quadrangle, then you can add or subtract layers from them by using the menu along the left-hand side of the image. You will also be given the option of downloading a free TerraGo Toolbar that allows you to measure distances, add comments, and merge products with Google Maps™ or your GPS. The older series of topographic maps are single-layer products scanned from paper maps.

Step 5: Printing Tips. You can print all or parts of the maps and orthoimages that you display in US Topo. The GeoPDF® files are very large (10–20 MB), so be patient and allow time for them to load, display, and print. To print an entire map or orthoimage on letter-size paper, be sure to set your printer to the "shrink to fit" setting. If you have a snipping tool on your computer, then you can also snip the images as low-resolution JPEG files.

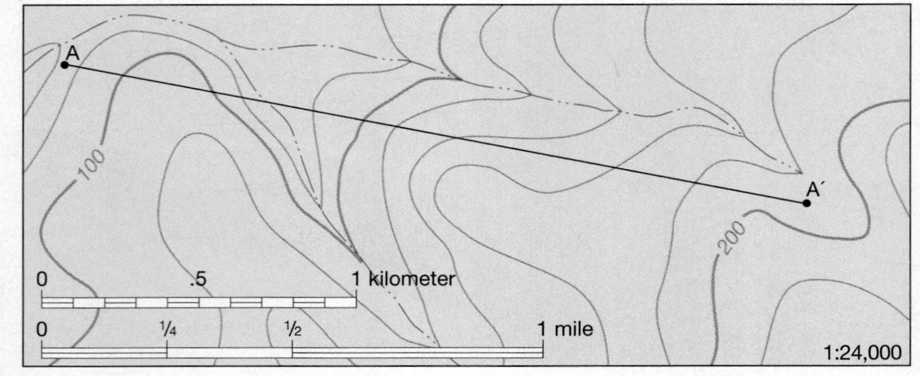

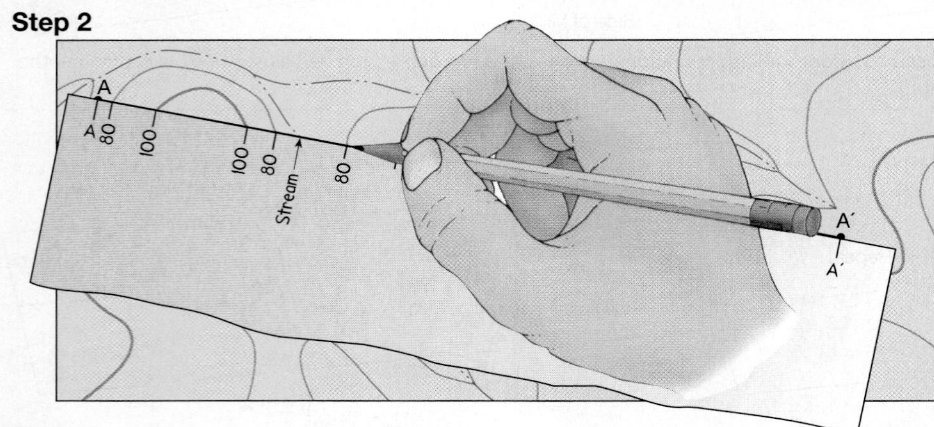

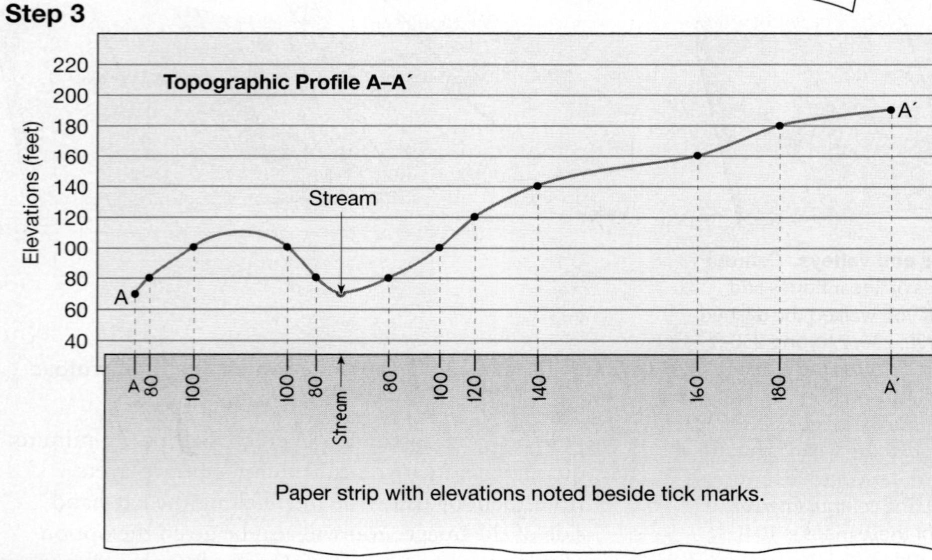

Step 4 Vertical Exaggeration

On most topographic profiles, the vertical scale is exaggerated (stretched) to make landscape features more obvious. One must calculate how much the vertical scale (V) has been exaggerated in comparison to the horizontal scale (H).

The horizontal scale is the map's scale. This map has an H ratio scale of 1:24,000, which means that 1 inch on the map equals 24,000 inches of real elevation. It is the same as an H fractional scale of 1/24,000.

On the vertical scale of this topographic profile, one inch equals 120 feet or 1440 inches (120 feet x 12 inches/foot). Since one inch on the vertical scale equals 1440 inches of real elevation, the topographic profile has a V ratio scale of 1:1440 and a V fractional scale of 1/1440.

The vertical exaggeration of this topographic profile is calculated by either method below:

Method 1: Divide the horizontal ratio scale by the vertical ratio scale.

$$\frac{\text{H ratio scale}}{\text{V ratio scale}} = \frac{1:24{,}000}{1:1440} = \frac{24{,}000}{1440} = 16.7\times$$

Method 2: Divide the vertical fractional scale by the horizontal fractional scale.

$$\frac{\text{V fractional scale}}{\text{H fractional scale}} = \frac{1/1440}{1/24{,}000} = \frac{24{,}000}{1440} = 16.7\times$$

FIGURE 9.16 Topographic profile construction and vertical exaggeration. Shown are a topographic map (Step 1), topographic profile constructed along line **A–A′** (Steps 2 and 3), and calculation of vertical exaggeration (Step 4). **Step 1**—Select two points (**A, A′**), and the line between them (line **A–A′**), along which you want to construct a topographic profile. **Step 2**—To construct the profile, the edge of a strip of paper was placed along line **A–A′** on the topographic map. A tick mark was then placed on the edge of the paper at each point where a contour line and stream intersected the edge of the paper. The elevation represented by each contour line was noted on its corresponding tick mark. **Step 3**—The edge of the strip of paper (with tick marks and elevations) was placed along the bottom line of a piece of lined paper, and the lined paper was graduated for elevations (along its right margin). A black dot was placed on the profile above each tick mark at the elevation noted on the tick mark. The black dots were then connected with a smooth line to complete the topographic profile. **Step 4**—*Vertical exaggeration* of the profile was calculated using either of two methods. Thus, the vertical dimension of this profile is exaggerated (stretched) to 16.7 times greater than it actually appears in nature compared to the horizontal/map dimension.

ACTIVITY 9.1 Map and Google Earth™ Inquiry

Name: ______________________ **Course/Section:** ______________________ **Date:** ____________

A. Imagine that a friend has asked you, "Where are you located right now?" Give three different answers below:

1.

2.

3.

B. REFLECT & DISCUSS One of the oldest ways of describing a location on our spherical Earth is by using a coordinate system of latitude and longtude like the one below. Latitude and longitude are both measured in degrees (°). Latitude is measured from 0° at the Equator to 90°N (the North Pole) or 90°S (the South Pole). Longitude is measured in degrees east or west of the prime meridian, an imaginary line that runs on Earth's surface from the North Pole to the South Pole through Greenwich, England. When viewed from above, locations between 0 and 180° east of the prime meridian are in the Eastern Hemisphere, and locations between 0 and 180° west are in the Western Hemisphere. On the map below, notice how the red location is identified as 30°N, 150°W. For points between labeled lines, the position can be estimated as done for the blue location.

1. On the map below, lightly color in the country where you live. Refer to the reference map at the back of the lab manual as needed.
2. Name a place of your choice to locate on the map (your current location, home, school, place of work, or other location). This will be referred to as your "**home place**" in the inquiry items below.

3. Place a dot on the map to show approximately where your home place is located, then estimate its latitude and longitude as already done for the red and blue points on the map.

   Latitude: ______________________ Longitude: ______________________

4. Notice that the lines of latitude and longitude outline somewhat rectangular areas called **quadrangles**, like the one shaded pale red on the map. Describe the location of the red quadrangle based on its latitude and longitude.

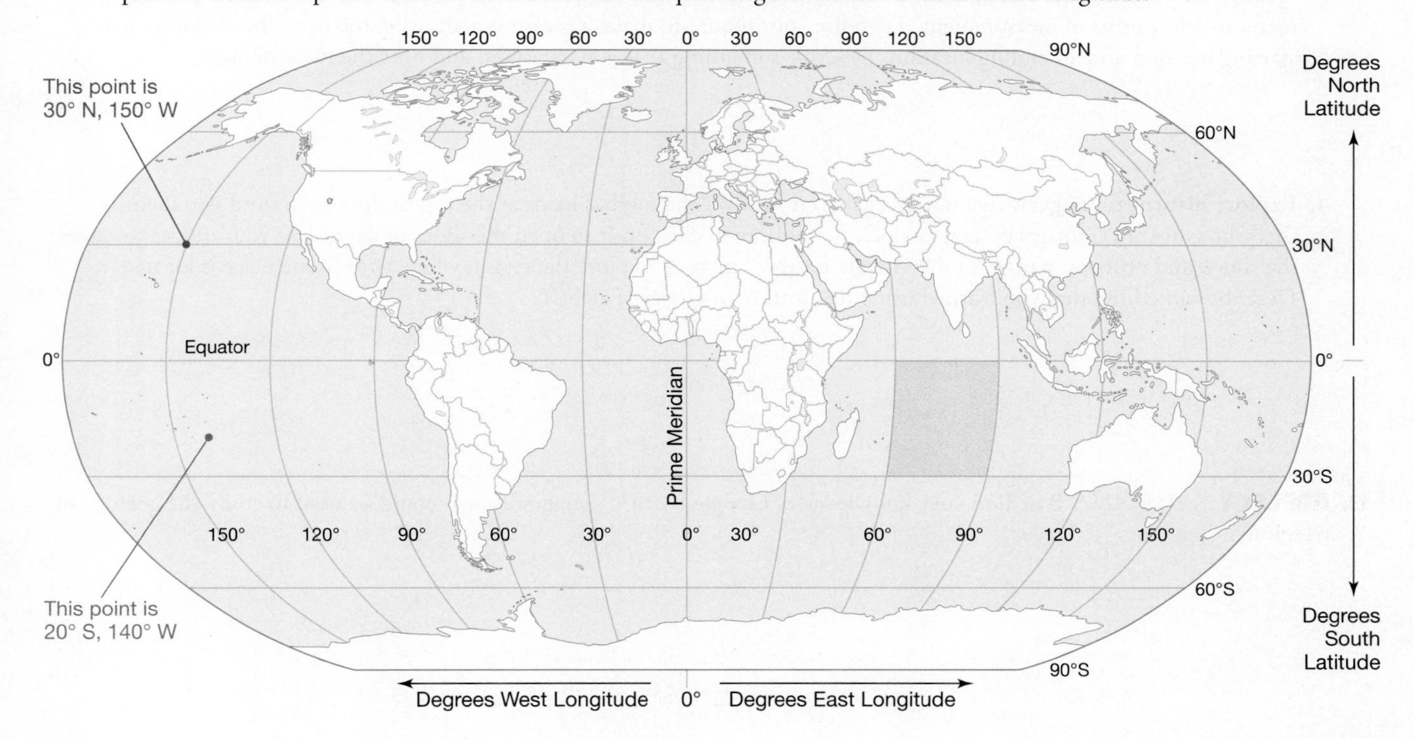

C. Use Google Earth™ to locate places on Earth. If you do not have Google Earth™ loaded on your computer, then do so now. Go to **http://www.google.com/earth/index.html** and click on "Dowload Google Earth". Read the Google Maps/Earth Terms of Service (and uncheck the Google Chrome browser option of you do not want to install it), then click "Agree and Download". The installation will begin immediately and takes just a few minutes to complete on most computers. Open Google Earth™ and do the following:

1. Determine exactly the latitude, longitude, and elevation of your home place.

a. Start by using the Google Earth™ "Search" option located in the upper lefthand corner of the screen. You can type in the name (e.g., Washington Monument) or address of your home place, then click on the "Search". Google Earth™ will then zoom in on the location. You can then change the zoom (in or out) using your mouse wheel or the plus-minus slide bar in the upper righthand corner of the screen.

b. Next, place your cursor over your home place to locate its latitude and longitude more exactly. As you move the cursor, notice along the bottom righthand edge of your screen that the latitude and longitude coordinates of the cursor location are identified. (If you do not see degrees of latitude North or South and longitude East or West, then go to the top of the Google Earth™ screen and choose "Tools", then "Options". On that menu, under "Show Lat/Long", choose "Degrees, minutes, seconds". Then click on "Apply" and "OK".) Notice that more than just degrees of latitude and longitude are indicated. For finer measurements of latitude and longitude, each degree can be subdivided (like a clock) into 60 subdivisions called minutes ('), and the minutes can be divided into 60 equal subdivisions called seconds ("). Record the exact location of your home place below in degrees, minutes, and seconds.

Latitude: ________________________ Longitude: ________________________

c. Notice that elevation above sea level (elev) is indicated to the right of latitude and longitude at the bottom righthand edge of the Google Earth™ screen. What is the elevation of your home place? ____________________

2. Experiment with the Layers of Google Earth™ to learn more about your home place. From the menu on the lefthand side of the Google Earth™ screen, open "Layers". Experiment with turning layers on and off to see how it affects what Google Earth™ displays. For example, you may want to start by choosing Roads, Borders and Labels, Gallery, or More to see what is available to display. When you are done, describe something that you learned about your home place or the region around it by experimenting with the Layers.

3. Measure a distance in Google Earth™. Move your cursor over the toolbar icons at the top of the screen until you identify the "Show Ruler" icon, then click on it to open the Ruler menu. Select the "Line" tab, and use the pull down menu to select units of measurement. Then use your mouse to make a measurement using the ruler (by clicking on a starting location and an ending location. Describe something that you measured and note the measurement.

4. Explore historical imagery. Again, move your cursor over the toolbar icons at the top of the screen until you identify the "Show historical imagery" icon (a clock symbol), then click on it to open the slider of dates. Use your mouse to move the slider, and observe changes in the Google Earth™ images. Explore the region where your home place is located. Describe something that you learned using this feature of Google Earth™.

D. REFLECT & DISCUSS Based on your knowledge of Google Earth™, suggest how it could be used to study the geology of a region.

B. These are 2011 US Topo Series products. The digital map is matched with an orthoimage of the same area. (Courtesy of USGS)

1. The orthoimage reveals that SP Mountain is a volcano (cinder cone) with a very visible closed depression (crater) at its summit. The image also reveals an older, reddish volcano (cinder cone) that is very eroded (worn down). Draw a dashed line on the orthoimage and map to show the outline of this older volcano.

2. Draw a solid line on the orthoimage and map to show the outline of the crater of the older volcano. Why is it not shown with hachure lines on the map, like the crater for SP Mountain?

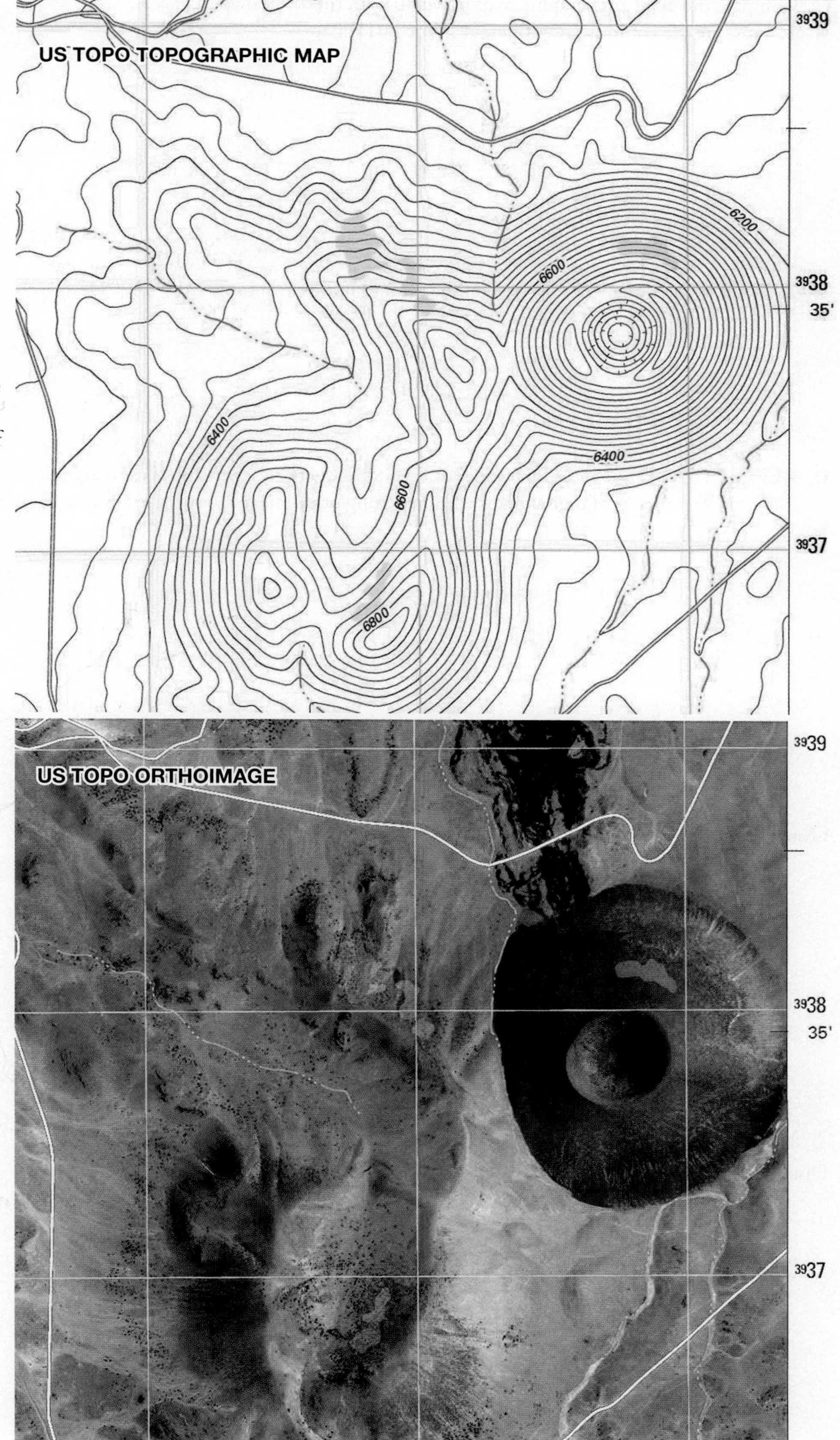

C. Compare the 2011 topographic map in part B with the 1989 map in part A.

1. How are the contour lines different in the 2011 map?

2. What is one thing that changed on the ground in this area from 1989 to 2011?

D. **REFLECT & DISCUSS** Which series of USGS products do you think is more useful: the 1989 paper maps series or the 2011 US Topo series of digital maps with matching orthoimages? List some advantages and disadvantages of your choice.

Advantages:

Disadvantages:

LABORATORY 10

Geologic Structures, Maps, and Block Diagrams

CONTRIBUTING AUTHORS

Michael J. Hozik • *Stockton College of New Jersey*

William R. Parrott, Jr. • *Stockton College of New Jersey*

Raymond W. Talkington • *Stockton College of New Jersey*

Satellite image of deformed rock layers, Anti-Atlas Mts., Morocco. False colors highlight layers of different rock types. (NASA/GSFC/METI/ERSDAC/JAROS and U.S./Japan ASTER Science Team)

BIG IDEAS

Some of Earth's rocky landscapes expose originally horizontal layers of rock that have been visibly tilted, fractured, dislocated, folded, or otherwise deformed into complex geologic structures. The structures are often too large to see in one place. To visualize them, geologists make regional geologic maps and block diagrams with field data assembled from many sites. The maps and diagrams are then used to reveal the structures and determine how they formed.

FOCUS YOUR INQUIRY

THINK About It | How are deformed rocks identified and classified?

ACTIVITY 10.1 Geologic Structures Inquiry *(p. 260)*

THINK About It | What kinds of stress relationships cause geologic structures to develop?

ACTIVITY 10.2 Visualizing How Stresses Deform Rocks *(p. 260)*

THINK About It | How do geologists map geologic structures on and beneath Earth's surface?

ACTIVITY 10.3 Map Contacts and Formations *(p. 261)*

ACTIVITY 10.4 Determine Attitude of Rock Layers and a Formation Contact *(p. 262)*

THINK About It | How are three-dimensional models used to visualize geologic structures?

ACTIVITY 10.5 Cardboard Model Analysis and Interpretation *(p. 263)*

ACTIVITY 10.6 Block Diagram Analysis and Interpretation *(p. 263)*

THINK About It | How do geologists define, analyze, and interpret geologic structures using images of landscapes?

ACTIVITY 10.7 Nevada Fault Analysis Using Orthoimages *(p. 263)*

THINK About It | How do geologists visualize geologic structures using geologic maps?

ACTIVITY 10.8 Appalachian Mountains Geologic Map *(p. 263)*

ACTIVITY

10.1 Geologic Structures Inquiry

THINK About It | **How are deformed rocks identified and classified?**

OBJECTIVE Classify types of rock deformation based on an analysis of deformed and underformed rocks.

PROCEDURES

1. **Before you begin**, do not look up definitions and information. Use your current knowledge, and complete the worksheet with your current level of ability. Also, this is **what you will need** to do the activity:

   ____ Activity 10.1 Worksheet (p. 273) and pencil
2. **Complete the worksheet in a way that makes sense to you.**
3. **After you complete the worksheet**, be prepared to discuss your observations and classification with other geologists.

ACTIVITY

10.2 Visualizing How Stresses Deform Rocks

THINK About It | **What kinds of stress relationships cause geologic structures to develop?**

OBJECTIVE Identify four kinds of stresses in everyday objects and actions, and then infer how they cause rock deformation.

PROCEDURES

1. **Before you begin**, read the Introduction and Stress and Strain below. Also, this is **what you will need:**

   ____ Activity 10.2 Worksheet (p. 274) and pencil
2. **Then follow your instructor's directions** for completing the worksheets.

Introduction

When two cars collide, their kinetic energy is converted to compressional stresses that force ductile metal and plastic to crumple and brittle glass to fracture and shatter. When lithospheric plates collide, pull apart, or slide past one another, a similar crumpling, fracturing, and shattering of rock occurs. However, the process is accompanied by hazardous earthquakes, may last for tens-of-millions of years, and can affect rocks over hundreds to thousands of square kilometers. Structural geology is the study of rock **deformation** (change in position, volume or shape of a body of rock) and **geologic structures** (fractures, faults, and folds that result from deformation).

Structural geology relies on the Laws of Original Horizontality and Lateral Continuity, which state that sediment and lava are deposited in relatively flat, continuous, horizontal layers called *beds* or *strata* (plural of stratum) like a layer cake. Wherever strata are folded (no longer relatively flat), tilted (no longer horizontal), or fractured apart (no longer continuous), they have been deformed. Structural geologists must decipher the shapes and internal characteristics of geologic structures to help locate the mineral, energy, and water resources hidden within them and to be sure that dams, power plants, and other structures are safely constructed on stable ground. Interpreting geologic structures requires knowledge of stress–strain relationships and the nature of the rocks themselves.

Stress and Strain

Stress is the amount of pressure or force acting on a unit surface area, like pounds per square inch or kilograms per square meter. Strain (deformation) occurs when the rock yields (gives in) to the stress. In other words, stress (pressure, force) acts on a body of rock and causes it to deform, or strain.

Confining pressure is pressure (stress, force) applied equally in all directions and shown by the red arrows in FIGURE 10.1. It is like the water pressure you feel all over your body when you dive to the bottom of a swimming pool. The deeper you go, the more confining is the water pressure you feel. The same thing happens with rocks. Confining pressure is equal in all directions and increases with depth below Earth's surface. Under low confining pressure, rocks remain *brittle*, which means that they tend to fracture and shatter when they yield to stress. Under high confining pressure, they undergo *dilation* (decrease in volume). This is what compacts sediment into more dense sedimentary rock (and eventually metamorphic rock). Rocks under high confining pressure also lose their brittle nature and, instead, become *ductile*—capable of plastic flow and folding if the confining pressure is accompanied by directed pressure.

Directed pressure occurs when the stress is unequal—greater in one direction than another (while the rocks are also under some amount of confining pressure). The vector directions of directed stress are indicated with arrows in FIGURE 10.1. Stress arrows pointed directly at one another indicate **compressional stresses**—opposing stresses that push and squeeze rocks (shorten them parallel to the arrows). Ductile rocks fold, and brittle rocks develop

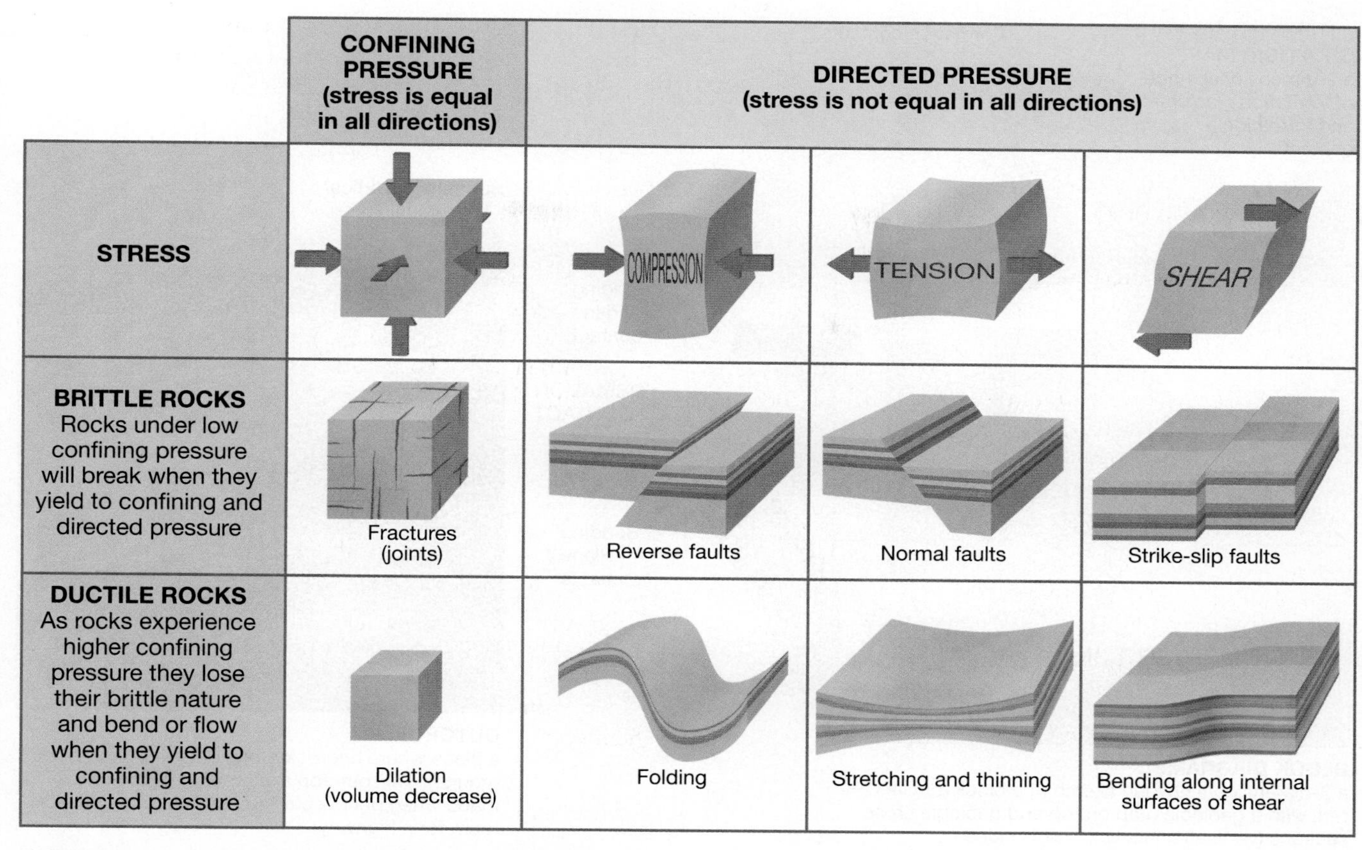

FIGURE 10.1 Stress and strain effects on rocks.

fractures and faults as the rocks yield to compressional stress. Stress arrows pointing directly away from one another indicate **tensional stresses**—stresses that pull rocks apart (lengthen them parallel to the arrows). Brittle rocks develop faults and ductile materials stretch (become elongated along the axis of the tensional stresses). Stress arrows pointing past one another indicate **shear stresses**—stresses that cause parts of the body of rock to slide past one another in opposite directions and parallel to the surface between them. Brittle rocks develop faults and ductile materials bend.

ACTIVITY 10.3 Map Contacts and Formations

THINK About It **How do geologists map geologic structures on and beneath Earth's surface?**

OBJECTIVE Map contacts and formations using images and a topographic map.

PROCEDURES

1. **Before you begin**, read Formations, Geologic Maps, and Block Diagrams next. Also, this is **what you will need:**
   ____ red and blue colored pencils, plus a dark pen (red or blue)
   ____ Activity 10.3 Worksheet (p. 275) and pencil
2. **Then follow your instructor's directions** for completing the worksheets.

Formations, Geologic Maps, and Block Diagrams

Geologists can see how bodies of bedrock or sediment are positioned three dimensionally where they *crop out* (stick out of the ground as an outcrop) at Earth's surface. The outcrops are classified into mappable units, called *formations.*

Formations

Formations are mappable rock units (**FIGURE 10.2**). This means that they can be distinguished from one another "in the field" and are large enough to appear on geologic maps (which usually cover a $7\frac{1}{2}$-minute quadrangle). The surfaces between formations are called **formation contacts** and appear as black lines on geologic maps and cross sections. Formations may be subdivided into mappable *members* composed of *beds* (individual strata, layers of rock or sediment). **Bedding plane contacts** are surfaces between individual beds within a formation.

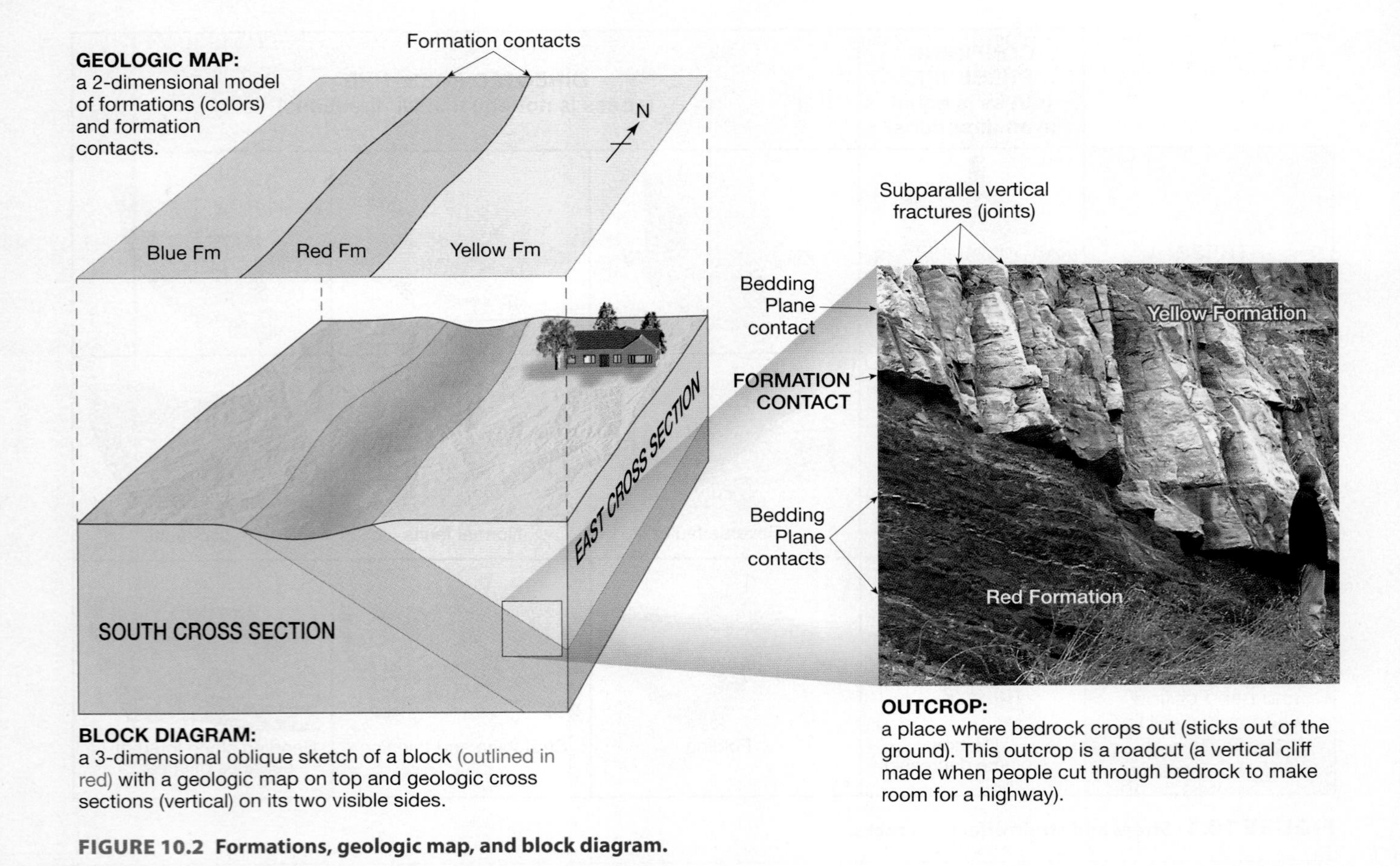

FIGURE 10.2 Formations, geologic map, and block diagram.

Individual beds/strata are rarely mapped because they are not wide enough to show up on a typical $7\frac{1}{2}$-minute quadrangle map (where a pencil line equals about 6 meters, 20 ft). Geologists assign each formation a formal name, which is capitalized (e.g., Yellow Formation or Yellow Fm in FIGURE 10.2) and published with a description of its distinguishing features and a "type locality" upon which the name and distinguishing features are based. The formal name can include the word "formation" or the name of the rock type that makes up the formation. For example, the Dakota Formation is also formally called the Dakota Sandstone.

Geologic Maps

Geologic maps are flat (two-dimensional, like a sheet of paper or computer display) models of Earth's surface, viewed from directly above, that use different colors and symbols to represent the locations of formations (FIGURE 10.2). You can search for U.S. geologic maps and formation descriptions with the National Geologic Map Database and Geologic Names Lexicon (**http://ngmdb.usgs.gov**). This site also has links to state geologic maps.

Block Diagrams

The widths of formations vary in outcrops and on maps because of variations in formation thickness, angle of tilting, and the angle of the land surface at which they

ACTIVITY

10.4 Determine Attitude of Rock Layers and a Formation Contact

THINK About It **How do geologists map geologic structures on and beneath Earth's surface?**

OBJECTIVE Determine the attitude of rock layers, draw geologic cross sections, and infer a formation contact.

PROCEDURES

1. **Before you begin**, read Attitude—Strike and Dip, Constructing Geologic Cross Sections, Fractures and Faults, Folded Structures, Unconformities, and Cardboard Models next. Also, this is **what you will need:**

   ___ dark pen (red or blue)
   ___ Activity 10.4 Worksheet (p. 276) and pencil

2. **Then follow your instructor's directions** for completing the worksheets.

crop out. To visualize this, geologists use block diagrams. A **block diagram** is an oblique sketch of a block of Earth's lithosphere, like a block of cake cut from a sheet cake and viewed from one corner, just above the level of the table on which the cake is sitting. It has a geologic map on top and a geologic cross section on each of its visible sides (**FIGURE 10.2**). Notice how the block diagram in **FIGURE 10.2** gives you a three-dimensional perspective of how the formations are oriented.

ACTIVITY 10.5 Cardboard Model Analysis and Interpretation

THINK About It | **How are three-dimensional models used to visualize geologic structures?**

OBJECTIVE Visualize, analyze, and interpret geologic structures using cardboard models.

PROCEDURES

1. **If you have not already done so**, then first read Attitude—Strike and Dip, Constructing Geologic Cross Sections, Fractures and Faults, Folded Structures, Unconformities, and Cardboard Models next. Also, this is **what you will need:**
   ___ scissors and protractor and ruler cut from GeoTools (at the back of manual)
   ___ Activity 10.5 Worksheets (pp. 277–278) and pencil
2. **Then follow your instructor's directions** for completing the worksheets.

ACTIVITY 10.6 Block Diagram Analysis and Interpretation

THINK About It | **How are three-dimensional models used to visualize geologic structures?**

OBJECTIVE Visualize, analyze, and interpret geologic structures in block diagrams.

PROCEDURES

1. **If you have not already done so**, then first read Attitude—Strike and Dip, Constructing Geologic Cross Sections, Fractures and Faults, Folded Structures, Unconformities, and Cardboard Models next. Also, this is **what you will need:**
   ___ Activity 10.6 Worksheets (pp. 279–280) and pencil
2. **Then follow your instructor's directions** for completing the worksheets.

ACTIVITY 10.7 Nevada Fault Analysis Using Orthoimages

THINK About It | **How do geologists define, analyze, and interpret geologic structures using images of landscapes?**

OBJECTIVE Analyze and interpret Nevada faults using orthoimages.

PROCEDURES

1. **If you have not already done so**, then first read Attitude—Strike and Dip, Constructing Geologic Cross Sections, Fractures and Faults, Folded Structures, Unconformities, and Cardboard Models next. Also, this is **what you will need:**
   ___ dark pen (red or blue)
   ___ Activity 10.7 Worksheet (p. 281) and pencil
2. **Then follow your instructor's directions** for completing the worksheets.

ACTIVITY 10.8 Appalachian Mountains Geologic Map

THINK About It | **How do geologists d efine, analyze, and interpret geologic structures using geologic maps?**

OBJECTIVE Construct and interpret a geologic cross section of folded rocks.

PROCEDURES

1. **If you have not already done so**, then first read Attitude—Strike and Dip, Constructing Geologic Cross Sections, Fractures and Faults, Folded Structures, Unconformities, and Cardboard Models next. Also, this is **what you will need:**
   ___ protractor and ruler cut from GeoTools at the back of the lab manual
   ___ Activity 10.8 Worksheet (p. 282) and pencil
2. **Then follow your instructor's directions** for completing the worksheets.

Attitude—Strike and Dip

The geologic map and block diagram in **FIGURE 10.2** shows where each formation occurs but it does not yet include any information about its three-dimensional orientation. The block diagram in **FIGURE 10.2** shows the three-dimensional orientation of each formation, but it is an oblique view in which angles are distorted. Therefore, geologists must measure the orientation of formations, and then record the orientation data on maps and block diagrams using symbols.

What Is Attitude?

Attitude is the orientation of a rock unit, surface (contact), or line relative to horizontal and/or a compass direction. Geologists have devised a system of strike-and-dip for measuring and describing the attitude of tilted rock layers or surfaces, so they can visualize how they have been deformed from their original horizontality (**FIGURES 10.3**). Strike and dip are usually measured directly from an outcrop using a compass and clinometer (device for measuring the angle of inclined surfaces). However, they can be measured or estimated by the shapes of landforms observed from a distance or on aerial photographs, orthoimages, and satellite images (**FIGURE 10.4**).

What Is Strike?

Strike is the *compass bearing* (line of direction or trend) of a line formed by the intersection of a horizontal plane, such as the surface of a lake, and an inclined surface (contact) or rock layer such as a bed, stratum, or formation (**FIGURE 10.3**).

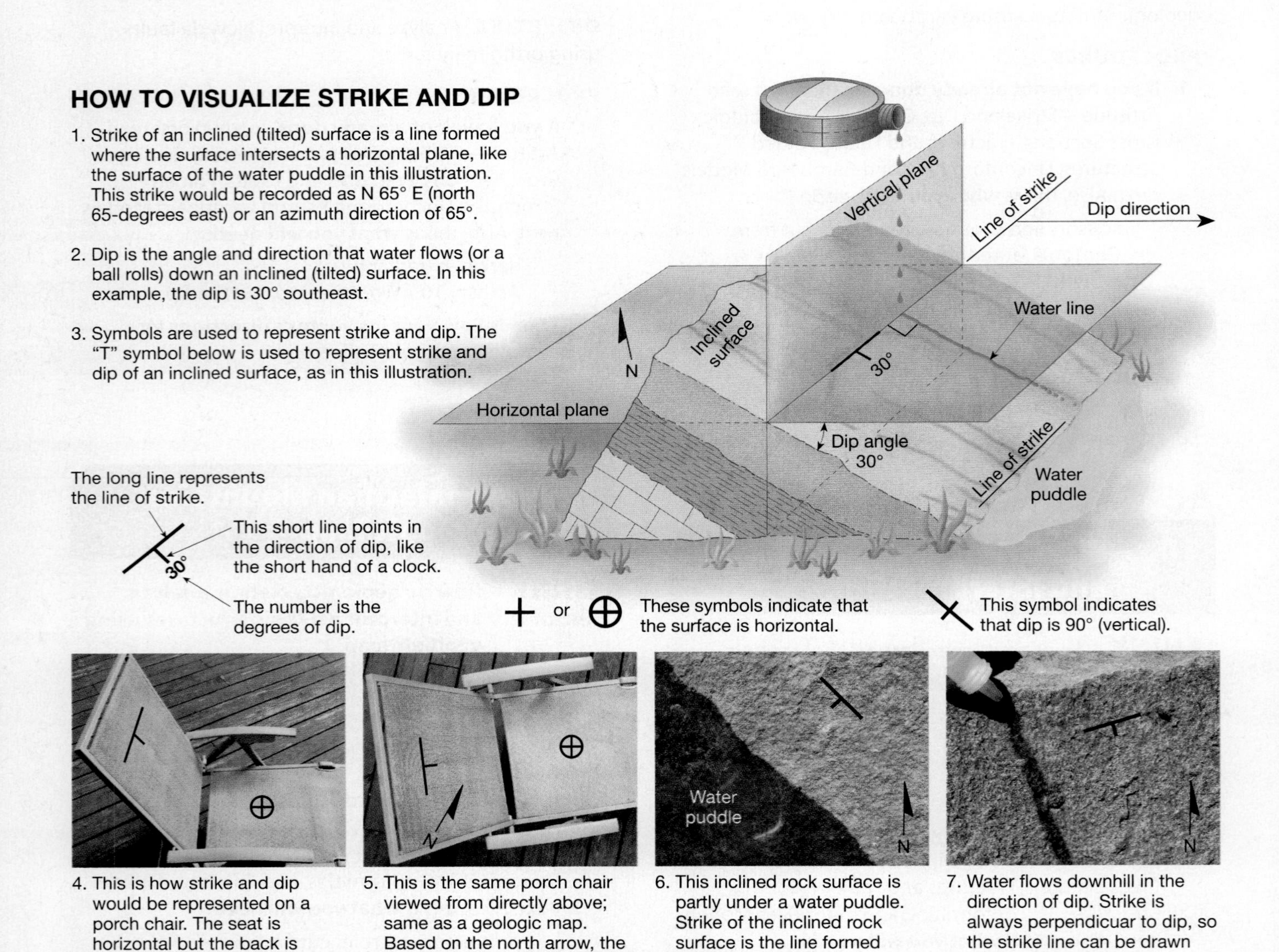

FIGURE 10.3 How to visualize strike and dip. Also refer to **FIGURE 10.4**.

STRIKE AND DIP ON MAPS AND IMAGES OF LANDSCAPES

Examples of how to read strike-and-dip symbols.
Notice how strike is normally expressed relative to north:

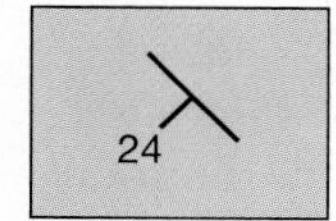

Quadrant: North 45° West (or South 45° East), 24° Southwest
Azimuth: Strike = 315° (or Strike = 135°), Dip = 24° @ 245°

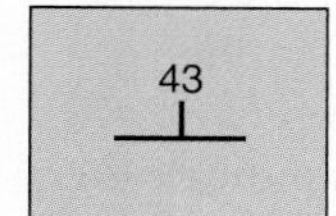

Quadrant: North 90° East (or South 90° West), 43° North
Azimuth: Strike = 090° (or Strike = 270°), Dip = 43° @ 000°

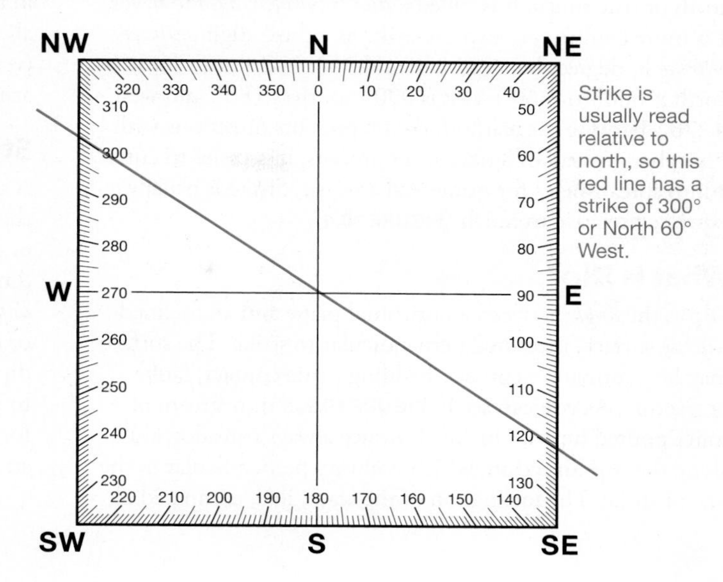

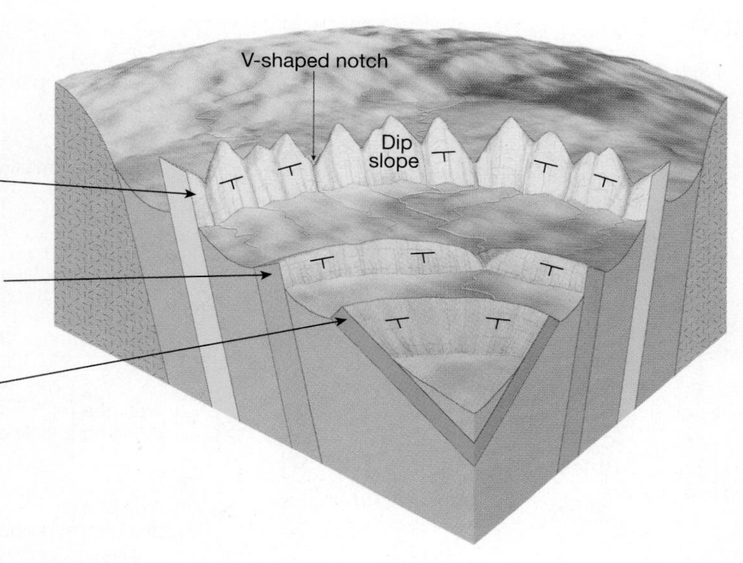

Flatiron–a triangular ridge of steeply dipping resistant rock between two V-shaped notches (cut by streams) and resembling the flat pointed end of a clothing iron or a triangular roof (above). A jagged ridge of flatirons is parallel to strike, and the flatiron surfaces are dip slopes.

Hogback–a sharp-crested ridge of resistant rock that slopes equally on both sides, so it resembles the back of a razorback hog. The ridge crest is parallel to strike and dip is > 30°.

Cuesta–a ridge or hill of resistant rock with a short steep slope on one side (scarp) and a long gentle dip slope on the other side. The ridge is parallel to strike and the long gentle slope is a dip slope. Dip is < 30°.

RULE OF Vs FOR FINDING DIP DIRECTION

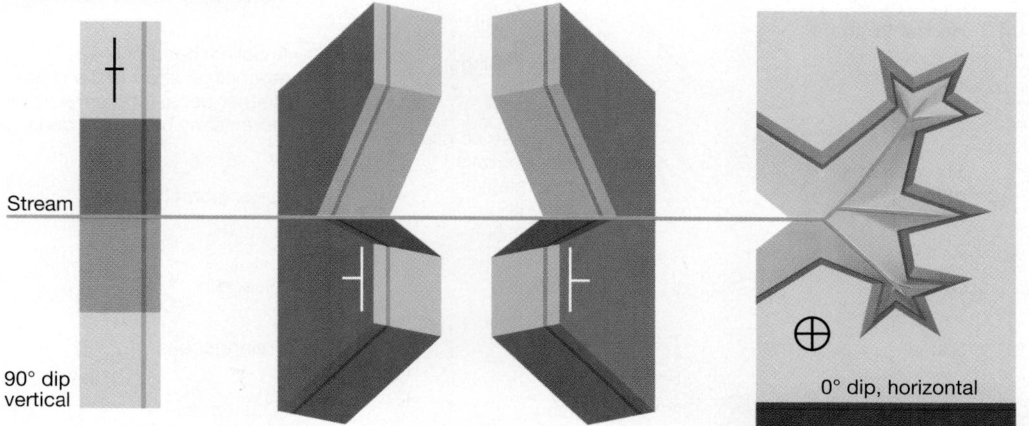

Vertical strata: No V-shapes in the rock layers or contacts can be seen on orthoimages and maps.

Tilted strata: Streams cut V-shapes into the rock layers and contacts that point in the direction of dip (except in rare cases when the slope of the stream bed is greater than the dip of the strata).

Horizontal strata: Streams cut V-shapes into the rock layers and contacts that point upstream and form a characteristic dendritic drainage (streams branching like a plant).

FIGURE 10.4 Strike and dip on maps and images of landscapes. Note how to read strike-and-dip symbols, plot strike using a protractor, and estimate strike and dip when viewing aerial photographs, orthoimages, and satellite images of landscapes.

When the strike is expressed in degrees east or west of true north or true south, it is called a *quadrant bearing*. However, it is more common to express strike as a three-digit *azimuth bearing* in degrees between 000 and 360. In azimuth form, north is 000° (or 360°), east is 090°, south is 180°, and west is 270°. Because the azimuth data represents directions with a number, instead of letters and numbers, it is easier to enter it into spreadsheets for numerical analysis. Strike is usually expressed relative to north (**FIGURE 10.4**).

What Is Dip?

Dip is the *angle* between a horizontal plane and an inclined (tilted) surface, measured perpendicular to strike. The surface may be a formation contact, bedding plane contact, fault, or fracture. As you can see in **FIGURE 10.3**, a thin stream of water poured onto an inclined surface always runs downhill along the **dip direction**, which is always perpendicular to the line of strike. The inclination of the water line, compared to a horizontal plane, is the **dip angle**. Dip is always expressed in terms of its dip angle and dip direction. The dip angle is always expressed in degrees of angle from 0 (horizontal) to 90 (vertical). The dip direction can be expressed as a three-digit azimuth direction or as a quadrant direction.

Strike-and-Dip Symbols

A strike-and-dip symbol consists of a long line showing the orientation of strike, plus a short line for the direction of dip (see **FIGURES 10.3, 10.4, 10.5**). Note that the dip direction is always perpendicular to strike and points *downdip*—the direction that drops of water would flow or a ball would roll. Accompanying numerals indicate the dip angle in degrees. See **FIGURE 10.4** for examples of how to read and express strike and dip in quadrant or azimuth form. Also note that special symbols are used for horizontal strata (rock layers) and vertical strata (**FIGURE 10.3, 10.5**).

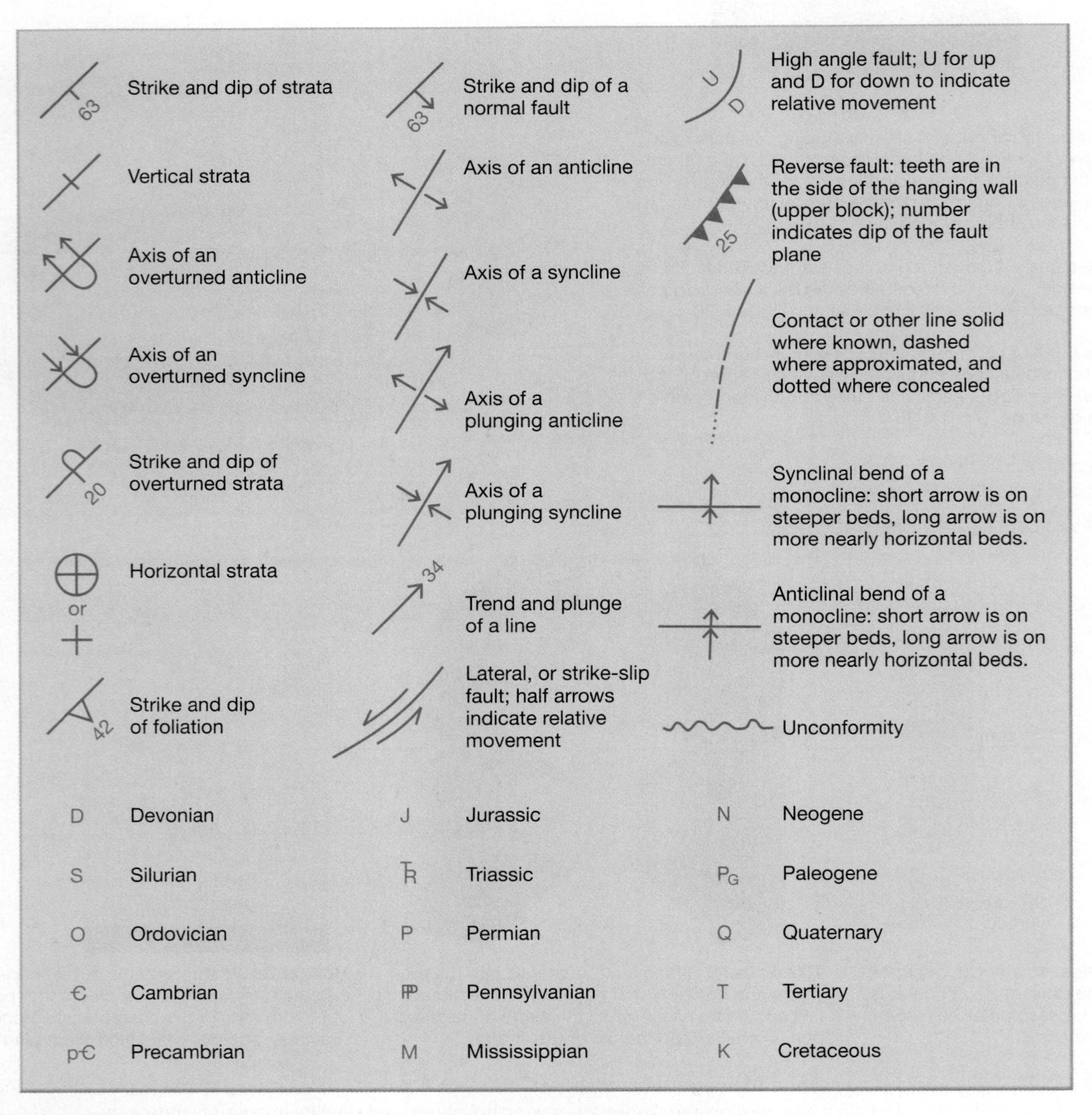

FIGURE 10.5 Structural geology symbols and abbreviations used on geologic maps.

Constructing Geologic Cross Sections

Geologic maps contain evidence of the surface locations and orientations of formations and the structures into which they have been deformed. To help visualize the geologic structures, geologists convert this surface information into vertical geologic cross sections like the sides of the block diagram in FIGURE 10.2.

Geologic cross sections are often drawn perpendicular to strike, so you can see the dip of the rocks more exactly. Most are drawn beneath a topographic profile (FIGURE 10.6), so you can see the topographic expression of the formations and geologic structures. However, some geologic cross sections are just rectangular cross sections that do not show the topography. Once you have constructed a topographic profile (or drawn a rectangular space) for the map line segment of the cross section (FIGURE 10.6), then follow the directions in FIGURE 10.6 to add the geologic information. You will need to use a pencil (with a good eraser), protractor, ruler, and colored pencils and be very neat and exact in your work.

Fractures and Faults

Brittle deformation is said to occur when rocks **fracture** (crack) or **fault** (slide in opposite directions along a crack in the rock). Motion and scraping of brittle rocks along the fault surfaces causes development of *slickensides*, polished surfaces with lineations and steplike linear ridges that indicate the direction of movement along the fault (FIGURE 10.7). If you gently rub the palm of your hand back and forth along the slickensides, then one direction will seem smoother (down the step like ridges) than the other. That is the relative direction of the side of the fault represented by your hand.

Faults form when brittle rocks experience one of these three kinds of directed pressure (stress): *tension* (pulling apart or lengthening), *compression* (pushing together, compacting, and shortening), or *shear* (smearing or tearing). The three kinds of stress produce three different kinds of faults: normal, reverse/thrust, and strike-slip (FIGURES 10.1, 10.7).

Normal and reverse/thrust faults both involve vertical motions of rocks. These faults are named by noting the *sense of motion* of the top surface of the fault (top block) relative to the bottom surface (bottom block), regardless of which one actually has moved. The top surface of the fault is called the **hanging wall** and is the base of the **hanging wall** (top) **block** of rock. The bottom surface of the fault is called the **footwall** and forms the top of the **footwall block**. Whenever you see a fault in a vertical cross section, just imagine yourself walking on the fault surface. The surface that your feet would touch is the footwall.

Normal Faults

Normal faults are caused by tension (rock lengthening). As tensional stress pulls the rocks apart, gravity pulls down the hanging wall block. Therefore, normal faulting gets its name because it is a normal response to gravity. You can recognize normal faults by recognizing the motion of the hanging wall block relative to the footwall block. First, imagine that the footwall block is stable (has not moved). If the hanging wall block has moved downward in relation to the footwall block, then the fault is a normal fault.

Reverse Faults

Reverse faults are caused by compression (rock shortening). As compressional stress pushes the rocks together, one block of rock gets pushed atop another. You can recognize reverse faults by recognizing the motion of the hanging wall block relative to the footwall block. First, imagine that the footwall block is stable (has not moved). If the hanging wall block has moved upward in relation to the footwall block, then the fault is a reverse fault. **Thrust faults** are reverse faults that develop at a very low angle and may be very difficult to recognize (FIGURE 10.7). Reverse faults and thrust faults generally place older strata on top of younger strata.

Strike–Slip Faults

Strike–slip faults (lateral faults) are caused by shear and involve horizontal motions of rocks (FIGURE 10.7). If you stand on one side of a strike–slip fault and look across it, then the rocks on the opposite side of the fault will appear to have slipped to the right or left. Along a *right-lateral (strike–slip) fault*, the rocks on the opposite side of the fault appear to have moved to the right. Along a *left-lateral (strike–slip) fault*, the rocks on the opposite side of the fault appear to have moved to the left.

Folded Structures

Folds are upward, downward, or sideways bends of rock layers. **Synclines** are "downfolds" or "concave folds," with the *youngest* rocks in the middle (FIGURE 10.8A). **Anticlines** are "upfolds" or "convex folds" with the *oldest* rocks in the middle (FIGURE 10.8B).

In a fold, each stratum (rock layer) is bent around an imaginary axis, like the crease in a piece of folded paper. This is the **fold axis** (or **hinge line**). For all strata in a fold, the fold axes lie within the **axial plane** of the fold (FIGURE 10.8A–D). The axial plane divides the fold into two **limbs** (sides, FIGURE 10.8B). For symmetric anticlines and synclines, the fold axis is vertical, but most anticlines and synclines are asymmetric. The axial plane of asymmetric folds is leaning to one side or the other, so one limb is steeper and shorter than the other.

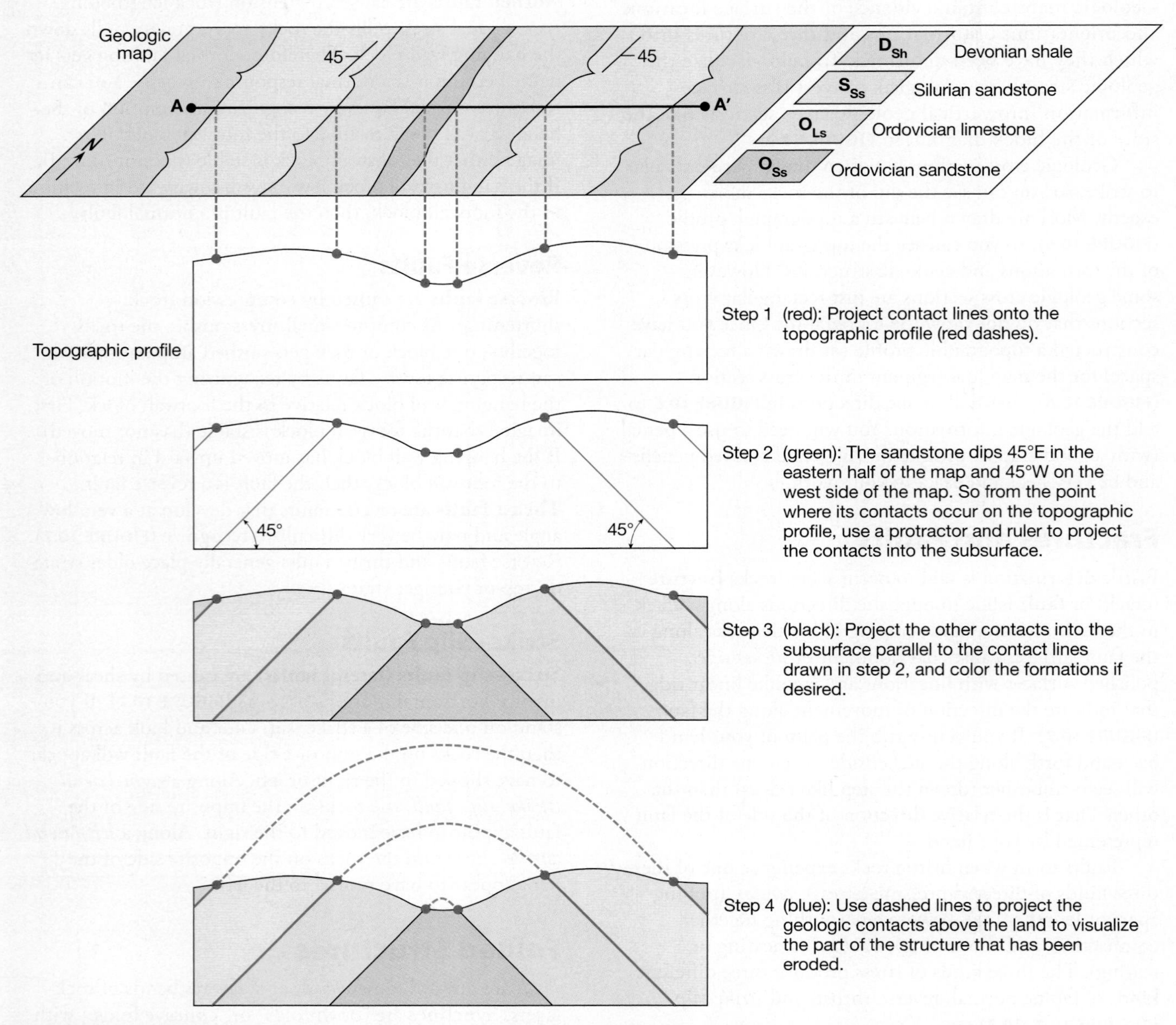

FIGURE 10.6 Geologic cross section construction. Follow the four steps in the illustration to construct a cross section.

The fold axis may not be horizontal, but rather it may plunge into the ground. This is called a **plunging fold** (FIGURE 10.8C, D). **Plunge** is the angle between the fold axis and horizontal. The **trend** of the plunge is the bearing (compass direction), measured in the direction that the axis is inclined downward. You can also think of the trend of a plunging fold as the direction a marble would roll if it were rolled down the plunging axis of the fold.

If a fold is tilted so that one limb is upside down, then the entire fold is called an **overturned fold** (FIGURE 10.8H). **Monoclines** have two axial planes that separate two nearly horizontal limbs from a single, more steeply inclined limb (FIGURE 10.8G).

Domes and **basins** (FIGURE 10.8E, F) are large, somewhat circular structures formed when strata are warped upward, like an upside-down bowl (dome) or downward, like a bowl (basin). Strata are oldest at the center of a dome, and youngest at the center of a basin.

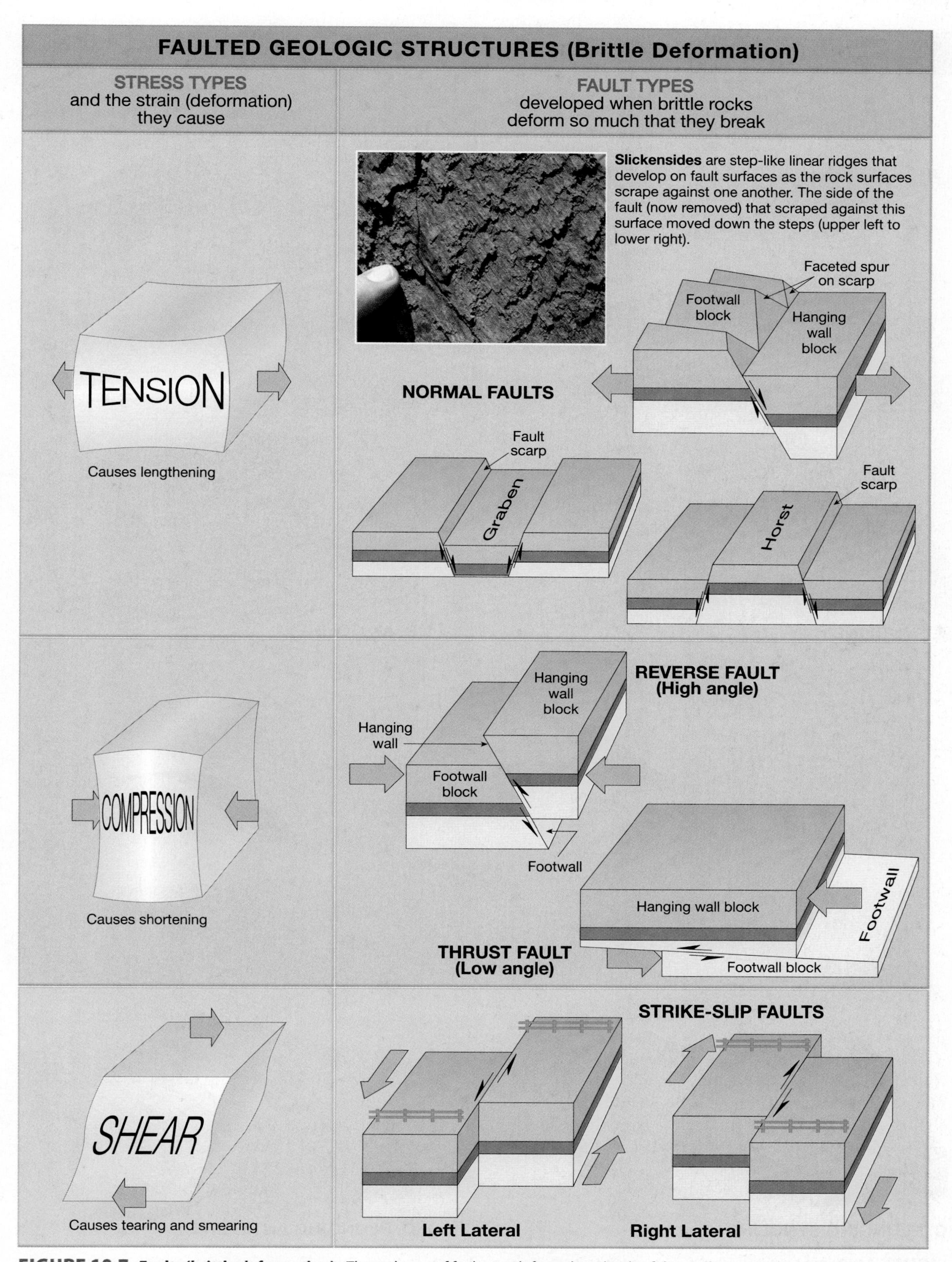

FIGURE 10.7 Faults (brittle deformation). Three classes of faults result from three kinds of directed pressure (stress: tension, compression, shear) applied to brittle rocks.

FOLDED GEOLOGIC STRUCTURES (Ductile Deformation)

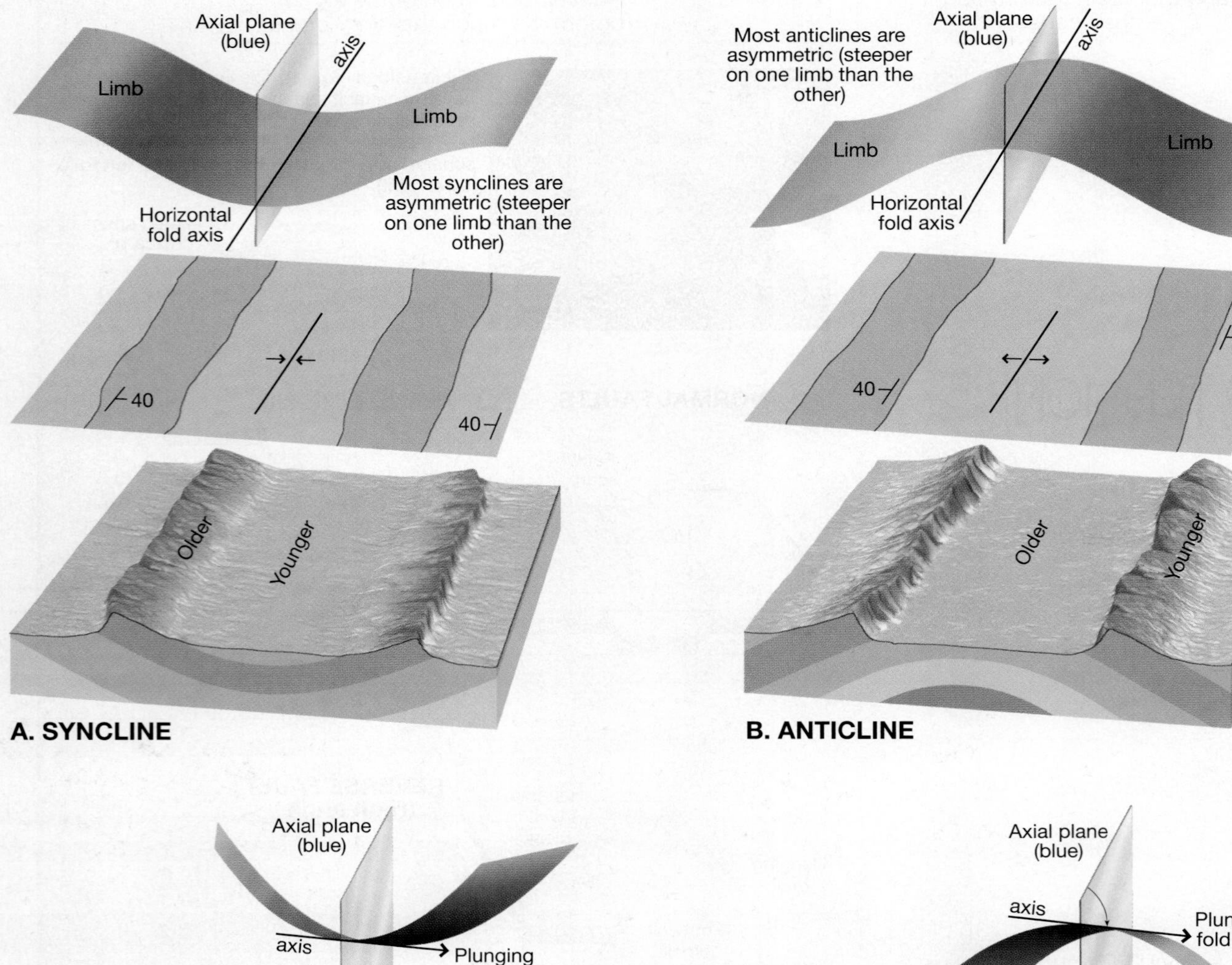

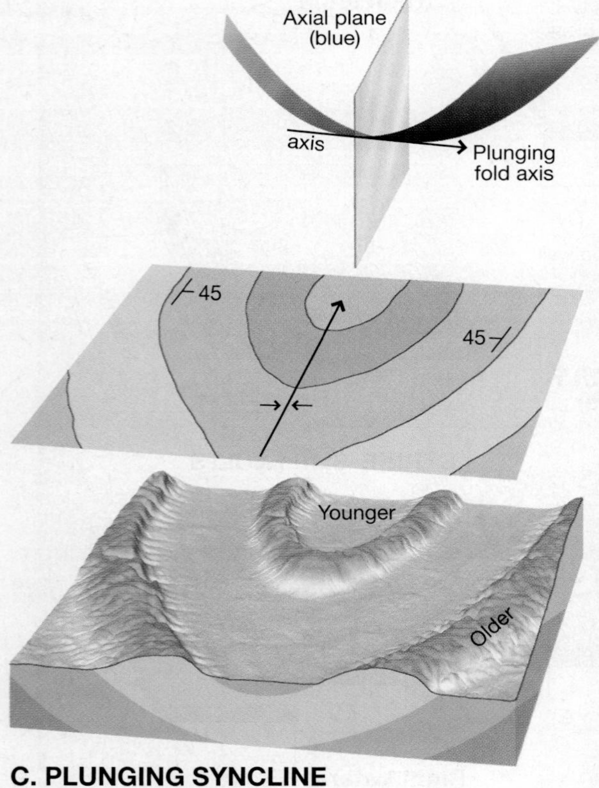

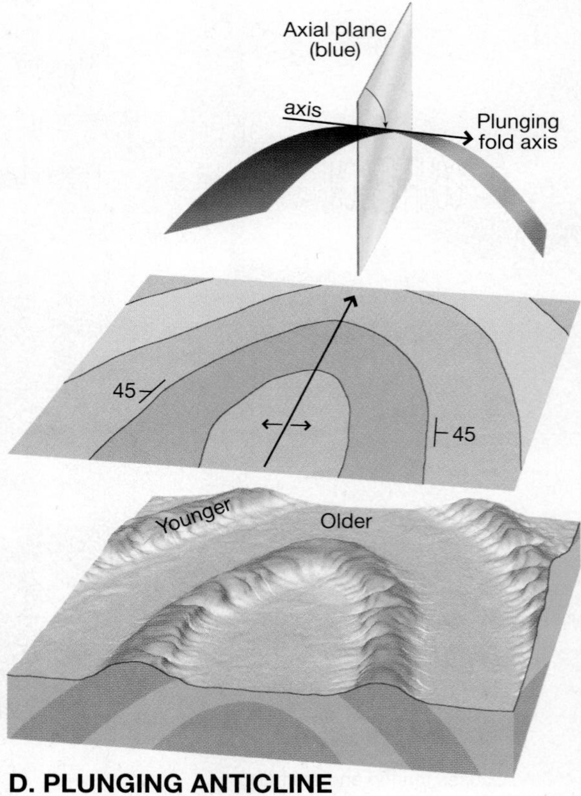

FIGURE 10.8 Folds (ductile deformation) in block diagrams and geologic maps. Note how and where symbols from **FIGURE 10.5** are used on the geologic maps.

FOLDED GEOLOGIC STRUCTURES (Ductile Deformation)

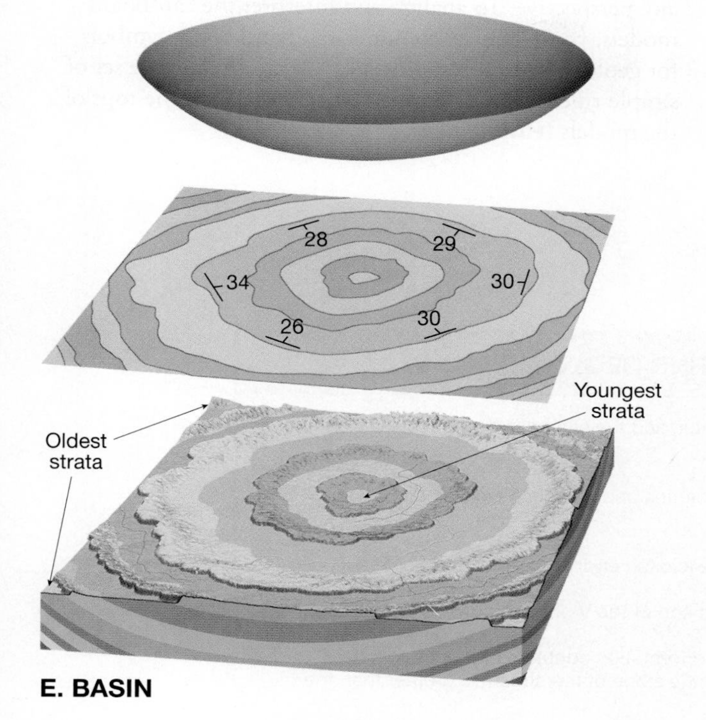

E. BASIN

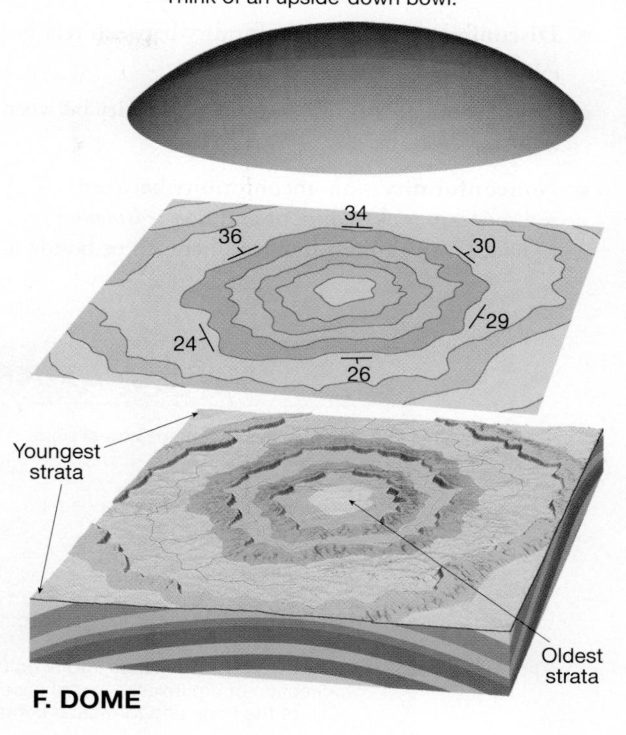

F. DOME

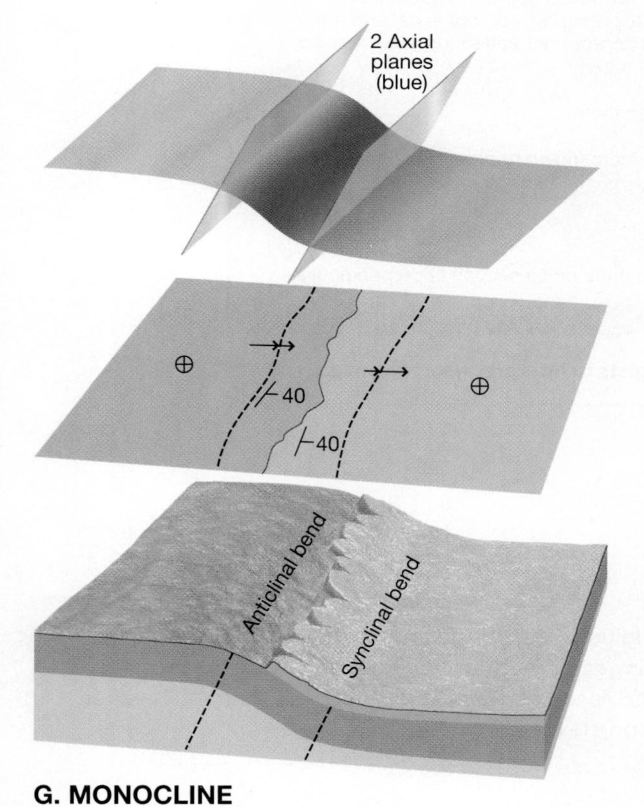

G. MONOCLINE

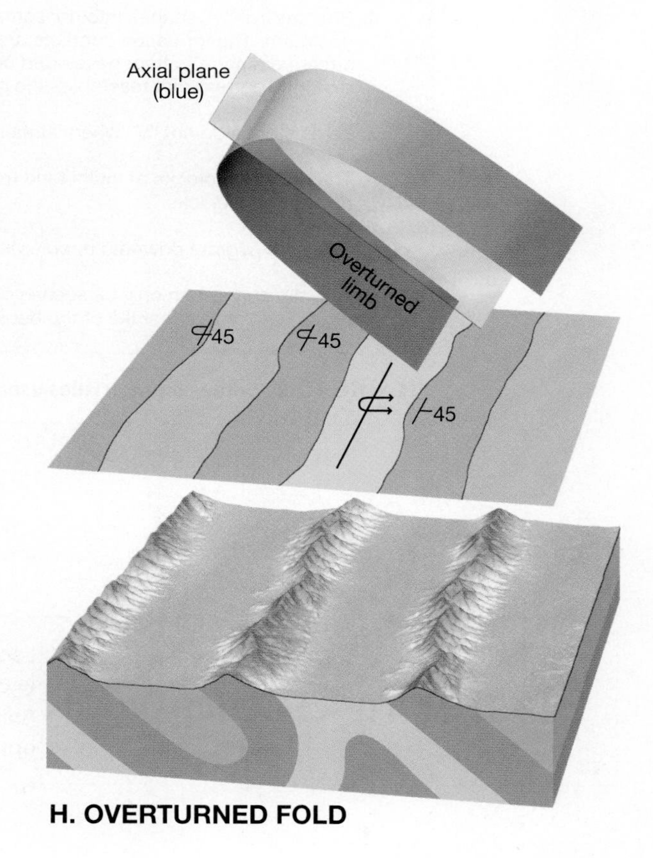

H. OVERTURNED FOLD

Unconformities

Structural geologists must locate, observe, and interpret many different structures. Fundamentally, these include unconformities, faults, and folds. There are three common types of *unconformities* (see **FIGURE 8.1**, p. 209):

- **Disconformity**—an unconformity between relatively *parallel* strata.
- **Angular unconformity**—an unconformity between *nonparallel* strata.
- **Nonconformity**—an unconformity between sedimentary rock/sediment and *non-sedimentary* (igneous or metamorphic) rock beneath or beside it.

Cardboard Models

Six cardboard block diagrams (Cardboard Models 1–6) are provided at the back of this laboratory manual. Unlike illustrated block diagrams, these actually are three-dimensional models that you can analyze from any perspective. To analyze and interpret the cardboard models, you will need to understand and apply symbols for geologic structures (**FIGURE 10.5**) and follow the set of simple rules for interpreting geologic maps on the tops of the models (**FIGURE 10.9**).

RULES FOR INTERPRETING GEOLOGIC MAPS

1. Anticlines have their oldest beds in the center, and their limbs (sides) dip away from the fold axis.
2. Synclines have their youngest beds in the center, and their limbs (sides) dip toward the fold axis.
3. Plunging anticlines plunge toward the nose (closed end) of the V-shaped outcrop belt.
4. Plunging synclines plunge toward the open end of the V-shaped outcrop belt.
5. Streams cut "V" shapes into tilted beds and formation contacts that point in the direction of dip (except in rare cases when the slope of the stream is greater than the dip of the beds and formation contacts).
6. Streams cut "V" shapes into horizontal beds and formation contacts that point upstream. The formation contacts are parallel to topographic contour lines, and the stream drainage system developed on horizontal and/or unstratified formations has a dendritic pattern that resembles the branching of a tree.
7. Vertical beds do not "V" where streams cut across them.
8. The upthrown blocks of faults tend to be eroded more (down to older beds) than downthrown blocks.
9. Contacts migrate downdip upon erosion.
10. True dip angles can only be seen in cross section if the cross section is perpendicular to the fault or to the strike of the beds.

FIGURE 10.9 Some common rules used by geologists to interpret geologic maps.

ACTIVITY 10.1 Geologic Structures Inquiry

Name: ______________________ **Course/Section:** ______________________ **Date:** ____________

A. Recall the Law of Original Horizontality, which states that layers of sedimentary rock and lava tend to accumulate as relatively horizontal layers. The layers may remain undisturbed or they may be deformed—have their original shape or form disturbed in some way. Analayze each image below. Beneath each image answer these two questions:

- Are the rocks deformed or not deformed?
- If deformed, then describe how and why you think the rocks were deformed.

1. Grand Canyon, Arizona rock layers.

2. Cliff face about 400 m tall, south-central Alaska (USGS photograph by N.J. Silberling).

3. Quartzite, Maria Mts, Riverside County, California (USGS photograph by W.B. Hamilton).

4. Sandstone on a steep wall about 100 m tall, Little Colorado River Gorge, Navajo Nation, Arizona

B. **REFLECT & DISCUSS** Based on your analysis of deformed rocks above, classify rock deformation into two categories, and note what images above would be in each category. Be prepared to explain your classification to other geologists.

ACTIVITY 10.2 Visualizing How Stresses Deform Rocks

Name: ______________________ **Course/Section:** ______________________ **Date:** ____________

Recall that Earth materials can deform when stress (amount of force acting on a unit surface area) exerts pressure on them. *Confining pressure* occurs when stress is equal in all directions. *Directed pressure* occurs when the stress is greater in one direction than another—described as *compression, tension, or shear.*

A. Stress in everyday objects and actions. Place checks in all of the boxes that describe the stress involved in each object and action described below.

Object or Action	Confining Pressure	Directed Pressure	Compression	Tension	Shear
1. A cardboard box collapses when you sit on it.					
2. An unopened bottle of soft drink (soda, pop).					
3. You stretch out a rubber band by pulling its ends apart.					
4. You rub your hands together, and back and forth, to keep warm.					

B. Stress in the Geosphere. Think about how tectonic plates deform along convergent, divergent, and transform plate boundaries. *Draw arrows* on each of the bottom six illustrations of plate deformation to show the directions of directed stress (directed pressure) that are causing the deformation, then *identify the stresses as compression, tension, or shear.*

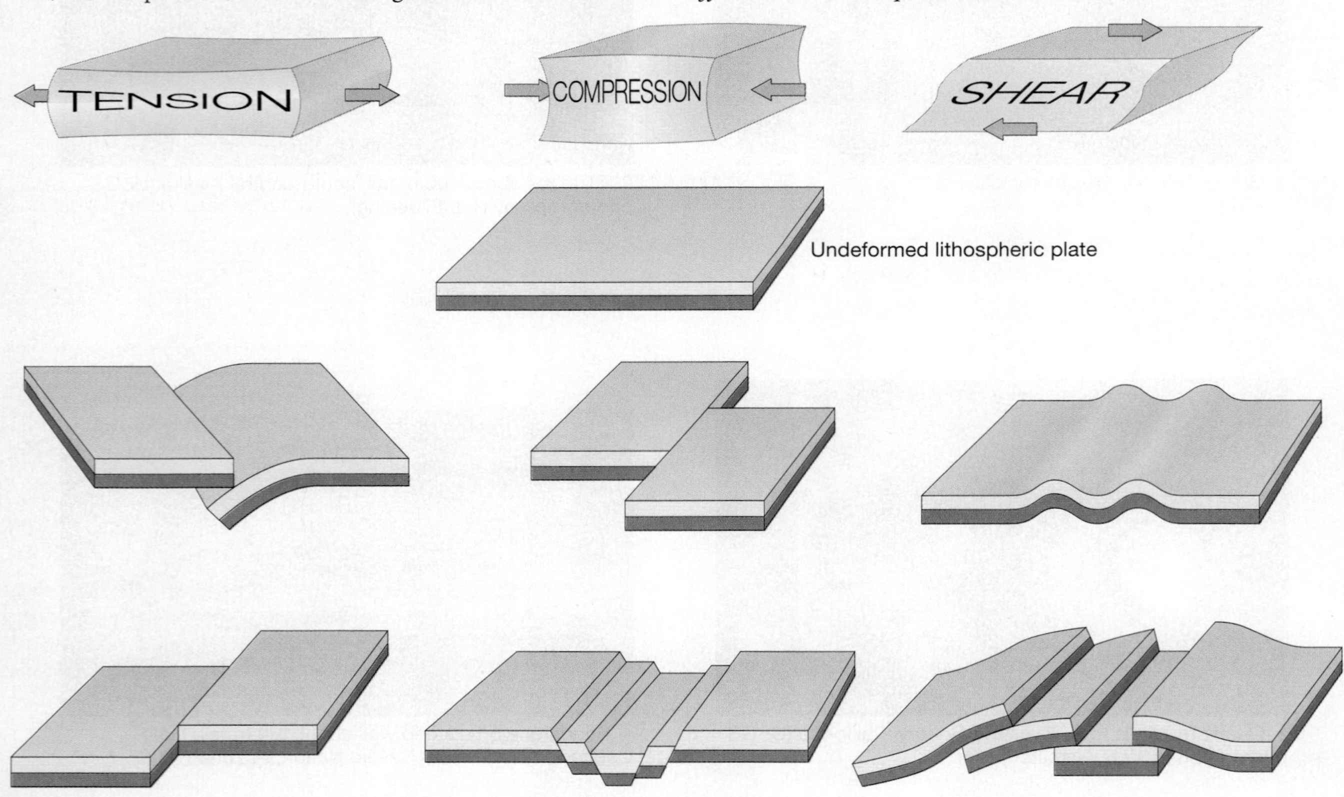

C. REFLECT & DISCUSS The examples of deformation above (part B) are primarily caused by directed stresses that are pressing on rocks in the horizontal dimension. What kinds of stress can you think of that are pushing or pulling on rocks in the vertical dimension, and what kind of deformation do they cause or aid?

ACTIVITY 10.3 Map Contacts and Formations

Name: ______________________ **Course/Section:** ______________________ **Date:** ____________

A. Map and interpret a bedding plane contact. Analyze this USGS photograph by M.R. Mudge showing outcrops of a resistant rock unit on the south end of Scapegoat Mountain, Montana. Notice the many prominent bedding plane contacts.

1. Using a pen, trace the labeled bedding plane contact across the photograph.
2. What kind of geologic structure is indicated by the shape of the contact, and what kind of stress caused it to form (**FIGURE 10.1**)?

Trace this bedding plane contact to the other side of the photograph

B. Mapping contacts and formations in the Grand Canyon, Arizona. On the orthoimage, notice that the Cambrian Muav Limestone Formation forms the floor of the canyon and has been labeled with a red "**M**". The location of the Pennsylvanian Watahomigi Formation has also been labeled with blue "**W**" symbols in several places where it crops out on the walls of the Grand Canyon. Based on their topographic expression on the orthoimage and map, *map both formations onto the topographic map* by coloring (as neatly and exactly as you can) the Muav red and the Watahomigi blue everywhere they appear. (Courtesy of USGS)

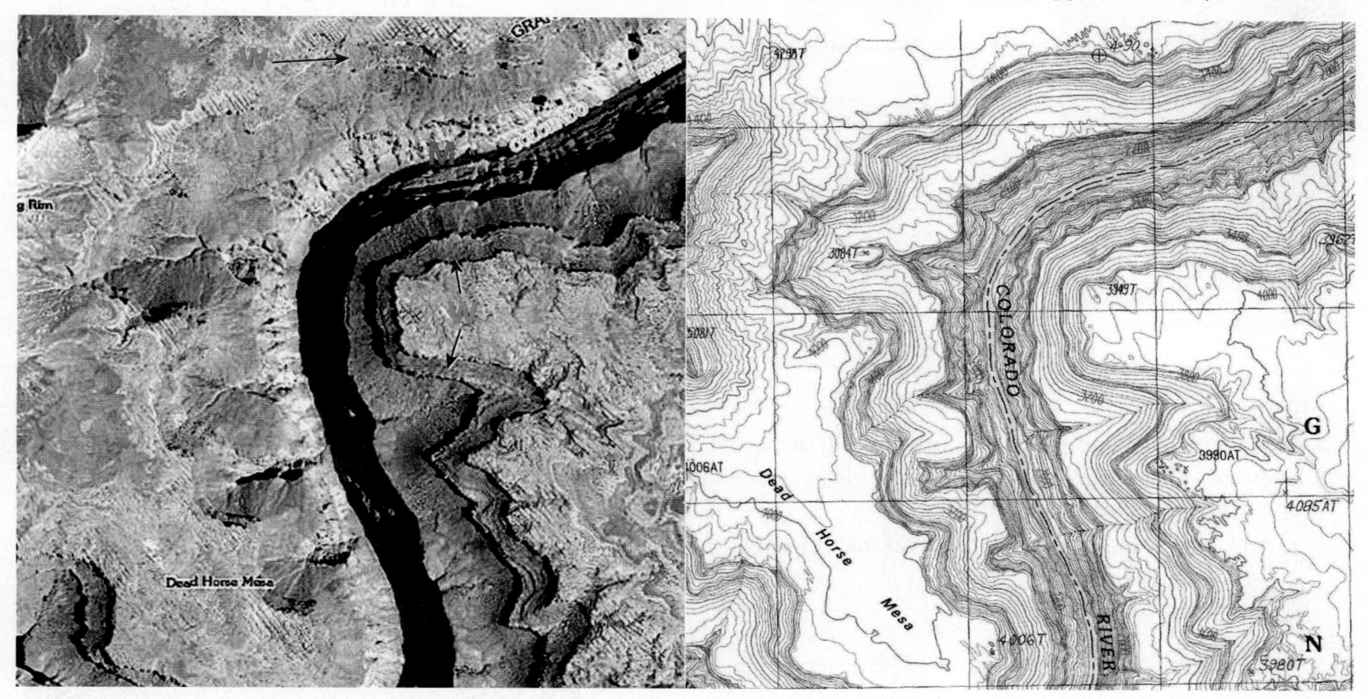

C. REFLECT & DISCUSS Analyze your geologic map, and picture in your mind how the Watahomigi Fm once extended across the canyon (before the river eroded it). Do you think the Watahomigi Formation is deformed into a geologic structure or is it relatively undeformed? Why?

ACTIVITY 10.4 Determine Attitude of Rock Layers and a Formation Contact

Name: ______________________ **Course/Section:** ______________________ **Date:** ____________

A. **On the flat surface below**, draw a symbol (**FIGURE 10.3**) that indicates its horizontal attitude (orientation). **On each of the inclined surfaces**, draw a strike-and-dip symbol (**FIGURE 10.3**) to indicate exactly how its strike and dip are oriented. Based on the north arrows and your strike-and-dip symbols, use a protractor from the GeoTools at the back of the lab manual to measure exactly and record the strike of surfaces in both quadrant and azimuth form (Refer to **FIGURES 10.3** and **10.4**).

1. Flat top surface of a rock wall

2. Inclined rock surface

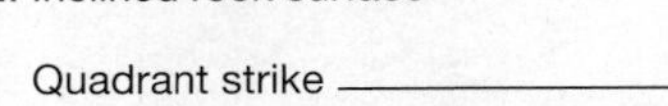

Quadrant strike ____________

Azimuth strike ____________

3. Inclined rock surface

Quadrant strike ____________

Azimuth strike ____________

B. For each map below, record the strike of the formations and their contacts. Then use the Rule of Vs (**FIGURE 10.4**) to determine the dip direction and sketch formation contacts on the geologic cross section (**FIGURE 10.6**, steps 1 and 2).

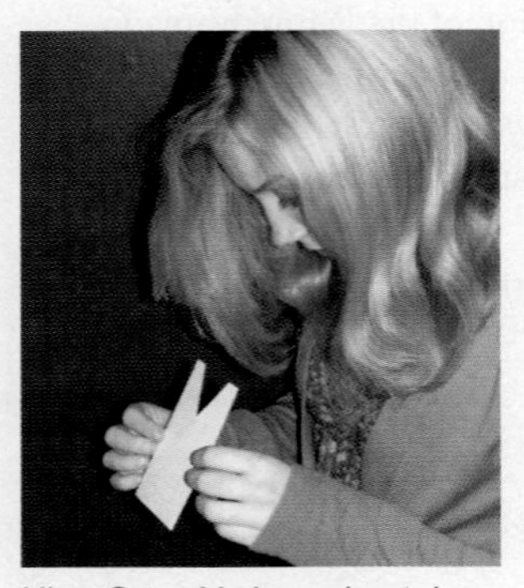

Hint: Cut a V-shaped notch into the top edge of a 3 x 5 card. Hold card vertically and look down on edge. See Vs at different dips of the card.

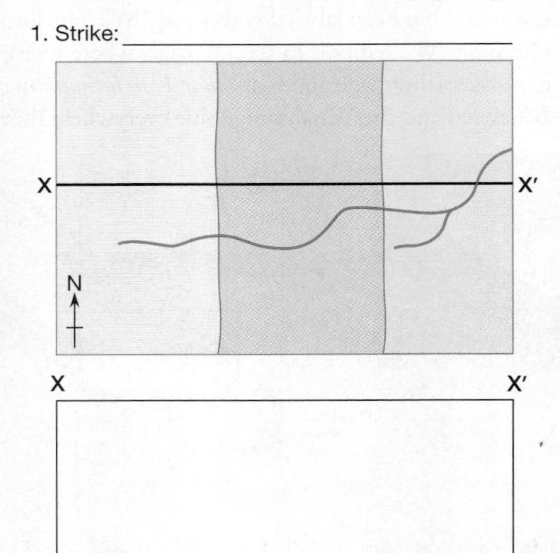

2. Strike: ____________

Y′

N

Y

Y Y′

C. This is an aerial photograph taken from directly above the landscape. The view is 1 km wide.

1. Circle a V-shaped notch.

2. Add a strike and dip symbol to one of the flatirons.

3. What is the strike? ____________

4. What is the dip direction? ____________

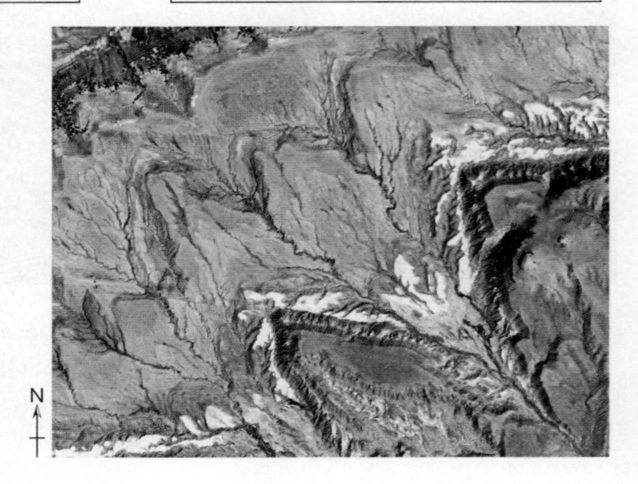

D. **REFLECT & DISCUSS** Draw (pen) a possible formation contact on the image in part C and explain why you drew it there.

ACTIVITY 10.5 Cardboard Model Analysis and Interpretation

Name: ____________________ **Course/Section:** ____________________ **Date:** __________

Tear Cardboard Models 1–6 from the back of your lab manual. Cut and fold them as noted in red on each model.

A. Cardboard Model 1

This model shows Ordovician (green), Silurian (light gray), Devonian (bule-gray), Mississippian (dark gray), Pennsylvanian (yellow), and Permian (salmon) formations striking due north and dipping 24° to the west. Provided are a complete geologic map (the top of the diagram) and three of the four vertical cross sections (south, east, and west sides of the block diagram).

1. Finalize Cardboard Model 1 as follows. First construct the vertical cross section on the north side of the block so it shows the formations and their attitudes (dips). On the map, draw a strike and dip symbol on the Mississippian sandstone that dips 24° to the west (see **FIGURES 10.3** or **10.5** for the strike and dip symbol).
2. Explain the sequence of events that led to the existence of the formations and the relationships that now exist among them in this block diagram.

B. Cardboard Model 2

This model is slightly more complicated than the previous one. The geologic map is complete, but only two of the cross sections are available. Letters **A–G** are ages from oldest (**A**) to youngest (**G**).

1. Finalize Cardboard Model 2 as follows. First, complete the north and east sides of the block. Notice that the rock units define a fold. This fold is an anticline, because the strata are convex upward and the oldest formation (**A**) is in the center (inside) of the fold. It is symmetric (non-plunging), because its axis is horizontal. (Refer back to **FIGURE 10.8** for the differences between plunging and non-plunging folds if you are uncertain about this.) On the geologic map, draw strike and dip symbols to indicate the attitudes of formation **E** (gray formation) at points **I, II, III,** and **IV.** Also draw the proper symbol on the map (top of model) along the axis of the fold (refer to **FIGURE 10.5**).
2. How do the strikes at all four locations compare with each other?
3. How does the dip direction at points **I** and **II** compare with the dip direction at points **III** and **IV?** *In your answer, include the dip direction at all four points.*

C. Cardboard Model 3

This cardboard model has a complete geologic map. However, only one side and part of another are complete. Letters **A–E** are ages from oldest (**A**) to youngest (**E**).

1. Finalize Cardboard Model 3 as follows. Complete the remaining two-and-a-half sides of this model, using as guides the geologic map on top of the block and the one-and-a-half completed sides. On the map, draw strike and dip symbols showing the orientation of formation **C** at points **I, II, III,** and **IV.** Also draw the proper symbol along the axis of the fold (refer to **FIGURE 10.5**).
2. How do the strikes of all four locations compare with each other?
3. How does the dip direction (of formation **C**) at points **I** and **II** compare with the dip direction at points **III** and **IV?** *Include the dip direction at all four points in your answer.*
4. Is this fold plunging or non-plunging? ______________ 5. Is it an anticline or a syncline? ______________
6. On the basis of this example, how much variation is there in the strike at all points in a non-plunging fold?

D. Cardboard Model 4

Letters **A–H** are ages from oldest (**A**) to youngest (**H**). This model shows a plunging anticline.. The anticline plunges to the north, following the general rule that *anticlines plunge in the direction in which the fold closes.*

1. Finalize Cardboard Model 4 as follows. Complete the north and east sides of the block. Draw strike and dip symbols on the map at points **I, II, III, IV,** and **V.** Draw the proper symbol on the map along the axis of the fold, including its direction of plunge. Also draw the proper symbol on the geologic map to indicate the orientation of beds in formation **J.**
2. How do the directions of strike and dip differ from those in Model 3?

E. Cardboard Model 5

Letters **A–H** are ages from oldest (**A**) to youngest (**H**). This model shows a plunging syncline. Two of the sides are complete and two remain incomplete.

1. Finalize Cardboard Model 5 as follows. Complete the north and east sides of the diagram. Draw strike and dip symbols on the map at points **I, II, III, IV,** and **V** to show the orientation of layer **G.** *Synclines plunge in the direction in which the fold opens.* Draw the proper symbol along the axis of the fold (on the map) to indicate its location and direction of plunge.
2. In which direction (bearing, trend) does this syncline plunge?

F. Cardboard Model 6

This model shows a fault that strikes due west and dips 45° to the north. Three sides of the diagram are complete, but the east side is incomplete.

1. Finalize Cardboard Model 6 as follows. At point **I,** draw a symbol from **FIGURE 10.7** to show the *orientation of the fault.* On the west edge of the block, draw arrows parallel to the fault, indicating relative motion. Label the hanging wall and the footwall. Complete the east side of the block. Draw half-arrows (**FIGURE 10.7**) parallel to the fault, to indicate its relative motion. Now look at the geologic map and at points **II** and **III.** Write **U** on the side that went up and **D** on the side that went down. At points **IV** and **V,** draw strike and dip symbols for formation **B.**
2. Is the fault in this model a normal fault or a reverse fault? Why?

3. On the geologic map, what happens to the contact between units **A** and **B** where it crosses the fault?

4. Could the same offset along this fault have been produced by strike-slip motion?

5. **REFLECT & DISCUSS** There is a general rule that, as erosion of the land proceeds, *contacts migrate downdip.* Is this true in this example? Explain.

ACTIVITY 10.6 Block Diagram Analysis and Interpretation

Name: ______________________ **Course/Section:** ______________________ **Date:** ____________

For each block diagram: **1.** Complete the diagram so that contact lines between rock formations are drawn on all sides. **2.** Add symbols (**FIGURES 10.5, 10.7–10.9**) to indicate the attitudes of all structures. **3.** On the line provided, write the exact name of the geologic structure represented in the diagram.

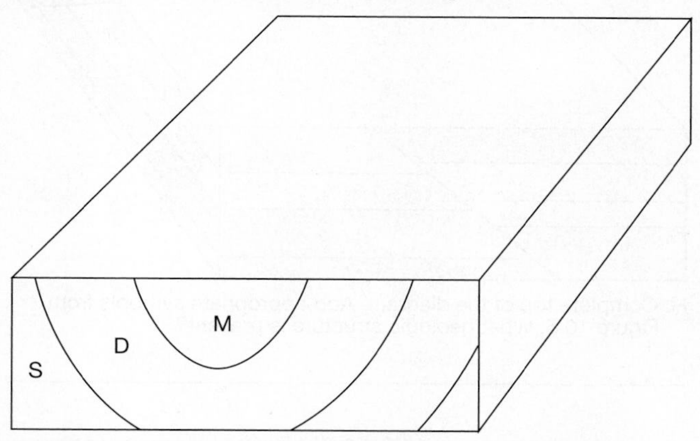

A. Complete top and side. Add appropriate symbols from Figure 10.5. What geologic structure is present?

____

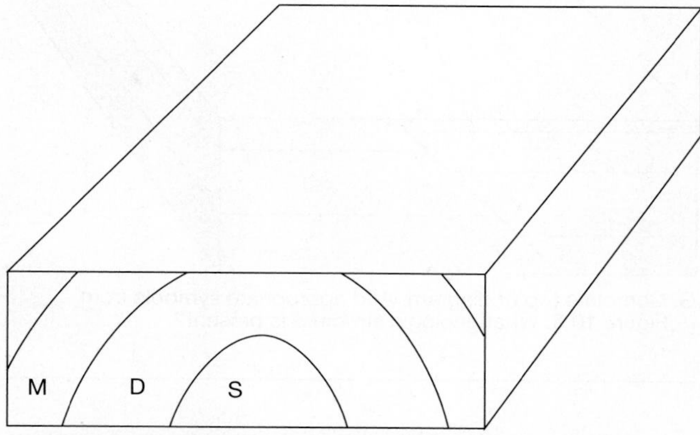

B. Complete top and side. Add appropriate symbols from Figure 10.5. What geologic structure is present?

____

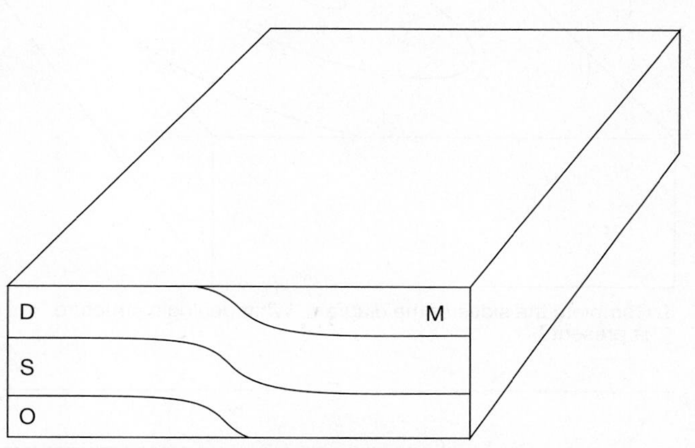

C. Complete top and side. Add appropriate symbols from Figure 10.5. What geologic structure is present?

____

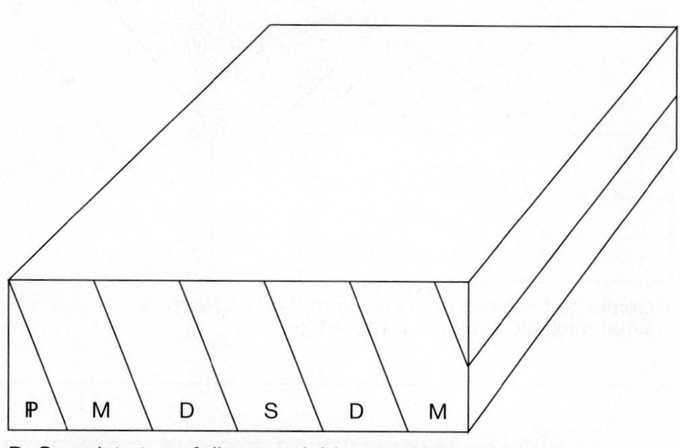

D. Complete top of diagram. Add appropriate symbols from Figure 10.5. What geologic structure is present?

____

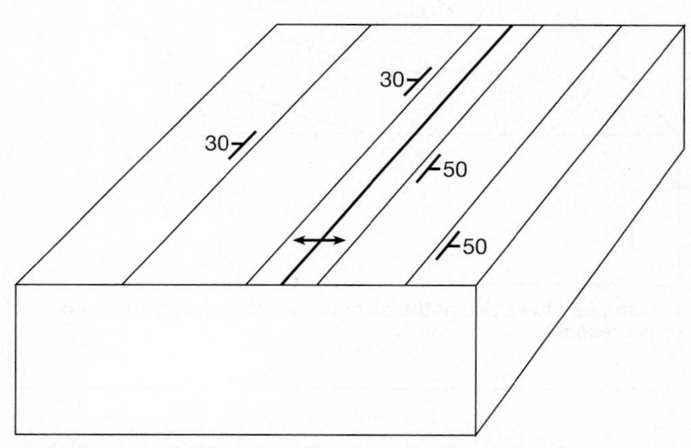

E. Complete the sides of the diagram. What geologic structure is present?

____

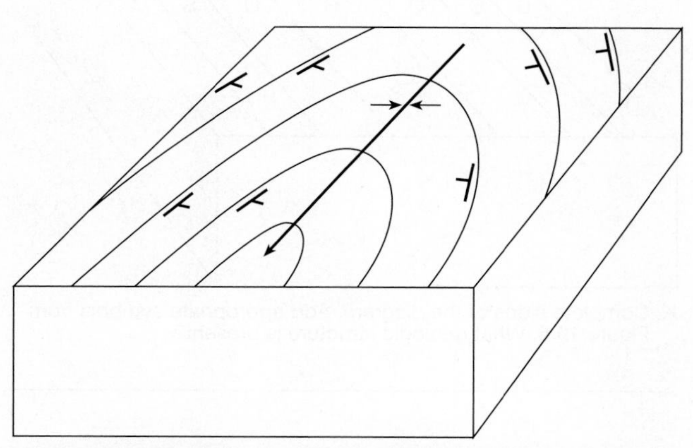

F. Complete the sides of the diagram. What geologic structure is present?

____

For each block diagram: **1.** Complete the diagram so that contact lines between rock formations are drawn on all sides. **2.** Add symbols (**FIGURES 10.5, 10.7–10.9**) to indicate the attitudes of all structures. **3.** On the line provided, write the exact name of the geologic structure represented in the diagram.

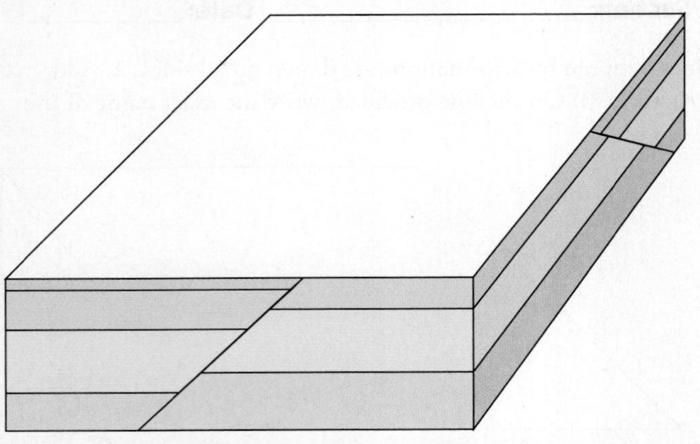

G. Complete top of diagram. Add appropriate symbols from Figure 10.5. What geologic structure is present?

____

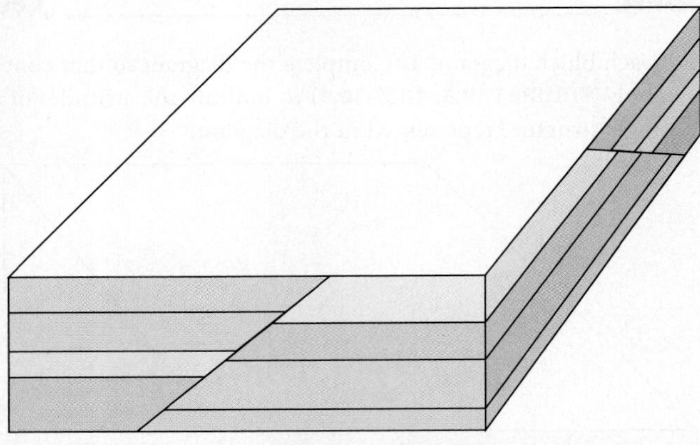

H. Complete top of the diagram. Add appropriate symbols from Figure 10.5. What geologic structure is present?

____

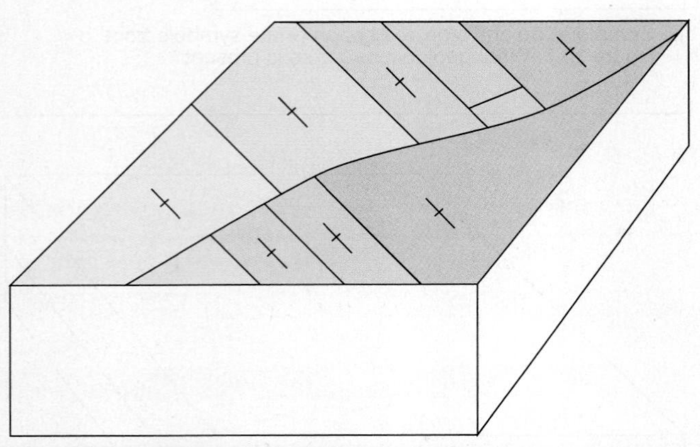

I. Complete the sides of the diagram. Add half-arrows. What geologic structure is present?

____

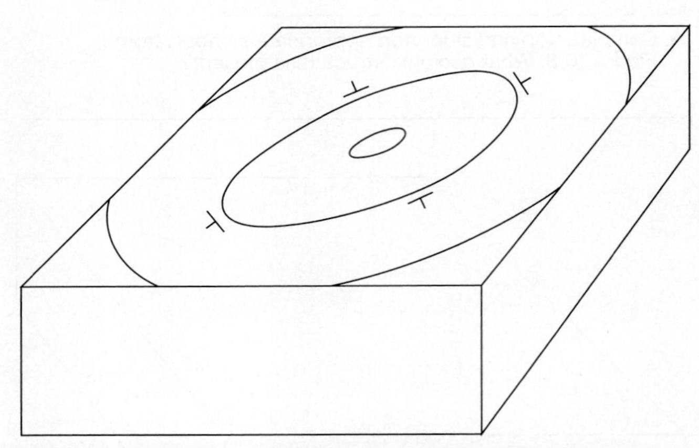

J. Complete the sides of the diagram. What geologic structure is present?

____

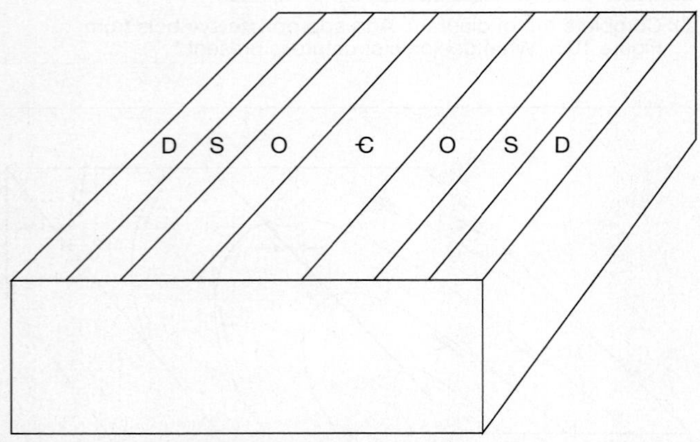

K. Complete sides of the diagram. Add appropriate symbols from Figure 10.5. What geologic structure is present?

____

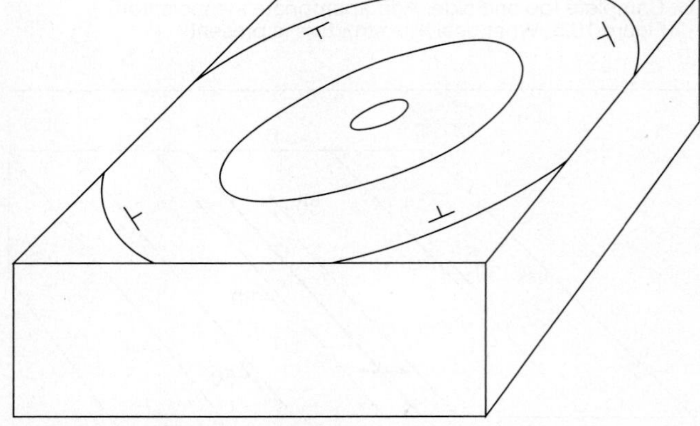

L. Complete the sides of the diagram. What geologic structure is present?

____

ACTIVITY 10.7 Nevada Fault Analysis Using Orthoimages

Name: ______________________ **Course/Section:** ______________________ **Date:** ____________

A. This orthoimage was acquired about 7 km east of Las Vegas, Nevada (US Topo: Frenchman Mtn. Quad.). The Nevada Geological Survey has determined that faulting occurred here 11–6 Ma. To view the region using Google Earth™, search 36 09 34.5N, 114 56 51.6W. Use a pen to neatly and precisely trace the fault line that crosses this area and add half-arrow symbols (**FIGURE 10.5**) to show its relative motion. Scale: 1 cm = 0.11 km.

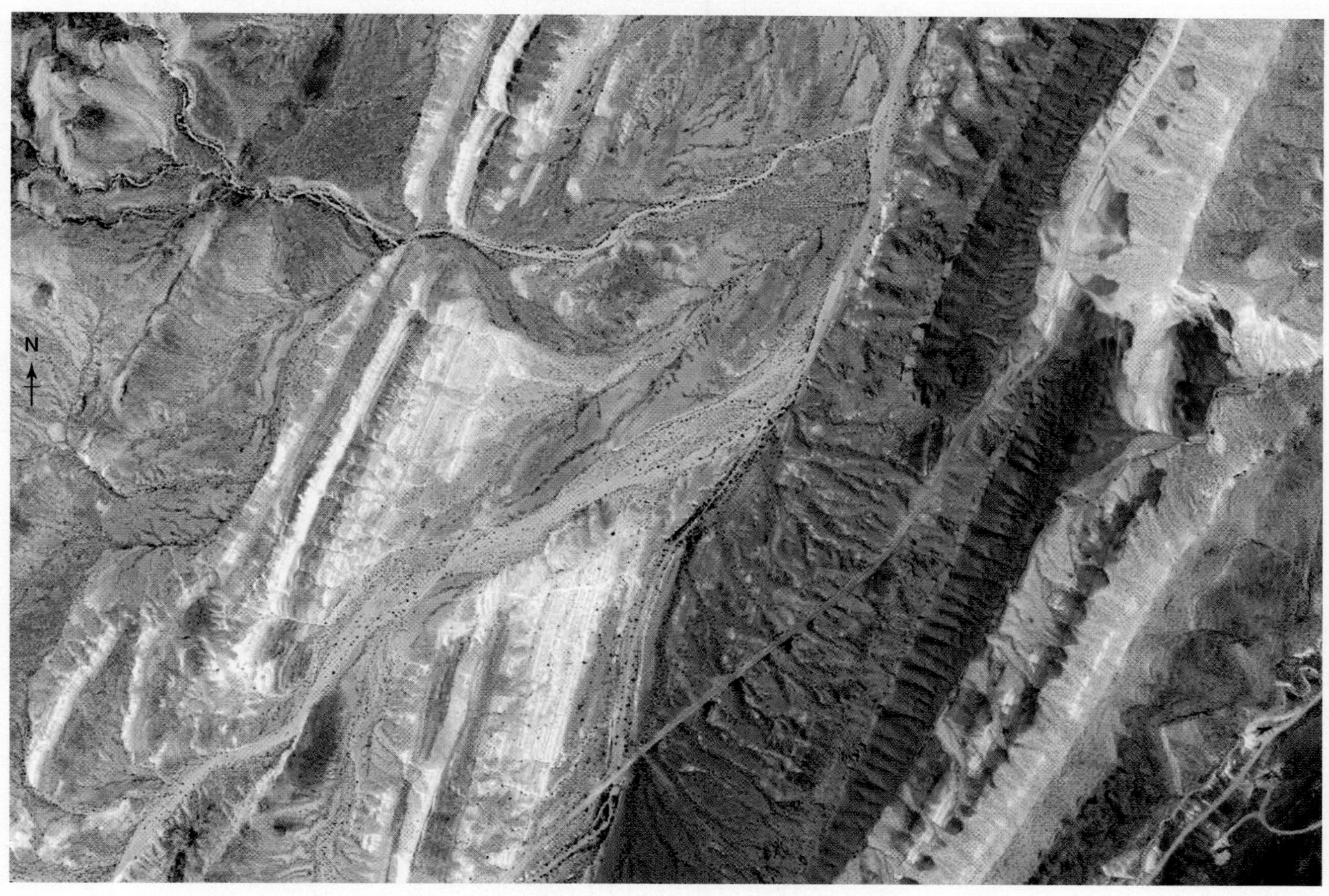

1. Exactly what kind of fault have you mapped above?

B. This image was acquired just west of the one above. You can view it in Google Earth™ at 36 08 19.4N, 114 58 41.8W.

1. Use a pen to neatly and precisely trace all of the faults that you can detect in the image, and add half-arrows to show their motion.
2. Exactly what kind(s) of fault(s) have you mapped?

C. REFLECT & DISCUSS Based on your work above, what kind of stress was this Las Vegas region experiencing 11–6 million years ago (when the faults formed), and why did faults form instead of folds? Refer to **FIGURES 10.1** and **10.7** as needed.

ACTIVITY 10.8 Appalachian Mountains Geologic Map

Name: ______________________ **Course/Section:** ______________________ **Date:** ____________

Directions: A. Complete the geologic cross section using the steps in **FIGURE 10.6**. **B.** Label the kind(s) of geologic structure(s) revealed by your work. Then add the appropriate symbols from **FIGURE 10.5** to the geologic map to show the axes of the folds. **C.** Add half-arrows to the fault near the center of the geologic map to show the relative motions of its two sides. Exactly what kind of fault is it?

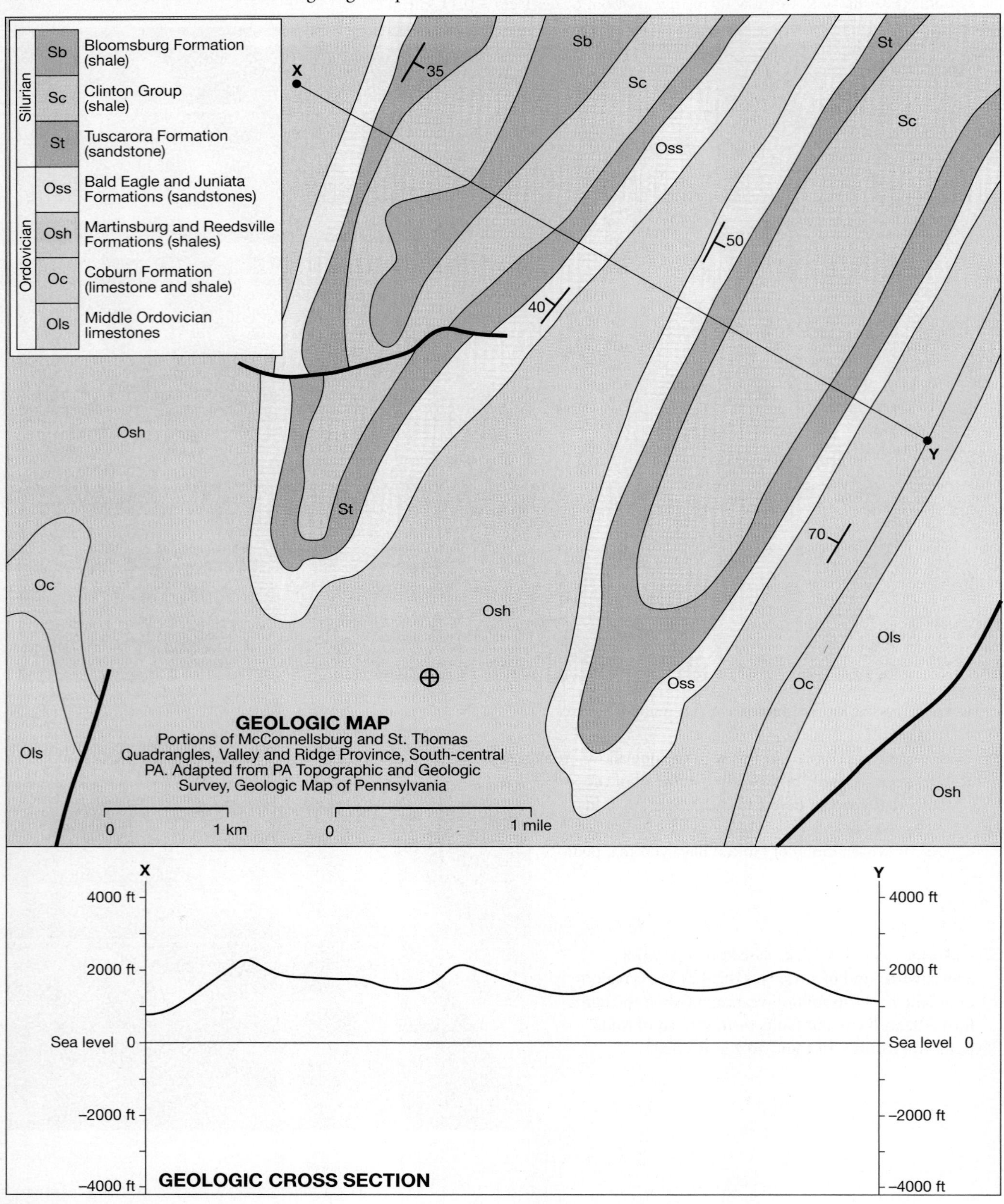

LABORATORY 11

Stream Processes, Landscapes, Mass Wastage, and Flood Hazards

CONTRIBUTING AUTHORS

Pamela J.W. Gore • *Georgia Perimeter College*

Richard W. Macomber • *Long Island University–Brooklyn*

Cherukupalli E. Nehru • *Brooklyn College (CUNY)*

The Middle Fork of the Salmon River flows for about 175 kilometers (110 miles) through a wilderness area in central Idaho. (Michael Collier)

BIG IDEAS

Streams shape the landscape and provide water to communities and agricultural systems. Flood hazards and mass wasting are also associated with streams. Tools and methods for determining flood hazards are provided by the Federal Emergency Management Agency (FEMA).

FOCUS YOUR INQUIRY

THINK About It How are you affected by streams?

ACTIVITY 11.1 Streamer Inquiry *(p. 284)*

THINK About It How does stream erosion shape the landscape?

ACTIVITY 11.2 Introduction to Stream Processes and Landscapes *(p. 284)*

ACTIVITY 11.3 Escarpments and Stream Terraces *(p. 284)*

ACTIVITY 11.4 Meander Evolution on the Rio Grande *(p. 284)*

ACTIVITY 11.5 Mass Wastage at Niagara Falls *(p. 292)*

THINK About It How do geologists determine the risk of flooding along rivers and streams?

ACTIVITY 11.6 Flood Hazard Mapping, Assessment, and Risk *(p. 295)*

Introduction

It all starts with a single raindrop, then another, and another. As water drenches the landscape, some soaks into the ground and becomes *groundwater*. Some flows over the ground and into streams and ponds of *surface water*. The streams will continue to flow for as long as they receive a water supply from additional rain, melting snow, or *base flow* (groundwater that seeps into a stream via porous rocks, fractures, and springs).

Perennial streams flow continuously throughout the year and are represented on topographic maps as blue lines. *Intermittent streams* flow only at certain times of the year, such as rainy seasons or when snow melts in the spring. They are represented on topographic maps as blue line segments separated by blue dots (three blue dots between each line segment). All streams, perennial and intermittent, have the potential to flood (overflow their banks). Floods damage more human property in the United States than any other natural hazard.

Streams are also the single most important natural agent of *land erosion* (wearing away of the land). They erode more sediment from the land than wind, glaciers, or ocean waves. The sediment is transported and eventually deposited, whereupon it is called *alluvium*. Alluvium consists of gravel, sand, silt, and clay deposited in floodplains, point bars, channel bars, deltas, and alluvial fans (**FIGURE 11.1**).

Therefore, stream processes (or *fluvial processes*) are among the most important agents that shape Earth's surface and cause damage to humans and their property.

ACTIVITY 11.1 Streamer Inquiry

THINK About It **Where does a stream near your community come from, where does it go, and why does it matter?**

OBJECTIVE Analyze where a community's stream water comes from and where it goes, then infer how a community may benefit from such knowledge.

PROCEDURES

1. **Before you begin**, do not look up definitions and information. Use your current knowledge, and complete the worksheet with your current level of ability. Also, this is **what you will need** to do the activity:
   ____ computer with Internet access
   ____ Activity 11.1 Worksheets (pp. 297–298) and pencil
2. **Complete the worksheet in a way that makes sense to you.**
3. **After you complete the worksheet**, be prepared to discuss your observations and ideas with others.

ACTIVITY 11.2 Introduction to Stream Processes and Landscapes

THINK About It **How does stream erosion shape the landscape?**

OBJECTIVE Analyze and interpret stream valley features using maps and an orthoimage, stream profile, and graph.

PROCEDURES

1. **Before you begin**, read Stream Processes and Landscapes below. Also, this is **what you will need:**
   ___ calculator
   ___ 15-inch (40 cm) piece of thin string or thread
   ___ Activity 11.2 Worksheets (pp. 299–303) and pencil
2. **Then follow your instructor's directions** for completing the worksheet.

ACTIVITY 11.3 Escarpments and Stream Terraces

THINK About It **How does stream erosion shape the landscape?**

OBJECTIVE Analyze and interpret escarpments and terraces along the Souris River using an orthoimage with topographic contours and a sketch of the river valley profile.

PROCEDURES

1. **Before you begin**, read Stream Processes and Landscapes below. Also, this is **what you will need:**
   ___ Activity 11.3 Worksheet (p. 304) and pencil
2. **Then follow your instructor's directions** for completing the worksheet.

ACTIVITY 11.4 Meander Evolution on the Rio Grande

THINK About It **How does stream erosion shape the landscape?**

OBJECTIVE Analyze a map of changes in the course of the Rio Grande to interpret the evolution of meanders.

PROCEDURES

1. **Before you begin**, read Stream Processes and Landscapes below. Also, this is **what you will need:**
   ___ calculator, ruler
   ___ Activity 11.4 Worksheet (p. 305) and pencil
2. **Then follow your instructor's directions** for completing the worksheet.

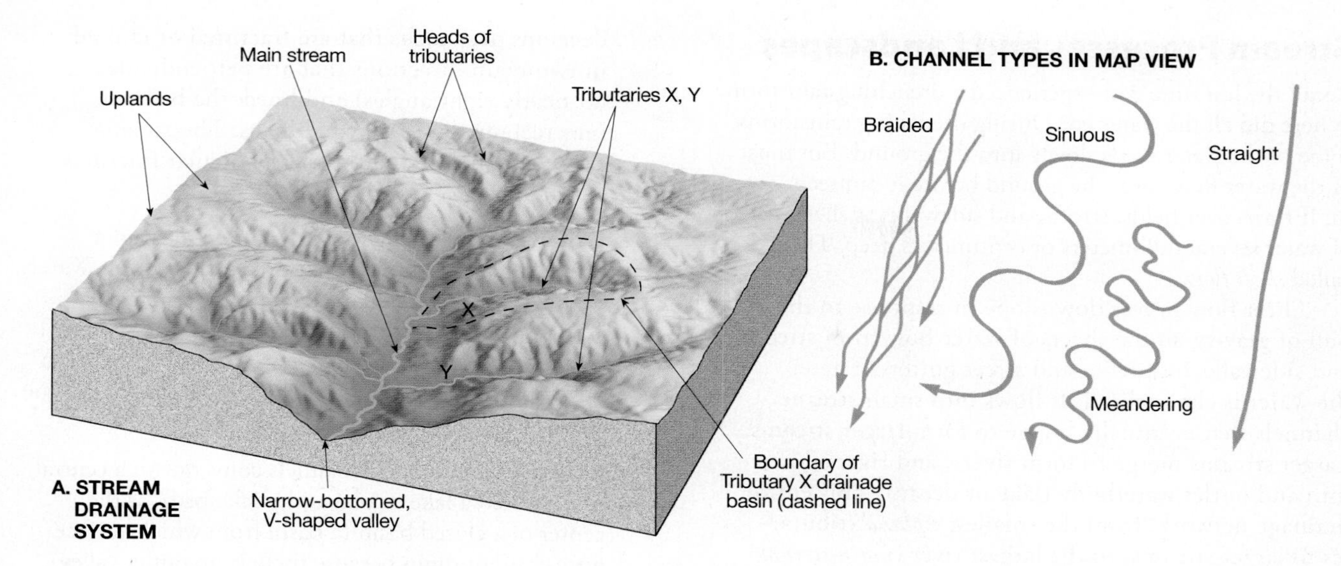

Sinuosity is a measure, below, of how much a stream channel meanders side-to-side.

$$\text{Sinuosity} = \frac{\text{Length A-B of stream channel measured along the path of water flow (blue arrows in drawing C, below)}}{\text{Length A-B measured along a straight line distance between A and B (red line in drawing C, below)}}$$

Straight channels have sinuosities less than 1.3, sinuous channels have sinuosities of 1.3 to 1.5, and meandering streams have sinuosities greater than 1.5.

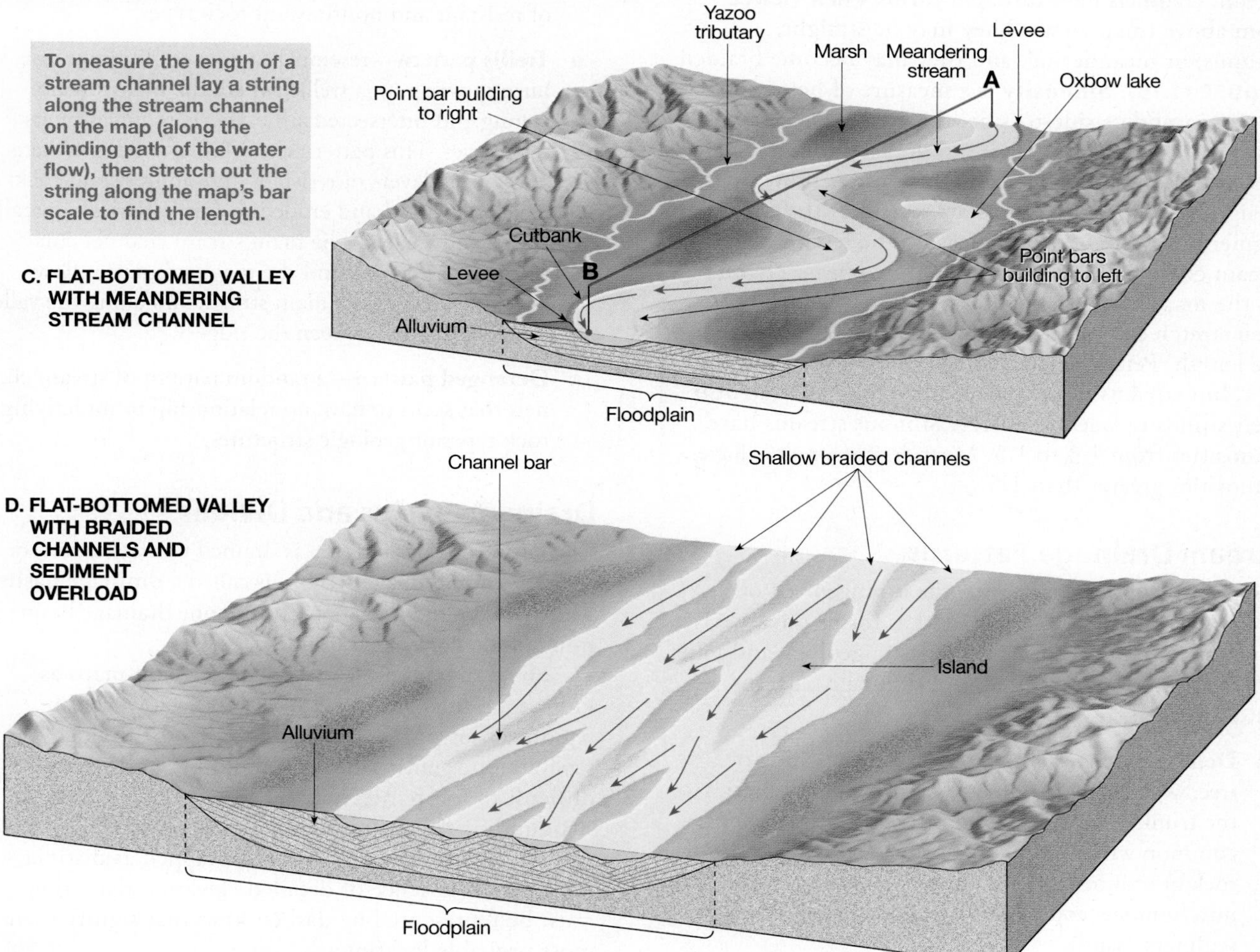

FIGURE 11.1 Drainage Basins, Streams, Channel Types, and Sinuosity. Arrows indicate current flow in the main channels of streams. A. Features of a stream drainage basin. B. Stream channel types as observed in map view. C. Features of a meandering stream valley. D. Features of a typical braided stream. Braided streams develop in sediment-choked streams. **To measure the length of a stream channel**, lay a string along the stream channel on the map (along the winding path of water flow), then stretch out the string along the bar scale to find the length.

Stream Processes and Landscapes

Recall the last time you experienced a drenching rainstorm. Where did all the water go? During drenching rainstorms, some of the water seeps slowly into the ground. But most of the water flows over the ground before it can seep in. It flows over fields, streets, and sidewalks as sheets of water several millimeters or centimeters deep. This is called *sheet flow.*

Sheet flow moves downslope in response to the pull of gravity, so the sheets of water flow from streets and sidewalks to ditches and street gutters. There, the water is channelized. It flows into small stream channels that eventually merge to form larger streams. Larger streams merge to form rivers, and rivers flow into and outlet waterbody (lake or ocean). This entire drainage network, from the smallest *upland* tributaries to larger streams, to the largest river (*main stream* or *main river*), is called a **stream drainage system** (**FIGURE 11.1A**).

Stream Channel Types and Their Sinuosity

Stream channels have different forms when viewed from above (map view). They may be straight, sinuous, or meandering, and they may become braided (**FIGURE 11.1B**). **Sinuosity** is a measure of how much a stream meanders side-to-side, the way a snake crawls. It can be calculated by dividing the length of a stream channel (along the winding path of water flow) by the straight-line distance from start to end of the stream segment (**FIGURE 11.1**). To measure the length of a stream channel, lay a string along the stream channel on the map (along the winding path of water flow), then stretch out the string along the bar scale to find the length. Perfectly straight channels have a sinuosity of 1, but streams in this lab are classified as straight if their sinuosity is less than 1.3. Sinuous streams have sinuosities from 1.3 to 1.5. Meandering streams have sinuosities greater than 1.5.

Stream Drainage Patterns

A **stream drainage pattern** is the arrangement of stream channels and tributaries that forms on a landscape as a result of its underlying geology and relief. These are some common stream drainage patterns (**FIGURE 11.2**):

- **Dendritic pattern**—resembles the branching of a tree. Water flow is from the branch-like tributaries to the trunk-like main stream or river. This pattern is common where a stream cuts into flat lying layers of rock or sediment. It also develops where a stream cuts into homogeneous rock (crystalline igneous rock) or sediment (sand).
- **Rectangular pattern**—a network of channels with right-angle bends that form a pattern of interconnected rectangles and squares. This pattern often develops over rocks that are fractured or faulted in two main directions that are perpendicular (at nearly right angles) and break the bedrock into rectangular or square blocks. The streams erode channels along the perpendicular fractures and faults.
- **Radial pattern**—channel flow outward from a central area, resembling the spokes of a wheel. Water drains from the inside of the pattern, where the "spokes" nearly meet, to the outside of the pattern (where the "spokes" are farthest apart). This pattern develops on conical hills, such as volcanoes and some structural domes.
- **Centripetal pattern**—channels converge on a central point, often a lake or playa (dry lake bed), at the center of a closed basin (a basin from which surface water cannot drain because there is no outlet valley).
- **Annular pattern**—a set of incomplete, concentric rings of streams connected by short radial channels. This pattern commonly develops on eroding structural domes and folds that contain alternating folded layers of resistant and nonresistant rock types.
- **Trellis pattern**—resembles a vine or climbing rose bush growing on a trellis, where the main stream is long and intersected at nearly right angles by its tributaries. This pattern commonly develops where alternating layers of resistant and nonresistant rocks have been tilted and eroded to form a series of parallel ridges and valleys. The main stream channel cuts through the ridges, and the main tributaries flow perpendicular to the main stream and along the valleys (parallel to and between the ridges).
- **Deranged pattern**—a random pattern of stream channels that seem to have no relationship to underlying rock types or geologic structures.

Drainage Basins and Divides

The entire area of land that is drained by one stream, or an entire stream drainage system, is called a **drainage basin**. The linear boundaries that separate one drainage basin from another are called **divides**.

Some divides are easy to recognize on maps as knife-edge ridge crests (**FIGURE 11.3**). However, in regions of lower relief or rolling hills, the divides separate one gentle slope from another and are more difficult to locate precisely (**FIGURE 11.1A**, dashed line surrounding the Tributary X drainage basin). For this reason, divides cannot always be mapped as distinct lines. In the absence of detailed elevation data, they must be represented by dashed lines that signify their most probable locations.

You may have heard of something called a *continental divide,* which is a narrow strip of land dividing surface waters that drain in opposite directions across the

STREAM DRAINAGE PATTERNS

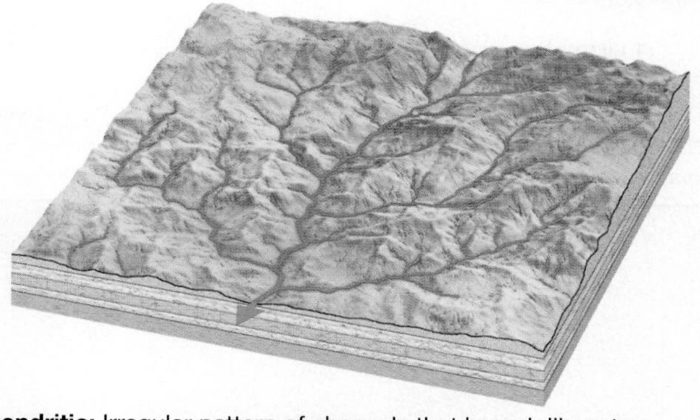

Dendritic: Irregular pattern of channels that branch like a tree. Develops on flat lying or homogeneous rock.

Rectangular: Channels have right-angle bends developed along perpendicular sets of rock fractures or joints.

Radial: Channels radiate outward like spokes of a wheel from a high point.

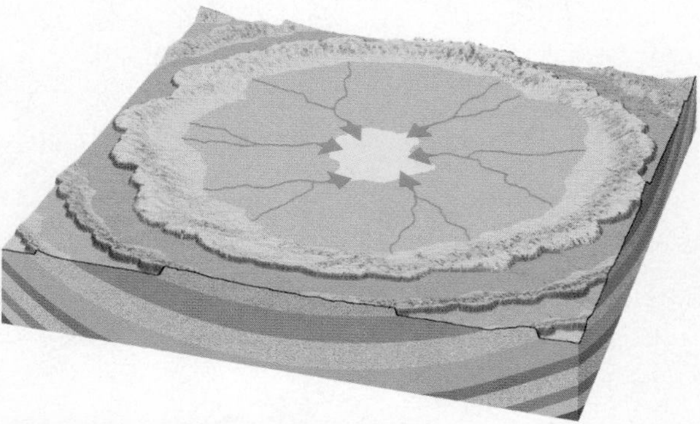

Centripetal: Channels converge on the lowest point in a closed basin from which water cannot drain.

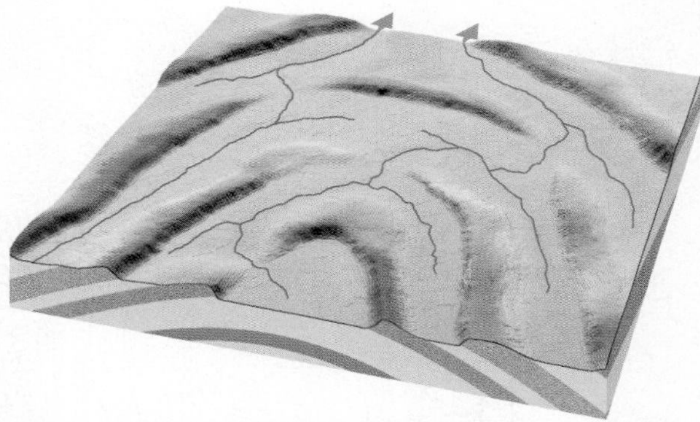

Annular: Long channels form a pattern of concentric circles connected by short radial channels. Develops on eroded domes or folds with resistant and nonresistant rock types.

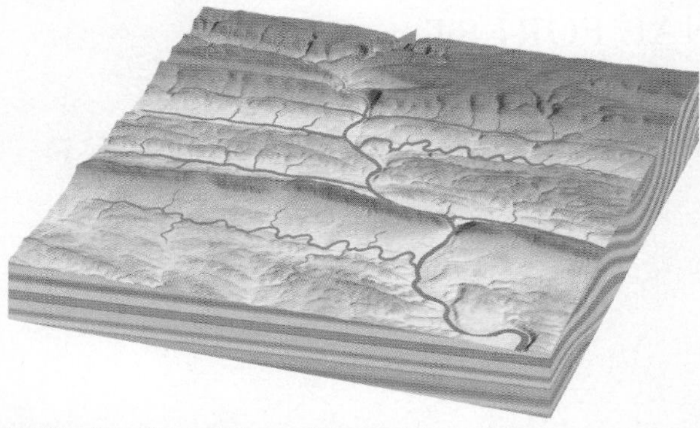

Trellis: A pattern of channels resembling a vine growing on a trellis. Develops where tilted layers of resistant and nonresistant rock form parallel ridges and valleys. The main stream channel cuts through the ridges, and the main tributaries flow along the valleys parallel to the ridges and at right angles to the main stream.

Deranged: Channels flow randomly with no relation to underlying rock types or structures.

FIGURE 11.2 Some stream drainage patterns. Note their relationship to bedrock geology.

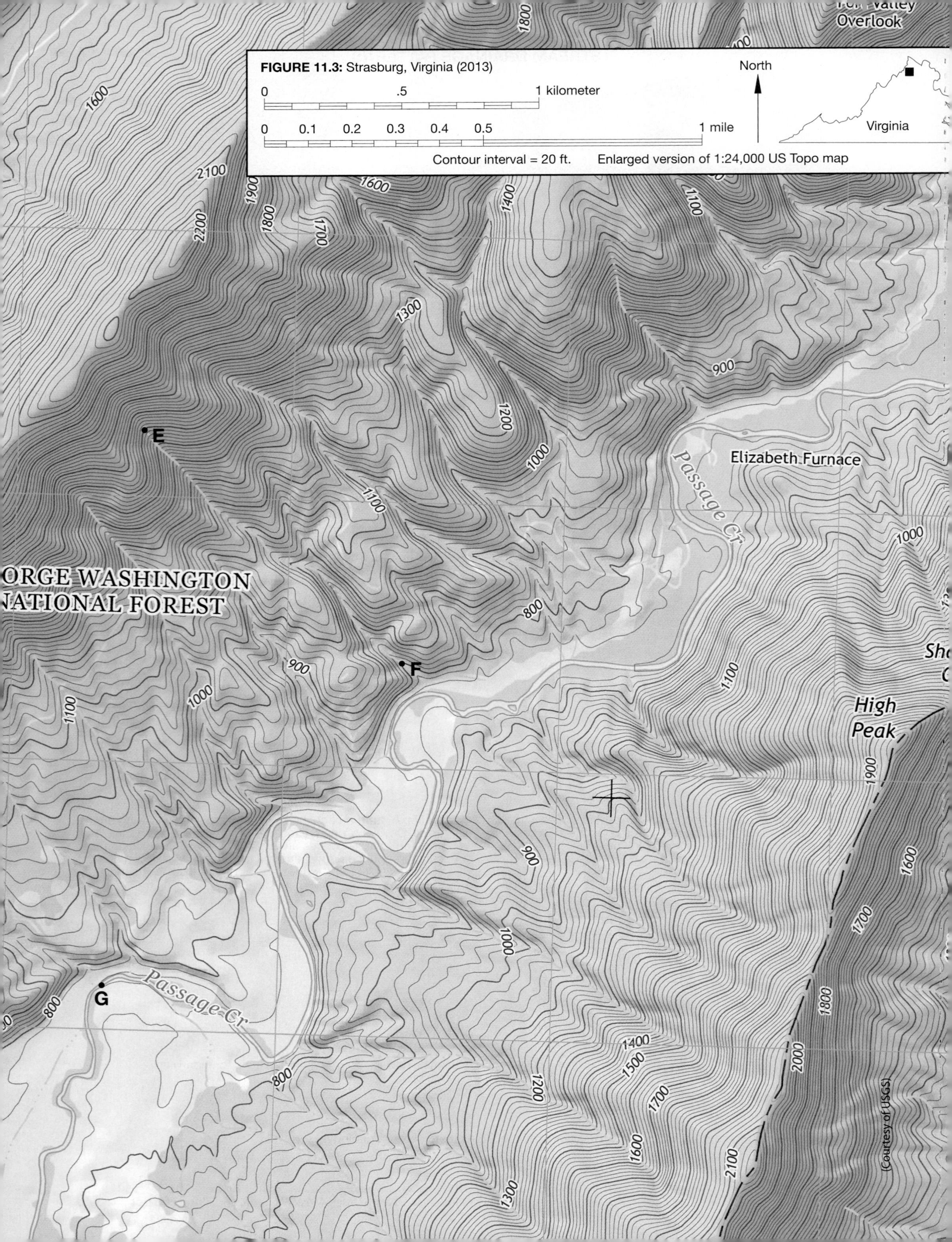

FIGURE 11.3: Strasburg, Virginia (2013)

continent. The continental divide in North America is an imaginary line along the crest of the Rocky Mountains (see red line on small map in **FIGURE 11.4**). Rainwater that falls east of the line drains eastward into the Atlantic Ocean, and rainwater that falls west of the line drains westward into the Pacific Ocean. Therefore, North America's continental divide is sometimes called "The Great Divide."

Stream Weathering, Transportation, and Deposition

Three main processes are at work in every stream. *Weathering* occurs where the stream physically erodes and disintegrates Earth materials and where it chemically decomposes or dissolves Earth materials to form sediment and aqueous chemical solutions. *Transportation* of these weathered materials occurs when they are dragged, bounced, and carried downstream (as suspended grains or chemicals in the water). *Deposition* occurs if the velocity of the stream drops (allowing sediments to settle out of the water) or if parts of the stream evaporate (allowing mineral crystals and oxide residues to form).

The smallest valleys in a drainage basin occur at its highest elevations, called **uplands** (**FIGURE 11.1**). In the uplands, a stream's (tributary's) point of origin, or **head**, may be at a spring or at the start of narrow runoff channels developed during rainstorms. Erosion (wearing away rock and sediment) is the dominant process here, and the stream channels deepen and erode their V-shaped channels uphill through time—a process called **headward erosion.** Eroded sediment is transported downstream by the tributaries.

Streams also weather and erode their own valleys along weaknesses in the rocks (fractures, faults), soluble nonresistant layers of rock (salt layers, limestone), and where there is the least resistance to erosion (see **FIGURE 11.2**). Rocks composed of hard, chemically resistant minerals are generally more resistant to erosion and form ridges or other hilltops. Rocks composed of soft and more easily weathered minerals are generally less resistant to erosion and form valleys. This is commonly called *differential erosion* of rock.

Headward tributary valleys merge into larger stream valleys, and these eventually merge into a larger river valley. Along the way, some new materials are eroded, and deposits (gravel, sand, mud) may form temporarily, but the main processes at work over the years in uplands are erosion (headward erosion and cutting V-shaped valleys) and transportation of sediment.

The end of a river valley is the **mouth** of the river, where it enters an outlet waterbody (lake, gulf, ocean) or a dry basin. At this location, the river water is dispersed into a wider area, its velocity decreases, and sediment settles out of suspension to form an alluvial deposit such as a delta (in water) or an alluvial fan (**FIGURE 11.5**). If the river water enters a dry basin, then it will evaporate and precipitate layers of mineral crystals and oxide residues (in a playa).

River Valley Forms and Processes

The form or shape of a river valley varies with these main factors:

- **Geology**—the bedrock geology over which the stream flows affects the stream's ability to find or erode its course (**FIGURE 11.2**).
- **Gradient**—the steepness of a slope—either the slope of a valley wall or the slope of a stream along a selected length (segment) of its channel (**FIGURE 11.6**). Gradient is generally expressed in *meters per kilometer* or *feet per mile*. This is determined by dividing the vertical rise or fall between two points on the slope (in meters or feet) by the horizontal distance (run) between them (in kilometers or miles). For example, if a stream descends 20 meters over a distance of 40 km, then its gradient is 20 m/40 km, or 0.5 m/km. You can estimate the gradient of a stream by studying the spacing of contours on a topographic map. Or, you can precisely calculate the exact gradient by measuring how much a stream descends along a measured segment of its course. Learn more about calculating slope (gradient) at this site featuring *The Math You Need, When You Need* It math tutorials for students in introductory geoscience courses: **http://serc.carleton.edu/mathyouneed/slope/index.html**
- **Base level**—the lowest level to which a stream can theoretically erode. For example, base level is achieved where a stream enters a lake or ocean. At that point, the erosional (cutting) power of the stream is zero and depositional (sediment accumulation) processes occur.
- **Discharge**—the rate of stream flow at a given time and location. Discharge is measured in water volume per unit of time, commonly *cubic feet per second* (ft^3/sec).
- **Load**—the amount of material (mostly alluvium, but also plants, trash, and dissolved material) that is transported by a stream. In the uplands, most streams have relatively steep gradients, so the streams cut narrow, V-shaped valleys. Near their heads, tributaries are quick to transport their load downstream, where it combines with the loads of other tributaries.

Therefore, the load of the tributaries is transferred to the larger streams and, eventually, to the main river. The load is eventually deposited at the mouth of the river, where it enters a lake, ocean, or dry basin.

From a stream's headwaters to its mouth, the gradient decreases, discharge generally increases, and valleys generally widen. Along the way, the stream's load may exceed the water's ability to carry it, so the solid particles accumulate as sedimentary deposits

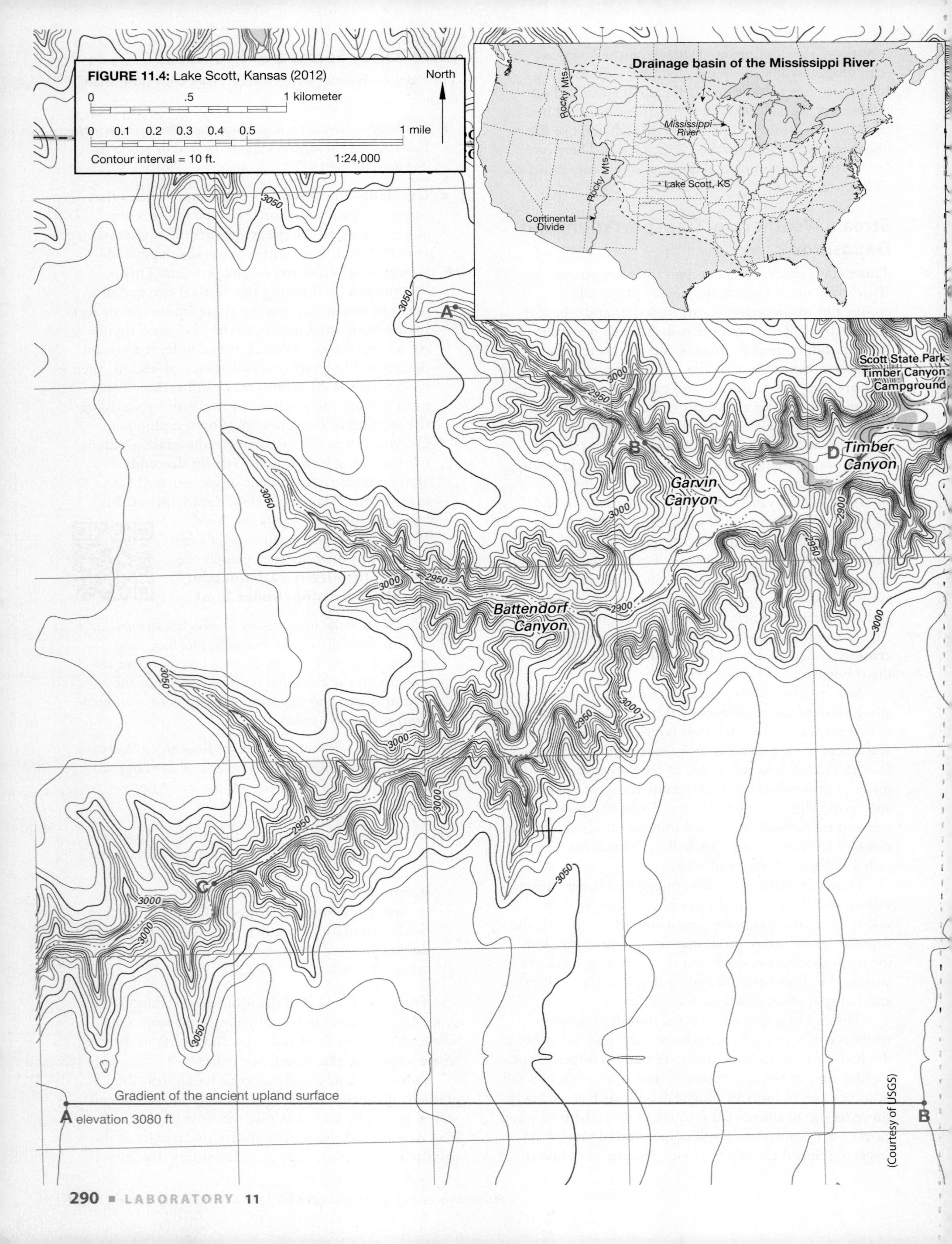

FIGURE 11.4: Lake Scott, Kansas (2012)
0
.5
1 kilometer
0
0.1
0.2
0.3
0.4
0.5
1 mile
North
Contour interval = 10 ft.
1:24,000
Drainage basin of the Mississippi River
Rocky Mts.
Mississippi River
Lake Scott, KS
Continental Divide
Scott State Park
Timber Canyon
Campground
A
B
C
D
Timber Canyon
Garvin Canyon
Battendorf Canyon
3050
3000
2950
2900
Gradient of the ancient upland surface
A elevation 3080 ft
B
(Courtesy of USGS)

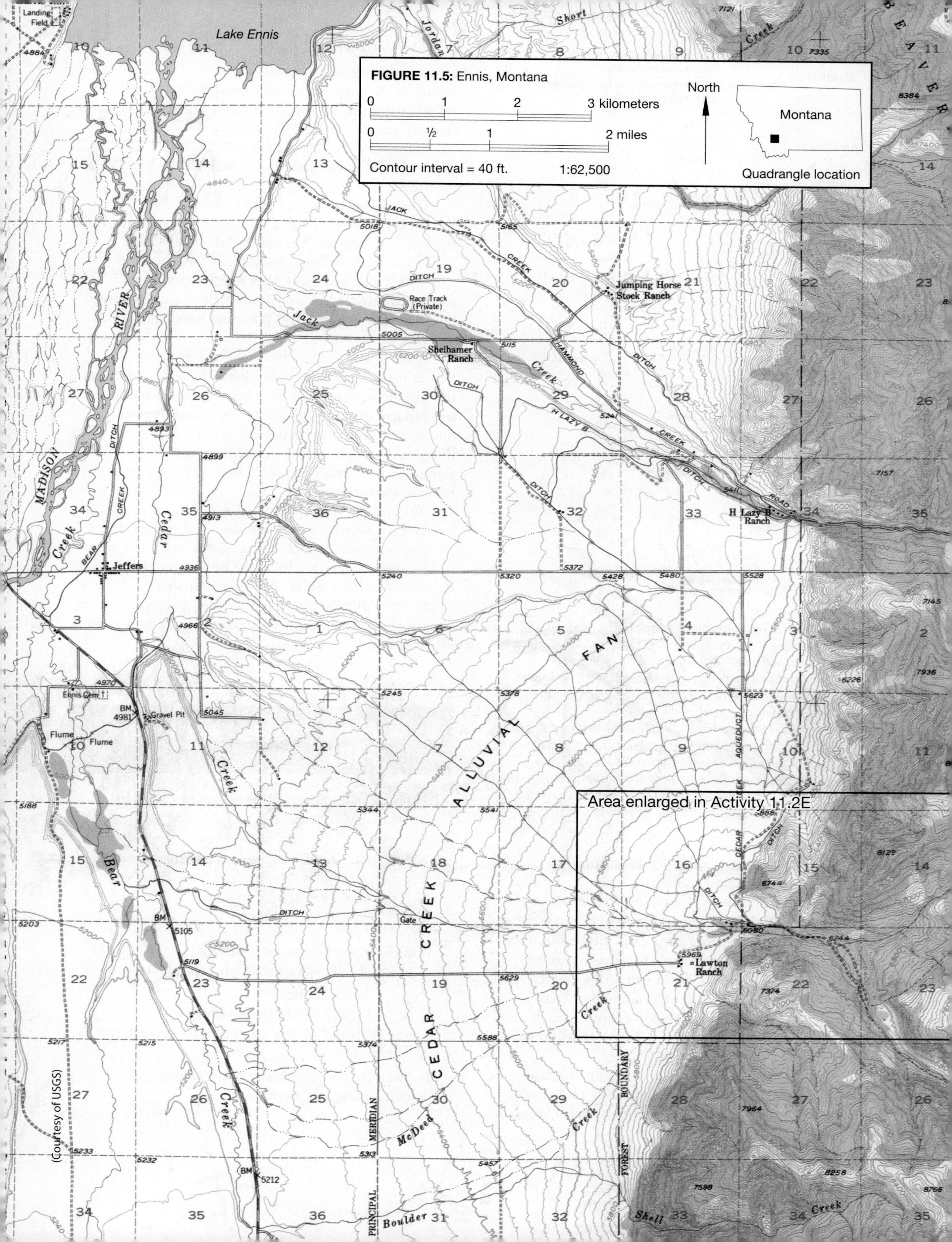
FIGURE 11.5: Ennis, Montana
0 1 2 3 kilometers
0 ½ 1 2 miles
Contour interval = 40 ft.
1:62,500
North
Montana
Quadrangle location
Area enlarged in Activity 11.2E
Lake Ennis
Landing Field
MADISON RIVER
Jack Creek
Race Track (Private)
Shelhamer Ranch
Jumping Horse Stock Ranch
HAMMOND DITCH
H LAZY B DITCH
H Lazy B Ranch
Jeffers
Ennis Cem
Gravel Pit
Flume
Bear Creek
Cedar Creek
CEDAR CREEK ALLUVIAL FAN
Lawton Ranch
AQUEDUCT
McDeed Creek
Boulder
Shell Creek
PRINCIPAL MERIDIAN
FOREST BOUNDARY
(Courtesy of USGS)

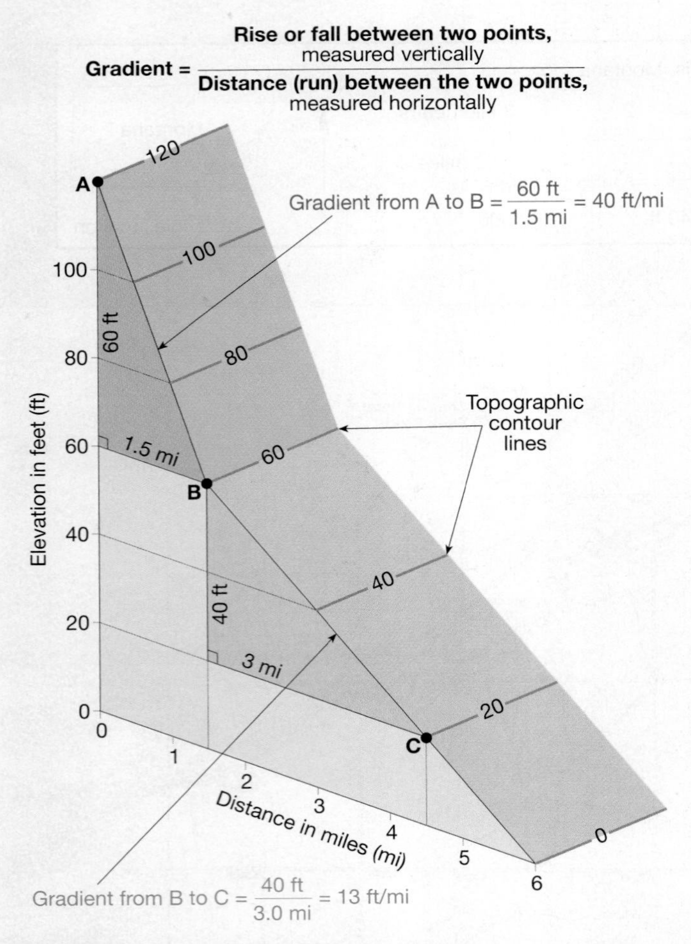

FIGURE 11.6 Stream gradient (slope). The gradient of a stream is a measure of the steepness of its slope. As above, gradient is usually determined by dividing the rise or fall (vertical relief) between two points on the map by the distance (run) between them. It is usually expressed as a fraction in feet per mile (as above) or meters per kilometer.

A second way to determine and express the gradient of a slope is by measuring its steepness in degrees relative to horizontal. Thirdly, gradient can be expressed as a percentage (also called *grade* of a slope). For example, a grade of 10% would mean a grade of 10 units of rise divided by (per) 100 units of distance (i.e., 10 in. per 100 in., 10 m per 100 m).

along the river margins, or banks. **Floodplains** develop when alluvium accumulates landward of the river banks, during floods (**FIGURES 11.1C** and **11.1D**). However, most flooding events do not submerge the entire floodplain. The more abundant minor flooding events deposit sediment only where the water barely overflows the river's banks. Over time, this creates natural **levees** (**FIGURE 11.1C**) that are higher than the rest of the floodplain. If a tributary cannot breach a river's levee, then it will become a **yazoo tributary** that flows parallel to the river (**FIGURE 11.1C**).

Still farther downstream, the gradient decreases even more as discharge and load increase. The stream valleys develop very wide, flat floodplains with sinuous channels. These channels may become highly sinuous, or **meandering** (see **FIGURES 11.1B, 11.7, 11.8**). Erosion occurs on the outer edge of meanders, which are called **cutbanks**. At the same time, **point bar** deposits (mostly gravel and sand) accumulate along the inner edge of meanders. Progressive erosion of cutbanks and deposition of point bars makes meanders "migrate" over time.

Channels may cut new paths during floods. This can cut off the outer edge of a meander, abandoning it to become a crescent-shaped **oxbow lake** (see **FIGURE 11.1C**). When low gradient/high discharge streams become overloaded with sediment, they may form **braided stream** patterns. These consist of braided channels with linear, underwater sandbars (**channel bars**) and islands (see **FIGURES 11.1B** and **11.1D**).

Some stream valleys have level surfaces that are higher than the present floodplain. These are remnants of older floodplains that have been dissected (cut by younger streams) and are called **stream terraces**. Sometimes several levels of stream terraces may be developed along a stream, resembling steps. The steep slopes or cliffs separating the relatively horizontal stream terraces are **escarpments**.

Where a stream enters a lake, ocean, or dry basin, its velocity decreases dramatically. The stream drops its sediment load, which accumulates as a triangular or fan-shaped deposit. In a lake or ocean, such a deposit is called a **delta**. A similar fan-shaped deposit of stream sediment also occurs where a steep-gradient stream abruptly enters a wide, dry plain, creating an **alluvial fan**.

ACTIVITY

11.5 Mass Wastage at Niagara Falls

THINK About It | **How does stream erosion shape the landscape?**

OBJECTIVE Describe erosional and mass wastage processes at Niagara Falls and calculate the rate at which the falls is retreating upstream.

PROCEDURES

1. **Before you begin**, read about Mass Wastage at Niagara Falls. Also, this is **what you will need**:
   ____ ruler, calculator
   ____ 30-cm (12-inch) length of string
   ____ Activity 11.5 Worksheet (p. 306) and pencil with eraser
2. **Then follow your instructor's directions** for completing the worksheet.

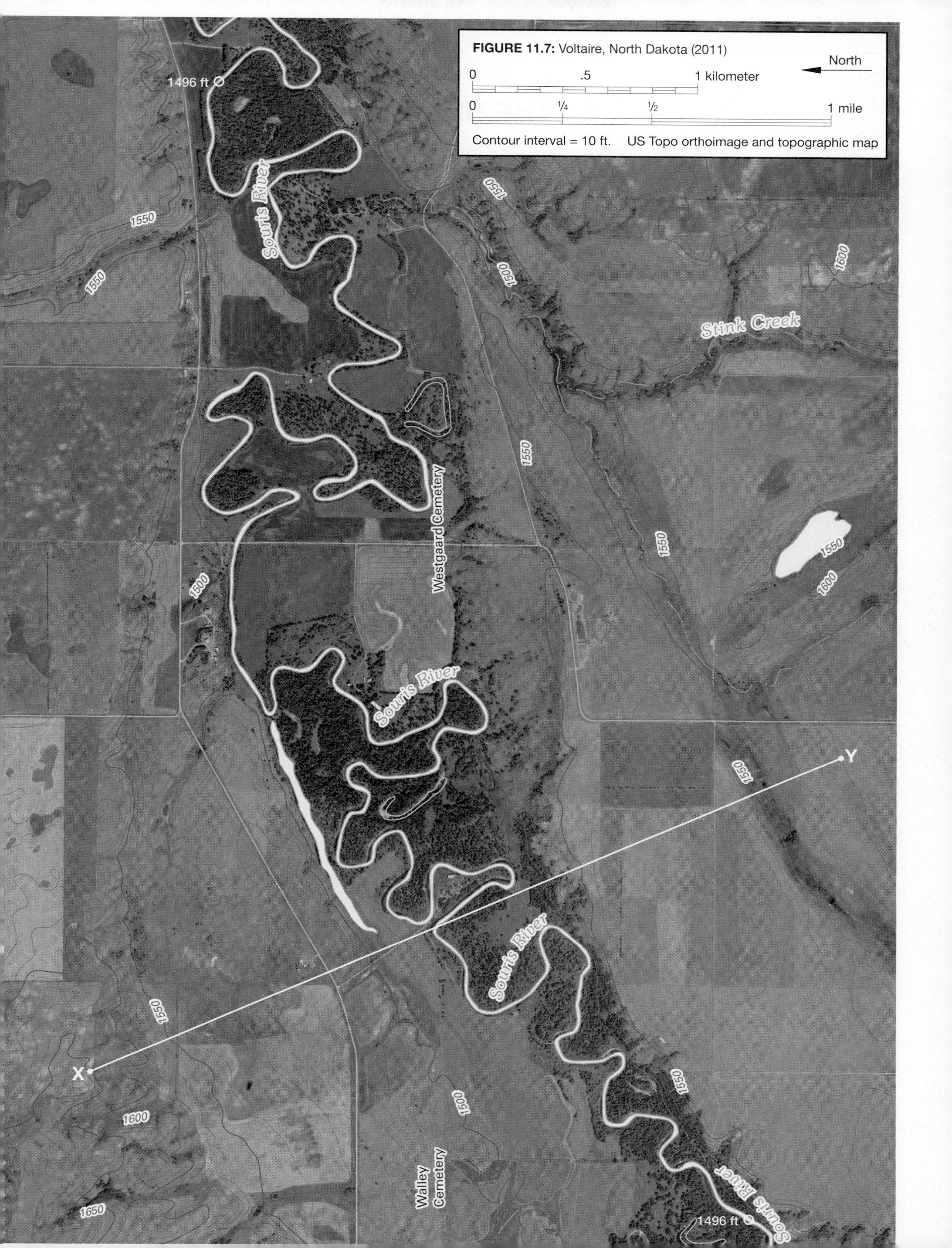

FIGURE 11.7: Voltaire, North Dakota (2011)

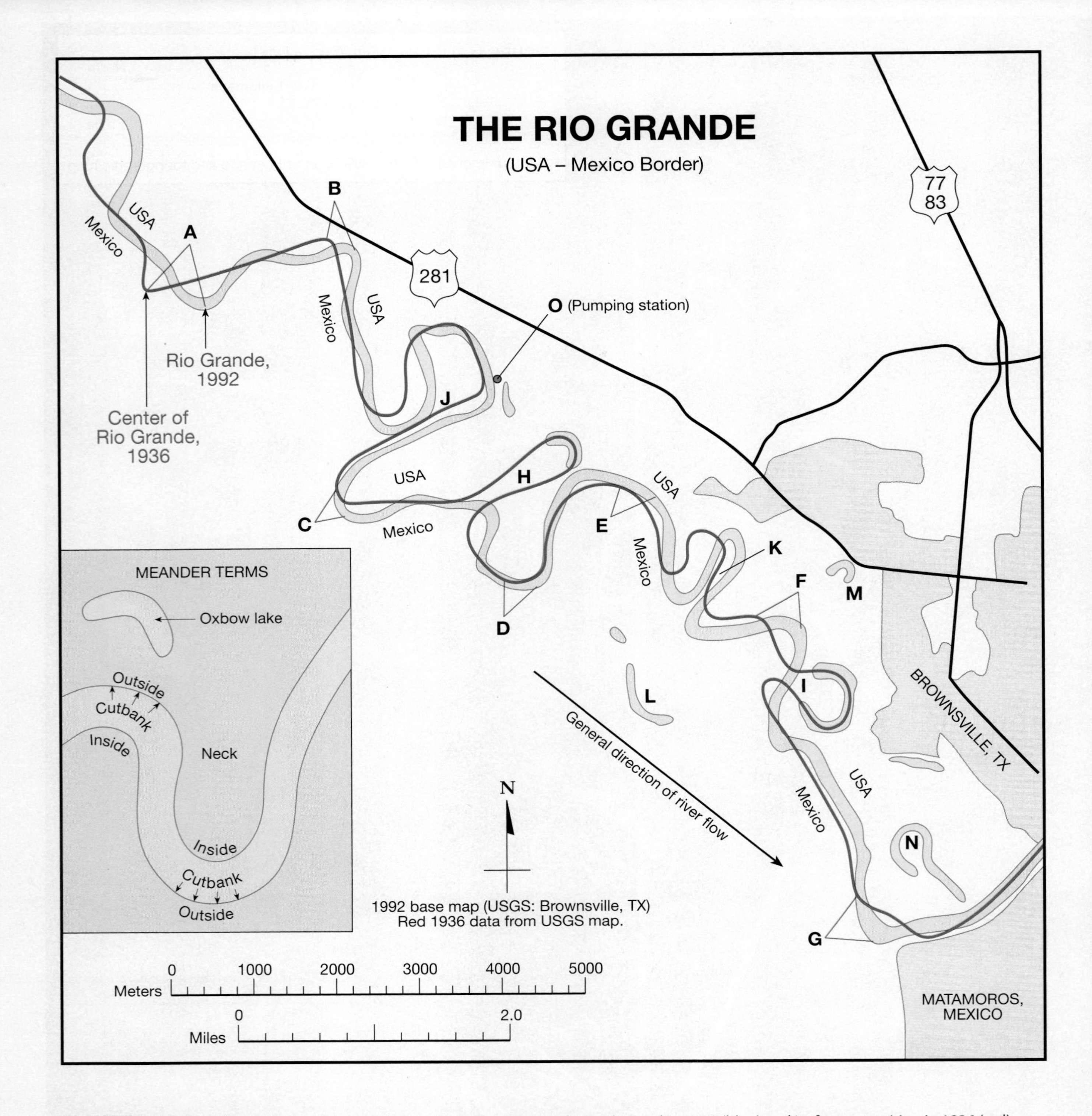

FIGURE 11.8 The Rio Grande. This map shows where the Rio Grande was located in 1992 (blue) and its former position in 1936 (red). The map is based on U.S. Geological Survey topographic maps (Brownsville, Texas, 1992; West Brownsville, Texas, 1936). The river flows east-southeast. Note the inset box of meander terms used to describe features of meandering streams.

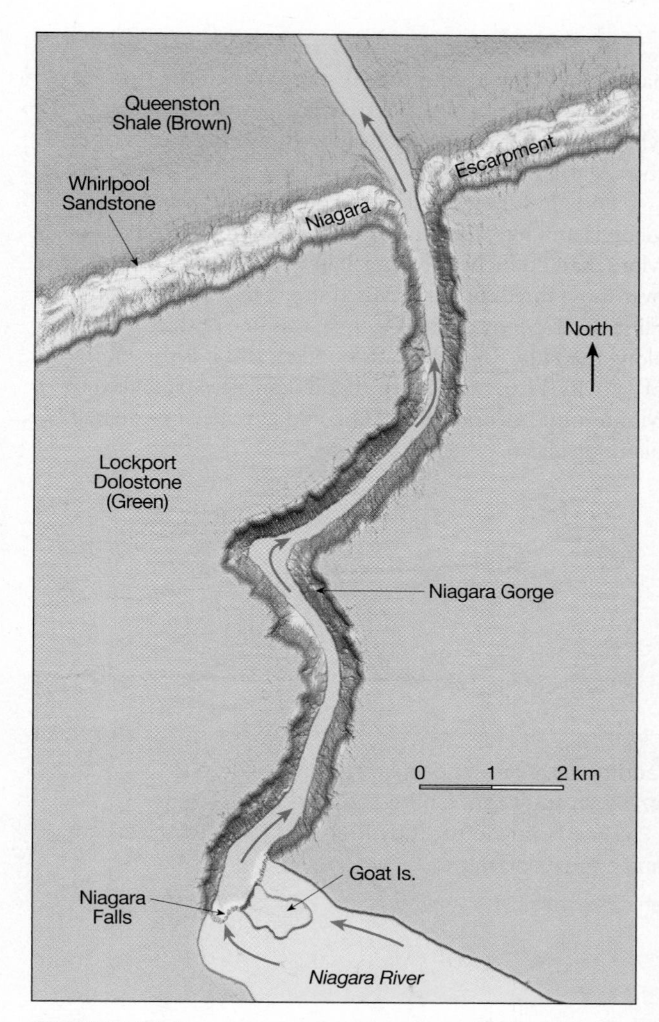

FIGURE 11.9 Geology of the Niagara Gorge Region. The Niagara River flows from Lake Erie north to Lake Ontario and forms the border between the United States and Canada. Niagara Falls is located on the Niagara River at the head of Niagara Gorge, about half way between the two lakes.

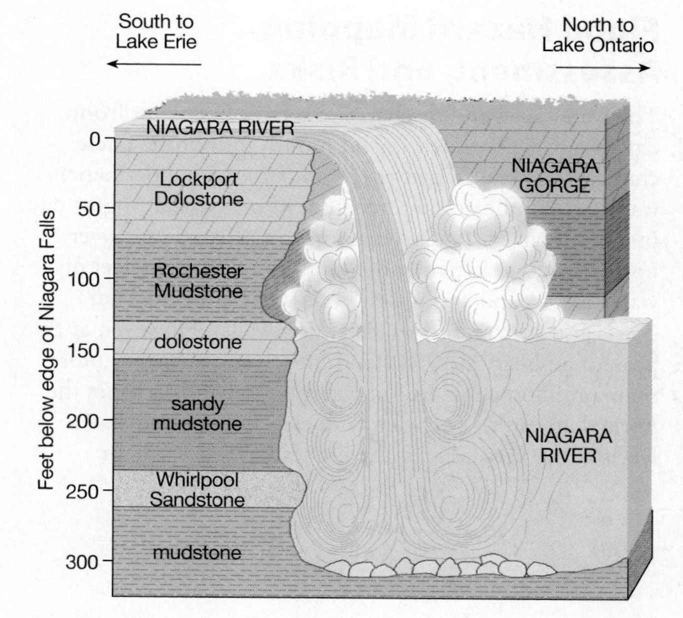

FIGURE 11.10 Cross Section of Niagara Falls. Niagara Falls exists because the named rock units beneath the falls vary in their hardness (resistance to erosion). As the hard dolostone caprock is undercut by erosion of the softer mudstone beneath it, pieces of the caprock break off and the falls moves upstream.

Mass Wastage at Niagara Falls

Mass wastage is the downslope movement of Earth materials such as soil, rock, and other debris. It is common along steep slopes, such as those created where rivers cut into the land. Some mass wastage occurs along the steep slopes of the river valleys. However, mass wastage can also occur in the bed of the river itself, as it does at Niagara Falls.

The Niagara River flows from Lake Erie to Lake Ontario (FIGURE 11.9). The gorge of the Niagara presents good evidence of the erosion of a caprock falls, Niagara Falls (FIGURE 11.10).

ACTIVITY

11.6 Flood Hazard Mapping, Assessment, and Risk

THINK About It | **How do geologists determine the risk of flooding along rivers and streams?**

OBJECTIVE Construct a flood magnitude/frequency graph, map floods, and flood hazard zones, and assess flood hazards along the Flint River, Georgia.

PROCEDURES

1. **Before you begin**, read Flood Hazard Mapping, Assessment, and Risks below. Also, this is **what you will need**:
   ___ calculator
   ___ Activity 11.6 Worksheets (pp. 307–310) and pencil with eraser
2. **Then follow your instructor's directions** for completing the worksheet.

Flood Hazard Mapping, Assessment, and Risks

The water level and discharge of a river fluctuates from day to day, week to week, and month to month. These changes are measured at *gaging stations,* with a permanent water-level indicator and recorder. On a typical August day in downtown St. Louis, Missouri, the Mississippi River normally has a discharge of about 130,000 cubic feet of water per second and water levels well below the boat docks and concrete *levees* (retaining walls). However, at the peak of an historic 1993 flood, the river discharged more than a million cubic feet of water per second (8 times the normal amount), swept away docks, and reached water levels at the very edge of the highest levees.

When the water level of a river is below the river's banks, the river is at a **normal stage**. When the water level is even with the banks, the river is at **bankfull stage**. And when the water level exceeds (overflows) the banks, the river is at a **flood stage**.

Early in July 1994, Tropical Storm Alberto entered Georgia and remained in a fixed position for several days. More than 20 inches of rain fell in west-central Georgia over those three days and caused severe flooding along the Flint River. Montezuma, Georgia, was one of the towns along the Flint River that was flooded, and it is the subject of Activity 11.6. Some of the flood damage experienced by Montezuma, Georgia, in 1994 could have been prevented by planning ahead.

ACTIVITY 11.1 Streamer Inquiry

Name: ______________________ **Course/Section:** ______________________ **Date:** ______________

Have you ever stood beside a stream and wondered where the water comes from or where it goes? *Streamer* is a map-based database of stream maps and information that allows you to find out. It is a new component of the U.S. National Atlas project, managed by the U.S. Geological Survey, that allows you to trace streams upstream to their sources or downstream to where they empty into larger streams or the ocean. To use Streamer, go to **http://nationalmap.gov/streamer/webApp/streamer.html** and click on the "Go to Map" panel. Then proceed below.

A. Where does the water come from?

1. Pick a community in the United States. What is the name of your community?

2. Locate your community on the *Streamer* map. You can double-click on the map to zoom in, or use your mouse to scroll in or out, to find your community. You can also type the location of the community (city, state) in the "Location Search" panel, then press "Enter," to locate your community.

3. Choose the largest stream located in or near your community to study as "**your stream**." Click on the "Trace Upstream" tab, then click on a point on the stream to display, in red, all of the streams that supply water to that point on the stream.

4. Now click on the "Trace Report" tab, and select "Detailed Report" to get a Stream Trace Detailed Report.

a. What stream did you study (Trace Origin Stream Name)?

b. What is the elevation [Trace Origin Elevation (feet)] and coordinates [Trace Origin (latitude, longitude)] of the point on the stream that you selected?

Elevation: ____________ feet above sea level. Latitude: ______________ Longitude: ______________

c. Through how many communities [Cities (count)] does the stream flow before it gets to this point? ______________________________

d. In how many named streams [Stream Names (count)] does the water flow to this point in the stream? ______________________________

e. What is the total length of the stream(s) named in Part **4d** [Total Length of Traced U.S. Streams (miles)]? ______________________________

f. Close the Stream trace Detailed Report (**not the map**) by clicking on the small gray "x" of the righthand tab at the top of the screen (it will say "National Atlas Streamer Detailed Trace Report" when you hover over it with your mouse), and proceed to part B below.

B. Where does the water go?

1. Click on the "Trace Downstream" tab. Then click on approximately the same point of the same stream that you studied in part A. to display, in red, where the water goes after that point on the stream.

2. Click on the "Identify" tab. Now when you click on any part of the red downstream trace of the water, it will identify the name of the stream at that point of the trace. List the names of all of the streams in the downstream trace of the water, from upstream (the starting red point in your map) to where it enters the "Outlet Waterbody" (downstream end of the red line on your map).

3. Click on the "Trace Report" tab, and select "Detailed Report" to get a new Stream Trace Detailed Report.

a. In how many named streams [Stream Names (count)] does the water flow downstream from this point? ______________________________

b. Through how many communities [Cities (count)] does the water flow downstream from this point? ______________________________

c. What is the total length of the stream(s) named in Part **3b** [Total Length of Traced U.S. Streams (miles)]? ______________________________

d. Into what "Outlet Waterbody" does the stream's water eventually empty? ______

e. What is the name of the last community or feature that the stream passes through before it enters the "Outlet Waterbody"? ______

f. Close the Stream trace Detailed Report (**not the map**), and proceed to part C below.

C. Your Stream Drainage System.

1. Click on the "Trace Downstream" tab again. Then click on the downstream end of the downstream trace that you identified in part B. It will be located near the place you identified in part B3e. This will display an entire stream drainage network, from the smallest upland tributaries to the largest river (main stream or main river). What do you think happens to the amount of water in the stream drainiange system, the width of the streams, and the slope of the streams as the water drains from small tributaries to the largest river?

D. **REFLECT & DISCUSS** Why would a community located on or near a stream want to know where its stream water comes from, and what else might they want to know about the water?

E. **REFLECT & DISCUSS** Why would a community located on or near a stream want to know where its stream water goes after passing their community?

Name: ______________________ **Course/Section:** ______________________ **Date:** ____________

A. Trout Run Drainage Basin: **1.** Complete items **a** through **h** below.

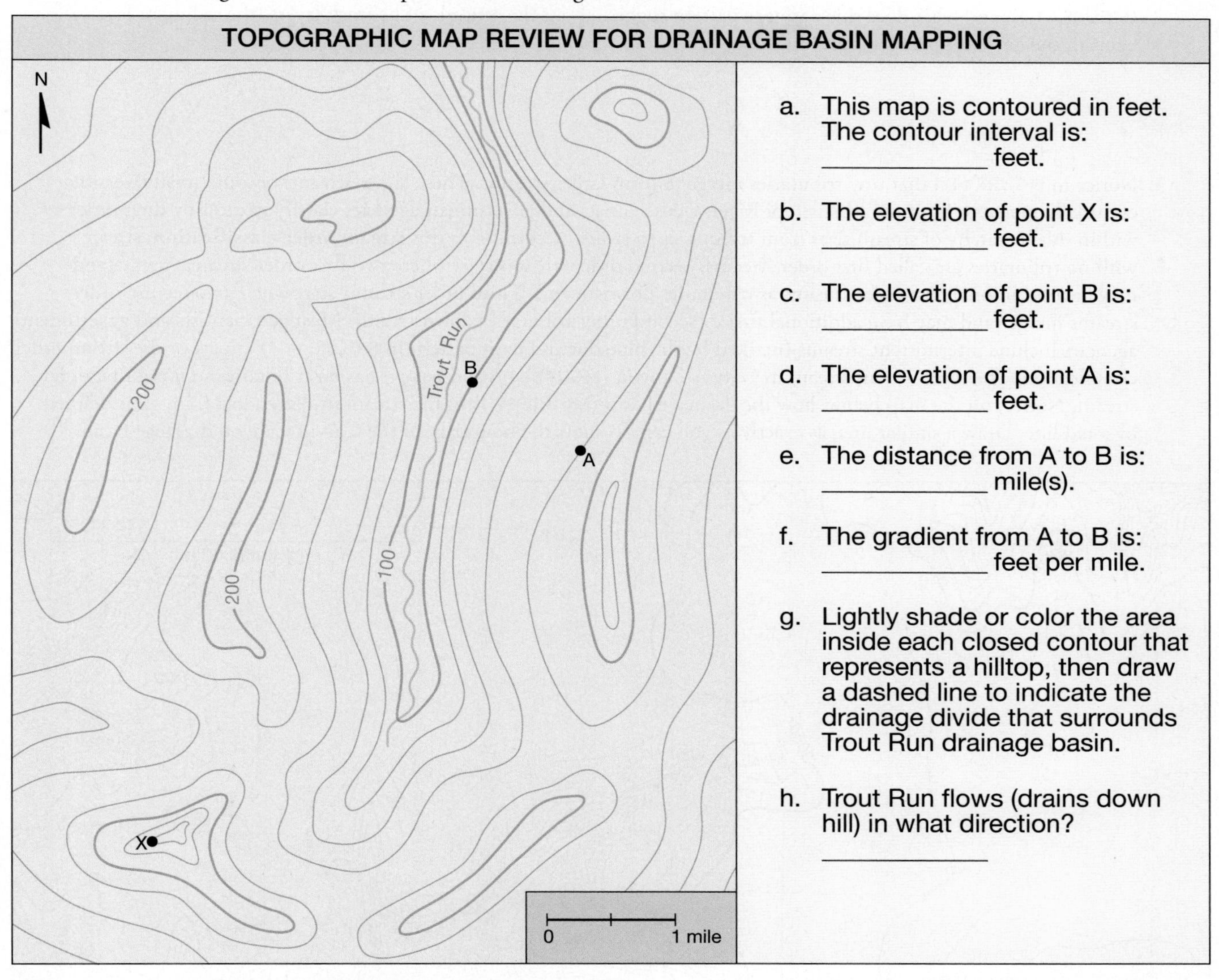

2. Imagine that drums of oil were emptied (illegally) at location **X** above. Is it likely that the oil would wash downhill into Trout Run? Explain your reasoning.

B. Refer to the topographic map of the Lake Scott quadrangle, Kansas (**FIGURE 11.4**). This area is located within the Great Plains physiographic province and the Mississippi River Drainage Basin. The Great Plains is a relatively flat grassland that extends from the Rocky Mountains to the Interior Lowlands of North America. It is an ancient upland surface that tilts eastward from an elevation of about 5500 feet along its western boundary with the Rocky Mountains to about 2000 feet above sea level in western Kansas. The upland is the top of a wedge of sediment that was weathered and carried eastward from the Rocky Mountains by a braided stream system that existed from Late Cretaceous to Pliocene time (65–2.6 Ma). Modern streams in western Kansas drain eastward across the Great Plains and cut channels into the ancient upland surface. Tiny modern tributaries merge to form larger streams that eventually flow into the Mississippi River. You can view this in *Streamer*: go to **http://nationalmap.gov/streamer/webApp/streamer.html** and click on the "Go to Map" panel. Then click on the "Trace Downstream" tab, zoom in to any stream in western Kansas (KS), and click on the stream. The red line will show how the stream is part of a stream drainage system the flows east across Kansas, on the ancient upland surface.

1. What is the gradient of the ancient upland surface in **FIGURE 11.4**? Show your work.

2. What is the name of the modern stream drainage pattern from **FIGURE 11.2** that is developed in the Lake Scott quadrangle (**FIGURE 11.4**), and what does this drainage pattern suggest about the attitude of bedrock layers (the sediment layers beneath the ancient upland surface) in this area?

3. Notice in **FIGURE 11.4** that tiny tributaries merge to form larger streams. These larger streams become small rivers that eventually merge to form the Mississippi River. Geoscientists and government agencies classify streams by their order within this hierarchy of stream sizes from tributaries to rivers. According to this **stream order classification**, streams with no tributaries are called first order streams. Second order streams start where two first order streams merge (and may have additional first order streams as tributaries downstream). Third order streams start where two second order streams merge (and may have additional first or second order tributaries downstream). Most geoscientists and government agencies include intermittent streams (marked by the blue dot and dash pattern in **FIGURE 11.4**) as part of the stream order classification. The intermittent stream in Garvin Canyon (**FIGURE 11.4** and below) has no tributaries so it is a first order stream. Notice, on the map below, how the drainage basin (**FIGURE 11.1**) of the stream in Battendorf Canyon is defined by a red line. Draw a similar line, as exactly as you can, to show the boundary of the Garvin Canyon drainage basin.

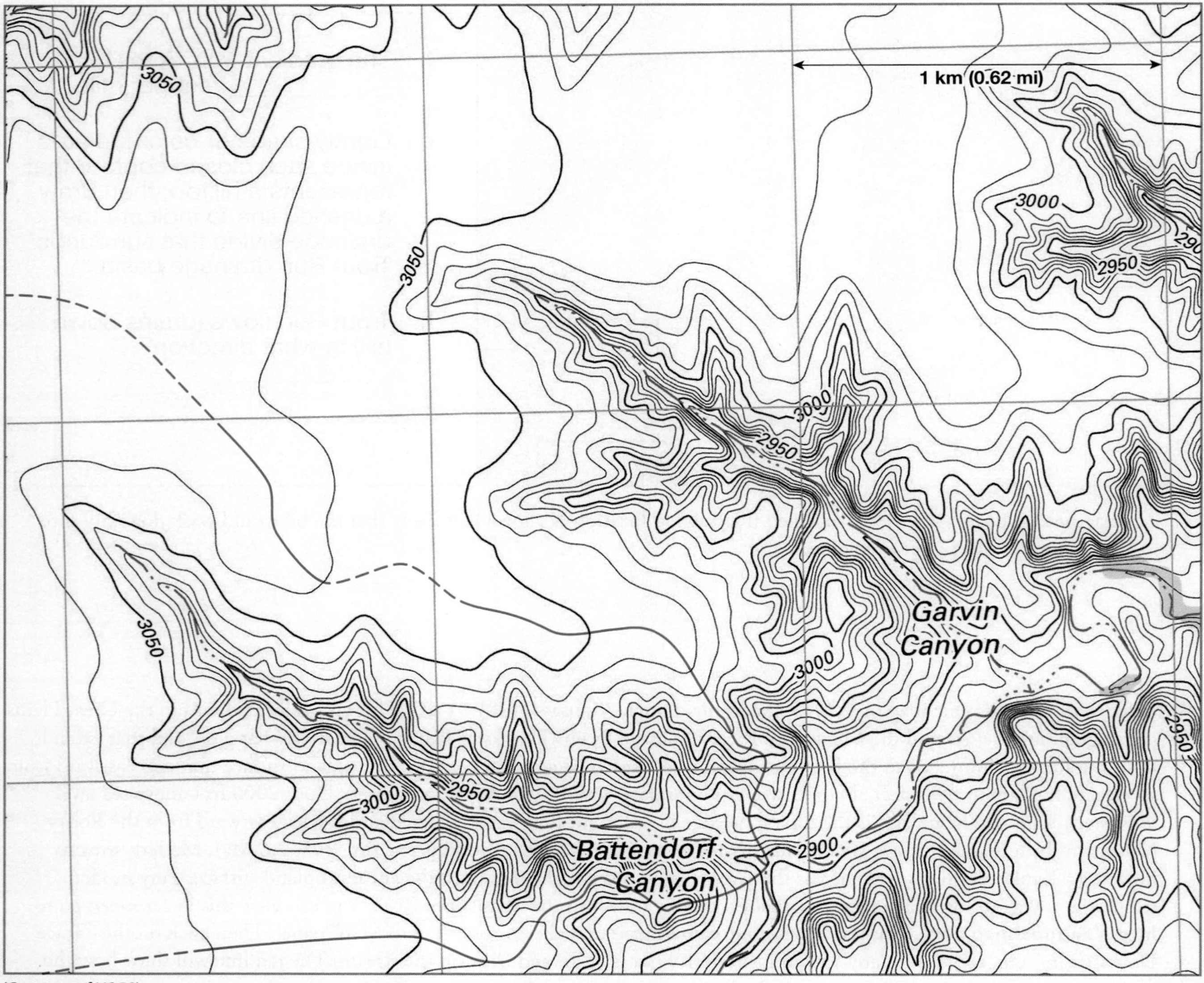

(Courtesy of USGS)

4. What is the gradient (ft/mi) and sinuosity, from A to B on **FIGURE 11.4**, of the first order stream in Garvin Canyon? (Refer to **FIGURES 11.1** and **11.6** for help measuring gradient and sinuosity.) Show your calculations. You will graph this data later in the activity.

   Gradient: ____________ft/mi Sinuosity: ____________

5. What is the stream order of the stream that occurs in Timber Canyon (**FIGURE 11.4**), and what is its gradient (ft/mi) and sinuosity from C to D? (Refer to **FIGURES 11.1** and **11.6** for help measuring gradient and sinuosity.) Show your calculations. You will graph this data later in the activity.

   Stream order: ________________

   Gradient: ____________ft/mi Sinuosity: ____________

6. The Mississippi River is a tenth order stream. Based on your answers to the two questions above, state what happens to the gradient of streams as they increase in order.

7. What do you think happens to the discharge of streams as they increase in order, and what effect do you think this would have on the relative number of fish living in each stream order within a basin?

C. Examine the enlarged part of the Strasburg, Virginia, quadrangle map in **FIGURE 11.3**.

1. What drainage pattern is developed in this area, and what does it suggest about the attitude of bedrock layers in this area? Explain your reasoning. (*Hint:* Refer to **FIGURE 11.2** and notice the stream pattern in relation to ridges and valleys.)

2. What is the gradient (ft/mi) and sinuosity of the small stream, from E to F? (Refer to **FIGURES 11.1** and **11.6** for help measuring gradient and sinuosity.) Show your calculations. You will graph this data later in the activity.

   Gradient: ____________ft/mi Sinuosity: ____________

3. What is the gradient (ft/mi) and sinuosity of Passage Creek from G to H? (Refer to **FIGURES 11.1** and **11.6** for help measuring gradient and sinuosity.) Show your calculations. You will graph this data later in the activity.

   Gradient: ____________ft/mi Sinuosity: ____________

D. Refer to the Ennis Montana 15' quadrangle in **FIGURE 11.5**. Some rivers are subject to large floods, either seasonal or periodic. In mountains, this flooding is due to snow melt. In drylands, it is caused by thunderstorms. During such times, rivers transport exceptionally large volumes of sediment. This causes characteristic features, two of which are braided channels and alluvial fans. Both features are relatively common in arid mountainous regions, such as the Ennis, Montana, area in **FIGURE 11.5**. (Both features also can occur wherever conditions are right, even at construction sites!)

1. What main stream channel types (shown in **FIGURE 11.1B**) are present on:

a. the streams in the forested southeastern corner of this map?

b. the Cedar Creek Alluvial Fan?

c. the valley of the Madison River (northwestern portion of **FIGURE 11.5**)?

2. Notice on **FIGURE 11.5** and the portion of that map enlarged below that Cedar Creek is the source of water that transports sediment onto Cedar Creek Alluvial Fan. Below, complete profile J-K of Cedar Creek by plotting and connecting the nine red elevation points (notice how points J and K have already been plotted).

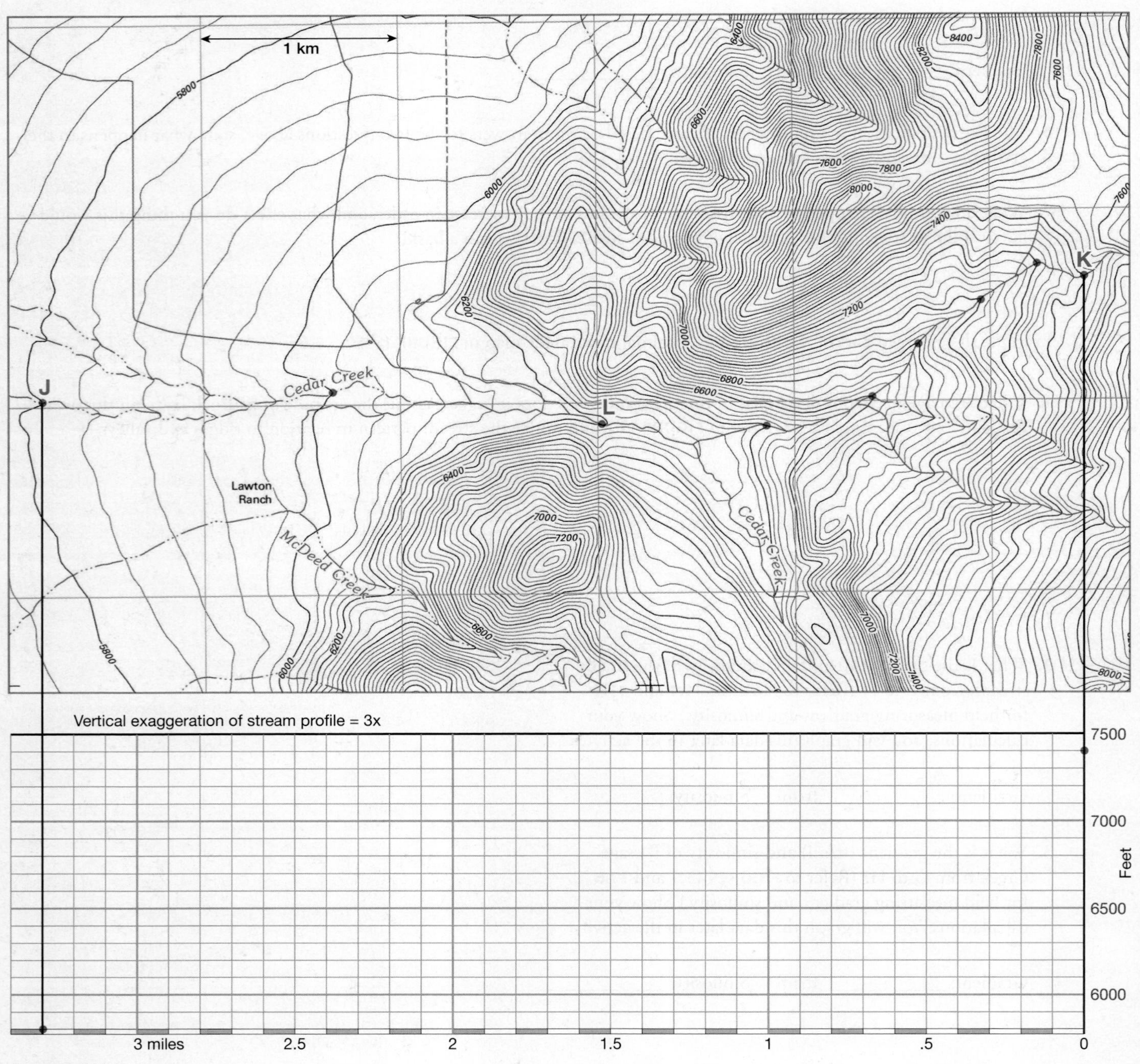

(Courtesy of USGS)

3. Observe the profile that you constructed in part D2. Label the part of the profile that is "bedrock eroded by Cedar Creek" and the part that is "sediment deposited from Cedar Creek."

4. Observe Cedar Creek on the map in part D2.

   a. What is its stream order classification at point L?

   b. What happens to that stream's gradient and order downstream, as it enters the alluvial fan, and how does this contribute to the formation of the alluvial fan?

E. On the semi-logarithmic graph paper provided at right, you can determine if a stream is linear, sinuous, or meandering by plotting a point based on the stream's gradient and sinuosity.

1. Plot points for the following streams, and draw a best-fit line through the points:
   - Stream segment A-B (Garvin Canyon stream) from part B4 of this activity
   - Stream segment C-D (Timber Canyon stream) from part B5 of this activity
   - Stream segment E-F (tributary of Passage Creek) from part C2
   - Stream segment G-H (Passage Creek) from part C3

2. Based on the summary graph that you just completed, is there a relationship between a stream's gradient and whether its channel is straight, sinuous, or meandering? If yes, then what is that relationship?

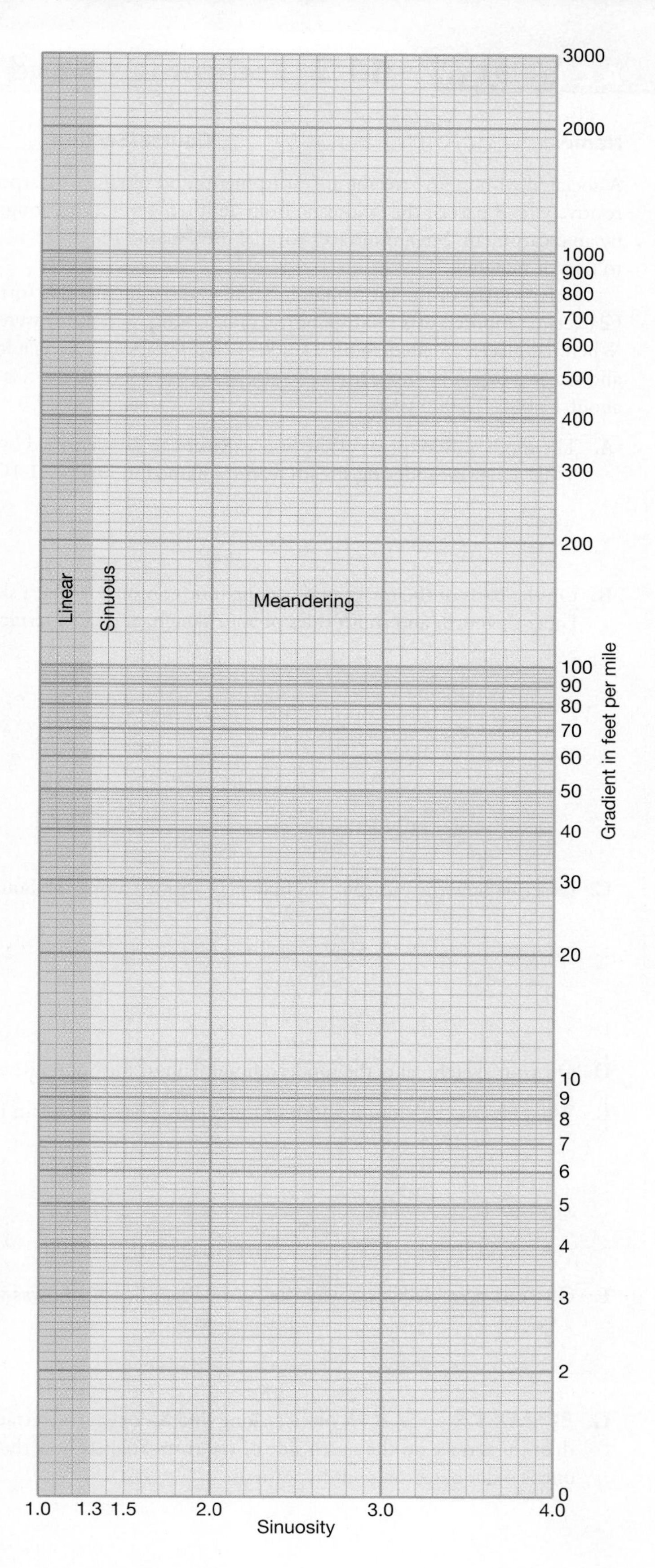

F. **REFLECT & DISCUSS** Compare the four landscapes that you studied in this activity. What factors determine the kind of drainage pattern that develops on a landscape and whether a stream is eroding bedrock or depositing sediment?

ACTIVITY 11.3 Escarpments and Stream Terraces

Name: ______________________ **Course/Section:** ______________________ **Date:** __________

Associated with many streams are escarpments and terraces. Escarpments are long cliffs or steep narrow slopes that separate one relatively level part of the landscape from another. Terraces are long, narrow or broad, level surfaces bounded on one or both sides by an escarpment. Stream terraces parallel the stream. The difference in elevation between two terraces can range from centimeters to tens of meters.

Refer to part of the the Voltaire, North Dakota quadrangle (orthoimage with topographic contours) in **FIGURE 11.7**. Glaciers (composed of a mixture of ice, gravel, sand, and mud) were present in this region at the end of the Pleistocene Ice Age. When the glaciers melted about 11,000–12,000 years ago, a thick layer of sand and gravel was deposited on top of the bedrock, and streams began forming from the glacial meltwater. Therefore, streams have been eroding and shaping this landscape for about 11,000–12,000 years.

A. The modern floodplain of the Souris River can be identified by its lush, dark green vegetation and blue meandering river. What other meandering stream features named in Figure 11.1C do you recognize in this image?

B. On the basis of the image and topographic contours, make a sketch, below, of a cross section of the landscape from **X** to **Y**. Label the north and south sides of your sketch, and label terraces with a "T" and escarpments with an "E."

C. Describe how the escarpments may have formed along the Souris River.

D. On your sketch, label the modern floodplain of the Souris River and record its width along line **X–Y**.

E. What was the maximum width of the Souris River floodplain in the past (measured along line **X–Y**) and how can you tell?

F. Give one possible reason why the Souris River floodplain was wider in the past.

G. **REFLECT & DISCUSS** Notice along line **X–Y** that the terrace on the south side of the Souris River is 30–40 feet higher than the terrace on the north side of the river. Suggest how these two different levels of terraces may have formed and which one is older based on your hypothesis.

ACTIVITY 11.4 Meander Evolution on the Rio Grande

Name: ______________________ **Course/Section:** ______________________ **Date:** __________

Refer to **FIGURE 11.8** showing the meandering Rio Grande, the river that forms the national border between Mexico and the United States. Notice that the position of the river changed in many places between 1936 (red line and leaders by lettered features) and 1992 (blue water bodies and leaders by lettered features). Study the meander terms provided in **FIGURE 11.8**, and then proceed to the questions below.

A. Study the meander cutbanks labeled **A** through **G.** The red leader from each letter points to the cutbank's location in 1936. The blue leader from each letter points to the cutbank's location in 1992. In what two general directions (relative to the meander, relative to the direction of river flow) have these cutbanks moved?

B. Study locations **H** and **I.**

1. In what country were **H** and **I** located in 1936?

2. In what country were **H** and **I** located in 1992?

3. Explain a process that probably caused locations **H** and **I** to change from meanders to oxbow lakes.

C. Based on your answer in item B3, predict how the river will change in the future at locations **J** and **K.**

D. What are features **L, M,** and **N,** and what do they indicate about the historical path of the Rio Grande?

E. What is the average rate at which meanders like **A** through **G** migrated here (in meters per year) from 1936 to 1992? Explain your reasoning and calculations.

F. **REFLECT & DISCUSS** Explain in steps how a meander evolves from the earliest stage of its history as a broad slightly sinuous meander to the stage when an oxbow lake forms.

ACTIVITY 11.5 Mass Wastage at Niagara Falls

Name: ______________________ **Course/Section:** ______________________ **Date:** __________

A. Geologic evidence indicates that the Niagara River began to cut its gorge (Niagara Gorge) about 11,000 years ago as the Laurentide Ice Sheet retreated from the area. The ice started at the Niagara Escarpment shown in **FIGURE 11.9** and receded (melted back) north to form the basin of Lake Ontario. The Niagara Gorge started at the Niagara Escarpment and retreated south to its present location. Based on this geochronology and the length of Niagara Gorge, calculate the average rate of falls retreat in cm/year. Show your calculations.

B. Name as many factors as you can that could cause the falls to retreat at a faster rate.

C. Name as many factors as you can that could cause the falls to retreat more slowly.

D. Niagara Falls is about 35 km north of Lake Erie, and it is retreating southward. If the falls was to continue its retreat at the average rate calculated in **A,** then how many years from now would the falls reach Lake Erie?

E. **REFLECT & DISCUSS** Look at the cross section of Niagara Falls in **FIGURE 11.10**. Describe how the process that formed the falls could have begun. (Hint: Use your knowledge of stream erosion and the effects of stream gradient.)

ACTIVITY 11.6 Flood Hazard Mapping, Assessment, and Risk

Name: ______________________ **Course/Section:** ______________________ **Date:** __________

A. On the Montezuma, Georgia topographic map below, locate the gaging station on Flint River in the map center. The gaging station is located at an elevation of 255.83 feet above sea level, and the river is considered to be at flood stage when it is 20 feet above this level (275.83 feet). A July 1994 flood established a record at 35.11 feet above the gaging station, or 289.94 feet above sea level. This corresponds to the 290-foot contour line on the map. Trace the 290-foot contour line on both sides of the Flint River and label the area within these contours (land lower than 290 feet) as "1994 Flood Hazard Zone."

(Courtesy of USGS)

B. Name two human structures that were submerged by the flood and tell what effect that would have had on the environment and human quality of life after the flood.

C. Notice line **X–Y** near the top center part of the topographic map in part A.

1. The map shows the Flint River at its normal stage. What is the width (in km) of the Flint River at its normal stage along line **X–Y?**

2. What was the width of the river (in km) along this line when it was at maximum flood stage (290 feet) during the July 1994 flood?

D. Notice the floodplain of the Flint River along line X–Y on the map in part A. It is the relatively flat (as indicated by widely spaced contour lines) marshy land between the river and the steep (as indicated by more closely spaced contour lines) walls of the valley (escarpments) that are created by erosion during floods.

1. What is the elevation (in feet above sea level) of the floodplain at point Z near line **X–Y?**

2. How deep (in feet above sea level) was the water that covered that floodplain at point Z during the 1994 flood? (Explain your reasoning or show your mathematical calculation.)

3. Did the 1994 flood (i.e., the highest river level ever recorded here) stay within the floodplain and its bounding valley slopes? Does this suggest that the 1994 flood was of normal or abnormal magnitude (severity) for this river? Explain your reasoning.

E. The USGS recorded annual high stages (elevation of water level) of the Flint River at the Montezuma gaging station for 99 years (1897 and 1905–2002). Parts of the data have been summarized in the Flood Data Table ahead on the next page.

1. The annual highest stages of the Flint River (S) were ranked in severity from S = 1 (highest annual high stage ever recorded; i.e., the 1994 flood) to S = 99 (lowest annual high stage). Data for 14 of these ranked years are provided in the Flood Data Table and can be used to calculate recurrence interval for each magnitude (rank, S). **Recurrence interval** (or **return period**) is the average number of years between occurrences of a flood of a given rank (S) or greater than that given rank. Recurrence interval for a rank of flood can be calculated as: RI = (n + 1)/S. Calculate the recurrence interval for ranks 1–5 and write them in the table. This has already been done for ranks of 20, 30, 40, 50, 60, 70, 80, 90, and 99.

2. Notice that a recurrence interval of 5.0 means that there is a 1-in-5 probability (or 20% chance) that an event of that magnitude will occur in any given year. This is known as a *5-year flood*. What is a *100-year flood?*

3. Plot (as exactly as you can) points on the flood magnitude/frequency graph (below the Flood Data Table) for all 14 ranks of annual high river stage in the Flood Data Table. Then use a ruler to draw a line through the points (and on to the right edge of the graph) so the number of, or distance to, points above and below the line is similar.

4. Your completed flood magnitude/frequency graph can now be used to estimate the probability of future floods of a given magnitude and frequency. A 10-year flood on the Flint River is the point where the line in your graph crosses the flood frequency (RI, return period) of 10 years. What is the probability that a future 10-year flood will occur in any given year, and what will be its magnitude (river elevation in feet above sea level)?

5. What is the probability for any given year that a flood on the Flint River at Montezuma, GA will reach an elevation of 275 feet above sea level?

F. Most homeowners insurance policies do not insure against floods, even though floods cause more damage than any other natural hazard. Homeowners must obtain private or federal flood insurance in addition to their base homeowners policy. The National Flood Insurance Program (NFIP), a Division of the Federal Emergency Management Agency (FEMA), helps communities develop corrective and preventative measures for reducing future flood damage. The program centers on floodplain identification, mapping, and management. In return, members of these communities are eligible for discounts on federal flood insurance. The rates are determined on the basis of a community's FIRM (Flood Insurance Rate Map), an official map of the community on which FEMA has delineated flood *hazard areas* and *risk premium zones* (with discount rates). The hazard areas on a FIRM are defined on a *base flood elevation* (BFE)—the computed elevation to which flood water is estimated to rise during a *base flood.* The regulatory-standard base flood elevation is the 100-year flood elevation. Based on your graph, what is the BFE for Montezuma, GA?

Flood Data Table

Recurrence Intervals for Selected, Ranked, Annual Highest Stages of the Flint River over 99 Years of Observation (1897 and 1905–2002) at Montezuma, Georgia (USGS Station 02349500, data from USGS)

Rank of annual highest river stage (S)	Year (*n = 99)	River elevation above gage, in feet	Gage elevation above sea level, in feet	River elevation above sea level, in feet	Recurrence interval** (RI), in years	Probability of occurring in any given year	Percent chance of occurring in any given year***
1 (highest)	1994	34.11	255.83	289.9		1 in 100	1%
2	1929	27.40	255.83	283.2		1 in 50	2%
3	1990	26.05	255.83	281.9		1 in 33.3	3%
4	1897	26.00	255.83	281.8		1 in 25	4%
5	1949	25.20	255.83	281.0		1 in 20	5%
20	1928	21.30	255.83	277.1	5.0	1 in 5	20%
30	1912	20.60	255.83	276.4	3.4	1 in 3.4	29%
40	1959	19.30	255.83	275.1	2.3	1 in 2.3	43%
50	1960	18.50	255.83	274.3	2.0	1 in 2	50%
60	1934	17.70	255.83	273.5	1.8	1 in 1.8	56%
70	1974	17.25	255.83	273.1	1.5	1 in 1.5	67%
80	1967	14.76	255.83	270.6	1.3	1 in 1.3	77%
90	1907	13.00	255.83	268.8	1.1	1 in 1.1	91%
99 (lowest)	2002	8.99	255.83	264.7	1.0	1 in 1	100%

*n = number of years of annual observations = 99
**Recurrence Interval (RI) = (n + 1) / S = average number of years between occurrences of an event of this magnitude or greater.
***Percent chance of occurrence = 1 / RI x 100.

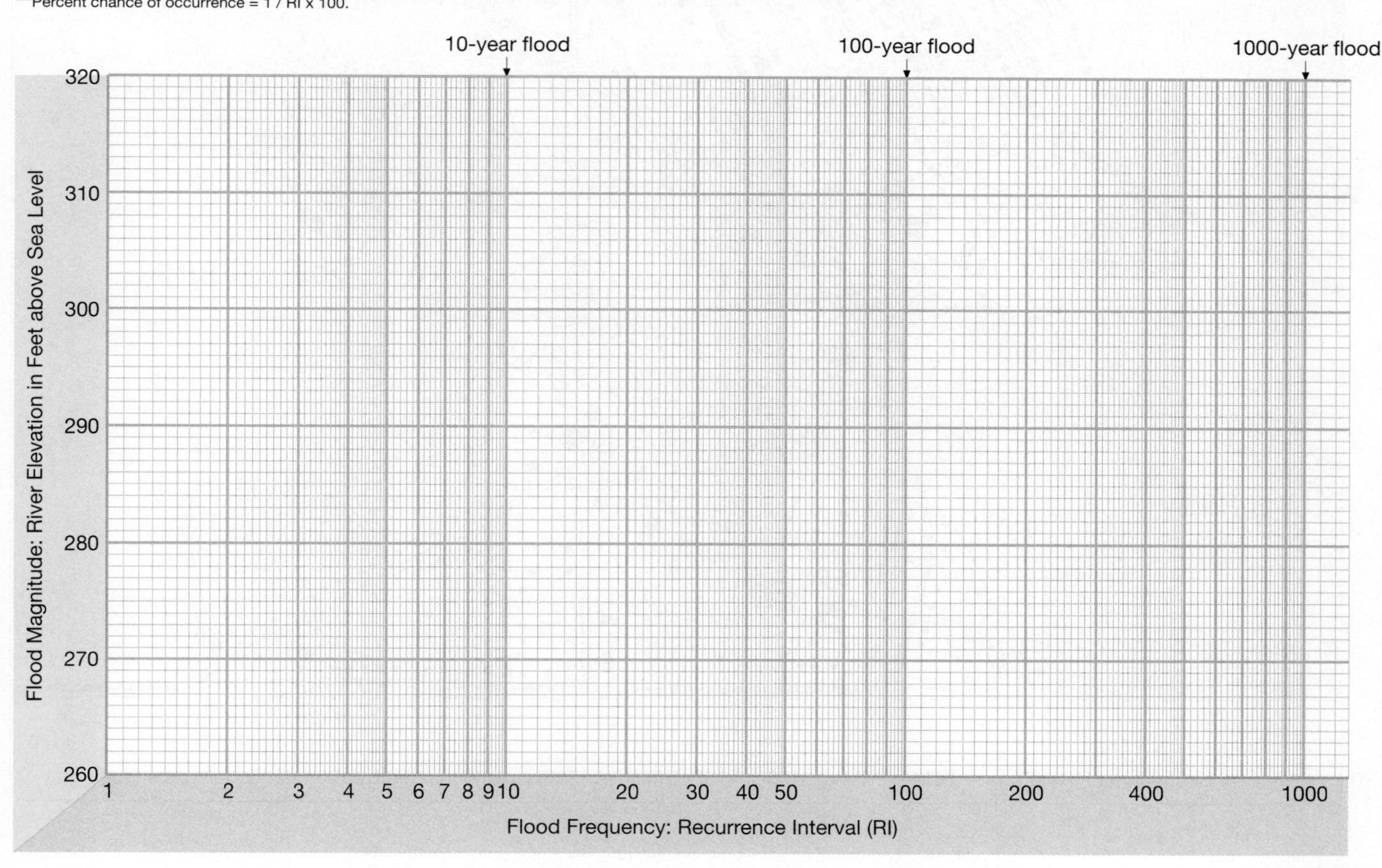

G. REFLECT & DISCUSS The 1996 FEMA FIRM for Montezuma, Georgia, shows hazard areas designated *zone A*. Zone A is the official designation for areas expected to be inundated by 100-year flooding even though no BFEs have been determined. The location of zone A (shaded gray) is shown on the Flood Hazard Map of a portion of the Montezuma, GA 7.5 minute map below. Your work above can be used to revise the flood hazard area. Place a dark line on this map (as exactly as you can) to show the elevation contour of the BFE for this community (your answer in item **F**). Your revised map reflects more accurately what area will be inundated by a 100-year flood. In general, how is the BFE line that you have plotted different from the boundary of zone A plotted by FEMA on its 1996 FIRM?

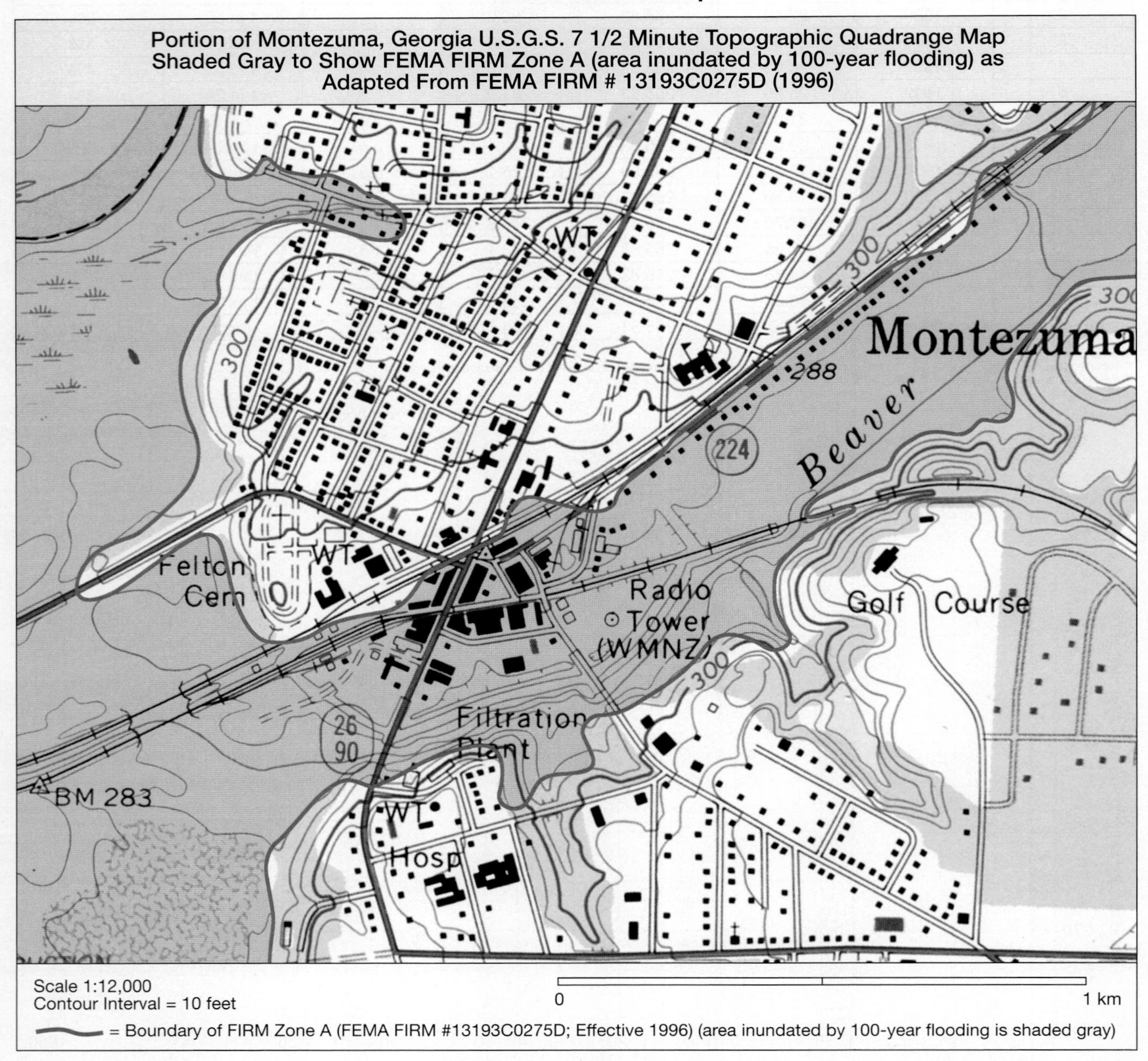

LABORATORY 12

Groundwater Processes, Resources, and Risks

CONTRIBUTING AUTHORS

Garry D. McKenzie • *Ohio State University*

Richard N. Strom • *University of South Florida, Tampa*

James R. Wilson • *Weber State University*

Stalactites hang from the ceiling of Luray Caverns, Virginia. Some merge with stalagmites forming on the cave floor.

PRE-LAB VIDEO

BIG IDEAS

Groundwater is subsurface water, beneath the landscape rather than on its surface. Most bodies of groundwater form when rainwater seeps into the ground under the influence of gravity and fills up (saturates) spaces in cracks and between grains. Some groundwater is unconfined and must be pumped from the ground to be used. Confined groundwater is under pressure and will flow on its own if a well is drilled to its location. Karst topography and rapid movement of water can occur when groundwater dissolves caves in soluble rocks, and land subsidence can occur when humans withdraw groundwater faster than it can be replenished.

FOCUS YOUR INQUIRY

THINK About It How does groundwater behave underground?

ACTIVITY 12.1 Groundwater Inquiry *(p. 312)*

THINK About It What is karst topography and how does water flow beneath it?

ACTIVITY 12.2 Karst Processes and Topography *(p. 312)*

ACTIVITY 12.3 Floridan Limestone Aquifer *(p. 314)*

THINK About It What can happen if groundwater is withdrawn faster than it is replenished?

ACTIVITY 12.4 Land Subsidence from Groundwater Withdrawal *(p. 317)*

Introduction

Water that seeps into the ground is pulled downward by the force of gravity through spaces in the soil and *bedrock* (rock that is exposed at the land surface or underlies the soil). At first, the water fills just some spaces and air remains in the other spaces. This underground zone with water- and air-filled spaces is called the *zone of aeration* (FIGURE 12.1; also called the *unsaturated zone* or *vadose zone*). Eventually, the water reaches a zone below the zone of aeration, where all spaces are completely saturated with water. This water-logged zone is called the *zone of saturation,* and its upper surface is the **water table** (FIGURE 12.1). Water in the saturated zone is called **groundwater**, which can also be withdrawn from the ground through a **well** (a hole dug or drilled into the ground). Most wells are lined with *casing*, a heavy metal or plastic pipe. The casing is perforated in sections where water is expected to supply the well. Other sections of the casing are left impervious to prevent unwanted rock particles or fluids from entering the well.

ACTIVITY

12.1 Groundwater Inquiry

THINK About It **How does water behave underground?**

OBJECTIVE Experiment with water to determine its behavior in confined and unconfined spaces and in relation to shale and sandstone.

PROCEDURES

1. **Before you begin**, do not look up definitions and information. Use your current knowledge, and complete the worksheet with your current level of ability. Also, this is **what you will need** to do the activity:
   ____ empty plastic drink bottle (2 liter), water, tape
   ____ nail, drill, or other object that can be used to safely make small holes in a plastic bottle
   ____ Activity 12.1 Worksheet (p. 321) and pencil
2. **Complete the worksheet in a way that makes sense to you.**
3. **After you complete the worksheet**, be prepared to discuss your observations and ideas with others.

ACTIVITY

12.2 Karst Processes and Topography

THINK About It **What is karst topography and how does water flow beneath it?**

OBJECTIVE Explore and evaluate the topographic features, groundwater movements, and hazards associated with karst topography.

PROCEDURES

1. **Before you begin**, read the Introduction and Caves and Karst below. Also, this is **what you will need:**
   ____ calculator, colored pencils
   ____ Activity 12.2 Worksheet (p. 323) and pencil
2. **Then follow your instructor's directions** for completing the worksheet.

Recall the last time that you consumed a drink from a fast-food restaurant (a paper cup containing ice and liquid that you drink using a plastic straw). The mixture of ice and liquid (no air) at the bottom of the cup was a zone of saturation, and your straw was a well. Each time you sucked on the straw, you withdrew liquid from the drink container just as a homeowner withdraws water from a water well. After you drank some of the drink, the cup contained both a zone of saturation (water and ice in the bottom of the cup) and a zone of aeration (ice and mostly air in the upper part of the cup). The boundary between these two zones was a water table. In order to continue drinking the liquid, you had to be sure that the bottom of your straw was within the zone of saturation, below the water table. Otherwise, sucking on the straw produced only a slurping sound, and you obtained mostly air. Natural water wells work the same way. The wells must be drilled or dug to a point below the water table (within the zone of saturation), so that water can flow or be pumped out of the ground.

Porosity and Permeability

The volume of void space (space filled with water or air) in sediment or bedrock is termed *porosity*. The larger the voids, and the greater their number, the higher is the porosity. If void spaces are interconnected, then fluids (water and air) can migrate through them (from space to space), and the rock or sediment is said to be *permeable*. Sponges and paper towels are household items that are permeable, because liquids easily flow into and through them. Plastic and glass are *impermeable* materials, so they are used to contain fluids.

Aquifers

Permeable bedrock materials make good **aquifers**, or rock strata that conduct water. Some examples are sandstones and limestones. Impermeable bedrock materials prevent the flow of water and are called **confining beds** (or **aquitards**). Some examples are layers of clay, mudstone, shale, or dense igneous and metamorphic rock. But how does groundwater move through aquifers?

Confined aquifers are sandwiched between two confining beds, the groundwater fills them from confining bed to confining bed, so there is no water table. The weight of the groundwater (being pulled downward by gravity) in a confined space creates water pressure, like the pressure inside of a garden hose or kitchen sink faucet. If a confined aquifer is penetrated by a well, then water flows naturally from the well. When aquifers are not confined (i.e., they are **unconfined aquifers**), the groundwater establishes a water table just beneath the surface of the land (**FIGURE 12.1**). For this reason, unconfined aquifers are also called *water table aquifers.* If an unconfined aquifer is penetrated by a well, then the water must be pumped from the ground using a submersible pump lowered into the well on a cable. An electric line runs from the top of the well to the submersible pump, and a water hose runs from the submersible pump to the top of the well.

Hydraulic Gradient

Groundwater in an unconfined (water table) aquifer is pulled down by gravity and spreads out through the ground until it forms the water table surface (such as the one in the drink cup full of crushed ice described previously). You can see the water table where it leaves the ground and becomes the level surface of a lake (**FIGURE 12.1A**) or springs flowing from a hillside. However, because groundwater is

Water Table Contours and Flow Lines

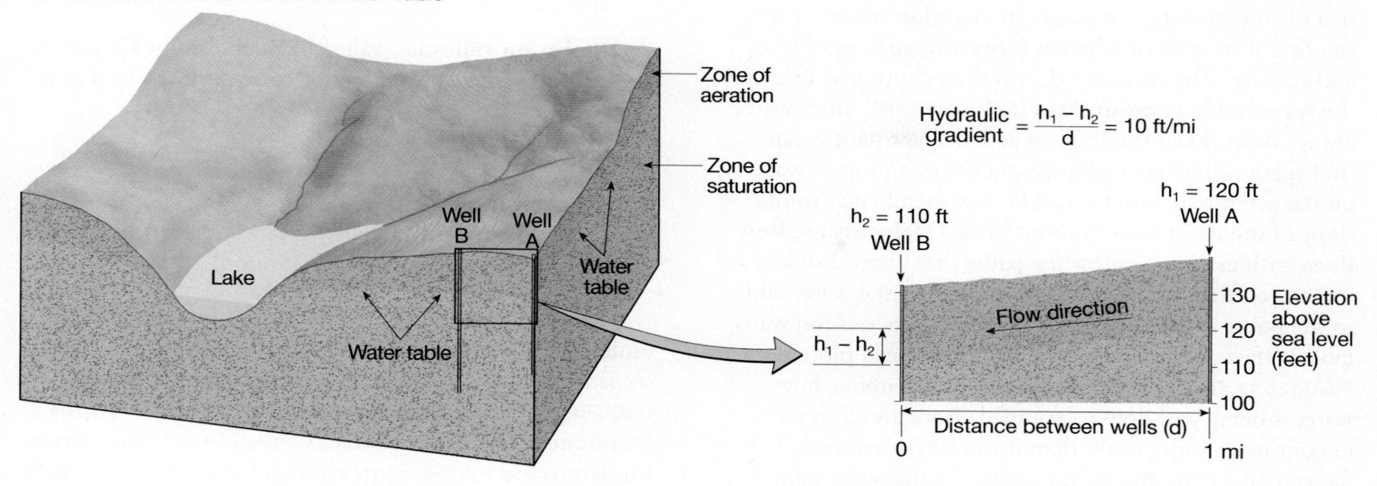

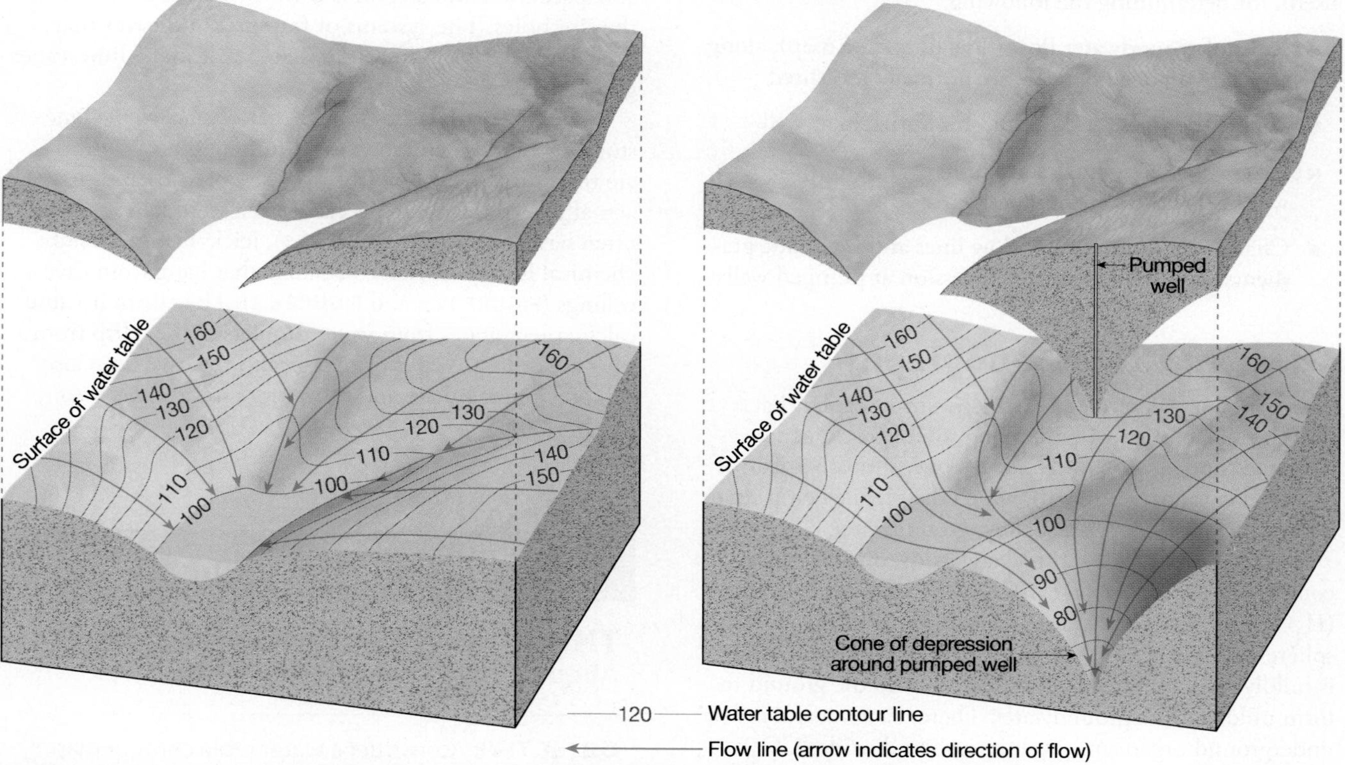

FIGURE 12.1 Water movement through an unconfined aquifer. A. Rainwater seeps into the *zone of aeration* (unsaturated zone, vadose zone), where void spaces are filled with air and water. Below it is the *zone of saturation,* where all void spaces are filled with water. Its upper surface is the water table. Water in the saturated zone is called groundwater, which always flows down the hydraulic gradient in unconfined aquifers. B. A water table surface is rarely level. Contour lines (contours) are used to map its topography and identify flow lines—paths traveled by droplets of water from the points where they enter the water table to the points where they enter a lake or stream. Flow lines with arrows run perpendicular to contour lines, converge or diverge, but never cross. C. A pumped well is being used to withdraw water faster than it can be replenished, causing development of a cone of depression in the water table and a change in the groundwater flow lines.

continuously being replenished (recharged) upslope, and it takes time for the water to flow through the ground, the water table is normally not level. It is normally higher uphill, where water flows into the ground, and lower downhill, where water seeps out of the ground at a lake or springs. The slope of the water table surface is called the **hydraulic gradient** (FIGURE 12.1A)—the difference in elevation between two points on the water table (observed in wells or surfaces of lakes and ponds) divided by the distance between those points.

Mapping Water Table Topography

To better understand the topography of the water table in a region, geologists measure its elevation wherever they can find it in wells or where it forms the surfaces of lakes and streams. The elevation data is then contoured to map the **water table contour lines** (FIGURE 12.1B). Since water always flows down the shortest and steepest path it can find (path of highest hydraulic gradient), a drop of water on the water table surface will flow perpendicular to the slope of the water table contour lines. Geologists use **flow lines with arrows** to show the paths that water droplets will travel from the point where they enter the water table to the point where they reach a lake, stream, or level water table surface. Notice how flow lines have been plotted on FIGURES 12.1B and 12.1C. In FIGURE 12.1C, notice how water is being withdrawn (pumped) from a well in an unconfined aquifer faster than it can be replenished. This has caused a cone-shaped depression in the water table (**cone of depression**) and a change in the regional flow of the groundwater. Thus, water table contour maps are useful for determining the following:

- Paths of groundwater flow (flow lines on a map), along which hydraulic gradients are normally measured
- Where the water comes from for a particular well
- Paths (flow lines) that contaminants in groundwater will likely follow from their source
- Changes to groundwater flow lines and hydraulic gradients caused by cones of depression at pumped wells

Caves and Karst Topography

The term **karst** describes a distinctive topography that indicates dissolution of underlying soluble rock, generally limestone (FIGURE 12.2). Limestone is mostly made of calcite (a carbonate mineral), which dissolves when it reacts with acidic rainwater and shallow groundwater.

Rainwater may contain several acids, but the most common is carbonic acid (H_2CO_3). It forms when water (H_2O) and carbon dioxide (CO_2) combine in the atmosphere ($H_2O + CO_2 = H_2CO_3$). All natural rainwater is mildly acidic (pH of 5–6) and soaks into the ground to form mildly acidic groundwater. There, bacteria and other underground organisms produce carbon dioxide (CO_2) as a waste product of their respiration (metabolic process whereby they convert food and oxygen into energy, plus water and carbon dioxide waste). This carbon dioxide makes the groundwater even more acidic, so it easily dissolves the calcite making up the limestone by this reaction:

$$\underset{\text{Calcite}}{CaCO_3} + \underset{\text{Carbonic acid}}{H_2CO_3} = \underset{\text{Calcium ions dissolved in groundwater}}{Ca^{+2}} + \underset{\text{Bicarbonate ions dissolved in groundwater}}{2HCO_3^{-1}}$$

A typical karst topography has these features, which are illustrated in FIGURE 12.2 and visible on the US Topo orthoimage of the Park City, Kentucky Quadrangle in FIGURE 12.3.

- (**Sinkholes**)—surface depressions formed by the collapse of caves or other large underground void spaces.
- (**Solution valleys**)—valley-like depressions formed by a linear series of sinkholes or collapse of the roof of a linear cave.
- (**Springs**)—places where water flows naturally from the ground (from spaces in the bedrock).
- (**Disappearing streams**)—streams that terminate abruptly by seeping into the ground.

Much of the drainage in karst areas occurs underground rather than by surface runoff. Rainwater seeps into the ground along fractures in the bedrock (FIGURE 12.4), whereupon the acidic water dissolves the limestone around it. The cracks widen into narrow **caves** (underground cavities large enough for a person to enter), which may eventually widen into huge cave galleries. Sinkholes develop where the ceilings of these galleries collapse, and lakes or ponds form wherever water fills the sinkholes. The systems of fractures and caves that typically develop in limestones are what make limestones good aquifers.

Eventually, the acidic water that was *dissolving* limestone becomes so enriched in calcium and bicarbonate that it turns alkaline (the opposite of acid) and may actually begin *precipitating* calcite. Caves in karst areas often have *stalactites* (FIGURE 12.5), icicle-like masses of chemical limestone made of calcite that hang from cave ceilings (FIGURE 12.5 and FIGURE 6.8). They form because calcite precipitates from water droplets as they drip from the cave ceiling. Water dripping onto the cave floor also can precipitate calcite and form more stout *stalagmites*.

ACTIVITY 12.3 Floridan Limestone Aquifer

THINK About It | **What is karst topography and how does water flow beneath it?**

OBJECTIVE Construct a water table contour map and determine the rate and direction of groundwater movement.

PROCEDURES

1. **Before you begin**, read the Introduction and Caves and Karst (above, if you have not already done so) and the Floridan Aquifer (below). Also, this is **what you will need:**
   ___ calculator
   ___ Activity 12.3 Worksheet (p. 325) and pencil
2. **Then follow your instructor's directions** for completing the worksheet.

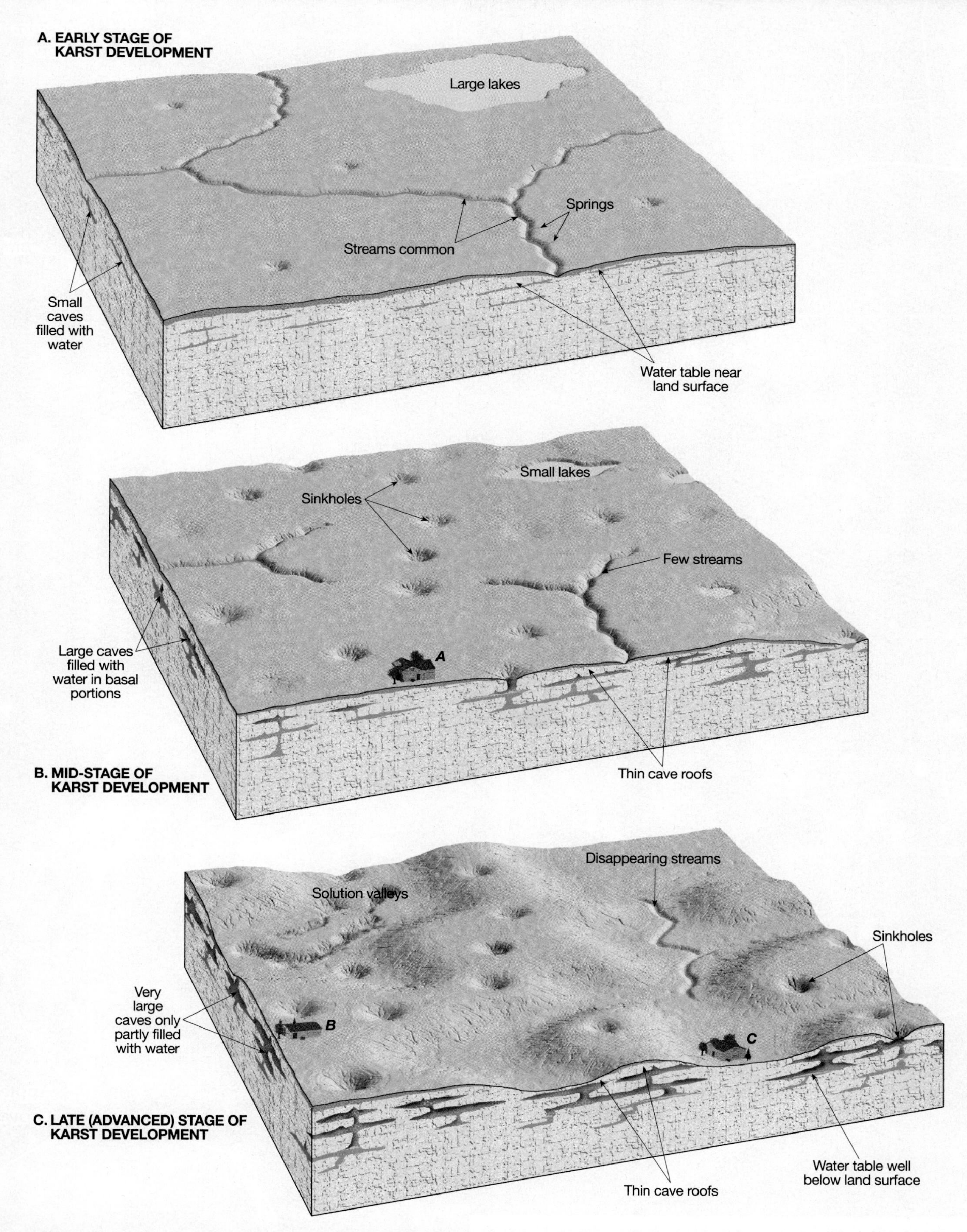

FIGURE 12.2 Stages in the evolution of karst topography. Karst topography is the result of dissolution of soluble bedrock (usually limestone).

FIGURE 12.3: Park City, Kentucky 7.5′ Quad (US Topo, 2010)

FIGURE 12.4 Water flow through fractures. Looking east toward the Arkansas River from Vap's Pass, Oklahoma (15 miles northeast of Ponca City). The Fort Riley Limestone bedrock *crops out* (is exposed at the surface) here. There is no soil, but plants have grown naturally along linear features in the bedrock.

FIGURE 12.5 Stalactites. These stalactites formed on part of the ceiling of Cave of the Winds, which has formed in Paleozoic limestones near Manitou Springs, Colorado.

The Floridan Aquifer

FIGURE 12.6 shows karst features developed in the Floridan Limestone Aquifer in the northern part of Tampa, Florida. Notice the abundant lakes and ponds. They are mostly sinkholes, which are filled with water and surrounded by hachured contour lines (contours with small tick marks that point inward, indicating a closed depression). By determining and mapping the elevations of water surfaces in the lakes, you can determine the slope of the water table and the direction of flow of groundwater here (as in FIGURE 12.1B).

ACTIVITY 12.4 Land Subsidence from Groundwater Withdrawal

THINK About It **What can happen if groundwater is withdrawn faster than it is replenished?**

OBJECTIVE Evaluate how groundwater withdrawal can cause subsidence (sinking) of the land.

PROCEDURES

1. **Before you begin**, read Land Subsidence Hazards Caused by Groundwater Withdrawal below. Also, this is **what you will need:**

   ____ calculator

   ____ Activity 12.4 Worksheet (p. 327) and pencil

2. **Then follow your instructor's directions** for completing the worksheet.

Land Subsidence Hazards Caused by Groundwater Withdrawal

Land subsidence caused by human withdrawal of groundwater is a serious problem in many places throughout the world. For example, in the heart of Mexico City, the land surface has gradually subsided up to 7.6 m (25 ft). At the northern end of California's Santa Clara Valley, 17 square miles of land have subsided below the highest tide level in San Francisco Bay and now must be protected by earthworks. Other centers of subsidence include Houston, Tokyo, Venice, and Las Vegas. With increasing withdrawal of groundwater and more intensive use of the land surface, we can expect the problem of subsidence to become more widespread.

Subsidence induced by withdrawal of groundwater commonly occurs in areas underlain by stream-deposited (alluvial) sand and gravel that is interbedded with lake-deposited (lacustrine) clays and clayey silts (FIGURE 12.7A). The sand-and-gravel beds are aquifers, and the clay and clayey silt beds are confining beds.

Subsidence in Unconfined and Confined Aquifers

In FIGURE 12.8, the water in the lower aquifer ("sand and gravel") is confined between impermeable beds of clay and silt and is under pressure from its own weight. Thus, water in wells **A** and **C** rises naturally from the confined aquifer to the *potentiometric* (water-pressure) *surface*. Such wells are termed **artesian wells** (water flows naturally from the top of the well). The sand in the water table aquifer (FIGURE 12.8) contains water that is not confined under pressure, so it is an **unconfined aquifer** (also called a *water table aquifer*). The water in well **B** stands at the level of the water table and must be pumped up to the land surface.

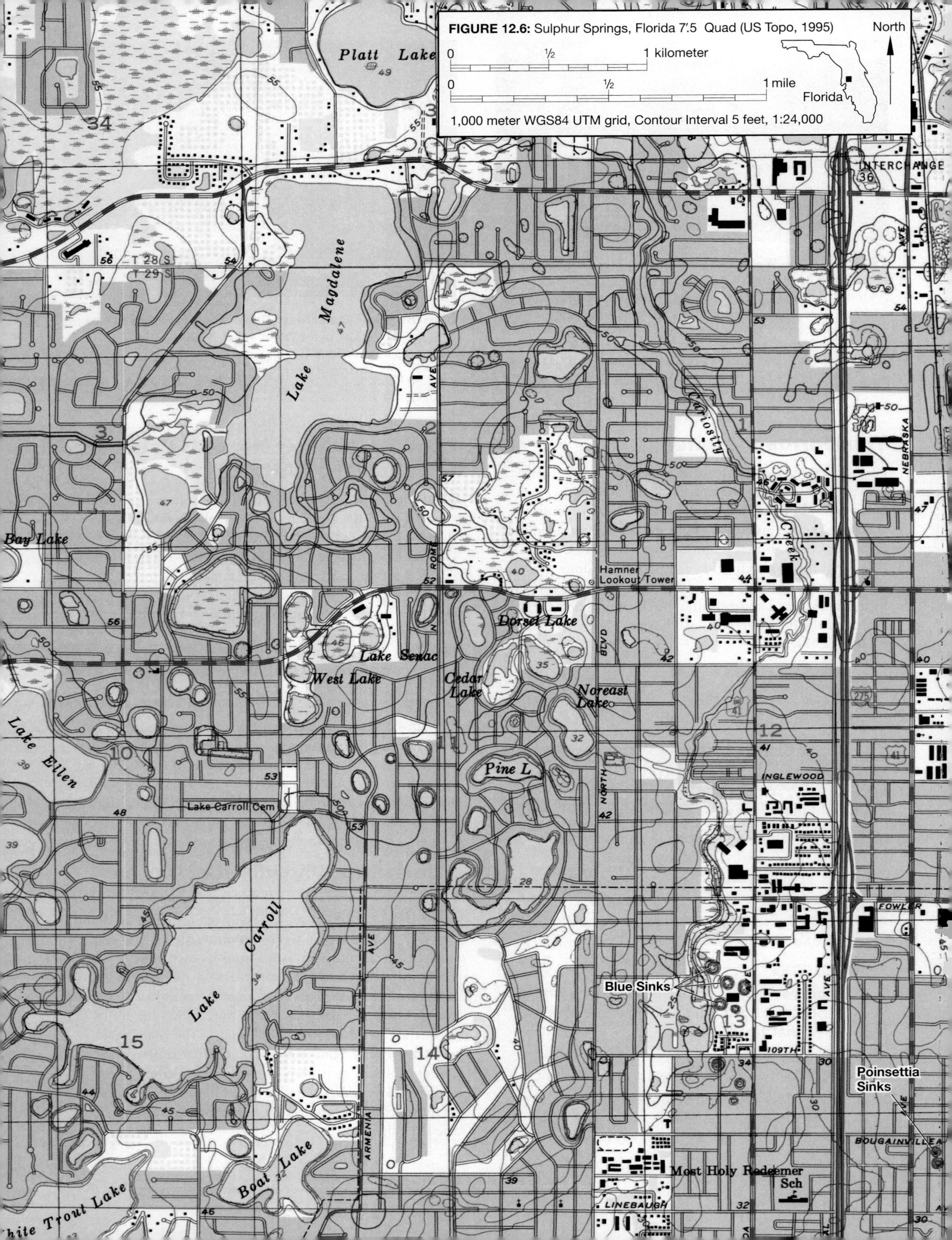

FIGURE 12.6: Sulphur Springs, Florida 7.5′ Quad (US Topo, 1995)

1,000 meter WGS84 UTM grid, Contour Interval 5 feet, 1:24,000

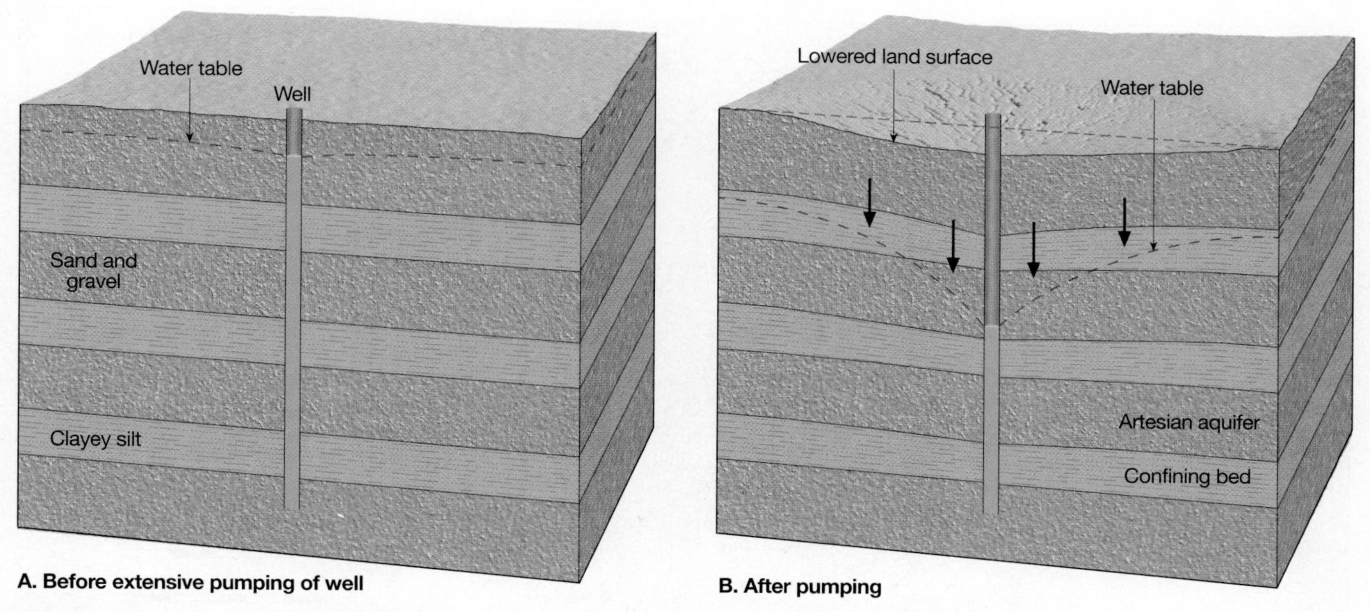

FIGURE 12.7 Before (A) and after (B) extensive pumping of a well. Note in **B** the lowering of the water-pressure surface, compaction of confining beds between the aquifers, and resulting subsidence of land surface. Arrows indicate the direction of compaction caused by the downward force of gravity, after the opposing water pressure was reduced by excessive withdrawal (discharge) of groundwater from the well.

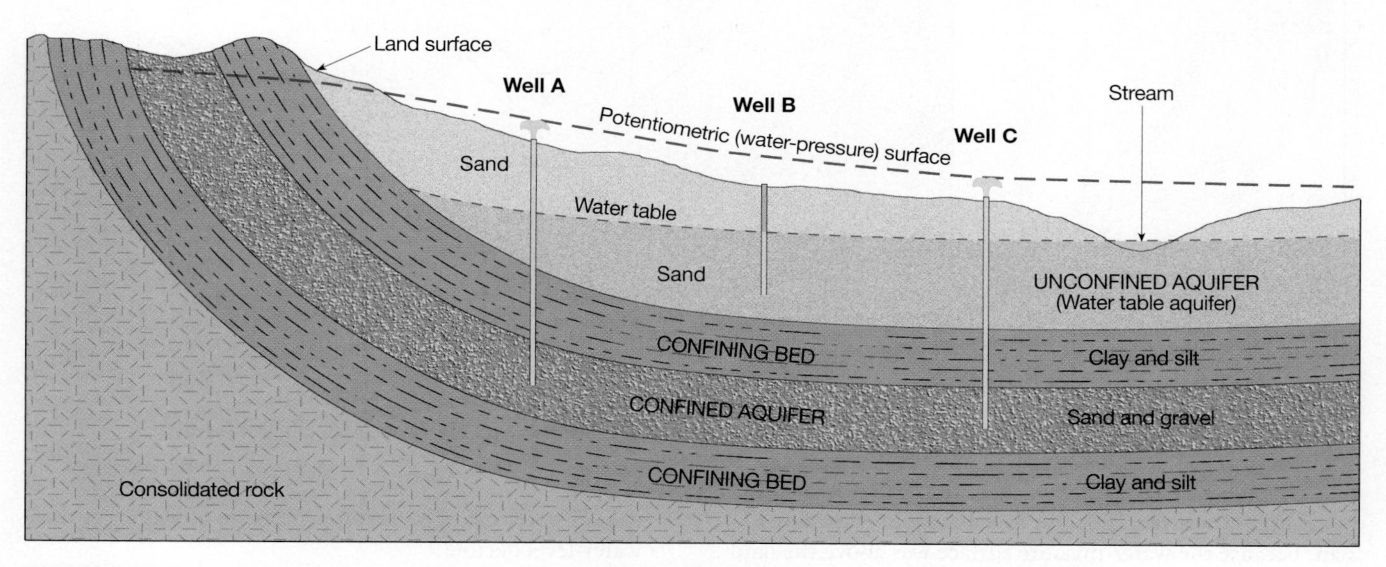

FIGURE 12.8 Geologic cross section of unconfined and confined aquifers. Vertical scale is exaggerated.

Land subsidence (FIGURE 12.7B) is related to the compressibility of water-saturated sediments. Withdrawing water from wells not only removes water from the system, but it also lowers the potentiometric surface and reduces the water pressure in the confined artesian aquifers. As the water pressure is reduced, the aquifer is gradually compacted and the ground surface above it is gradually lowered. The hydrostatic pressure can be restored by replenishing (or **recharging**) the aquifer with water. But the confining beds, once compacted, will not expand to their earlier thicknesses.

Subsidence in the Santa Clara Valley

The Santa Clara Valley (FIGURE 12.9) of California is a very important center of agriculture that depends on groundwater for irrigation. It was one of the first areas in the United States where land subsidence due to withdrawal of groundwater was recognized. The Santa Clara Valley is a large structural trough filled with alluvium (river sediments) more than 460 m (1500 ft) thick. Sand-and-gravel aquifers predominate near the valley margins, but the major part of the alluvium is silt and clay. Below a depth of 60 m (200 ft), the groundwater is confined by layers of clay, except near the margins.

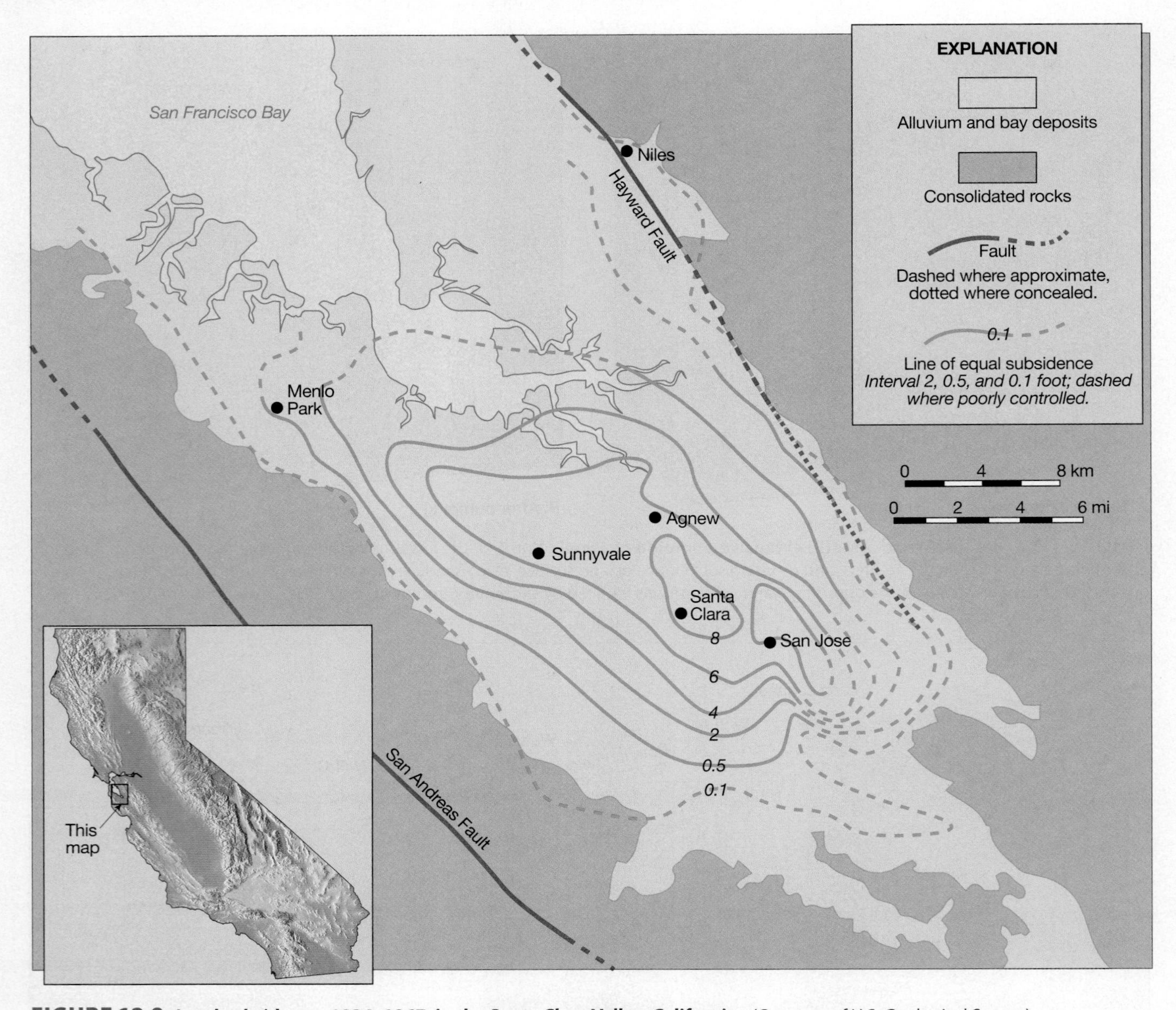

FIGURE 12.9 Land subsidence, 1934–1967, in the Santa Clara Valley, California. (Courtesy of U.S. Geological Survey)

Initially, wells as far south as Santa Clara were artesian, because the water-pressure surface was above the land surface. However, pumping them for irrigation lowered the water-pressure surface 40–60 m (150–200 ft) by 1965. This decline was not continuous. Natural recharge of the aquifer occurred between 1938 and 1947. As of 1971, the subsidence had been stopped due to a reversal of the water-level decline.

Most wells tapping the artesian system are 150–300 m (500–1000 ft) deep, although a few reach 365 m (1200 ft). Well yields in the valley are 500–1500 gallons per minute (gpm), which is very high.

ACTIVITY 12.1 Groundwater Inquiry

Name: ______________________ **Course/Section:** ______________________ **Date:** ____________

Consider the following experiments to understand how groundwater flows or is confined. Do not look up information. Just consider the experiments and complete the worksheet as well as you can.

A. EXPERIMENT 1: Analyze what happened when water was dropped on the rocks.

1. In general, what effect do you think layers of shale would have on groundwater movement or storage? Why?

2. In general, what effect do you think layers of sandstone would have on groundwater movement or storage? Why?

EXPERIMENT 1: What happens when a drop of water is applied to shale and sandstone?

Procedure: A drop of water was placed on four different rocks. This is what happened after 5 seconds.

Shale

Sandstone

Shale

Sandstone

B. EXPERIMENT 2: Analyze the procedures in the image on the next page, then plot the data on the graph.

1. How is the distance of the water jet related to the height of water in the bottle? Why?

2. Some water wells flow like a water hose and are called **artesian**. The water actually flows up and out of the well on its own. In non-artesian wells, the water must be pumped from the ground, because the water will not flow up and out of the well on its own.

a. Label the part of your Experiment 2 graph where the hole in the bottle can be considered artesian, and explain your reasoning used to label that part of the graph.

b. Label the part of your Experiment 2 graph where the hole in the bottle would be considered non-artesian, and explain why you decided to label that part of the graph.

EXPERIMENT 2: What happens when water drains from a hole in the side of a bottle?

Procedure:

1. A small hole was punched in the side of a 2-liter plastic bottle, 6 cm above its bottom.
2. Tape was placed over the hole.
3. The bottle was filled with water to a height of 22 cm.
4. The tape was removed, and a jet of water shot out of the hole. The distance that the water jet shot from the bottle to the table top was recorded for specific water heights in the bottle.
5. Plot the data on the graph paper to see if there is a trend.

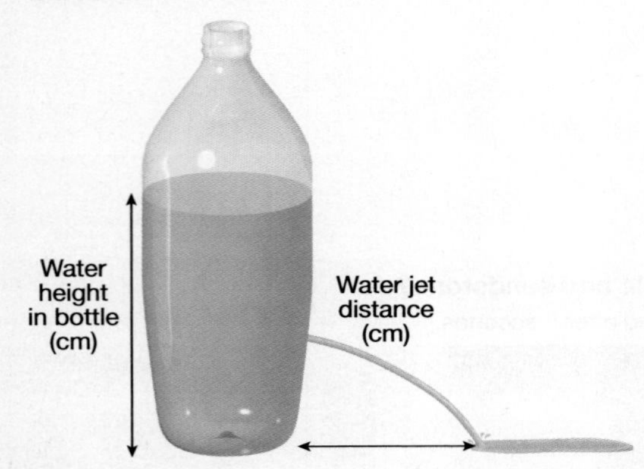

Water height in bottle (cm)	Water jet distance (cm)
22.0	10.0
18.5	9.5
16.5	9.0
14.0	8.0
13.0	7.5
11.0	6.0
10.0	5.5
9.0	4.5
8.0	3.5
7.5	2.5
7.0	2.0
6.5 (hole height)	0

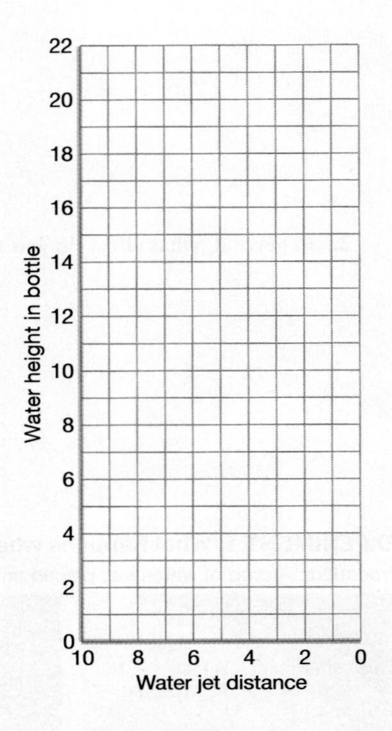

C. REFLECT & DISCUSS In Experiment 2, the jet of water was horizontal for a few centimeters, and then gravity exceeded the horizontal force and pulled the jet of water down onto the table. Water wells are normally vertical. Make a sketch of how you would re-arrange the materials from Expeiment 2 in order to get the jet of water to flow in more of a vertical position, like a natural artesian well, and explain how such a situation could occur among layers of sandstone and shale underground.

ACTIVITY 12.2 Karst Processes and Topography

Name: ____________________ **Course/Section:** ____________________ **Date:** __________

A. Analyze **FIGURES 12.4** and **12.5**.

1. In the area photographed in **FIGURE 12.4**, there is no soil developed on the limestone bedrock surface, yet abundant plants are growing along linear features in the bedrock. What does this indicate about how water travels through bedrock under this part of Oklahoma?

2. If you had to drill a water well in the area pictured in **FIGURE 12.4**, where would you drill (relative to the pattern of plant growth) to find a good supply of water? Why?

3. **REFLECT & DISCUSS** How is **FIGURE 12.5** related to **FIGURE 12.4**?

B. It is common for buildings to sink into newly formed sinkholes as they develop in karst regions. Consider the three new-home construction sites (labeled **A, B,** and **C**) in **FIGURE 12.2**, relative to sinkhole hazards.

1. Which new-home construction site (**A, B,** or **C**) is the **most** hazardous? Why?

2. Which new-home construction site (**A, B,** or **C**) is the **least** hazardous? Why?

3. **REFLECT & DISCUSS** Imagine that you are planning to buy a new-home construction site in the region portrayed in **FIGURE 12.2**. What could you do to find out if there is a sinkhole hazard in the location where you are thinking of building your home?

C. Study the orthoimage of the Park City (Kentucky) topographic map in **FIGURE 12.3**. Almost all of this area is underlain by limestone. The limestone is overlain by sandstone in the small northern part of this image (Bald Knob, Opossum Hollow) that is covered by dense dark green trees.

1. How can you tell the area on this orthoimage where limestone crops out at Earth's surface?

2. Recall that on a topographic map, a depression is shown by a contour line, with hachures (tic marks), that forms a closed loop. Describe the pattern of depressions on the topo map of Park City. Why do some of the depressions contain ponds, while others do not?

3. **REFLECT & DISCUSS** Notice that there are many naturally formed circular ponds in the northwest half of the image. (The triangular ponds are surface water impounded behind dams constructed by people.) How could you use the elevations of the surfaces of the ponds to determine how groundwater flows through this region?

D. Refer to the map on the back of this page, a topographic map of the orthoimaged area in **FIGURE 12.3**.

1. Compare the map and orthoimage, then draw a contact (line) on the map that separates limestone with karst topography from forested, more resistant sandstone. Color the sandstone bedrock with a colored pencil.
2. Gardner Creek is a *disappearing stream*. Place arrows along all parts of the creek to show its direction of flow, then circle the location where it disappears underground. Circle the disappearing end of two other disappearing streams.
3. Notice that there are nine different springs that flow from the east-west trending hill on which Apple Grove is located. Label the elevation of each spring (where it starts a stream), then use the elevation points to draw a flow line with a large arrow to show the direction that water travels down the hydraulic gradient within the hill.
4. Find and label a solution valley anywhere on the map.
5. **REFLECT & DISCUSS** Notice that a pond has been constructed on the sandstone bedrock on top of Bald Knob and filled with water from a well. If the well is located on the dark blue edge of the pond, then how deep below that surface location was the well drilled just to reach the water table? Show your work.

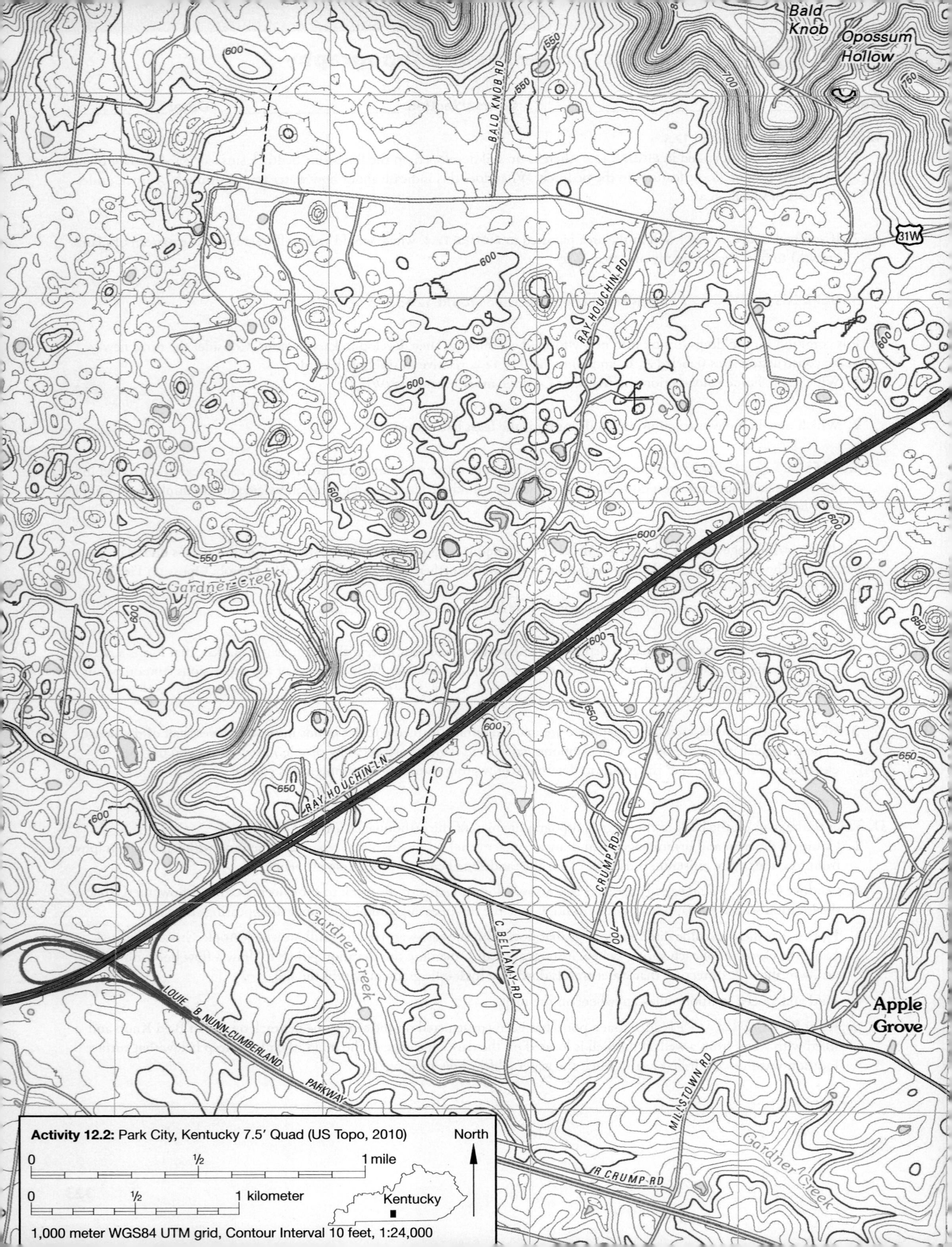

Bald Knob
Opossum Hollow
BALD KNOB RD
31W
RAY HOUCHIN RD
Gardner Creek
RAY HOUCHIN LN
CRUMP RD
C BELLAMY RD
LOUIE B NUNN-CUMBERLAND PARKWAY
MILLSTOWN RD
R CRUMP RD
Gardner Creek
Gardner Creek
Apple Grove
Activity 12.2: Park City, Kentucky 7.5′ Quad (US Topo, 2010)
North
0
½
1 mile
0
½
1 kilometer
Kentucky
1,000 meter WGS84 UTM grid, Contour Interval 10 feet, 1:24,000

ACTIVITY 12.3 Floridan Limestone Aquifer

Name: ______________________ **Course/Section:** ______________________ **Date:** ____________

Refer to **FIGURE 12.6** (Sulphur Springs Quadrangle) and the "sketch map" on page 326.

A. On the sketch map, mark the elevations of water levels in the lakes (obtain this information from **FIGURE 12.6**). The elevations of Lake Magdalene and some lakes beyond the boundaries of the topographic map already are marked for you.

B. Contour the water table surface (use a 5-foot contour interval) on the sketch map. Draw only contour lines representing whole fives (40, 45, and so on). Do this in the same manner that you contoured land surfaces in the topographic maps lab.

C. The flow of shallow groundwater in the sketch map is at right angles to the contour lines. The groundwater flows from high elevations of the hydraulic gradient to lower elevations, just like a stream. Draw three or four flow lines with arrows on the sketch map to indicate the direction of shallow groundwater flow in this part of Tampa. The southeastern part of **FIGURE 12.6** shows numerous closed depressions but very few lakes. What does this indicate about the level of the water table in this region?

D. Note the Poinsettia Sinks, a pair of sinkholes with water in them in the southeast corner of the topographic map (see **FIGURE 12.6**). Note their closely spaced hachured contour lines. Next, find the cluster of five similar sinkholes, called Blue Sinks, about 1 mile northwest of Poinsettia Sinks (just west of the WHBO radio tower). Use asterisks (*) to mark their locations on **FIGURE 12.8**, and label them "Blue Sinks."

E. On the sketch map, draw a straight arrow (vector) along the shortest path between Blue Sinks and Poinsettia Sinks. The water level in Blue Sinks is 15 feet above sea level, and the water level in Poinsettia Sinks is 10 feet above sea level. Calculate the hydraulic gradient (in ft/mi: show your work below) along this arrow and write it next to the arrow on the sketch map. (Refer to the hydraulic gradient in **FIGURE 12.1** if needed.)

F. On **FIGURE 12.6**, note the stream and valley north of Blue Sinks. This is a fairly typical disappearing stream. Draw its approximate course onto the sketch map. Make an arrowhead on one end of your drawing of the stream to indicate the direction that water flows in this stream. How does this direction compare to the general slope of the water table?

G. In March 1958, fluorescent dye was injected into the northernmost of the Blue Sinks. It was detected 28 hours later in Sulphur Springs, on the Hillsborough River to the south (see sketch map). Use these data to calculate (show your work) the approximate velocity of flow in this portion of the Floridan Aquifer:

1. in feet per hour: ____________ **2.** in miles per hour: ____________ **3.** in meters per hour: ____________

H. The velocities you just calculated are quite high, even for the Floridan Aquifer. But this portion of Tampa seems to be riddled with solution channels and caves in the underlying limestone. Sulphur Springs has an average discharge of approximately 44 cubic feet per second (cfs), and its maximum recorded discharge was 165 cfs (it once was a famous spa). During recent years, the discharge at Sulphur Springs has decreased. Water quality has also worsened substantially.

1. Examine the human-made structures on **FIGURE 12.6**. Note especially those in red, the color used to indicate new structures. Why do you think the discharge of Sulphur Springs has decreased in recent years?

2. Why do you think the water quality has decreased in recent years?

I. **REFLECT & DISCUSS** Name two potential groundwater-related hazards to homes and homeowners in the area that you can think of.

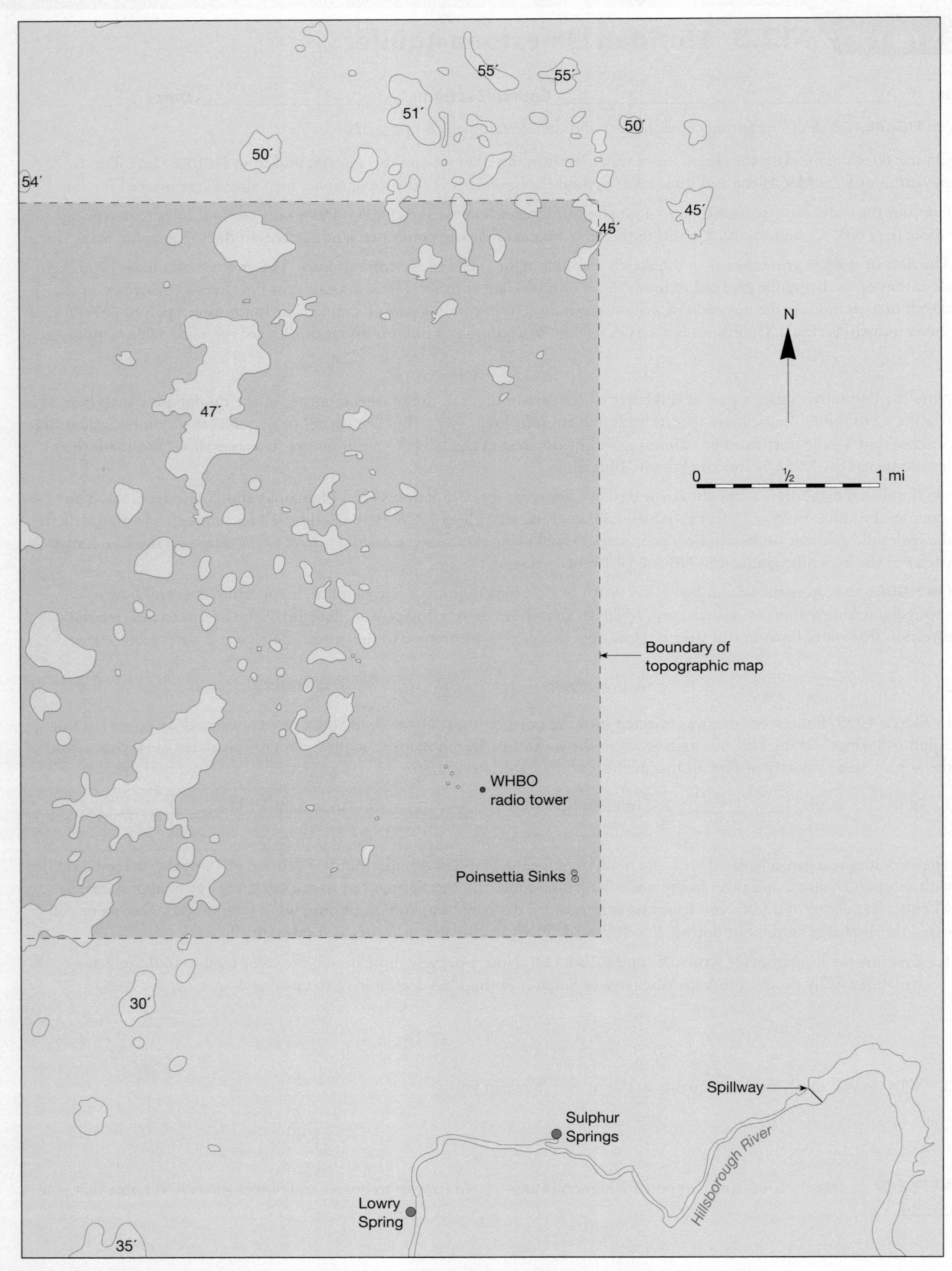

Sketch map of the area shown in **FIGURE 12.6** (Sulphur Springs Quadrangle) and surrounding region.

ACTIVITY 12.4 Land Subsidence from Groundwater Withdrawal

Name: ______________________ **Course/Section:** ______________________ **Date:** ____________

A. Santa Clara Valley, California.

1. In **FIGURE 12.9** on page 320, where are the areas of greatest subsidence in the Santa Clara Valley?

2. What was the total subsidence at San Jose (**FIGURE12.10**) from 1934 to 1967 in feet?

Year	Total Subsidence (feet) from 1912 level
1912	0.0
1920	0.3
1934	4.6
1935	5.0
1936	5.0
1937	5.2
1940	5.5
1948	5.8
1955	8.0
1960	9.0
1963	11.0
1967	12.7

FIGURE 12.10 Subsidence at benchmark P7 in San Jose, California.

3. What was the average annual rate of subsidence for the period of 1934 to 1967 in feet per year?

4. Analyze **FIGURE 12.9**. At what places in the Santa Clara Valley would subsidence cause the most problems? Explain your reasoning.

5. Would you expect much subsidence to occur in the darker shaded areas of **FIGURE 12.9**? Explain.

6. By 1960, the total subsidence at San Jose had reached 9.0 feet (**FIGURE 12.10**). What was the average annual rate of subsidence (in feet per year) for the seven-year period from 1960 through 1967? (Show your work.)

7. Refer to Figure 12.11 on the back of this page. What was the level of the water in the San Jose well in:

   a. 1915? __________ feet   b. 1967? __________ feet

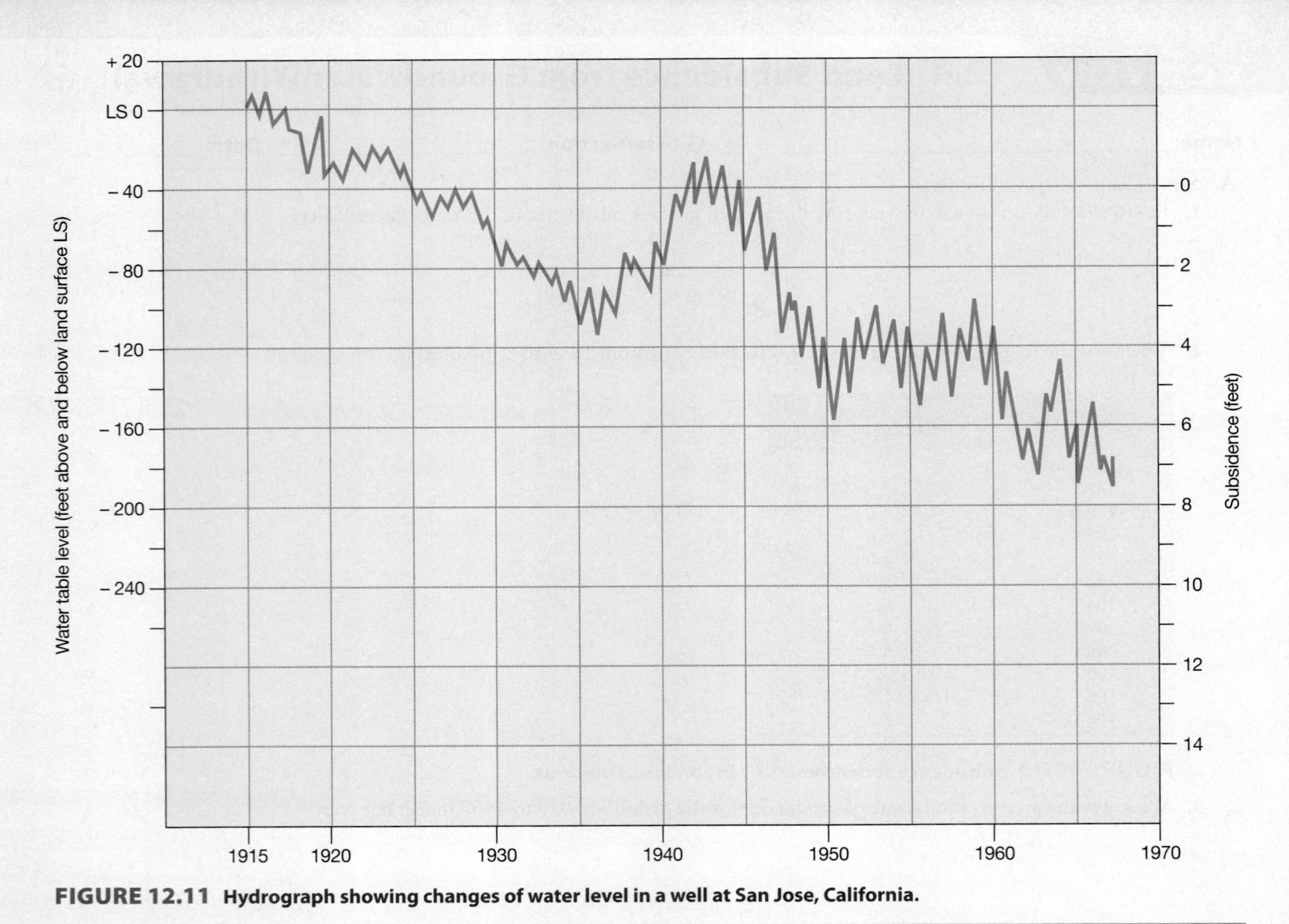

FIGURE 12.11 Hydrograph showing changes of water level in a well at San Jose, California.

8. During which years would the San Jose well have been a flowing artesian well?

9. How can you explain the minor fluctuations in the hydrograph (**FIGURE 12.12**) like those between 1920 and 1925?

10. In **FIGURE 12.11**, the slope of a line joining the level of the land surface in 1915 with subsidence that had occurred by 1967 gives the average rate of subsidence for that period. How did the rate of subsidence occurring between 1938 and 1948 differ from earlier rates?

B. **REFLECT & DISCUSS** Adolf Hitler came into power as head of the National Socialist German Workers' Party (Nazi Party) in 1933 and German Troops invaded Austria in 1938 and Poland in 1939 to initiate World War II. Japan invaded China in 1932, withdrew, and then launched a full-scale invasion of China in 1937. The United States officially entered World War II in 1941 (when Japan attacked Pearl Harbor). Explain how these world events could have caused the change in subsidence rates noted in Question 10.

C. **REFLECT & DISCUSS** Subsidence was stopped by 1971. What measures might have been taken to accomplish this?

PRE-LAB VIDEO

LABORATORY 13

Glaciers and the Dynamic Cryosphere

CONTRIBUTING AUTHORS

Sharon Laska • *Acadia University*

Kenton E. Strickland • *Wright State University–Lake Campus*

Nancy A. Van Wagoner • *Acadia University*

Kennicott Glacier, a long (43 km, 27 mi) valley glacier in Alaska. Mountains in the distance are where snow and ice accumulate and form the glacier. Down valley, dark medial moraines of rocky drift are deposited from melting ice. (Photo by Michael Collier)

BIG IDEAS

Earth's crysphere is its snow and ice (frozen water), including permafrost, sea ice, mountain glaciers, continental ice sheets, and the polar ice caps. The extent of snow and ice in any given area depends on how much snow and ice accumulates during winter months and how much snow and ice melts during summer months. Glaciers are one of the best known components of the cryosphere, because they are present on all continents except Australia and have created characteristic landforms and resources utilized by many people.

FOCUS YOUR INQUIRY

THINK About It What is the cryosphere, and how do changes in the cryosphere affect other parts of the Earth system?

ACTIVITY 13.1 Cryosphere Inquiry *(p. 330)*

THINK About It How do glaciers affect landscapes?

ACTIVITY 13.2 Mountain Glaciers and Glacial Landforms *(p. 330)*

ACTIVITY 13.3 Continental Glaciation of North America *(p. 330)*

THINK About It How is the cryosphere affected by climate change?

ACTIVITY 13.4 Glacier National Park Investigation *(p. 334)*

ACTIVITY 13.5 Nisqually Glacier Response to Climate Change *(p. 334)*

ACTIVITY 13.6 The Changing Extent of Sea Ice *(p. 335)*

Introduction

The **cryosphere** is all of Earth's snow and ice (frozen water). It all begins with a single snowflake falling from the sky or a single crystal of ice forming in a body of water. Over time, a visible body of snow or ice may form. Most snow and ice melts completely over summer months, providing much-needed water to communities. However, there are areas of Earth's surface where the annual amount of ice accumulation exceeds the annual amount of ice melting. Permanent masses of ice can exist there. These areas (**FIGURE 13.1**) range from places with permanently frozen ground (permafrost), to places

ACTIVITY

13.1 Cryosphere Inquiry

THINK About It **What is the cryosphere, and how do changes in the cryosphere affect other parts of the Earth system?**

OBJECTIVE Analyze global and regional components of the cryosphere, and then infer how they may change and ways that such change may affect other parts of the Earth system.

PROCEDURES

1. **Before you begin**, do not look up definitions and information. Use your current knowledge, and complete the worksheet with your current level of ability. Also, this is **what you will need** to do the activity:

    ____ pen
    ____ Activity 13.1 Worksheets (pp. 347–348) and pencil

2. **Complete the worksheet in a way that makes sense to you.**

3. **After you complete the worksheet**, be prepared to discuss your observations and classification with other geologists.

ACTIVITY

13.2 Mountain Glaciers and Glacial Landforms

THINK About It **How do glaciers affect landscapes?**

OBJECTIVE Analyze features of landscapes affected by mountain glaciation and infer how they formed.

PROCEDURES

1. **Before you begin**, read the Introduction, Glaciers, and Glacial Processes and Landforms. Also, this is **what you will need:**

    ____ ruler, calculator
    ____ Activity 13.2 Worksheets (pp. 349–350) and pencil

2. **Then follow your instructor's directions** for completing the worksheets.

where ice permanently covers the ground (glaciers and ice caps, ice sheets), to places where ice covers parts of the ocean (ice shelves, sea ice). The ice in your freezer may last for days or months, but ice in some of Earth's ice caps is thousands of years old.

ACTIVITY

13.3 Continental Glaciation of North America

THINK About It **How do glaciers affect landscapes?**

OBJECTIVE Analyze features of landscapes affected by continental glaciation and infer how they formed.

PROCEDURES

1. **Before you begin**, read the Introduction, Glaciers, and Glacial Processes and Landforms. Also, this is **what you will need:**

    ____ Activity 13.3 Worksheet (p. 351) and pencil

2. **Then follow your instructor's directions** for completing the worksheets.

Dynamic Cryosphere

The total amount of ice on Earth's surface is ever-changing due to annual variations in global patterns of air circulation and regional variations in things like ground temperature, ocean surface temperature, and the *weather* (daily to seasonal conditions of the atmosphere, such as air temperature and humidity, wind, cloud cover, and precipitation). Global and regional amounts of ice are also affected by *climate*—the set of atmospheric conditions (like air temperature, humidity, wind, and precipitaion) that prevails in a region over decades. A region's climate is generally determined by measuring the average conditions that exist there over a period of years or the conditons that normally exist in the region at a particular time of year.

Climate Change

A region's climate is based on factors like latitude, altitude, location relative to oceans (moisture sources), and location relative to patterns of global air and ocean circulation. **Climate change** refers to a significant change in atmospheric conditions of a region or the planet. This can occur due to natural factors like changing patterns of global air circulation, variations in volcanic activity, and changes in solar activity. It can also occur due to human factors like construction of regional urban centers (adding regional sources of heat energy) and deforestation (removing a transpiration source of atmospheric water vapor; adding soot and gases to the atmosphere as the forest is burned).

Map of Regional Variations in the Cryosphere

ICE SHELF: A sheet of ice attached to the land on one side but afloat on the ocean on the other side.

SEA ICE: A sheet of ice that originates from the freezing of seawater.

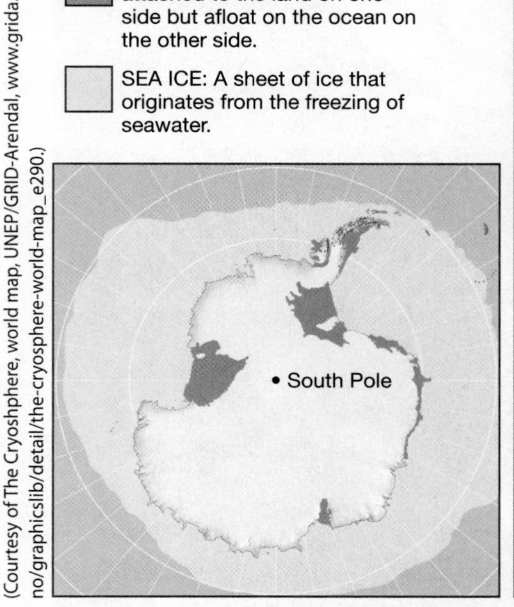

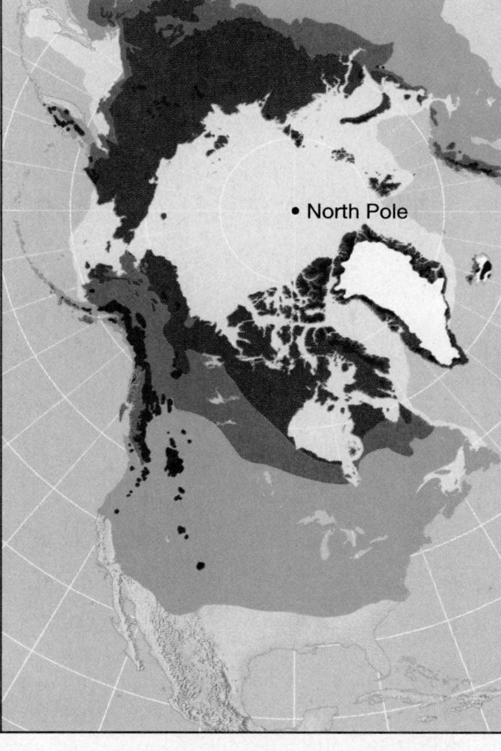

SEASONAL SNOW: Snow and ice may accumulate here in winter, but it melts over the following summer.

PERMAFROST CONTINUOUS: The ground is permanently frozen over this entire area.

PERMAFROST DISCONTINUOUS: The ground is permanently frozen in isolated patches within this area.

MOUNTAIN GLACIERS AND ICE CAPS: This area contains permanent patches of ice on mountain sides (cirques), river-like bodies of ice that flow down and away from mountains (valley and piedmont glaciers), and dome-shaped masses of ice and snow that cover the summits of mountains so that no peaks emerge (ice cap).

ICE SHEET: A pancake-like mound of ice covering a large part of a continent (more than 50,000 km^2).

(Courtesy of The Cryoshphere, world map, UNEP/GRID-Arendal, www.grida.no/graphicslib/detail/the-cryosphere-world-map_e290.)

FIGURE 13.1 Cryosphere components. You can also download a complete world map of cryosphere components from this UNEP (United Nations Environment Programme) website: **http://www.grida.no/graphicslib/detail/the-cryosphere-world-map_e290**

Glaciers

Glaciers are large ice masses that form on land areas that are cold enough and have enough snowfall to sustain them year after year. They form wherever the winter accumulation of snow and ice exceeds the summer ablation (also called *wastage*). *Ablation* (wastage) is the loss of snow and ice by melting and by *sublimation* to gas (direct change from ice to water vapor, without melting). Accumulation commonly occurs in *snowfields*—regions of permanent snow cover (FIGURE 13.2).

Glaciers can be divided into two zones, accumulation and ablation (FIGURE 13.2). As snow and ice accumulate in and beneath snowfields of the **zone of accumulation,** they become compacted and highly recrystallized under their own weight. The ice mass then begins to slide and flow downslope like a very viscous (thick) fluid. If you *slowly* squeeze a small piece of ice in the jaws of a vise or pair of pliers, then you can observe how it flows. In nature, glacial ice formed in the zone of accumulation flows and slides downhill into the **zone of ablation,** where it melts or sublimes (undergoes sublimation) faster than new ice can form. The *snowline* is the boundary between the zones of accumulation and ablation. The bottom end of the glacier is the **terminus.**

It helps to understand a glacier by viewing it as a river of ice. The "headwater" is the zone of accumulation, and the "river mouth" is the terminus. Like a river, glaciers *erode* (wear away) rocks, transport their load (tons of rock debris), and deposit their load "downstream" (down-glacier).

The downslope movement and extreme weight of glaciers cause them to abrade and erode (wear away) rock materials that they encounter. They also *pluck* rock material by freezing around it and ripping it from bedrock. The rock debris is then incorporated into the glacial ice and transported many kilometers by the glacier. The debris also gives glacial ice extra abrasive power. As the heavy rock-filled ice moves over the land, it scrapes surfaces like a giant sheet of sandpaper. Rock debris falling from valley walls commonly accumulates on the surface of a moving glacier and is transported downslope. Thus, glaciers transport huge quantities of sediment, not only *in,* but also *on* the ice.

When a glacier melts, it appears to retreat up the valley from which it flowed. This is called **glacial retreat,** even though the ice is simply melting back (rather than moving back up the hill). As melting occurs (FIGURE 13.3), deposits of rocky gravel, sand, silt, and clay accumulate where there once was ice. These deposits collectively are called **drift.** Drift that accumulates directly from the melting ice is unstratified (unsorted by size) and is called **till.** However, drift that is transported by the meltwater becomes more rounded, sorted by size, layered, and is called **stratified drift.** Wind also can transport the sand, silt, and clay particles from drift. This wind-transported sediment can form dunes or *loess* deposits (wind-deposited, unstratified accumulations of clayey silt).

FIGURE 13.2 Mountain glaciation. This is an ASTER infrared satellite image of a 20-by-20 km area in Alaska. Vegetation appears red, glacial ice is blue, and snow is white. (Image courtesy of NASA/GSFC/METI/ERSDAC/JAROS and U.S./Japan ASTER Science Team.)

There are five main kinds of glaciers based on their size and form.

- **Cirque glaciers**—small, semicircular to triangular glaciers that form on the sides of mountains. If they form at the head (up-hill end) of a valley and grow large enough, then they evolve into valley glaciers.
- **Valley glaciers**—long glaciers that originate at cirques and flow down stream valleys in the mountains.
- **Piedmont glaciers**—mergers of two or more valley glaciers at the foot (break in slope) of a mountain range.
- **Ice sheet**—a vast, pancake-shaped ice mound that covers a large portion of a continent and flows independent of the topographic features beneath it and covers an area greater than 50,000 km^2. The Antarctic Ice Sheet (covering the entire continent of Antarctica) and the Greenland Ice Sheet (covering Greenland) are modern examples.
- **Ice cap**—a dome-shaped mass of ice and snow that covers a flat plateau, island, or peaks at the summit of a mountain range and flows outward in all directions from the thickest part of the cap. It is much smaller than an ice sheet.

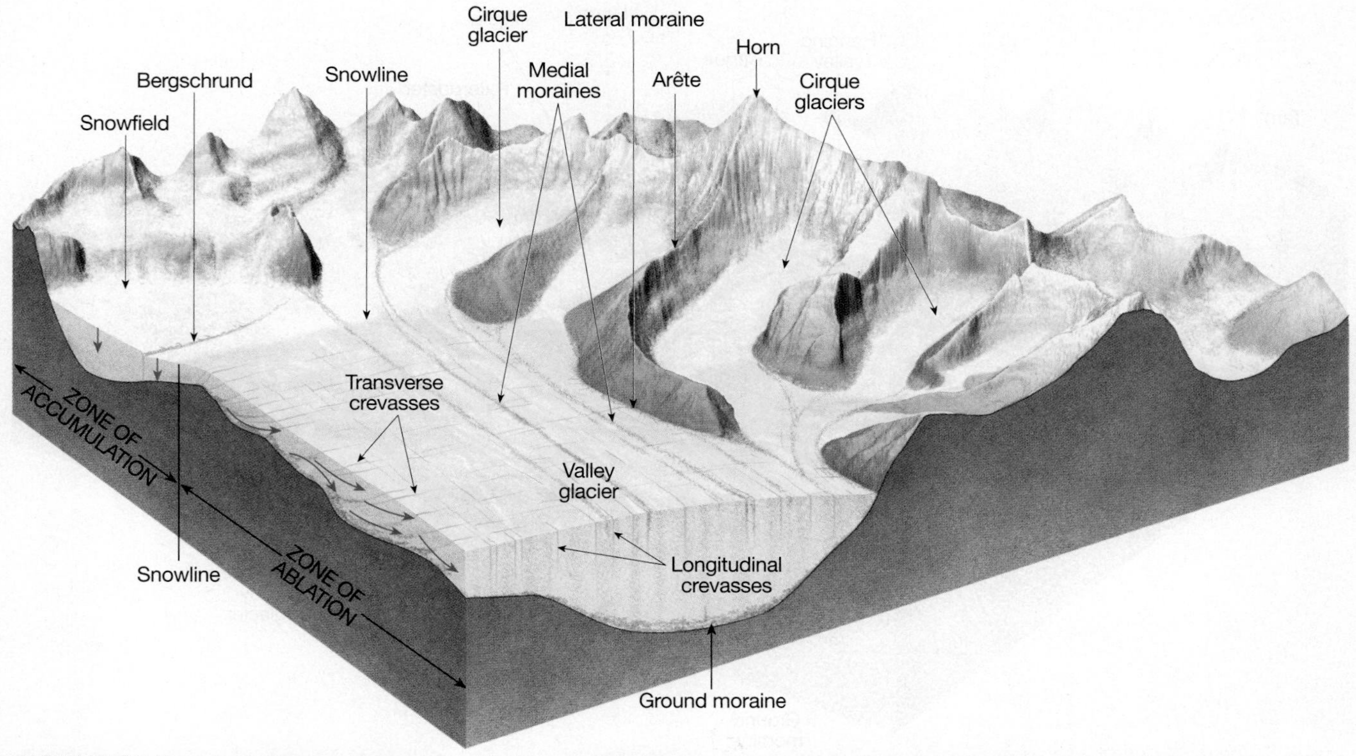

FIGURE 13.3 Active mountain glaciation, in a hypothetical region. Note the cutaway view of glacial ice, showing flow lines and direction (blue lines and arrows).

Glacial Processes and Landforms

Glaciated lands are affected by either local to regional "mountain glaciation" or more continent-wide "continental glaciation."

Mountain Glaciation

Mountain glaciation is characterized by cirque glaciers, valley glaciers, piedmont glaciers, and ice caps. Poorly developed mountain glaciation involves only cirques, but the best-developed mountain glaciation involves all three types. In some cases, valley and piedmont glaciers are so well developed that only the highest peaks and ridges extend above the ice. Ice caps cover even the peaks and ridges. FIGURE 13.2 shows a region with mountain glaciation. Note the extensive *snowfield* in the zone of accumulation. *Snowline* is the elevation above which there is permanent snow cover.

Also note that there are many cracks or fissures in the glacial ice of FIGURE 13.2. At the upper end of the glacier is the large *bergschrund* (German, "mountain crack") that separates the flowing ice from the relatively immobile portion of the snowfield. The other cracks are called **crevasses**—open fissures that form when the velocity of ice flow is variable (such as at bends in valleys). **Transverse crevasses** are perpendicular to the flow direction, and **longitudinal crevasses** are aligned parallel with the direction of flow.

FIGURE 13.3 shows the results of mountain glaciation after the glaciers have completely melted. Notice the characteristic landforms, water bodies, and sedimentary deposits. For your convenience, distinctive features of glacial lands are summarized in three figures: *erosional features* in FIGURE 13.4, *depositional features* in FIGURE 13.5, and *water bodies* in FIGURE 13.6. Note that some features are identical in mountain glaciation and continental glaciation, but others are unique to one or the other. Study the descriptions in these three figures and compare them with the visuals in FIGURES 13.2 and 13.3.

Continental Glaciation

During the Pleistocene Epoch, or "Ice Age," that ended 11,700 years ago, thick ice sheets covered most of Canada, large parts of Alaska, and the northern contiguous United States. These continental glaciers produced a variety of characteristic landforms (FIGURE 13.7, FIGURE 13.8).

Recognizing and interpreting these landforms is important in conducting work such as regional soil analyses, studies of surface drainage and water supply, and exploration for sources of sand, gravel, and minerals. The thousands of lakes in the Precambrian Shield area of Canada also are a legacy of this continental glaciation, as are the fertile soils of the north-central United States and south-central Canada.

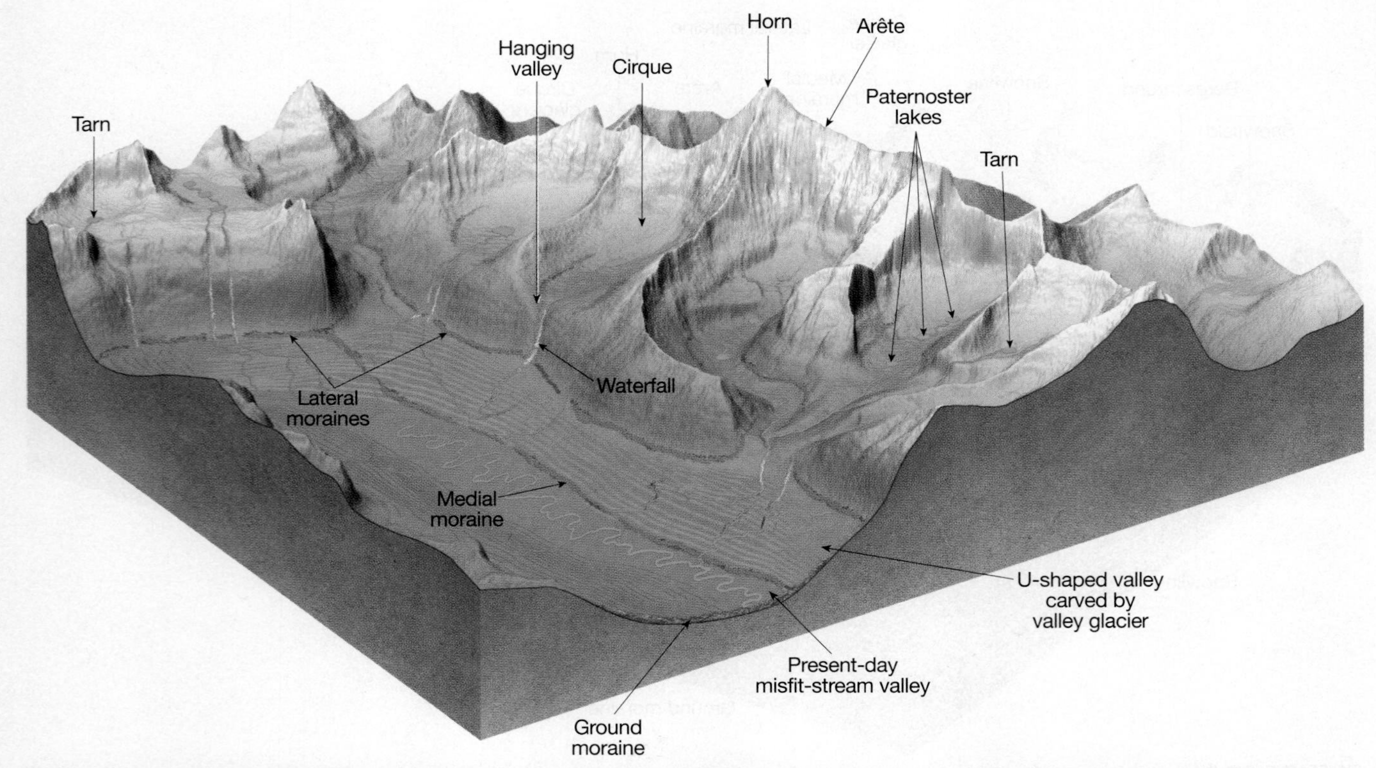

FIGURE 13.4 Erosional and depositional features of mountain glaciation. The same region as **FIGURE 13.3**, but showing erosion features remaining after total ablation (melting) of glacial ice.

ACTIVITY

13.4 Glacier National Park Investigation

THINK About It **How do glaciers affect landscapes? How is the cryosphere affected by climate change?**

OBJECTIVE Analyze glacial features in Glacier National Park and infer how glaciers there may change in the future.

PROCEDURES

1. **Before you begin**, read about Glacial National Park, Montana below. Also, this is **what you will need:**
   ____ calculator
   ____ Activity 13.4 Worksheet (p. 352) and pencil
2. **Then follow your instructor's directions** for completing the worksheets.

Glacier National Park, Montana

Glacier National Park is located on the northern edge of Montana, across the border from Alberta and British Columbia, Canada. Most of the erosional features formed by glaciation in the park developed during the Wisconsinan glaciation that ended about 11,700 years ago. Today, only small cirque glaciers exist in the park. Thirty-seven of them are named, and nine of those can be observed on the topographic map of part of the park in **FIGURE 13.14**.

ACTIVITY

13.5 Nisqually Glacier Response to Climate Change

THINK About It **How is the cryosphere affected by climate change?**

OBJECTIVE Evaluate the use of Nisqually Glacier as a global thermometer for measuring climate change.

PROCEDURES

1. **Before you begin**, read Nisqually Glacier—A Global Thermometer? Also, this is **what you will need:**
   ____ ruler
   ____ Activity 13.5 Worksheets (p. 353–354) and pencil
2. **Then follow your instructor's directions** for completing the worksheets.

EROSIONAL FEATURES OF GLACIATED REGIONS		MOUNTAIN GLACIATION	CONTINENTAL GLACIATION
Cirque	Bowl-shaped depression on a high mountain slope, formed by a cirque glacier	X	
Arête	Sharp, jagged, knife-edge ridge between two cirques or glaciated valleys	X	
Col	Mountain pass formed by the headward erosion of cirques	X	
Horn	Steep-sided, pyramid-shaped peak produced by headward erosion of several cirques	X	
Headwall	Steep slope or rock cliff at the upslope end of a glaciated valley or cirque	X	
Glacial trough	U-shaped, steep-walled, glaciated valley formed by the scouring action of a valley glacier	X	
Hanging valley	Glacial trough of a tributary glacier, elevated above the main trough	X	
Roche moutonnée	Asymmetrical knoll or small hill of bedrock, formed by glacial abrasion on the smooth stoss side (side from which the glacier came) and by plucking (prying and pulling by glacial ice) on the less-smooth lee side (down-glacier side)		X
Glacial striations and grooves	Parallel linear scratches and grooves in bedrock surfaces, resulting from glacial scouring	X	X
Glacial polish	Smooth bedrock surfaces caused by glacial abrasion (sanding action of glaciers analogous to sanding of wood with sandpaper)	X	X

FIGURE 13.5 Erosional features of mountain or continental glaciation.

Nisqually Glacier—A Global Thermometer?

Nisqually Glacier is one of many active valley glaciers that occupy the radial drainage of Mt. Rainier—an active volcano located near Seattle, Washington, in the Cascade Range of the western United States. Nisqually Glacier occurs on the southern side of Mt. Rainier and flows south toward the Nisqually River Bridge in **FIGURE 13.15**. The position of the glacier's terminus (downhill end) was first recorded in 1840, and it has been measured and mapped by numerous geologists since that time. The map in **FIGURE 13.15** was prepared by the U.S. Geological Survey in 1976 and shows where the terminus of Nisqually Glacier was located at various times from 1840 to 1997. (The 1994, 1997, and 2010 positions were added for this laboratory, based on NHAP aerial photographs and satellite imagery.) Notice how the glacier has more or less retreated up the valley since 1840.

Sea Ice

Sea ice is frozen ocean water. The largest masses of sea ice occur in the Arctic Ocean and around the continent of Antarctica (**FIGURE 13.16**). In both locations, the sea

ACTIVITY

13.6 The Changing Extent of Sea Ice

THINK About It | **How is the cryosphere affected by climate change?**

OBJECTIVE Measure how the extent of sea ice has changed annually in the past, predict how it may change in the future, and infer what benefits or hazards could result if Arctic sea ice continues to decline.

PROCEDURES

1. **Before you begin**, read Sea Ice. Also, this is **what you will need**:
   ____ 30 cm (12 in.) length of thread or thin string
   ____ ruler, calculator
   ____ Activity 13.7 Worksheets (pp. 355–356) and pencil
2. **Then follow your instructor's directions** for completing the worksheets.

DEPOSITIONAL FEATURES OF GLACIATED REGIONS		MOUNTAIN GLACIATION	CONTINENTAL GLACIATION
Ground moraine	Sheetlike layer (blanket) of till left on the landscape by a receding (wasting) glacier.	X	X
Terminal moraine	Ridge of till that formed along the leading edge of the farthest advance of a glacier.	X	X
Recessional moraine	Ridge of till that forms at terminus of a glacier, behind (up-glacier) and generally parallel to the terminal moraine; formed during a temporary halt (stand) in recession of a wasting glacier.	X	X
Lateral moraine	A body of rock fragments at or within the side of a valley glacier where it touches bedrock and scours the rock fragments from the side of the valley. It is visible along the sides of the glacier and on its surface in its ablation zone. When the glacier melts, the lateral moraine will remain as a narrow ridge of till or boulder train on the side of the valley.	X	
Medial moraine	A long narrow body of rock fragments carried in or upon the middle of a valley glacier and parallel to its sides, usually formed by the merging of lateral moraines from two or more merging valley glaciers. It is visible on the surface of the glacier in its ablation zone. When the glaciers melt, the medial moraine will remain as a narrow ridge of till or boulder train in the middle of the valley.	X	
Drumlin	An elongated mound or ridge of glacial till (unstratified drift) that accumulated under a glacier and was elongated and streamlined by movement (flow) of the glacier. Its long axis is parallel to ice flow. It normally has a blunt end in the direction from which the ice came and long narrow tail in the direction that the ice was flowing.		X
Kame	A low mound, knob, or short irregular ridge of stratified drift (sand and gravel) sorted by and deposited from meltwater flowing a short distance beneath, within, or on top of a glacier. When the ice melted, the kame remained.		X
Esker	Long, narrow, sinuous ridge of stratified drift deposited by meltwater streams flowing under glacial ice or in tunnels within the glacial ice		X
Erratic	Boulder or smaller fragment of rock resting far from its source on bedrock of a different type.	X	X
Boulder train	A line or band of boulders and smaller rock clasts (cobbles, gravel, sand) transported by a glacier (often for many kilometers) and extending from the bedrock source where they originated to the place where the glacier carried them. When deposited on different bedrock, the rocks are called erratics.	X	X
Outwash	Stratified drift (mud, sand and gravel) transported, sorted, and deposited by meltwater streams (usually muddy braided streams) flowing in front of (down-slope from) the terminus of the melting glacier.	X	X
Outwash plain	Plain formed by blanket-like deposition of outwash; usually an outwash braid plain, formed by the coalescence of many braided streams having their origins along a common glacial terminus.	X	X
Valley train	Long, narrow sheet of outwash (outwash braid plain of one braided stream, or floodplain of a meandering stream) that extends far beyond the terminus of a glacier.	X	
Beach line	Landward edge of a shoreline of a lake formed from damming of glacial meltwater, or temporary ponding of glacial meltwater in a topographic depression.		X
Glacial-lake deposits	Layers of sediment in the lake bed, deltas, or beaches of a glacial lake.		X
Loess	Unstratified sheets of clayey silt and silty clay transported beyond the margins of a glacier by wind and/or braided streams; it is compact and able to resist significant erosion when exposed in steep slopes or cliffs.		X

FIGURE 13.6 **Depositional features of mountain or continental glaciation.**

ice reaches its maximum thickness and extent during the winter months, then it melts back to a minimum extent and thickness during the summer months. In the northern hemisphere, Arctic sea ice reaches its minimum thickness and extent by September. Sea ice helps moderate Earth's climate, because its bright white surface reflects sunlight back into space. Without sea ice, the ocean absorbs the sunlight and warms up. Sea ice also provides the ideal environment for animals like polar bears, seals, and walruses to hunt, breed, and migrate as survival dictates. Some Arctic human populations rely on subsistence hunting of such species to survive.

WATER BODIES OF GLACIATED REGIONS		MOUNTAIN GLACIATION	CONTINENTAL GLACIATION
Tarn	Small lake in a cirque (bowl-shaped depression formed by a cirque glacier). A melting cirque glacier may also fill part of the cirque and may be in direct contact with or slightly up-slope from the tarn.	X	
Ice-dammed lake	Lake formed behind a mass of ice sheets and blocks that have wedged together and blocked the flow of water from a melting glacier and or river. Such natural dams may burst and produce a catastropic flood of water, ice blocks, and sediment.	X	X
Paternoster lakes	Chain of small lakes in a glacial trough.	X	
Finger lake	Long narrow lake in a glacial trough that was cut into bedrock by the scouring action of glacial ice (containing rock particles and acting like sand paper as it flows downhill) and usually dammed by a deposit of glacial gravel (end or recessional moraine).	X	X
Kettle lake or kettle hole	Small lake or water-saturated depression (10s to 1000s of meters wide) in glacial drift, formed by melting of an isolated, detached block of ice left behind by a glacier in retreat (melting back) or buried in outwash from a flood caused by the collapse of an ice-dammed lake.	X	X
Swale	Narrow marsh, swamp, or very shallow lake in a long shallow depression between two moraines.		X
Marginal glacial lake	Lake formed at the margin (edge) of a glacier as a result of accumulating meltwater; the upslope edge of the lake is the melting glacier itself.	X	X
Meltwater stream	Stream of water derived from melting glacial ice, that flows under the ice, on the ice, along the margins of the ice, or beyond the margins of the ice.	X	X
Misfit stream	Stream that is not large enough and powerful enough to have cut the valley it occupies. The valley must have been cut at a time when the stream was larger and had more cutting power or else it was cut by another process such as scouring by glacial ice.	X	X
Marsh or swamp	Saturated, poorly drained areas that are permanently or intermittently covered with water and have grassy vegetation (marsh) or shrubs and trees (swamp).	X	X

FIGURE 13.7 Water bodies resulting from mountain or continental glaciation.

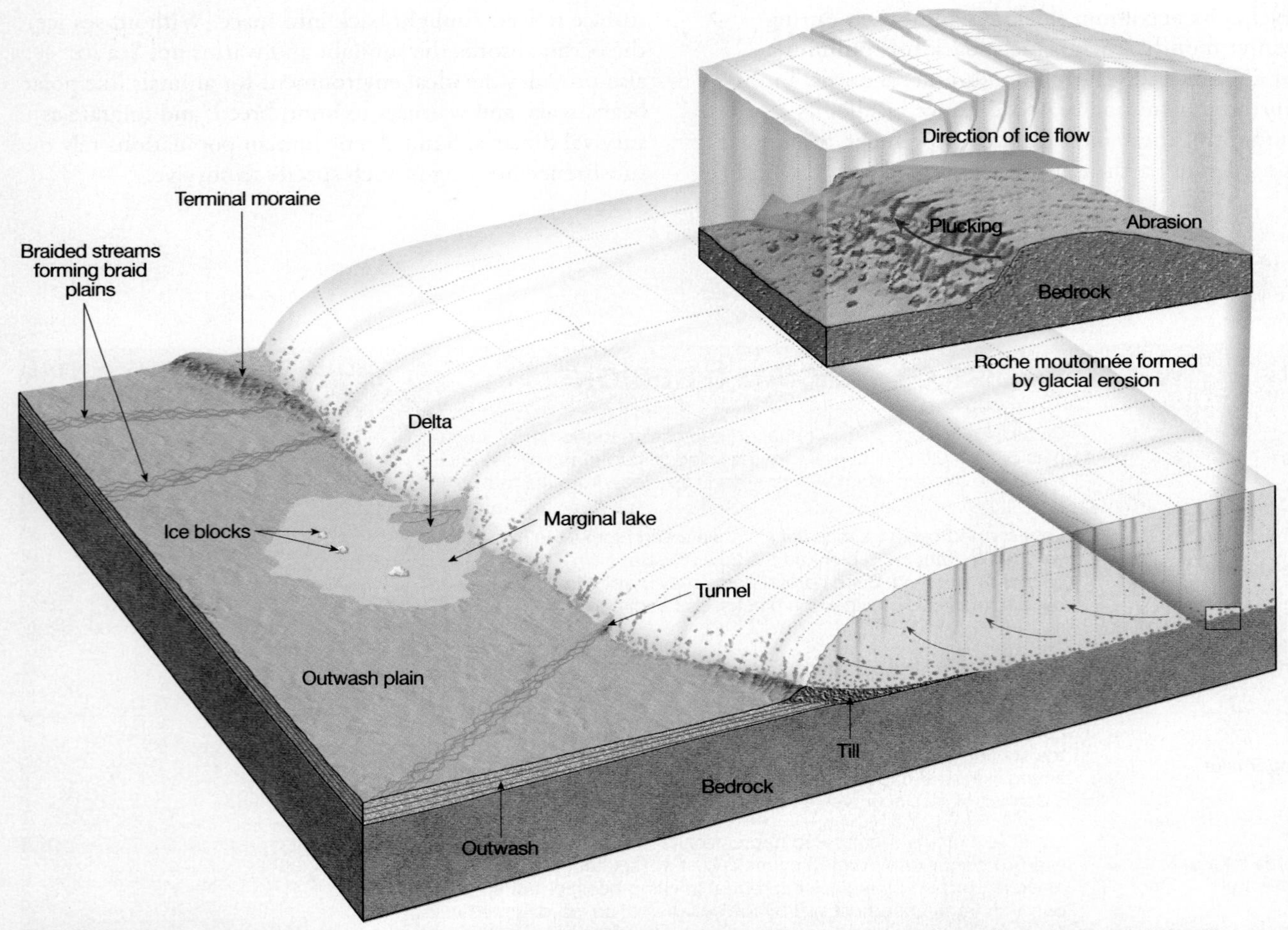

FIGURE 13.8 Continental glaciation in a hypothetical region. Continental glaciation produces these characteristic landforms at the beginning of ice wastage (decrease in glacier size due to severe ablation).

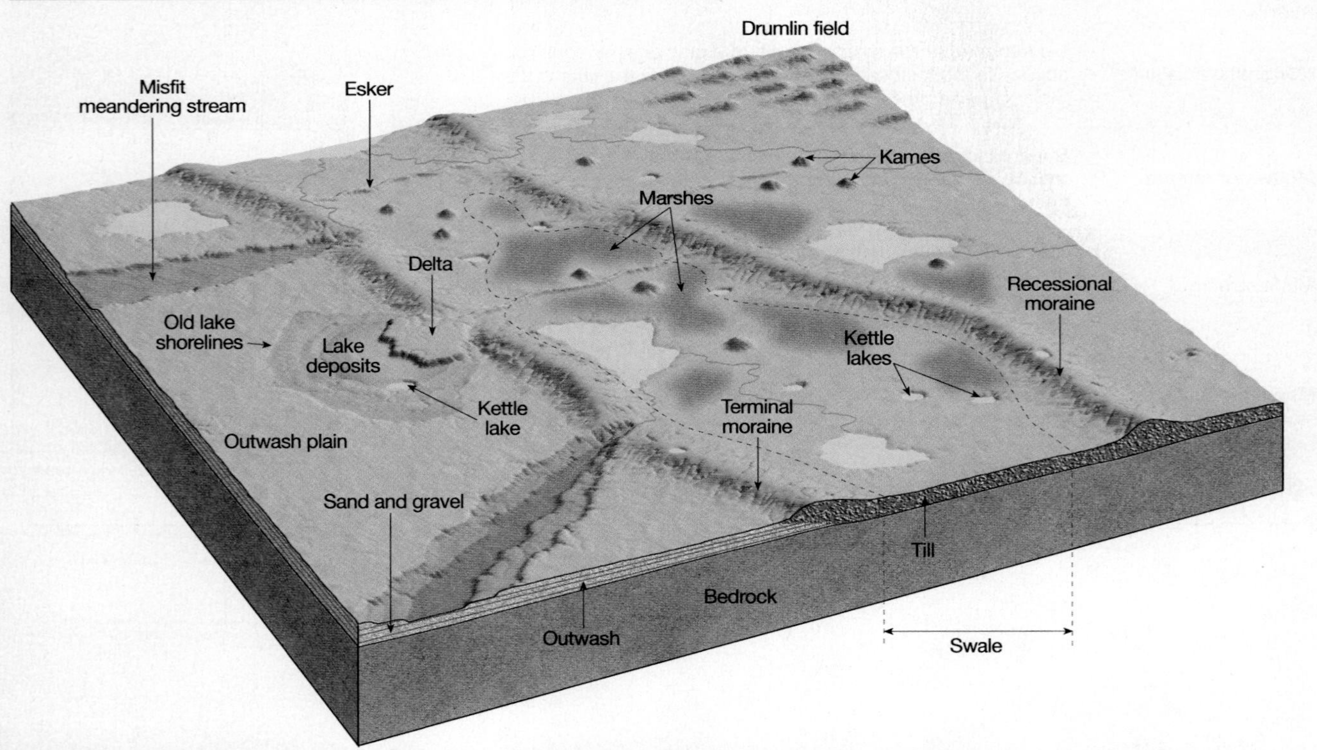

FIGURE 13.9 Erosional and depositional features of continental glaciation. Continental glaciation leaves behind these characteristic landforms after complete ice wastage. (Compare to **FIGURE 13.8.**)

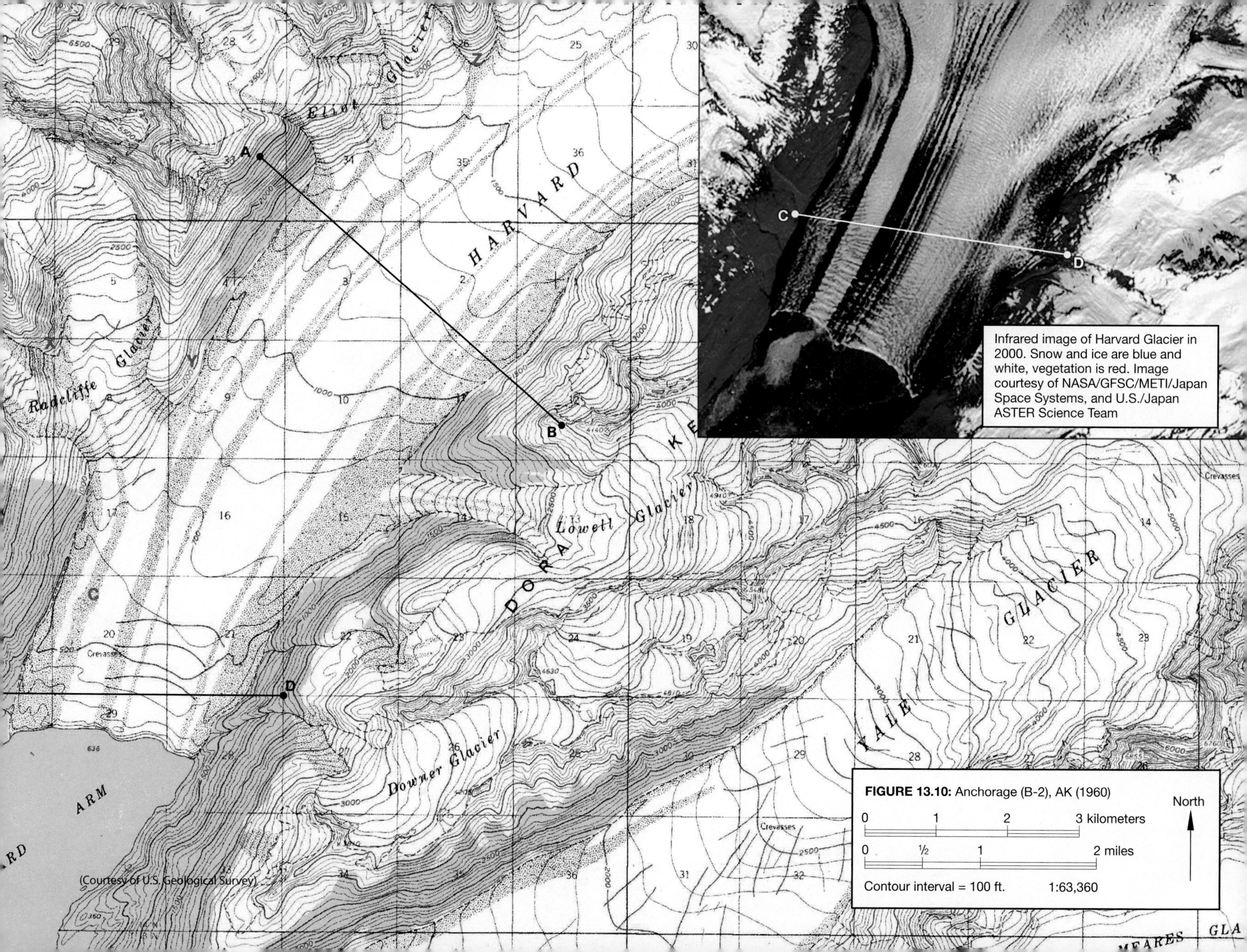

FIGURE 13.10: Anchorage (B-2), AK (1960)

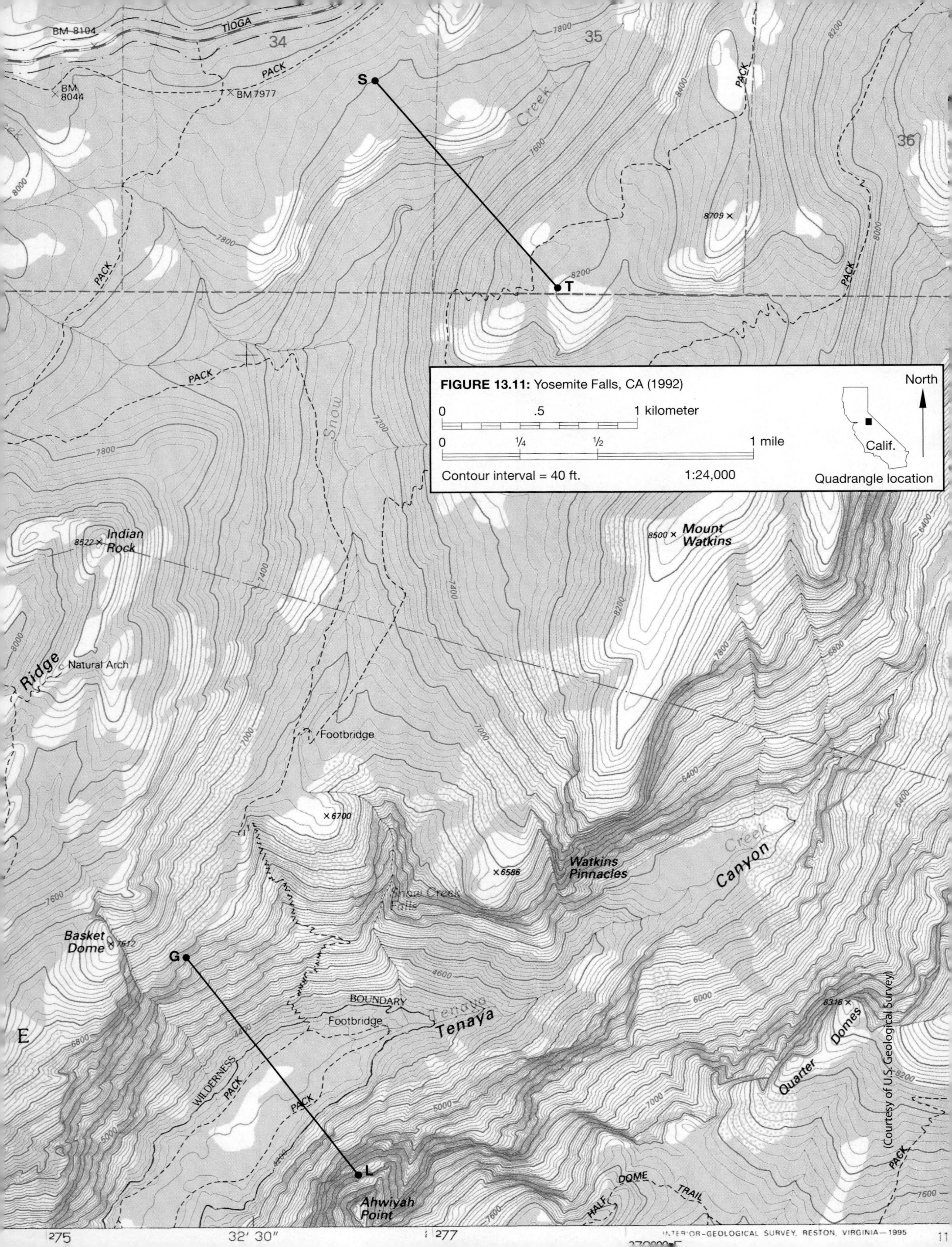
FIGURE 13.11: Yosemite Falls, CA (1992)
0
.5
1 kilometer
0
1/4
1/2
1 mile
Contour interval = 40 ft.
1:24,000
North
Calif.
Quadrangle location
(Courtesy of U.S. Geological Survey)
BM 8104
TIOGA
34
35
36
PACK
BM 8044
BM 7977
S
T
L
G
Creek
Snow
8709
8522
Indian Rock
Natural Arch
Ridge
8500
Mount Watkins
Footbridge
6700
6586
Watkins Pinnacles
Snow Creek Falls
Canyon
Creek
Basket Dome
7612
BOUNDARY
Footbridge
Tenaya
WILDERNESS
8318
Quarter Domes
Ahwiyah Point
HALF DOME TRAIL
E
275
32′ 30″
277
INTERIOR—GEOLOGICAL SURVEY, RESTON, VIRGINIA—1995

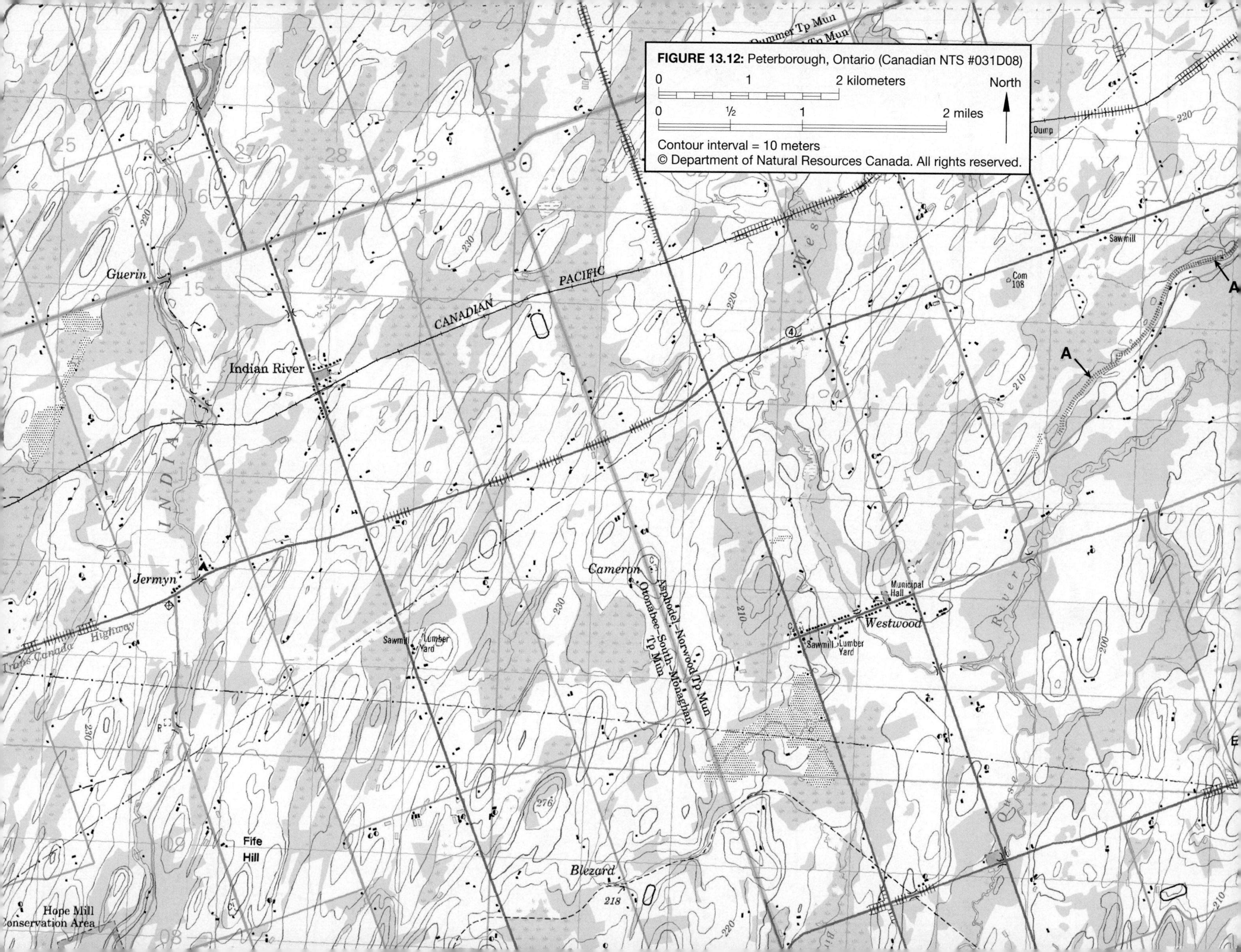

FIGURE 13.12: Peterborough, Ontario (Canadian NTS #031D08)

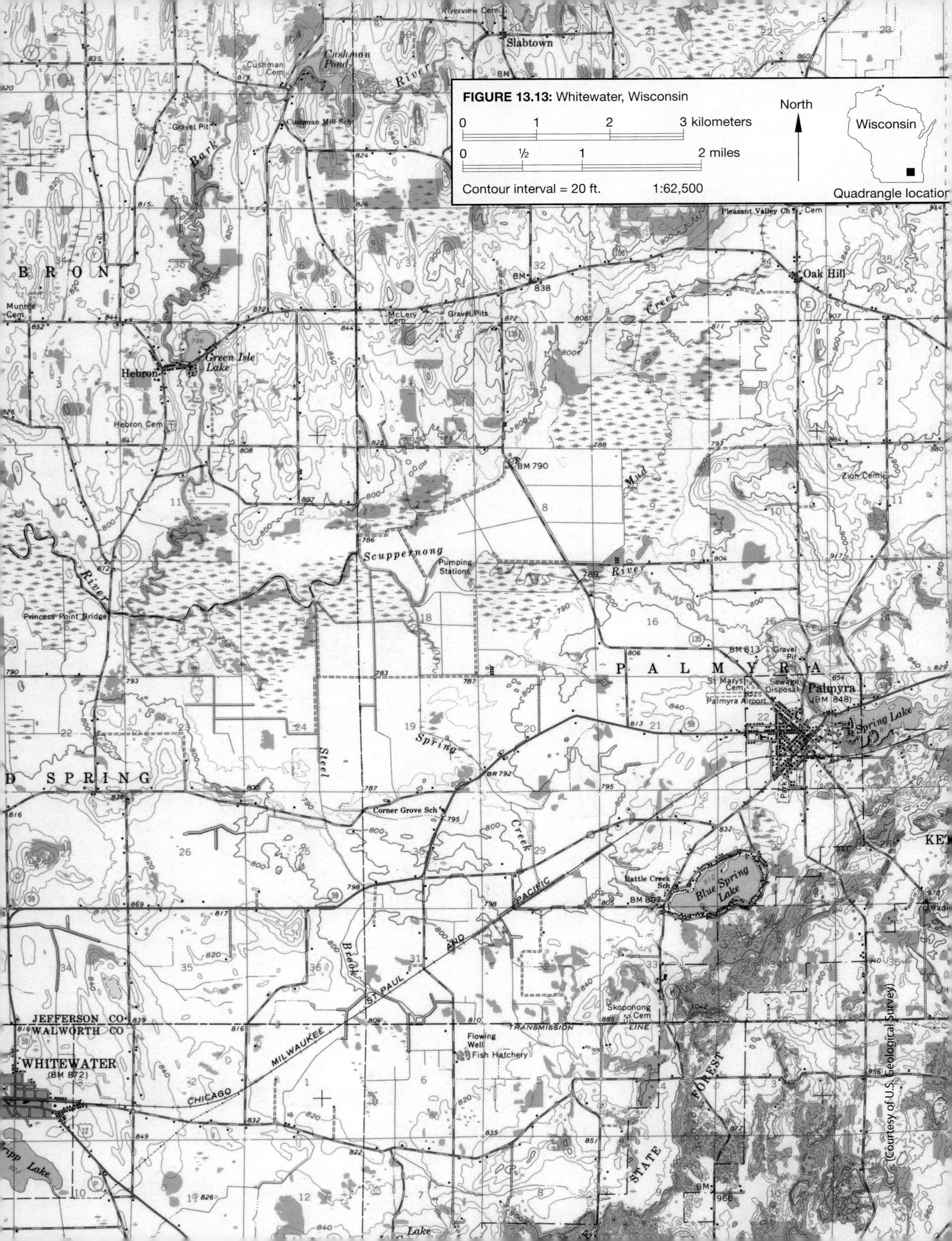

FIGURE 13.13: Whitewater, Wisconsin

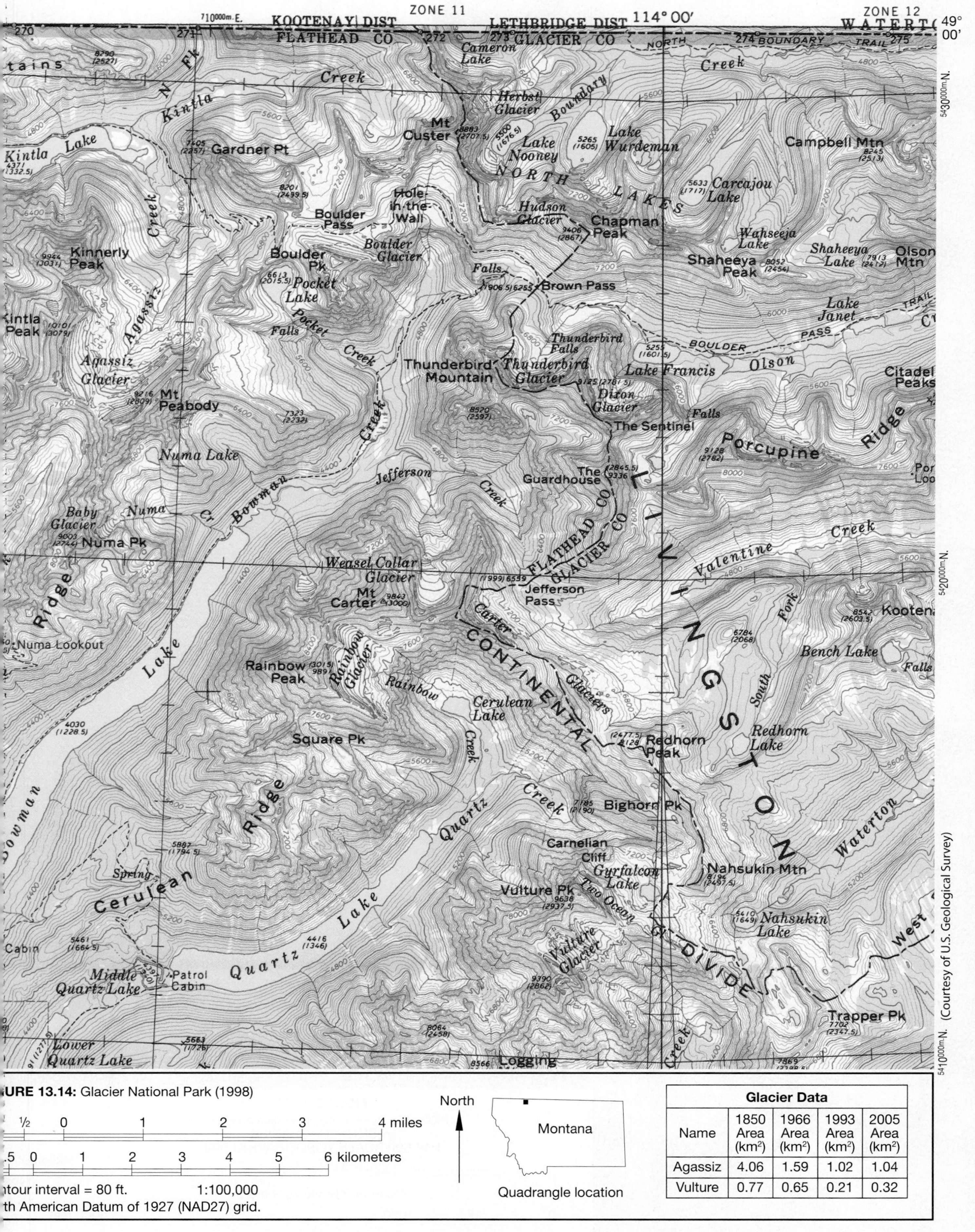

URE 13.14: Glacier National Park (1998)

Glacier Data

Name	1850 Area (km²)	1966 Area (km²)	1993 Area (km²)	2005 Area (km²)
Agassiz	4.06	1.59	1.02	1.04
Vulture	0.77	0.65	0.21	0.32

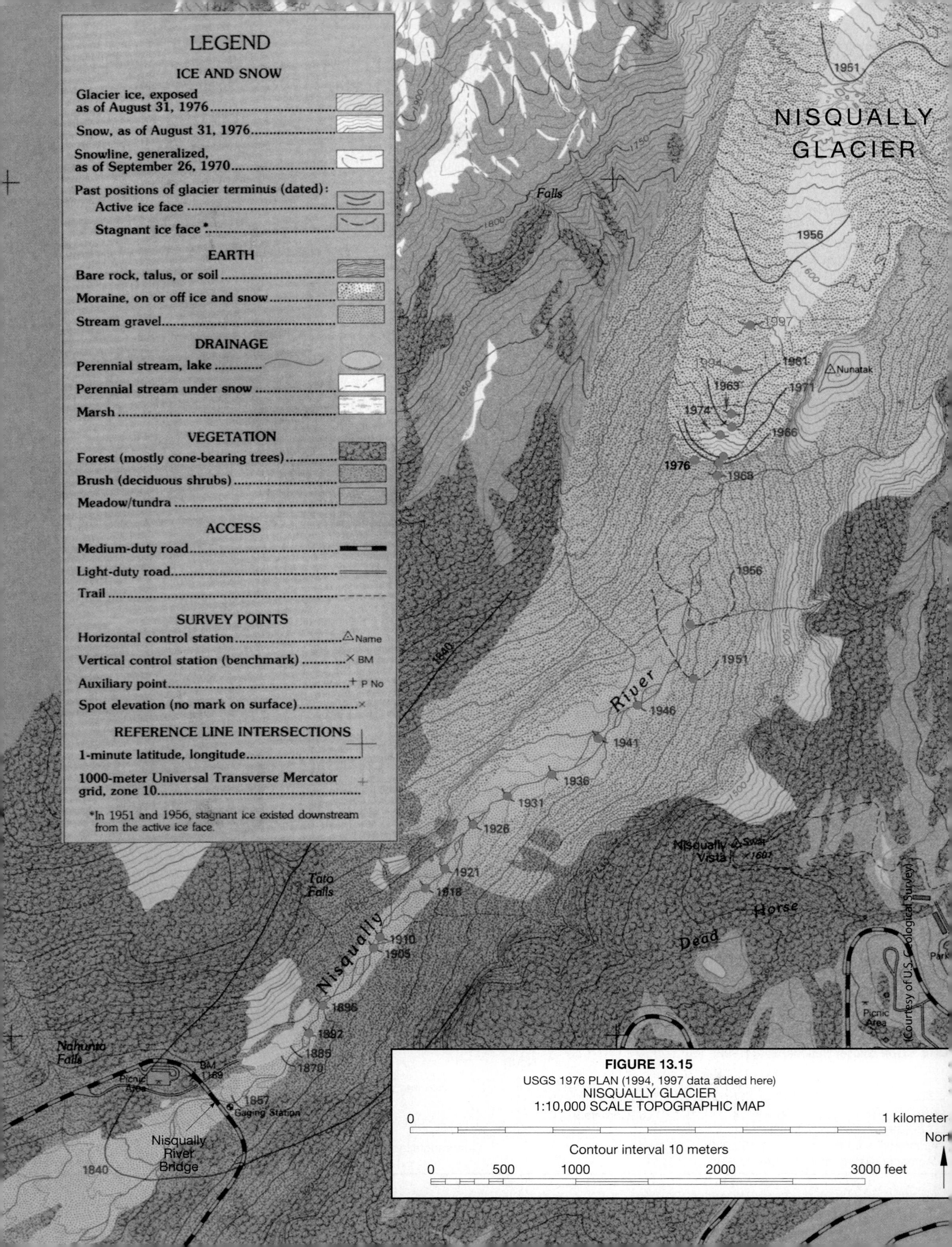

FIGURE 13.15
USGS 1976 PLAN (1994, 1997 data added here)
NISQUALLY GLACIER
1:10,000 SCALE TOPOGRAPHIC MAP

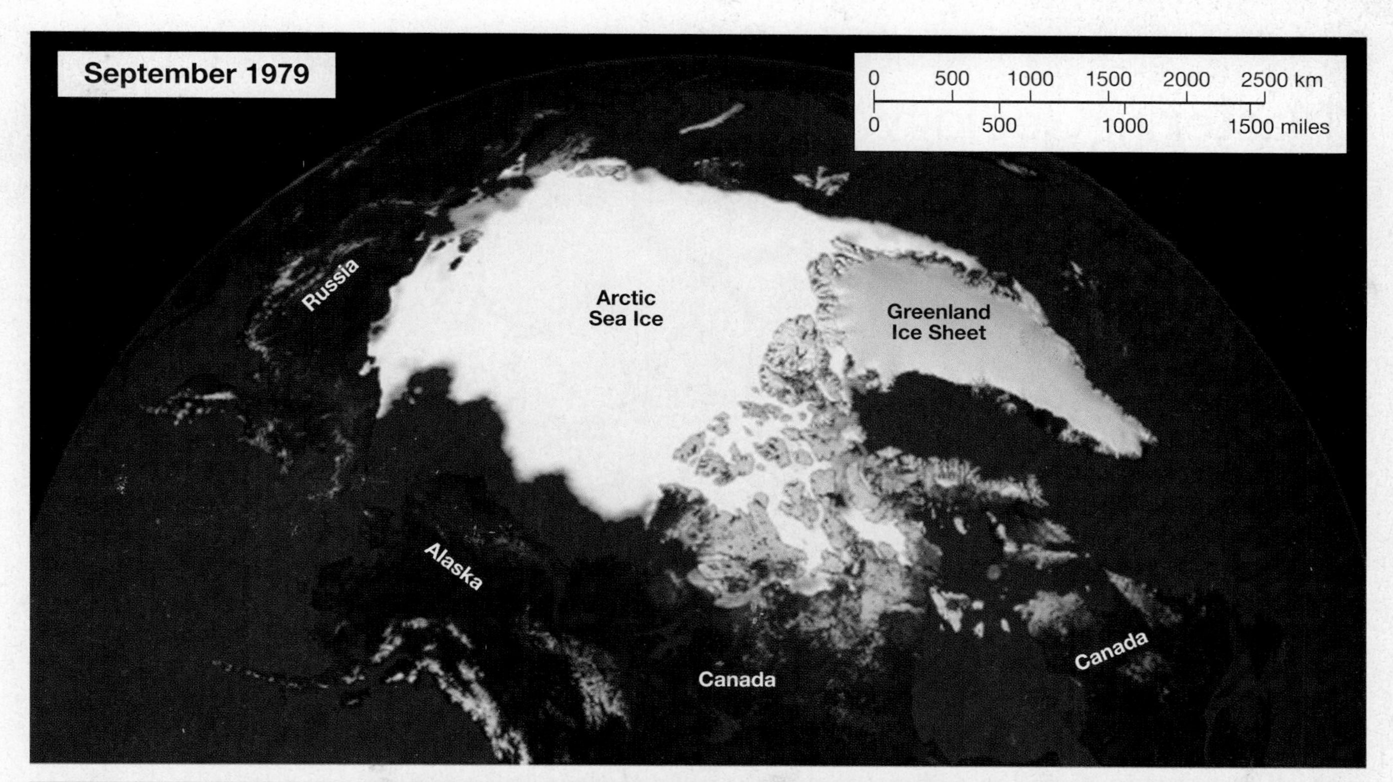

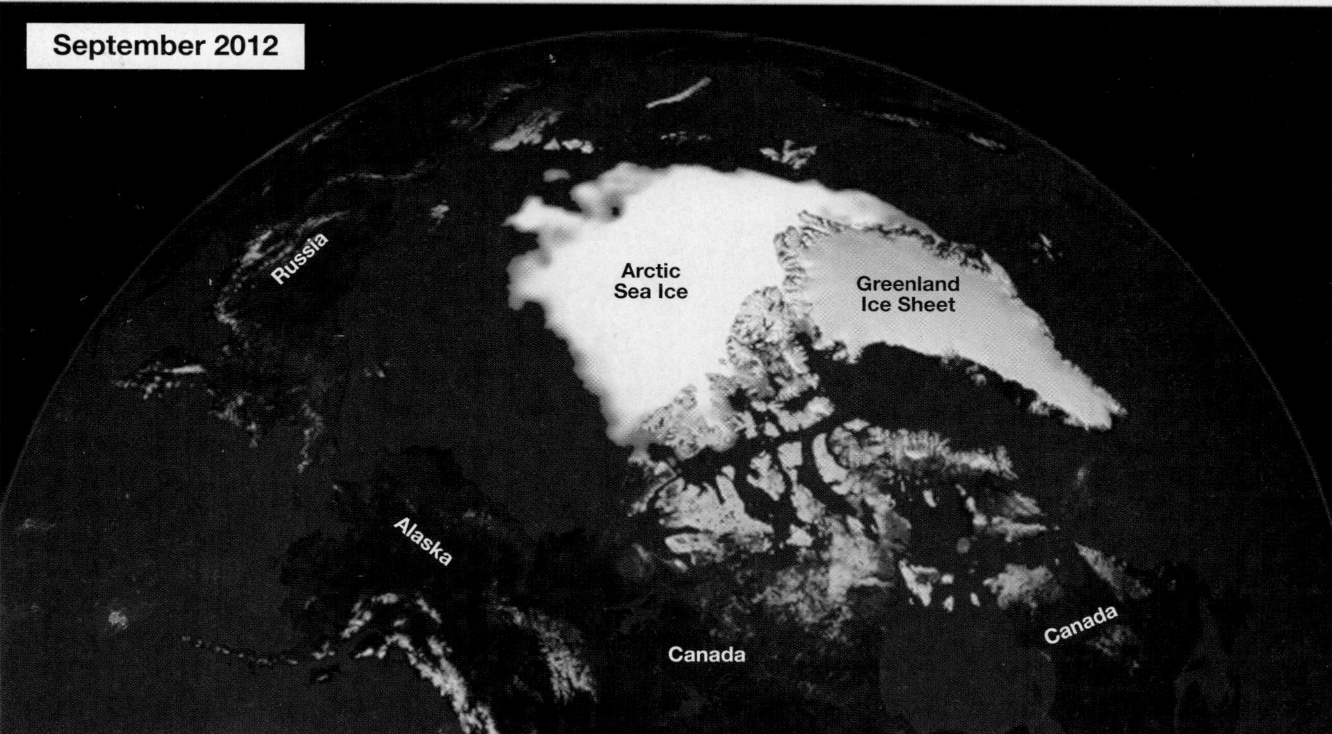

FIGURE 13.16 Extent of Arctic Sea Ice: 1979 and 2012. Sea ice covers essentially all of the Arctic Ocean in winter months, but it melts back to a minimum thickness and extent by the end of summer (September). These NASA satellite images reveal the minimum extent of Arctic sea ice at times 33 years apart. Dark blue areas are ocean; gray areas are mountain glaciers and the Greenland Ice Sheet. White and light blue areas are the Arctic sea ice.

ACTIVITY 13.1 Cryosphere Inquiry

Name: ______________________ **Course/Section:** ______________________ **Date:** ____________

A. The cryosphere is all of Earth's snow and ice.

1. In **FIGURE 13.1**, what is the sequence of cryosphere regions that you would encounter on the ground if you traveled from Mexico (a beige- to yellow-colored region with no snow or ice) to the North Pole?

2. Notice in **FIGURE 13.1** that mountain glaciers and ice caps occur in parts of Greenland, Canada, Russia, Alaska, and the western conterminous United States. Some mountain glaciers also exist very close to the equator (not shown in **FIGURE 13.1**). How do you think it is possible for glaciers to exist at the equator?

3. If the temperature of Earth's atmosphere were to rise, then how do you think it would affect the cryosphere, hydrosphere, and biosphere?

4. If the temperature of Earth's atmosphere were to cool, then how do you think it would affect the cryosphere, hydrosphere, and biosphere?

B. Snow and glaciers are two of the best known parts of the cryosphere. Notice the snow and glaciers in the satellite image on the next page. It is a perspective view, looking north, of part of the Himalayan Mountains and was made by draping an ASTER natural color satellite image over a digital elevation model (by the NASA/GSFC/METI/ERSDAC/JAROS and U.S./Japan ASTER Science Team). You can view the same region in Google Earth™ by searching for coordinates 28 09 38 N, 90 03 05 E.

The glaciers in the satellite image formed by compaction and recrystallization of snow at higher elevations. Then they flowed downhill, where they eventually melt. A glacier's **mass balance** is the difference between the mass of its ice that is accumulating and the mass of ice that is melting. If a glacier has more ice accumulating than melting, then it has a positive mass balance and will advance downhill. If a glacier melts faster than it accumulates ice, then it has a negative mass balance and will retreat (melt back).

1. The satellite image was acquired in summer of 2009, after most of the seasonal snow had melted. Using a pen, draw a line along the **snowline**—the line between areas with snow (higher elevations) and areas with no snow.
2. Place arrows on the glaciers to show their direction of flow, like a river of ice.
3. Label the "area of snow and ice accumulation" and two "areas of ablation" (glacial melting).
4. Label the area where the glaciers have "positive mass balance" and the areas where the glaciers have a "negative mass balance."
5. Is the mass balance of the snowline that you drew in part B1 positive, negative, or neither? Why?

C. Refer to **FIGURE 13.2**, an ASTER satellite image of a 20-by-20 km area of southern Alaska. It is an infrared image, so vegetation appears red, glacial ice is blue, and snow is white.

1. Where is the zone of ablation in this image, and how can you tell?

2. Name two resources (used by humans) that were created by the glaciers in **FIGURE 13.2**?

D. **REFLECT & DISCUSS** In what ways have the glaciers affected the landscape in the above image, and what does it suggest about how extensive these glaciers must have been in the past?

ACTIVITY 13.2 Mountain Glaciers and Glacial Landforms

Name: ______________________ **Course/Section:** ______________________ **Date:** __________

A. **FIGURE 13.10** is a topographic map of modern mountain glaciation near Anchorage, Alaska. **FIGURE 13.11** is a topographic map of the southeast part of the Yosemite Falls, California quadrangle (Yosemite National Park), which was shaped by Pleistocene glaciers (that have since melted away) and modern streams.

1. On the left-hand side of the graph below, construct and label a topographic profile for line ***S–T*** *across a valley on* **FIGURE 13.11** *that is being cut by a modern river.* Refer to **FIGURE 9.16** (Topographic Profile Construction), if needed.

2. On the left-hand side of the graph below, also construct and label a topographic profile for line ***G–L*** *across a valley on* **FIGURE 13.11** *that was scoured and shaped by a Pleistocene valley glacier.*

3. Which of the above cross sections (river or glacial) is "V" shaped: ______________________

Describe the erosional process that you think causes this shape.

4. Which of the above cross sections (river or glacial) is "U" shaped: ______________________

Describe the erosional process that you think causes this shape.

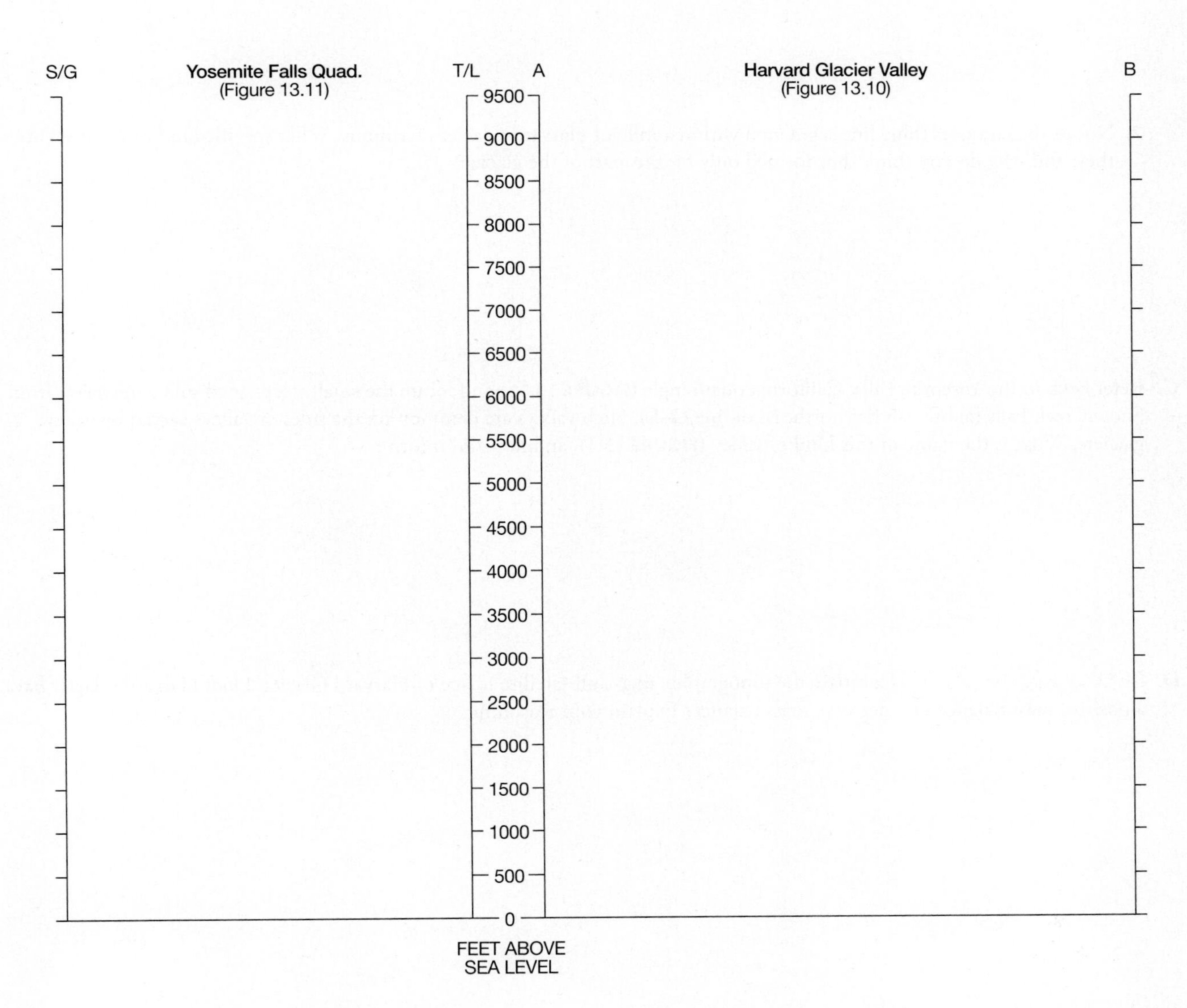

5. On the right-hand side of the graph on page 349, complete the topographic profile for line ***A–B*** *across the Harvard Glacier* (**FIGURE 13.10**).

   a. Label the part of the profile that is the top surface of the glacier.

   b. Using a dashed line, draw where you think the rock bottom of the valley is located under the Harvard Glacier. (Your drawing may extend slightly below the figure.)

   c. Based on the profile that you just constructed, what is the maximum thickness of Harvard Glacier along line **A–B?**

B. Refer back to **FIGURE 13.10**, a portion of the Anchorage (B-2), Alaska, quadrangle, for the following questions. In the southwestern corner, note the Harvard Arm of Prince William Sound. The famous *Exxon Valdez* oil spill occurred just south of this area (it did not affect Harvard Arm).

1. Lateral and medial moraines in/on the ablation zone of Harvard Glacier are indicated by the stippled (finely dotted) pattern on parts of the glacier. If a hiker found gold in rock fragments on the glacier at location **C,** then would you look for gold near location **X, Y,** or **Z?** Explain your reasoning.

2. Notice the crevasses (blue line segments) within a mile of Harvard Glacier's terminus. What specific kind of crevasses are they, and why do you think they formed only on this part of the glacier?

C. Refer back to the Yosemite Falls, California quadrangle (**FIGURE 13.11**) and locate the small steep-sided valley upstream from Snow Creek Falls (about 1.5 km northeast of line **G–L**). Such valleys are common on the sides of valleys carved by valley glaciers. What is the name of this kind of valley (**FIGURE 13.4**), and how did it form?

D. REFLECT & DISCUSS Compare the topographic map and satellite image of Harvard Glacier. Does Harvard Glacier have a positive mass balance or a negative mass balance? Explain your reasoning.

ACTIVITY 13.3 Continental Glaciation of North America

Name: ____________________ **Course/ection:** ____________________ **Date:** __________

A. Refer to **FIGURE 13.12**, part of the Peterborough, Ontario, quadrangle, for the following questions. This area lies north of Lake Ontario.

1. Study the size and shape of the short, oblong rounded hills. Fieldwork has revealed that they are made of till. What type of feature are they and how did they form?

2. Find the long narrow hill labeled **A.** It is marked by a symbol made of a long line of tiny pairs of brown dots. What would you call this linear feature, and how do you think it formed?

3. Towards what direction did the glacial ice flow here, and how can you tell?

B. The most recent glaciation of Earth is called the *Wisconsinan glaciation.* It reached its maximum development about 18,000 years ago, when a "*Laurentide Ice Sheet*" covered central and eastern Canada, the Great Lakes Region, and the northeastern United States. It ended by about 11,700 years ago, at the start of the Holocene Epoch. Refer to **FIGURE 13.13**, a portion of the Whitewater, Wisconsin, quadrangle.

1. List the features of glaciated regions from **FIGURES 13.8** and **13.9** that are present in this region.

2. Describe in what direction the ice flowed over this region. Cite evidence for your inference.

3. What kinds of lakes are present in this region, and how did they form? (Refer to **FIGURE 13.7**.)

4. In the southeastern corner of the map, the northwest-trending forested area is probably what kind of feature?

5. Note the swampy and marshy area running from the west-central edge of the map to the northeastern corner. Describe the probable origin of this feature (more than one answer is possible).

C. REFLECT & DISCUSS How are the glaciated areas of **FIGURES 13.12** and **13.13** different from areas affected by mountain glaciation and how are they they same?

ACTIVITY 13.4 Glacier National Park Investigation

Name: ______________________ **Course/Section:** ______________________ **Date:** __________

Refer to the map of Glacier National Park in **FIGURE 13.14.**

A. List the features of glaciation from **FIGURES 13.3, 13.4, 13.8,** or **13.9** that are present in **FIGURE 13.14.**

B. Locate Quartz Lake and Middle Quartz Lake in the southwest part of the map. Notice the Patrol Cabin located between these lakes. Describe the chain of geologic/glacial events (steps) that led to formation of Quartz Lake, the valley of Quartz Lake, the small piece of land on which the Patrol Cabin is located, and the cirque in which Rainbow Glacier is located today.

C. Based on your answers above, what kind of glaciation (mountain versus continental) has shaped this landscape?

D. Locate the Continental Divide and think of ways that it may be related to weather and climate in the region. Recall that weather systems generally move across the United States from west to east.

1. Describe how modern glaciers of this region are distributed in relation to the Continental Divide.

2. Based on the distribution you observed, describe the weather/climate conditions that may exist on opposite sides of the Continental Divide in this region.

E. Describe how the size (area in km^2) of Agassiz Glacier changed from 1850 to 2005.

F. Describe how the size (area in km^2) of Vulture Glacier changed from 1850 to 2005.

G. REFLECT & DISCUSS What do you expect the area (km^2) of Agassiz and Vulture Glaciers to be in 2020? Explain.

ACTIVITY 13.5 Nisqually Glacier Response to Climate Change

Name: ______________________ **Course/Section:** ______________________ **Date:** ____________

A. Refer to FIGURE 13.15 and fill in the Nisqually Glacier Data Chart below. To do this, use a ruler and the map's bar scale to measure the distance in kilometers from Nisqually River Bridge to the position of the glacier's terminus (red dot) for each year of the chart. Be sure to record your distance measurements to two decimal points (hundredths of km).

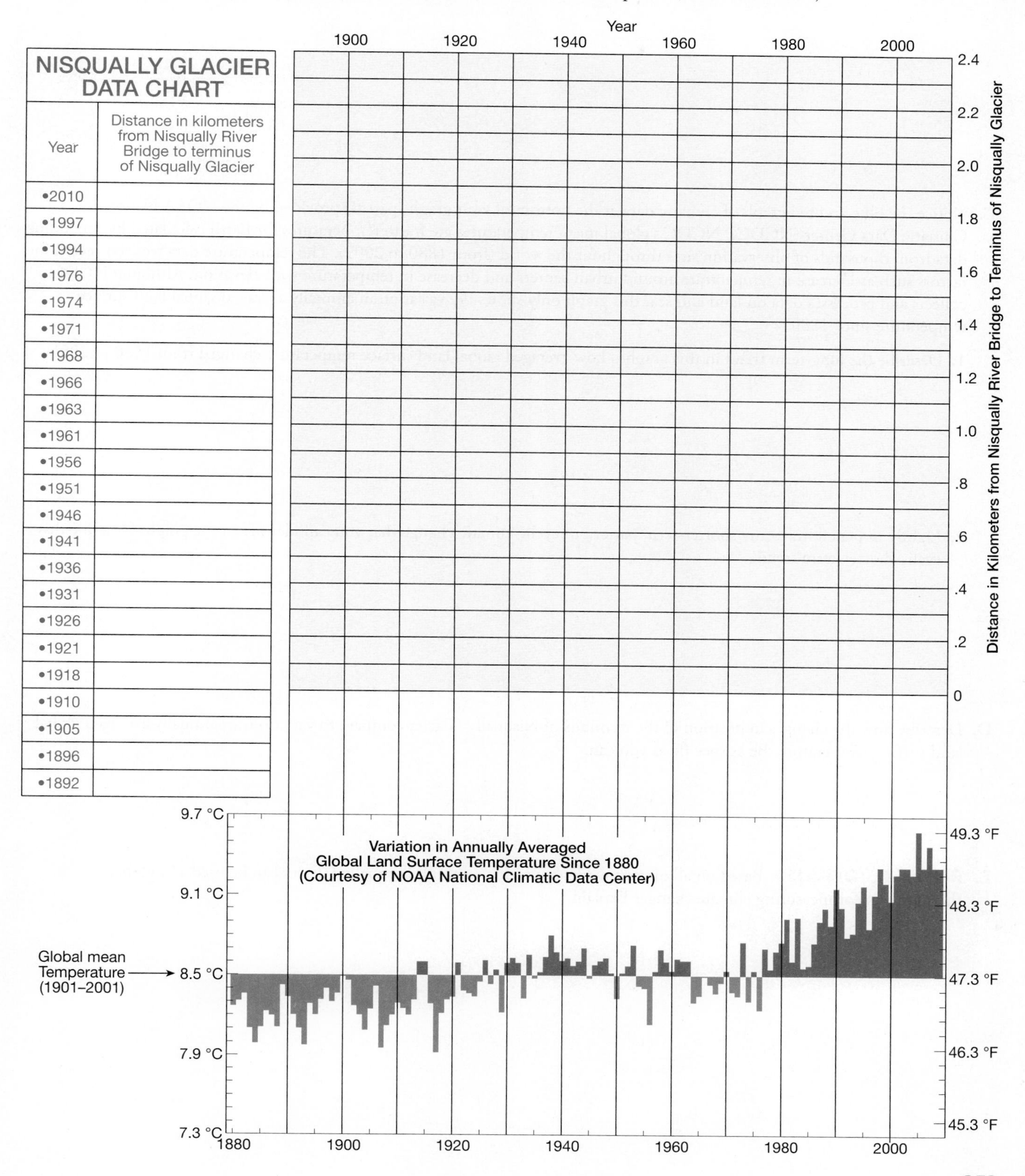

NISQUALLY GLACIER DATA CHART

Year	Distance in kilometers from Nisqually River Bridge to terminus of Nisqually Glacier
•2010	
•1997	
•1994	
•1976	
•1974	
•1971	
•1968	
•1966	
•1963	
•1961	
•1956	
•1951	
•1946	
•1941	
•1936	
•1931	
•1926	
•1921	
•1918	
•1910	
•1905	
•1896	
•1892	

B. Plot your data from part A (Nisqually Glacier Data Chart) in the graph to the right of the data chart. After plotting each point of data, connect the dots with a smooth, light pencil line. Notice that the glacier terminus retreated up the valley at some times, but advanced back down the valley at other times. Summarize these changes in a chart or paragraph, relative to specific years of the data.

C. Notice the blue and red graph of climatic data at the bottom of your graph (part B) provided by the NOAA National Climatic Data Center (NCDC). NCDC's global mean temperatures are mean temperatures for Earth calculated by processing data from thousands of observation sites throughout the world (from 1880 to 2009). The temperature data were corrected for factors such as increase in temperature around urban centers and decrease in temperature with elevation. Although NCDC collects and processes data on land and sea, this graph only shows the variation in annually averaged global land surface temperature since 1880.

1. Describe the long-term trend in this graph—how averaged global land surface temperature changed from 1880 to 2005.

2. Lightly in pencil, trace any shorter-term pattern of cyclic climate change that you can identify in the graph. Describe this cyclic shorter-term trend.

D. Describe how the changes in position of the terminus of Nisqually Glacier compare to variations in annually averaged global land surface temperature. Be as specific as you can.

E. **REFLECT & DISCUSS** Based on all of your work above, do you think Nisqually Glacier can be used as a global thermometer for measuring climate change? Explain.

ACTIVITY 13.6 The Changing Extent of Sea Ice

Name: ______________________ **Course/Section:** ______________________ **Date:** ____________

A. Refer to the satellite images of Arctic sea ice in **FIGURE 13.16** on page 345. These images were both taken in the month of September, when sea ice is at its minimum thickness and extent. "Extent" refers to how far the ice extends in all directions: its total area without regard for tiny ice free areas within the overall body of sea ice. It is easy to see that there was less sea ice in September of 2012 than in September of 1979, but how much less? To find out, you need to measure the extent of the sea ice in 1979 and 2012 by following these directions.

Step 1. Start by using a piece of thin string or thread about 30 cm (12 in.) long to measure the circumference of the sea ice. Carefully lay the string along the edge of the body of sea ice so that the string totally surrounds it as perfectly as possible. Then lay that length (segment) of string along the bar scale to determine the circumference of the sea ice in kilometers.

Step 2. Assume that the circumference of sea ice that you just measured is like the circumference of a circle. Circumference of a circle is equal to 2 times π (3.14) times radius, so radius (r) equals circumference divided by 2 times π. So, determine the radius of the ice sheet by dividing the circumference (Step 1) by 6.28.

Step 3. Area of a circle is equal to π (3.14) times the square of its radius. So determine the area of the ice sheet by multiplying the radius from step 2 by itself (to get radius squared), and then multiply that number by 3.14. Your answer will be in square kilometers (km^2).

1. Using the three steps above, what was the extent of Arctic sea ice in September of 1979, in millions of km^2? Show your work below.

2. Using the three steps above, what was the extent of Arctic sea ice in September of 2012, in millions of km^2? Show your work below.

3. Based on your limited set consisting of just two years of data, what has been the rate of Arctic sea ice decline from 1979 to 2012 (in km per year)? Show your work.

B. Scientists at the National Snow and Ice Data Center (NSIDC) have measured the annual September extent of Arctic and Antarctic sea ice in more exact ways. Arctic sea ice fills the Arctic Ocean, which is confined by land masses like Asia (Russia), North America, and Greenland (**FIGURE 13.1**). A table of the NSIDC data is provided on the next page.

1. Graph all of the data for extent of Arctic sea ice from 1979 to 2013, then use a ruler to draw a "best fit" line through the points so that the number of, and distance to, points above and below the line is similar. Label the line as "Arctic."

2. What was the average annual extent of Arctic sea ice from 2000 to 2007, in millions of km^2? Show your work.

3. What was the average annual extent of Arctic sea ice from 2008 to 2013, in millions of km^2? Show your work.

4. Based on your graph and calculations above, would you say that the annual amount of Arctic sea ice is decreasing, increasing, or staying about the same? Explain.

5. What do you predict the extent of Arctic sea ice will be in 2015?

6. Graph all of the data for extent of Antarctic sea ice from 1979 to 2013, then use a ruler to draw a "best fit" line through the points. Label the line as "Antarctic."

7. What was the average annual extent of Antarctic sea ice from 2000 to 2007, in millions of km^2? Show your work.

8. What was the average annual extent of Antarctic sea ice from 2008 to 2013, in millions of km^2? Show your work.

9. Based on your graph and calculations above, would you say that the annual amount of Antarctic sea ice is decreasing, increasing, or staying about the same? Explain.

Summer Extent of Sea Ice

Year	Arctic Extent in millions of square kilometers	Antarctic Extent in millions of square kilometers
2000	3.4	19.1
2001	3.5	18.4
2002	6.0	18.2
2003	6.2	18.6
2004	6.1	19.1
2005	5.6	19.2
2006	6.0	19.3
2007	4.3	19.2
2008	4.7	18.5
2009	5.4	19.1
2010	4.9	19.2
2011	4.6	18.9
2012	3.6	19.4
2013	5.4	19.8

(Courtesy of National Snow and Ice Data Center (NSIDC))

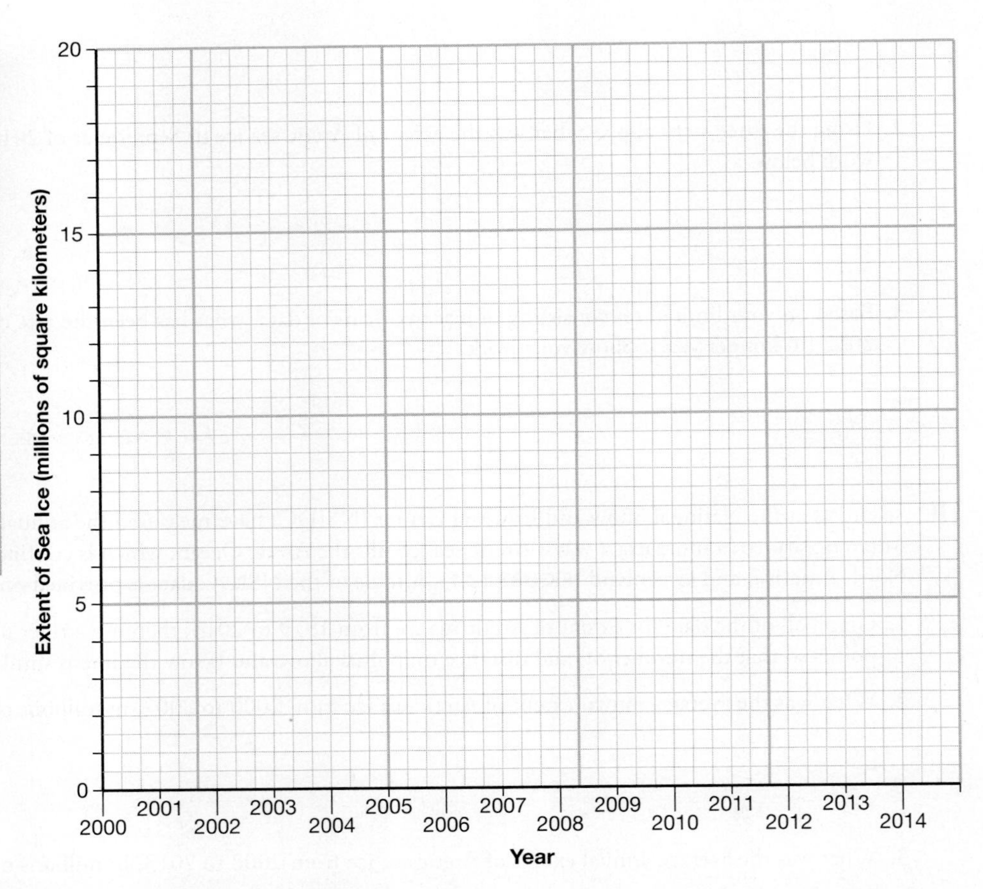

C. **REFLECT & DISCUSS** The changes in Arctic sea ice extent over time are not the same as the Antarctic changes. Why do you think the two bodies of sea ice are so different, and what benefits or hazards could result if the Arctic sea ice continues to decline?

PRE-LAB VIDEO

BIG IDEAS

Drylands are lands of arid-to-dry, subhumid climates that generally have sparse vegetation and receive precipitation just a few days or one season of the year. Even so, water is one of the primary agents that produces characteristic dryland landforms and flood hazards. Wind is also a factor in the erosion and transportation of sediment, especially dust and the sand that makes dunes. Although many people live in drylands, true deserts do not support any agriculture without irrigation or a well.

FOCUS YOUR INQUIRY

THINK About It | What are some characteristic processes, landforms, and hazards of drylands?

ACTIVITY 14.1 Dryland Inquiry *(p. 358)*

THINK About It | What can we learn from topographic maps and satellite images about dryland processes and landforms?

ACTIVITY 14.2 Mojave Desert, Death Valley, California *(p. 358)*

ACTIVITY 14.3 Sand Seas of Nebraska and the Arabian Peninsula *(p. 363)*

THINK About It | How can topographic maps and aerial photographs of drylands be used to interpret how their environments have changed?

ACTIVITY 14.4 Dryland Lakes of Utah *(p. 365)*

LABORATORY 14

Dryland Landforms, Hazards, and Risks

CONTRIBUTING AUTHORS

Charles G. Oviatt • *Kansas State University*

James B. Swinehart • *Institute of Agriculture & Natural Resources, University of Nebraska*

James R. Wilson • *Weber State University*

Rain is rare in drylands, so there are few plants to trap and bind loose rocks among their roots or aid in the development of soil that would absorb rainwater. When it does rain, flash floods cut channels and shape the landscape. (Photo by Michael Collier)

Introduction

Drylands are lands in arid, semi-arid, and dry-sub-humid climates. The United Nations Environment Programme (UNEP) estimates that drylands make up 41% of all land on Earth and that they support one-third of the world's human population. Sixteen percent of all existing drylands (about 6% of all land areas on Earth) are so dry that their biological productivity is too poor to support any type of agriculture (unless irrigation or wells are used). These regions are true **deserts.**

When people rely on land for farming or ranching, they must assess the potential for **land degradation**—a state of declining agricultural productivity due to natural and/or human causes. Humid lands (lands in humid climates) may undergo degradation from factors such as soil *erosion* (wearing away), farming without crop rotation or fertilization, overgrazing, or dramatic increases or decreases in soil moisture. However, degraded humid lands always retain the capability of some level of agricultural production. This is not true in drylands, where degradation may cause the land to become true desert with no agricultural value. This type of land degradation is called **desertification** (the process of land degradation toward drier, true desert conditions).

UNEP estimates that 70% of all existing drylands (about 25% of all land on Earth) are now experiencing the hazard of desertification from factors related to human population growth, climate change, poor groundwater use policies, overgrazing, and other poor land management practices. More than 100 nations now risk degradation of their productive cropland and grazeland to useless desert. For this reason, the Third World Academy of Sciences declared the 1990s as the "Decade of the Desert," and the United Nations General Assembly declared 2006 the International Year of Deserts and Desertification. In 2012, a United Nations Convention to Combat Desertification affirmed a goal of zero net land degradation.

ACTIVITY

14.1 Dryland Inquiry

THINK About It **What are some characteristic processes, landforms, and hazards of drylands?**

OBJECTIVE Analyze satellite images and photographs of American drylands and infer processes and hazards that occur there.

PROCEDURES

1. **Before you begin**, do not look up definitions and information. Use your current knowledge, and complete the worksheet with your current level of ability. Also, this is **what you will need** to do the activity:

    ____ Activity 14.1 Worksheet (p. 367) and pencil
2. **Complete the worksheet in a way that makes sense to you.**
3. **After you complete the worksheet**, be prepared to discuss your observations and classification with other geologists.

ACTIVITY

14.2 Mojave Desert, Death Valley, California

THINK About It **What can we learn from topographic maps and satellite images about dryland processes and landforms?**

OBJECTIVE Identify dryland landforms of Death Valley, California, and infer how the valley is forming.

PROCEDURES

1. **Before you begin**, read the Introduction and Eolian Processes, Dryland Landforms, and Desertification below. Also, this is **what you will need**:

    ____ colored pencils
    ____ Activity 14.2 Worksheet (p. 369) and pencil
2. **Then follow your instructor's directions** for completing the worksheets.

Eolian Processes, Dryland Landforms, and Desertification

Many drylands have specific landforms that result primarily from processes associated with degradation, erosion by streams and flash floods developed after infrequent rains (fluvial processes), or erosion and deposition associated with wind (eolian processes) (**FIGURES 14.1, 14.2, 14.3**). Fluvial and eolian processes erode dryland landscapes, transport Earth materials, and deposit sediments. Rocky surfaces, sparsely vegetated surfaces, sand dunes, and arroyos (steep-walled canyons with gravel floors) are typical of dryland landscapes. Therefore, humans living in drylands must adapt to these landforms, conditions, processes that have created the landforms, and the prospect of land degradation or even desertification.

Eolian Processes and Landforms

Water and ice are capable of moving large particles of sediment. The wind can move only smaller particles (**FIGURES 14.4** and **14.5**). For this reason, the eolian (wind-related) landforms may be subtle or even invisible on a topographic map. (However, they may be more evident on aerial photographs that have a higher resolution.) They may be superimposed on *fluvial* (stream-related) or *glacial* (ice- or glacier-related) features, particularly where recently exposed and unvegetated sediment occurs.

A lack of a dense vegetation cover is a prerequisite for significant wind erosion. This lack of vegetation can occur:

- On recently deposited sediment, such as floodplains and beaches.
- In areas where vegetation has been destroyed by fire, overgrazed, or removed by humans, or
- In true deserts, where the lack of water precludes substantial growth of vegetation.

When examining a topographic map, keep in mind that the green overprint represents only trees and shrubs. There could be an important soil-protecting grass cover present that is not indicated on the map. Your evaluation of the present climate of a topographic map area should consider surface water features, groundwater features, and the geographic location of the area.

Blowouts

The most common wind-eroded landform visible on a topographic map is usually a **blowout**—a shallow depression developed where wind has eroded and blown out the soil and fragmented rock (**FIGURE 14.6C**). Blowouts may

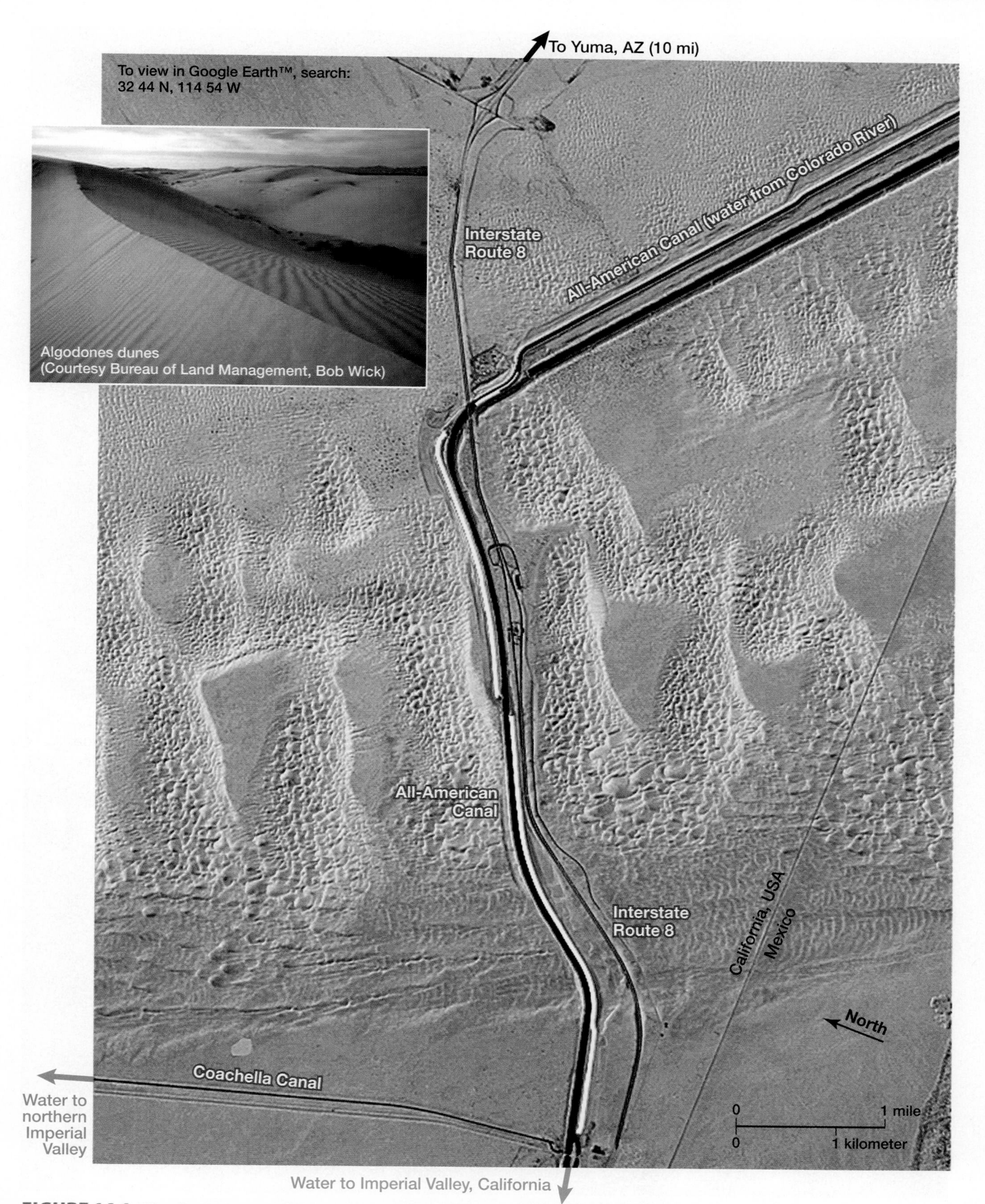

FIGURE 14.1 Algodones Dune Field, Sonoran Desert, near the California-Mexico border. Algodones Dune Field is about 10 km (6 mi) wide and 70 km (45 miles long). The long axis of the dune field is oriented northwest-southeast, along the direction of the prevailing northwest winds. Notice in this astronaut photograph how Interstate Route 8 and the All-American Canal cross the dune field. The 85-mile (53-km) canal and its branches carry water westward, from the Colorado River (at 26,000 cubic feet per second) to homes and farms of California's Imperial Valley. It is the largest irrigation canal in the world. Runoff from land irrigated by the canal system drains into the Salton Sea. (Astronaut photograph ISS018-E-24949, acquired 1-31-2009, courtsey of Image Science & Analysis Laboratory, NASA Johnson Space Center.)

FIGURE 14.2 Perspective view of Mojave Desert, Death Valley, California, looking northwest. Width of view is about 20 km (12.5 miles). The lowest point in North America (282 feet below sea level) is labeled in Badwater Basin at the bottom of the image. Tucki Mountain, at the top of the image, is 6732 feet above sea level. Death Valley is also the driest and hottest location in North America. Note the lack of vegetation. (ASTER true-color satellite image draped over an ASTER elevation model. Courtesy of NASA/GSFC/METI/ERSDAC/JAROS, and U.S./Japan ASTER Science Team.)

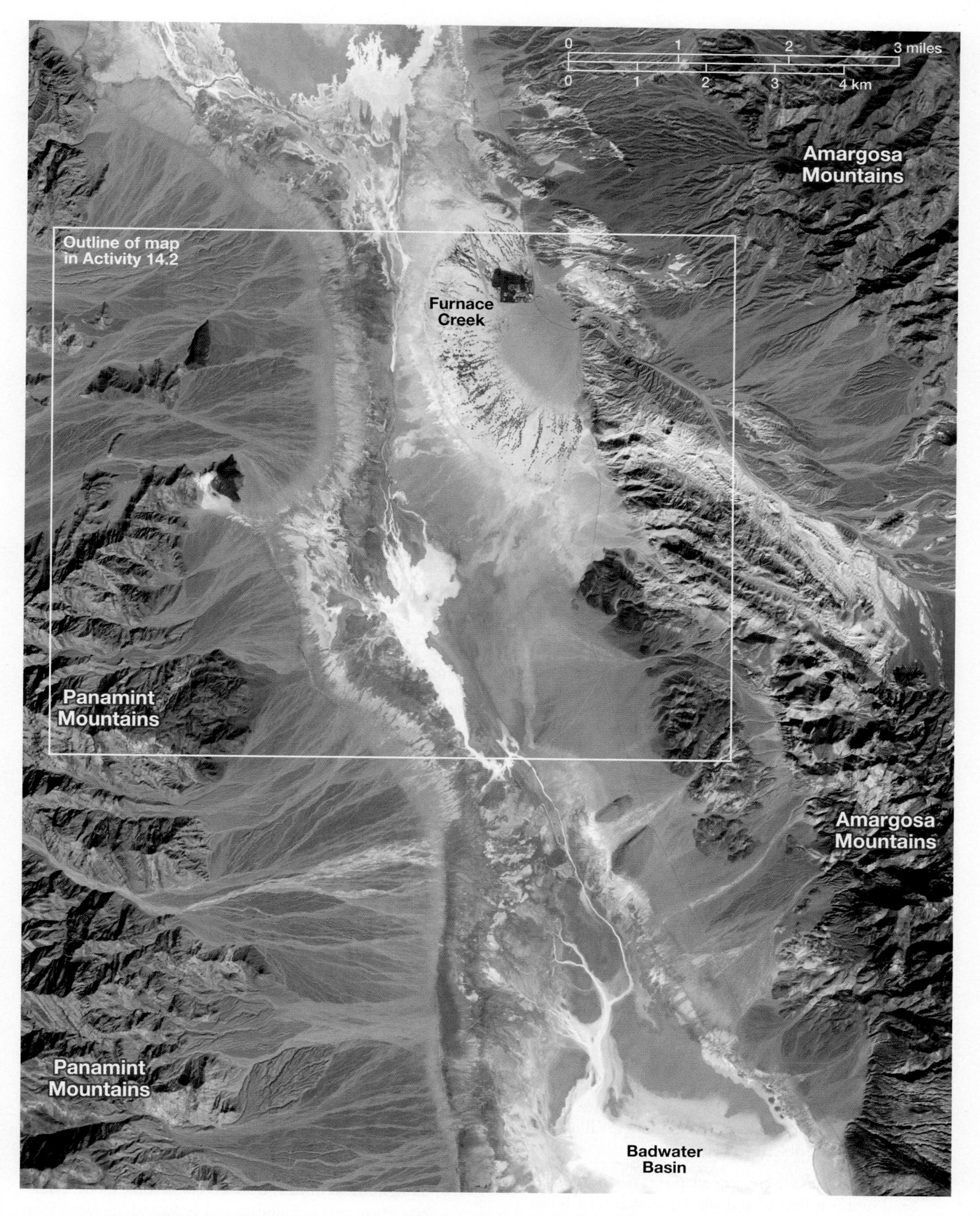

FIGURE 14.3 Overhead satellite view of Mojave Desert, Death Valley, California. Notice Badwater Basin at the bottom of the image. The water is very alkaline there, when present. Furnace Creek is the only location in the valley with potable water and significant vegetation. View in Google Earth™ by searching: 36 22 N, 116 53 W. (NASA Earth Observatory image processed by Jesse Allen and Robert Simmon using NASA EO-1 data.)

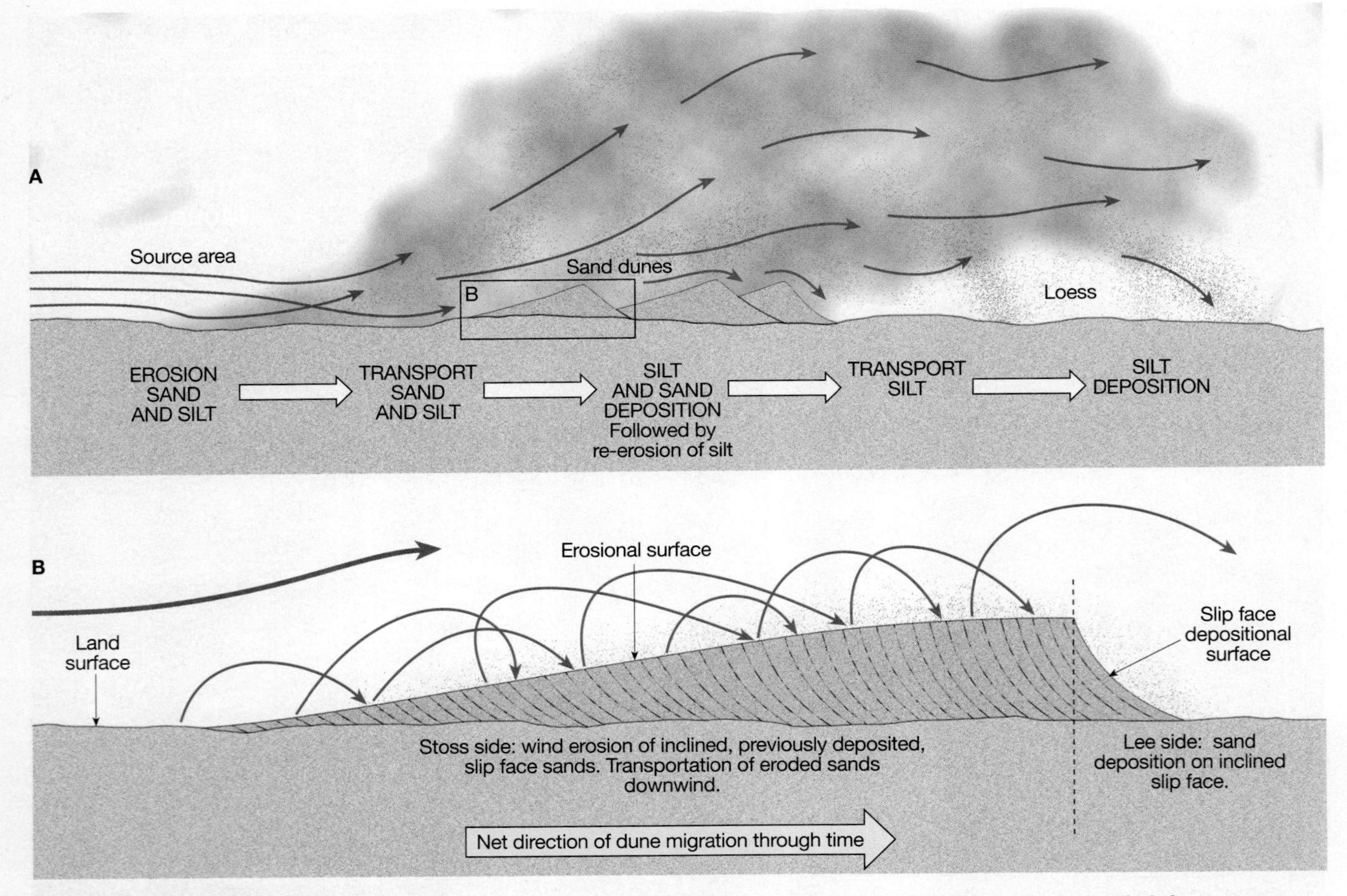

FIGURE 14.4 Eolian (wind-related) erosion, transportation, and deposition. A. Strong winds erode sand and silt from a source area and transport them to new areas. As the wind velocity decreases, the sand accumulates first (closest to the source) and the silt (loess) is carried further downwind. B. Hypothetical cross section through a sand dune. Wind erodes and transports sand up the *stoss side* (upwind side) of the dune. Sand bounces onto the slip face and accumulates, forming the *lee side* (downwind side) of the dune. This continuing process of wind erosion and transportation of sand on the stoss side of the dune, and simultaneous deposition of sand on the slip face of the dune, results in net downwind migration of the dune.

resemble sinkholes (depressions formed where caves have collapsed), or kettles (depressions formed where sediment-covered blocks of glacial ice have melted), but you can distinguish these different types of depressions in the context of other features observable on the map. Unlike sinkholes and kettles, blowouts usually have an adjacent sand dune or dunes that formed where sand-sized grains were deposited after being removed from the blowout. Blowouts also range in size from a few meters to a few kilometers in diameter.

Sand Dunes

When wind-blown (eolian) sediment accumulates, it forms sand dunes and silty loess deposits (see FIGURE 14.5). The process of dune and loess formation is shown in FIGURE 14.4. Some common types of dunes are illustrated in FIGURE 14.6 and described below:

- **Barchan dunes** are crescent shaped. They occur where sand supply is limited and wind direction is fairly constant. Barchans generally form around shrubs or large rocks, which serve as minor barriers to sand transportation. The *horns* (tips) of barchans point downwind.
- **Transverse dunes** occur where sand supply is greater. They form as long ridges perpendicular to the prevailing wind direction. The crests of transverse dunes generally are linear to sinuous.
- **Bachanoid ridge dunes** form when barchan dunes are numerous and the horns of adjacent barchan dunes merge into transverse ridges. The crests of barchanoid ridge dunes are chains of the short crescent-shaped segments that are the crests of individual barchan dunes. They can easily be distinguished from true transverse dunes that have long sinuous to very sinuous (like a snake) crests.
- **Parabolic dunes** somewhat resemble barchans. However, their horns point in the opposite direction—upwind. Parabolic dunes always form adjacent to blowouts, oval depressions from which come the sandy sediments that form the parabolic dunes.
- **Longitudinal (linear) dunes** occur in some modern deserts where sand is abundant and crosswinds merge to form these high, elongated dunes. They can be quite large, up to 200 km long and up to 100 m high. The crests of longitudinal dunes generally are straight to slightly sinuous.

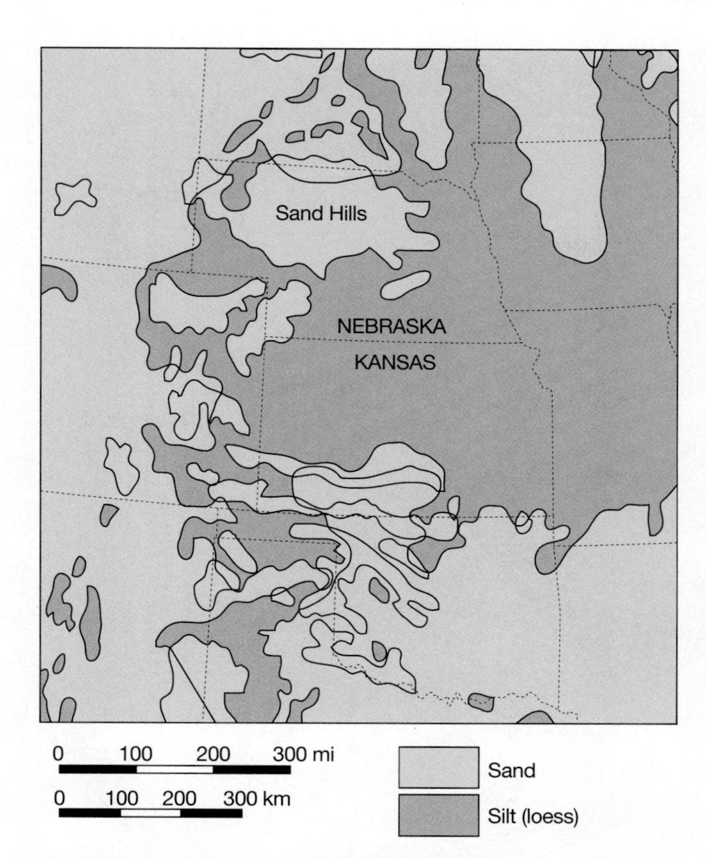

FIGURE 14.5 The Sand Hils of Nebraska. Map of part of the midwestern United States showing the location of Nebraska's Sand Hills, sand deposits, and silt (loess) deposits.

Dunes tend to migrate slowly in the direction of the prevailing wind (FIGURES 14.4 and 14.6). However, revegetation of exposed areas, due to changes in climate or mitigation, may stabilize them.

Water Erodes Drylands

Two characteristics of dryland precipitation combine to create some of the most characteristic dryland landforms other than blowouts and dunes. First, rainfall in drylands is minimal, so there are few plants to trap and bind loose rocks among their roots or aid in the development of soil that would absorb rainwater. Second, when rainfall does occur, it generally is in the form of violent thunderstorms. The high volume of water falling from such storms causes flash floods over dry ground. These floods develop suddenly, have high discharge, and last briefly. They carve steep-walled canyons, often floored with gravel that is deposited as the flow decreases and ends. Such steep-walled canyons with gravel floors commonly are called **arroyos** (or **wadis,** or **dry washes**).

Flash flooding in arid regions also erodes vertical cliffs along the edges of hills. When bedrock lies roughly horizontal, such erosion creates broad, flat-topped **mesas** bounded by cliffs. In time, the mesas can erode to stout, barrel-like rock columns, called **buttes.**

Mountainous Drylands

A variety of landforms are characteristic of drylands (FIGURE 14.7). They are primarily formed by the action of infrequent rain storms and flash floods that erode the landscape and transport and deposit sediment. However, these effects are enhanced in tectonically active regions, where there is greater relief of the land.

When it rains in mountainous drylands, the water simply runs off of the rocks because there is no soil to absorb it. This leads to development of severe flash floods, which have the cutting power to erode rock and transport sediment. These flash floods often develop into *mudflows* (sediment liquified with water, and having the consistency and density of concrete being poured from a "cement mixer" truck). Flash floods and mudflows do millions of dollars worth of damage to human properties each year and claim many lives. They also lead to development of **alluvial fans** (fan-shaped, delta-like deposits of sediment that develop where the flash floods and mudflows empty into a valley).

The southwestern United States (Great Basin) is one of many arid regions of the world where Earth's crust is being lengthened by tensional forces (pulled apart). This leads to **block faulting**—a type of regional rock deformation where Earth's crust is broken into fault-bounded blocks of different elevations. The higher blocks are called *horsts* and the lower blocks are *grabens* (see FIGURE 14.7). Steep slopes develop along faults, between the blocks. After severe thunderstorms, flash floods and mudflows commonly flow from the horsts into the graben valleys. Huge alluvial fans develop where the stream valleys of the flash floods and mudflows empty into the grabens, much as deltas develop where rivers empty into a lake or the ocean. In a humid climate, these basins might collect water in permanent lakes. But in a desert, precipitation usually is insufficient to fill and maintain permanent lakes. Many of the graben valleys are also closed basins, meaning that water has no outlet to flow from them. The only way that water can escape from such graben basins is by evaporation. Such ephemeral bodies of water are called **playas.** Chemicals

ACTIVITY 14.3 Sand Seas of Nebraska and the Arabian Peninsula

THINK About It **What can we learn from topographic maps and satellite images about dryland processes and landforms?**

OBJECTIVE Identify landforms, including types of sand dunes, in drylands and analyze drylands to determine their risk of desertification.

PROCEDURES

1. **Before you begin**, read Sand Seas of Nebraska and the Arabian Peninsula below. Also, this is **what you will need:**

   ____ colored pencils
   ____ Activity 14.3 Worksheet (p. 371) and pencil

2. **Then follow your instructor's directions** for completing the worksheets.

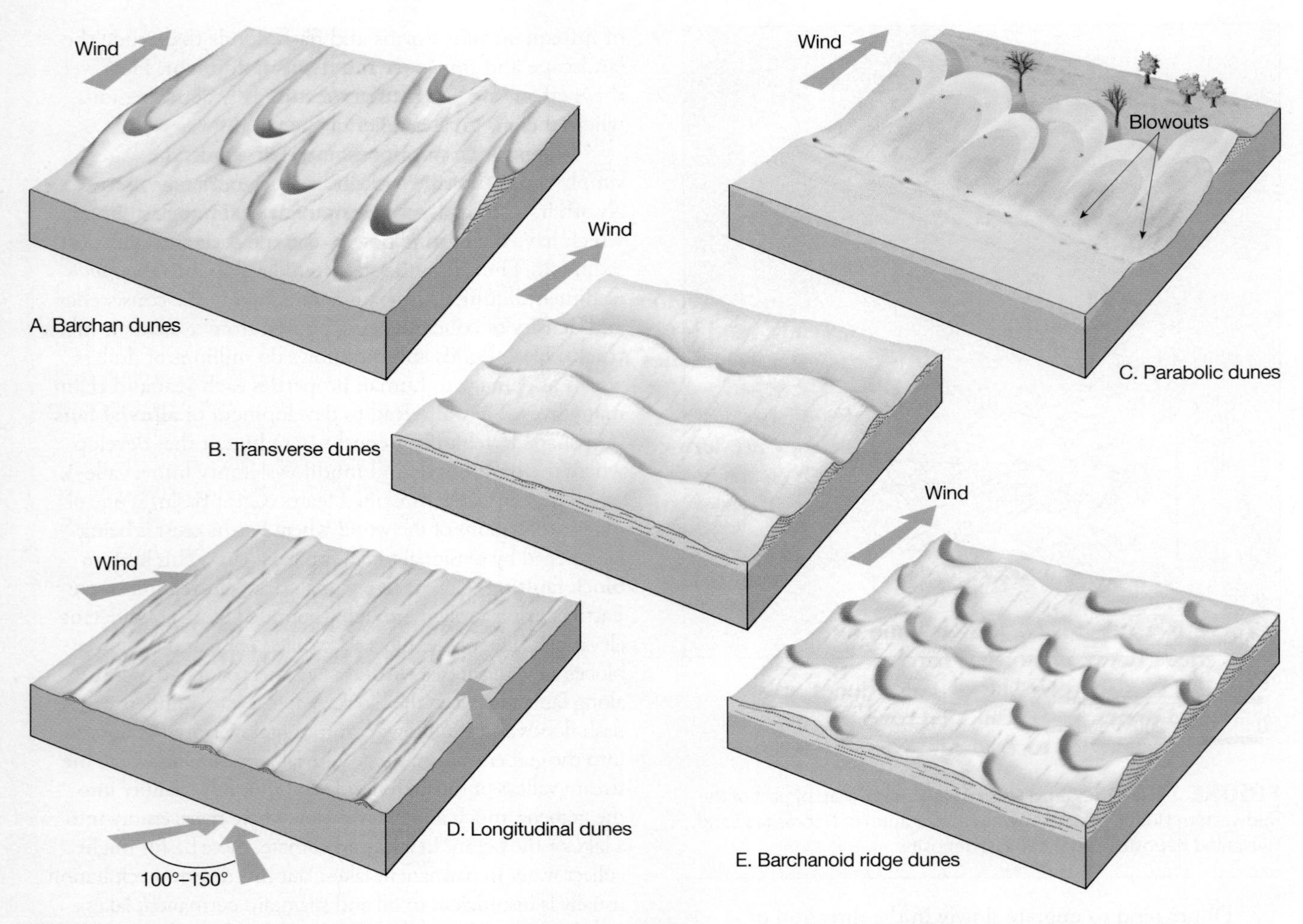

FIGURE 14.6 Common types of sand dunes. Note their basic morphology and internal stratification relative to wind direction.

dissolved in playa water become more and more concentrated as the water evaporates. Eventually, evaporite minerals (salts) precipitate from the water, and all that is left is a white, soggy or dry, **salt flat** (a level patch of land that is encrusted with salt).

Sand Seas of Nebraska and the Arabian Peninsula

Deserts can be rocky or sandy. An extensive sandy desert is called a sand sea, or **erg.** The largest erg on Earth is the Rub' al Khali (rūb al ka'lē), Arab for "the Empty Quarter," of the Arabian Peninsula. It covers an area of about 250,000 square kilometers (nearly the size of Oregon). Many kinds of active dunes occur there, and some reach heights of more than 200 meters. Rub' al Khali is a true desert (supports no agriculture) with rainfall less than 35 mm per year.

The Sand Hills of Nebraska (**FIGURES 14.5**) is only one-fifth the size of the Empty Quarter, or about 50,000 square kilometers of land, but it is the largest erg in the Western Hemisphere. This sand sea was active in Late Pleistocene and early Holocene time, but it has been inactive (i.e., the dunes are not actively forming or moving) for about the past 8000 years. This was determined by dating the radioactive carbon of organic materials that have been covered up by the large dunes. The large dunes are now covered with grass (short-grass prairie) that is suitable for limited ranching. About 17,000 people (mostly ranchers) now live in the Sand Hills.

Dryland Lakes

The amount of rain that falls on a particular dryland normally fluctuates over periods of several months, years, decades, centuries, or even millennia. Therefore, a dryland may actually switch back and forth between arid and semi-arid conditions, semi-arid and dry-subhumid conditions, arid and dry-subhumid conditions, and so on. Where lakes persist in the midst of drylands, their water levels fluctuate up and down in relation to such periodic changes in precipitation and climate. Periods of higher rainfall (or snow that eventually melts) and reduced aridity and evaporation create lakes that dry up during intervening periods of less rain and greater aridity and evaporation. Great Salt Lake, Utah, is an example.

Great Salt Lake is a closed basin, so water can escape from the lake only by evaporation. When it rains, or when snow melts in the surrounding hills, the water

ACTIVITY

14.4 Dryland Lakes of Utah

THINK About It **How can topographic maps and aerial photographs of drylands be used to interpret how their environments have changed?**

OBJECTIVE Analyze a stereogram and topographic map of the Utah desert to evaluate the history of Lake Bonneville.

PROCEDURES

1. **Before you begin**, read Dryland Lakes below. Also, this is **what you will need:**
   ____ colored pencils
   ____ Activity 14.4 Worksheet (p. 373) and pencil
2. **Then follow your instructor's directions** for completing the worksheets.

raises the level of the lake. Therefore, the level of Great Salt Lake has varied significantly in historic times over periods of months, years, and decades. During one dry period of many years, people ignored the dryland hazard of fluctuating lake levels and constructed homes, roads, farms, and even a 2.5-million-dollar resort, the Saltair, near the shores of Great Salt Lake. When a wet period occurred from 1982–87, many of these structures (including the resort) were submerged. The State of Utah installed huge pumps in 1987 to pump lake water into another valley, but the pumps were left high and dry during a brief dry period that lasted for 2 years (1988–89) after they were installed.

Geologic studies now suggest that the historic fluctuations of Great Salt Lake are minor in comparison to those that have occurred over millennia. Great Salt Lake is actually all that remains of a much larger lake that covered 20,000 square miles of Utah—Lake Bonneville. Lake Bonneville reached its maximum depth and geographic extent about 17,000 years ago as glaciers were melting near the end of the last Ice Age. One arm of the lake at that time extended into Wah Wah Valley, Utah, which is now a dryland (**FIGURE 14.8**).

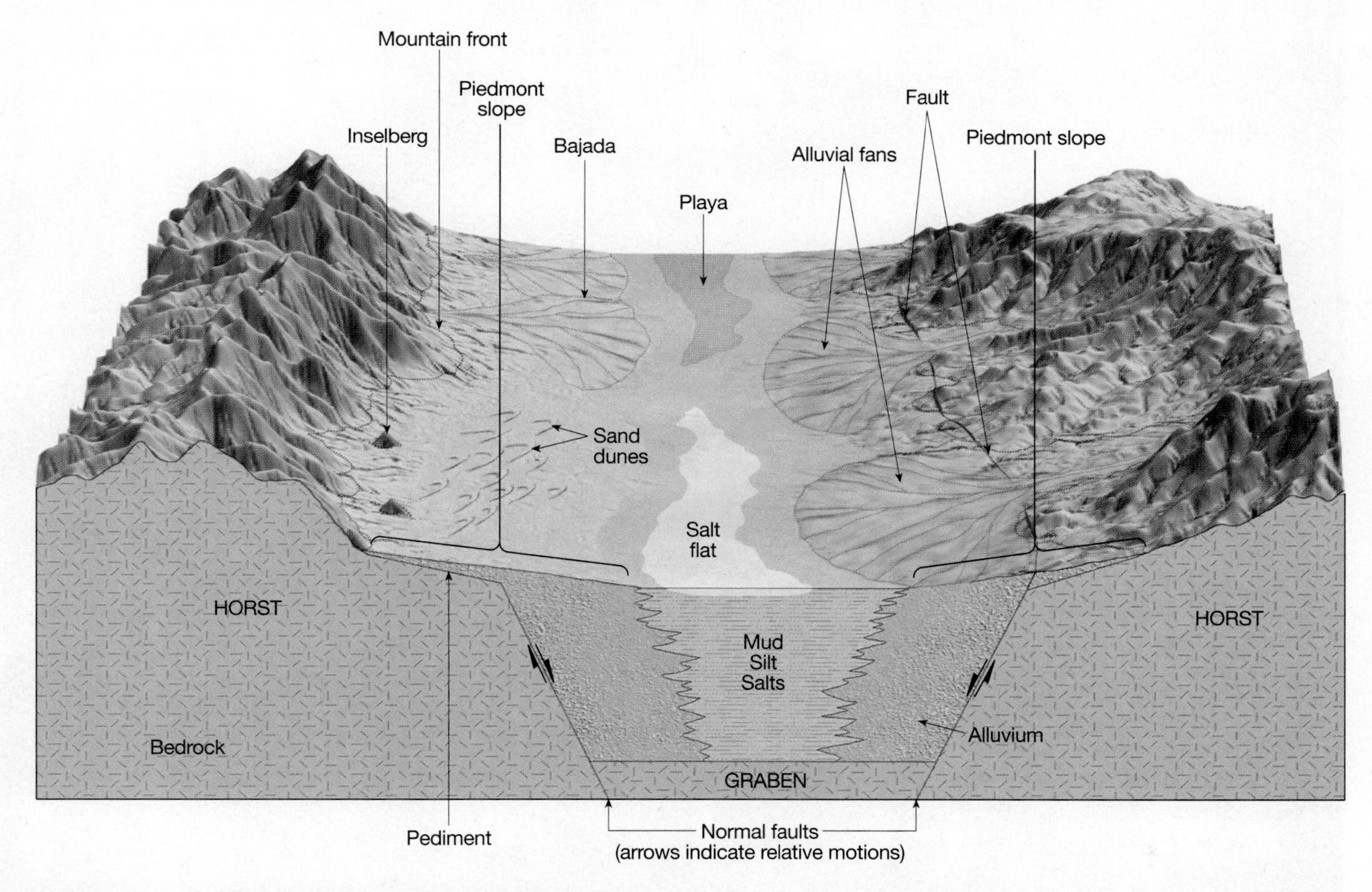

FIGURE 14.7 Landforms of mountainous drylands. These landforms are typical of arid mountainous deserts in regions of the southwestern United States, where Earth's crust has been lengthened by tensional forces (pulled apart). Mountain ranges and basins develop by **block faulting**—a type of regional rock deformation where Earth's crust is broken into fault-bounded blocks of different elevations. The higher blocks form mountains called *horsts* and the lower blocks form valleys called *grabens*. Note that the boundaries between horsts and grabens are typically normal faults. Sediment eroded from the horsts is transported into the grabens by wind and water. **Alluvial fans** develop from the mountain fronts to the valley floors. They may surround outlying portions of the **mountain fronts** to create **inselbergs** (island-mountains). The fans may also coalesce to form a **bajada.** In cases where there is no drainage outlet from the valley, the valley is a closed basin from which water can escape only by evaporation. Because rain is infrequent in drylands, the lakes that form are temporary (**playas**). When they evaporate, all that is left is a **salt flat** (a patch of level land that is encrusted with salt.). Wind blowing over the valleys can form **sand dunes** made of salt crystals or muineral grains eroded from bedrock (usually quartz sand).

FIGURE 14.8 Stereogram of the Wah Wah Valley, Utah, area: 1:58,000 scale. The stereogram is constructed with two 1991, National high-altitude photograph (NHAP) aerial photographs (Courtesy of U.S. Geological Survey). View this stereogram using a pocket stereoscope (**FIGURE 9.9**, page 241) or by crossing your eyes. Hold the stereogram at a comfortable distance (one foot or so) from your eyes with your nose facing the line between the images. Cross your eyes until you see four photographs (two stereograms), then relax your eyes to let the two center photographs merge into one stereo image.

ACTIVITY 14.1 Dryland Inquiry

Name: ______________________ **Course/Section:** ______________________ **Date:** ____________

A. When most people think of drylands and deserts, they imagine hot sandy landscapes. Most of the southwestern United States is desert, including the Sonoran Desert of southern California and Arizona. However, most of the Sonoran Desert is rocky landscapes. Sandy areas are present, but limited, like the Algodones Dune Field in **FIGURE 14.1**. This is the location where Star Wars producers filmed desert scenes of the film's planet Tatooine.

1. Notice the sand dunes of the Algodones Dune Field in **FIGURE 14.1**. Why do you think there are no plants growing on the dunes?

2. Winds in the Algodones Dune Field can reach velocities up to 60 miles per hour. This can create hazardous conditions and the need for maintanence on the canal and Interstate Route 8. What would be the hazard, and what maintenance would be needed periodically on the canal and Interstate Route 8 as a result of the hazard?

3. This region is managed by the U.S. Bureau of Land Management as Imperial Sand Dunes Recreation Area. Portions of the dunes are available for operating off-road vehicles. What effect do you think the operation of off-road vehicles here would have on plant growth and the hazards you described above?

B. Death Valley, occurs in the Mojave Desert. Analyze the images of the Death Valley region in **FIGURES 14.2** and **14.3**.

1. Notice in **FIGURE 14.2** that steep mountainous slopes occur on both sides of Death Valley. Also notice that there is almost no soil or vegetation on the slopes. Describe what you think conditions would be like in the river valleys on these mountain slopes when a heavy rain falls on them.

2. Notice the delta-like landforms that form at the mouths of the rivers in **FIGURES 14.2** and **14.3**, where the rivers enter the valley. Explain how you think these landforms form.

3. Notice that there is no standing water in Death Valley, even though you can see that water sometimes flows into the valley from the mountains. It is a closed basin (meaning that water has no way to drain from it). It is also the hottest and driest place in North America. When there is water on the floor of the vally, it is alkaline to salty and not potable (drinkable). How do you think the water gets so alkaline and salty?

4. Suppose you could walk down to the white patches on the floor of Death Valley in **FIGURES 14.2** and **14.3** and examine them. Predict what materials and conditions you would find there.

5. Residents of Furnace Creek have grassy lawns, trees, and potable water to drink. Why do you think their water is potable?

C. Open Google Earth™. Type coordinates "33 44 15 N, 116 25 W" into the search box and press enter to go to the location. You should arrive at Rancho Mirage, California, a resort community located in a desert region.

1. Notice that Rancho Mirage has a triangular shape. Based on your work in part **B**, how was the landform beneath Rancho Mirage formed?

2. Using your mouse, hover over the icons at the top of the Google Earth™ screen to find and select the "Show historical imagery" feature. Use the slider to go back in time and view how Rancho Mirage has changed. Describe how it has changed since 1996.

3. In July of 1979, a violent storm developed here and nearly six inches of rain fell on the San Jacinto Mountains, uphill from Rancho Mirage. At that time, flood control at Rancho Mirage included a system of earthen channels and concrete walls that were overwhelmed by a flash flood. Many homes were damaged and two lives were lost in the event. Analyze, recent images of Rancho Mirage in Google Earth™ and describe any evidence of flood control measures.

4. Would you feel safe living at Rancho Mirage? Explain.

D. REFLECT & DISCUSS The United Nations Convention to Combat Desertification (UNCCD) points out that many people live in drylands, and that more than half of the world's productive land is dryland. The challenge is to manage the land so that it does not degrade to useless desert that supports no agriculture, and to live safely. Based on your work above, make a list of challenges that people face when they live in drylands, and some ways that the land could be managed to make it safer and more productive.

ACTIVITY 14.2 Mojave Desert, Death Valley, California

Name: ______________________ **Course/Section:** ______________________ **Date:** ____________

Death Valley is located in the USGS 15-minute Furnace Creek, California quadrangle (provided on the back of this activity sheet). It is the large valley (graben) in the middle of the map and the lowest valley in the United States (**FIGURES 14.2, 14.3**). The mountains on each side of the valley are *horsts* (**FIGURE 14.7**).

A. Obtain your set of colored pencils and do the following on the topographic map on the back of this page.

1. Neatly and precisely color alluvial fan **A** yellow, including the two arroyos at the top (upslope end) of the fan.
2. Color the inselbergs red in the vicinity of location **B**.
3. Color alluvial fan **C** yellow.
4. Color alluvial fan **D** yellow.
5. Color the 00 (sea level) topographic contours blue on both sides of the valley.
6. Make a green line along the downhill edge of the *mountain front* (**FIGURE 14.7**) on both sides of the valley and label it "mountain front."
7. What is the elevation of the lowest contour line on the map? ______________ feet
8. What is the elevation of the lowest point on the map? ______________ feet

B. Notice the intermittent stream that drains from the upstream end of the alluvial fan/arroyo system **A** (that you have already colored yellow) to the playa at **E**. How would the grain size of the sediments along this stream change as you walk downslope from the high arroyo to the playa **(E)**? Why?

C. The floor of most grabens is tilted, because fault movement is usually greater on one side of the graben than the other. There are also half grabens, valleys developed along only one normal fault. Carefully examine the map on the back of this page for evidence of faults on either or both sides of Death Valley. Draw a dark dashed line (with a normal pencil or a black colored pencil or pen) wherever you think a normal fault may be present on either or both sides of the valley. Based on your work, do you think that Death Valley is a graben or a half graben?

D. REFLECT & DISCUSS Notice that people chose to build a ranch on alluvial fan **C**, even though this entire region is dryland. What do you think was the single most important reason why those people chose alluvial fan **C** for their ranch instead of one of the other fans?

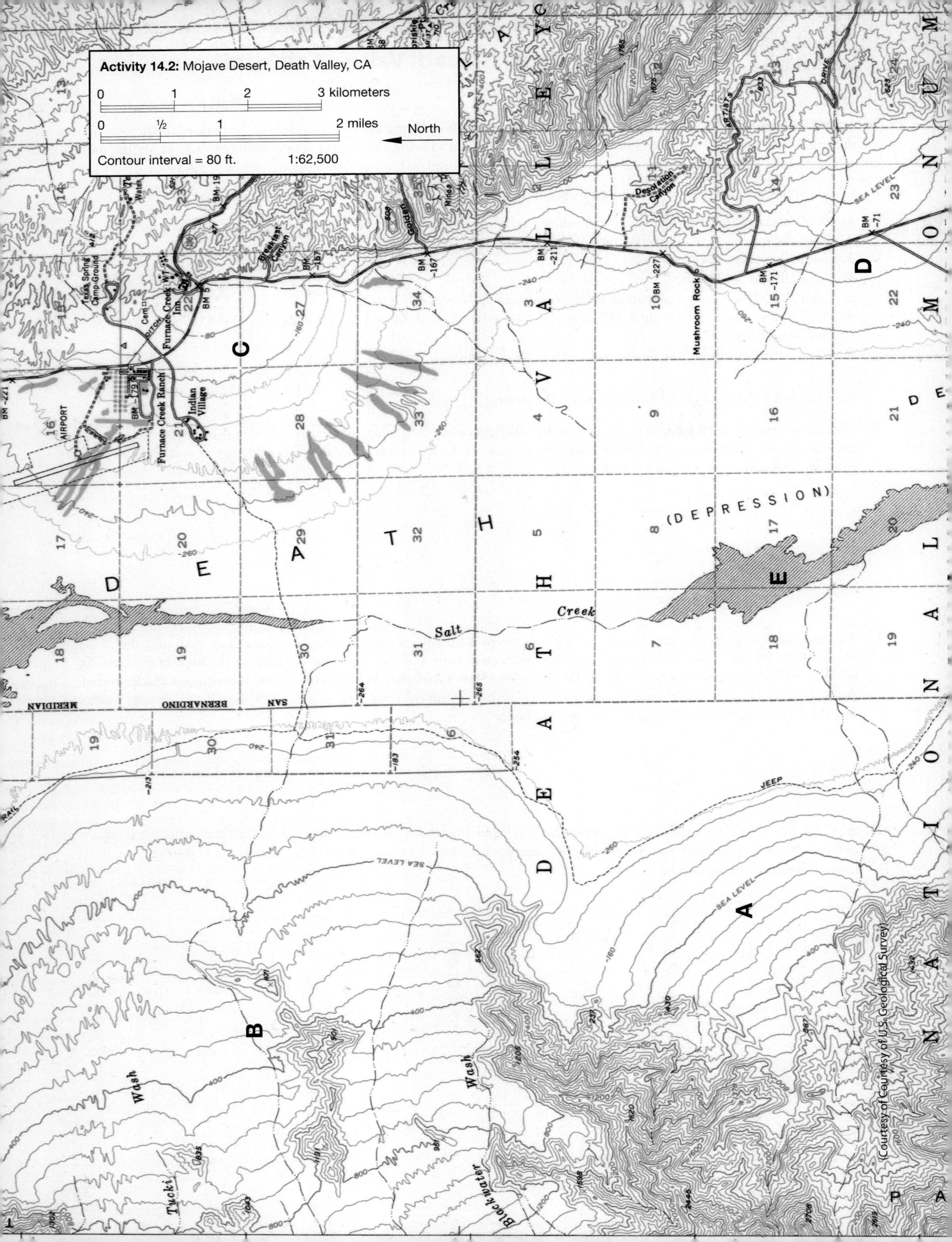
Activity 14.2: Mojave Desert, Death Valley, CA
0 1 2 3 kilometers
0 ½ 1 2 miles
North
Contour interval = 80 ft.
1:62,500
Furnace Creek Ranch
Indian Village
Furnace Creek Inn
Texas Spring Camp Ground
AIRPORT
Breakfast Canyon
Golden
Desolation Canyon
Mushroom Rock
(DEPRESSION)
Salt Creek
SAN BERNARDINO MERIDIAN
Tucki Wash
Blackwater Wash
SEA LEVEL
A
B
C
D
E
(Courtesy of Courtesy of U.S. Geological Survey)

ACTIVITY 14.3 Sand Seas of Nebraska and the Arabian Peninsula

Name: ______________________ **Course/Section:** ______________________ **Date:** ______________

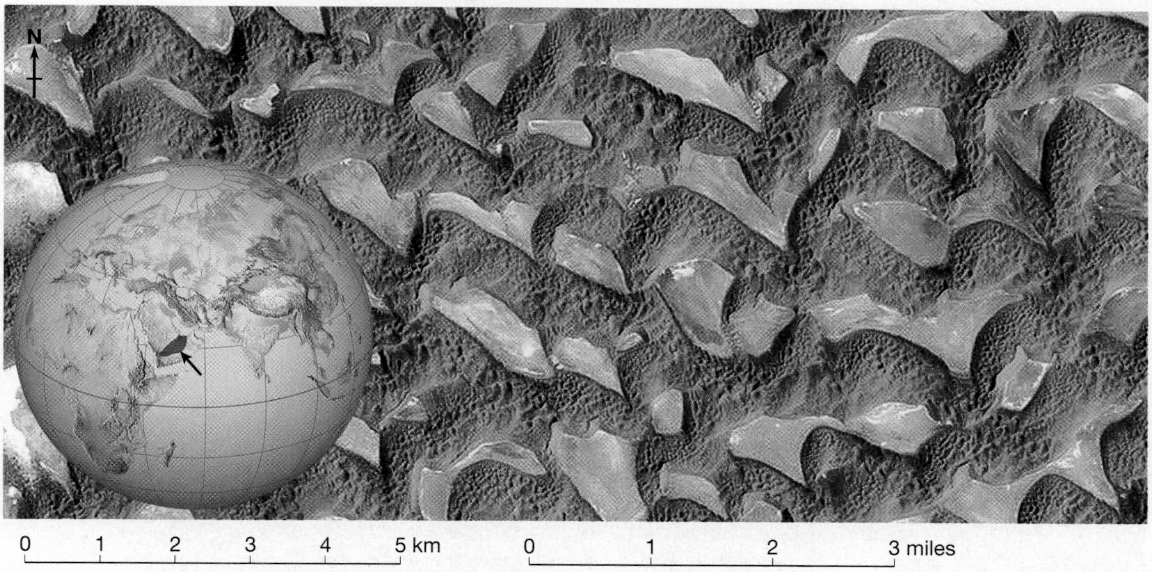

N

0 1 2 3 4 5 km   0 1 2 3 miles

A. Analyze the satellite image above of the Empty Quarter Desert. This is a sandy desert. The sand is mostly quartz (with a reddish hematite coating) transported south by strong winds from Jordan, Syria, and Iraq. These winds, called *shamals*, can reach speeds of nearly 50 mph that last for days. The dunes sit on gray and brown clay (ancient lake beds). White patches are salt flats (mostly gypsum).

1. What kind of dunes are developed in this image?
2. The dominant winds here are northerly (i.e., come from the north and blow south). How can you tell?

B. Analyze the USGS orthoimage of part of Nebraska's semi-arid Sand Hills region on the next page. Rainwater quickly drains through the porous sand, so the hilltops are dry and support only sparse grass. There is a shallow water table, so there are lakes, marshes, and moist fields between the hills.

1. How are the Sand Hills sand dunes similar to the sand dunes of the Empty Quarter Desert?

2. Locate Star Ranch in the northwest corner of the orthoimage. This ranch was present in the 1930s during the famous Dust Bowl days, when dry windy conditions persisted for years and dust storm after dust storm scoured the land here. Star Ranch was not covered by any advancing dunes during the Dust Bowl. What does this suggest about how much desertification must occur here in order for the large dunes of the Sand Hills to once again become an active sand sea?

3. The hills (Sand Hills) in the orthoimage are either large barchan dunes (called *megabarchans*) or barchanoid ridges (**FIGURE 14.6**). Using a highlighter or colored pencil color an isolated megabarchan yellow and a barchanoid ridge green (color it from one side of the map to the other). What is the relief, in feet, of the megabarchan that you colored?
4. According to the orientations of the megabarchans and the barchanoid ridges, the winds that made these dunes were coming *from* what direction?

5. What evidence is there in **FIGURE 14.5** that the source of sand for the Sand Hills was located northwest of Nebraska and the Sand Hills? Explain your reasoning.

C. The Sand Hills region now receives an average of 580 mm (about 2 in.) of rain per year, and large active dunes tend to occur in regions that receive less than 250 mm of rain per year (less than an inch). Winds capable of moving sand already occur in this region, but the rainfall and moisture from evapotranspiration is presently enough to sustain grasses that hold the sand in place in the dunes. If you were a rancher in the Sand Hills, what grazing practices would you follow to decrease the risk of desertification there?

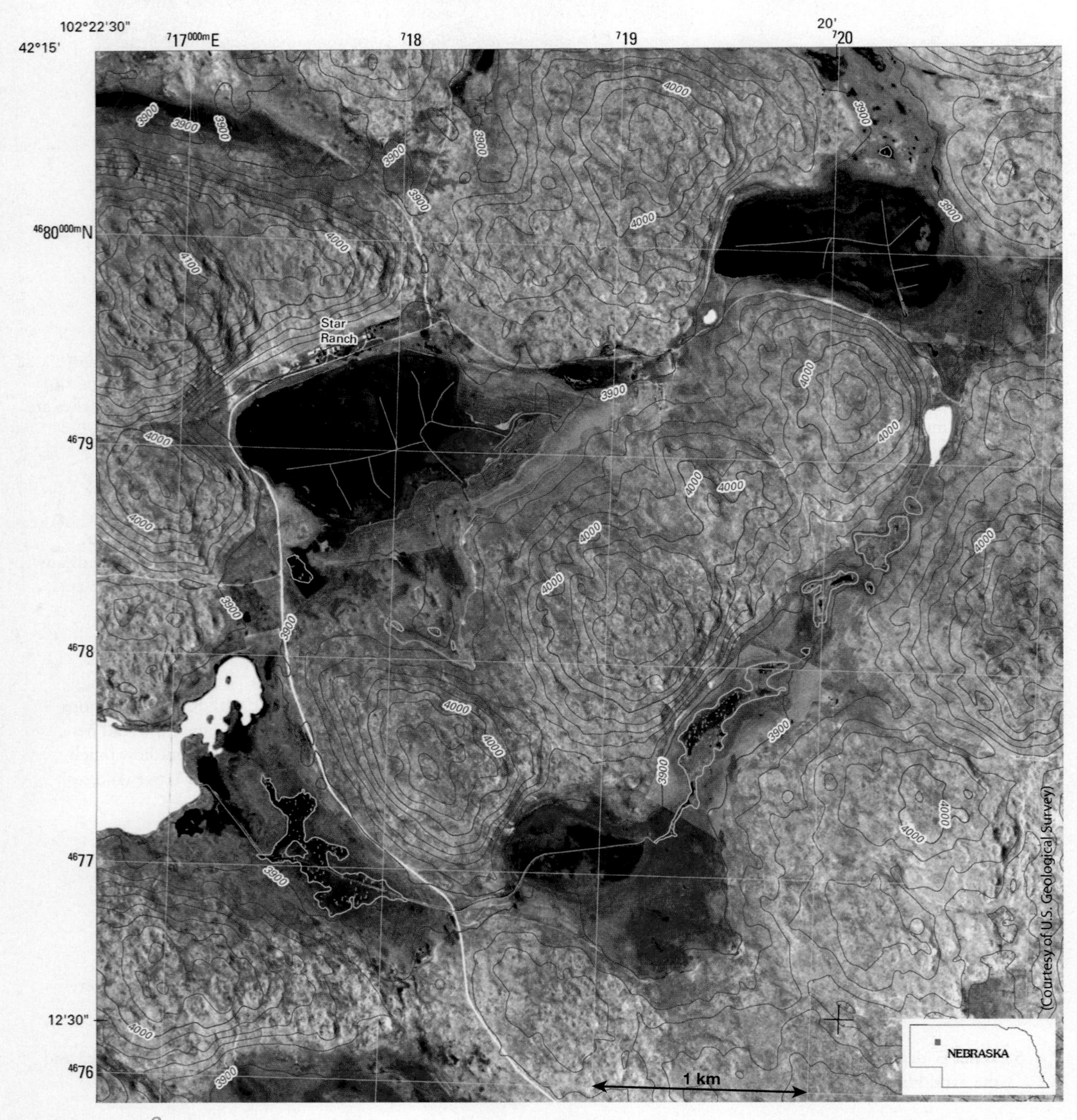

D. **REFLECT & DISCUSS** Many cities in central and eastern Nebraska rely on groundwater for consumption, industry, and pleasure. As these cities continue to grow, and their use of groundwater increases, what effect might this have on the environments and people of the Sand Hills?

ACTIVITY 14.4 Dryland Lakes of Utah

Name: ______________________ **Course/Section:** ______________ **Date:** __________

Refer to **FIGURE 14.8** (stereogram) and the Frisco Peak topographic map on the back of this page.

A. What specific type of feature is the Wah Wah Valley Hardpan?

B. If the Wah Wah Valley Hardpan were to fill with water, then how deep could the lake become before it overflows to the northeast (along the jeep trail, near the red number 27)? Show your work.

C. On **FIGURE 14.8**, notice the lines beneath the letters **X** and **Y**. They are low, steplike terraces that go all around the valley (like bathtub rings). How do you think these terraces formed?

D. Also on **FIGURE 14.8**, notice that there is a line of small deltas upslope and downslope from letter **Y**. They are also visible on the topographic map. How did they form in a line from upslope to downslope?

E. On the stereogram (**FIGURE 14.8**) and topographic map, what evidence can you identify for a former deeper lake (an arm of Lake Bonneville) in Wah Wah Valley (deeper than your answer in part D) at location **X** and what was its elevation?

F. Are the alluvial fans at **Z** in **FIGURE 14.8** older or younger than the shoreline you identified in **B**? Why?

G. On the topographic map, use a blue colored pencil to draw the position (line) of the shoreline of ancient Lake Bonneville where it reached its highest elevation (answer **G**). Then shade (blue) the area that was submerged (i.e., color in the area that was the lake).

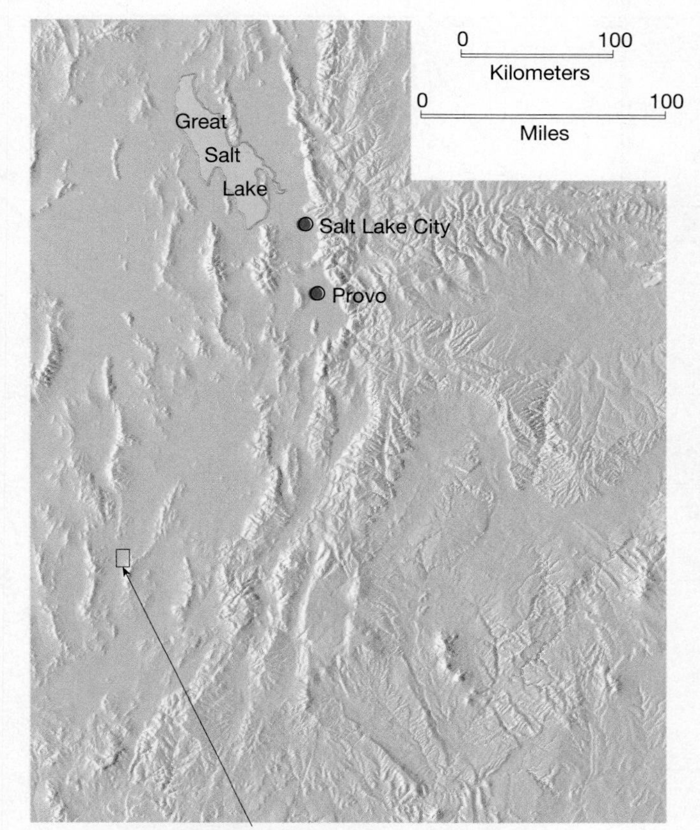

Frisco Peak Quadrangle: Figure 14.8 (stereogram) and map on back of this page

H. Studies by geologists of the Utah Geologic Survey and U.S. Geological Survey indicate that ancient Lake Bonneville stabilized in elevation at least three times before present: 5100 ft about 17,000 years ago, 4800 ft about 16,000 years ago, and 4300 ft about 12,000 years ago.

1. What is the age of the lake level that you identified in **G**?

2. Modern Great Salt Lake has an elevation of about 4200 ft and is 30 ft deep. How deep was the Great Salt Lake location at the time you identified above (part **H1**)?

I. REFLECT & DISCUSS Explain how the climate must have changed in Utah over the past 17,000 years to explain the fluctuations in levels of Lake Bonneville investigated above. In your answer, consider the times identified in part **H** and conditions in Utah today.

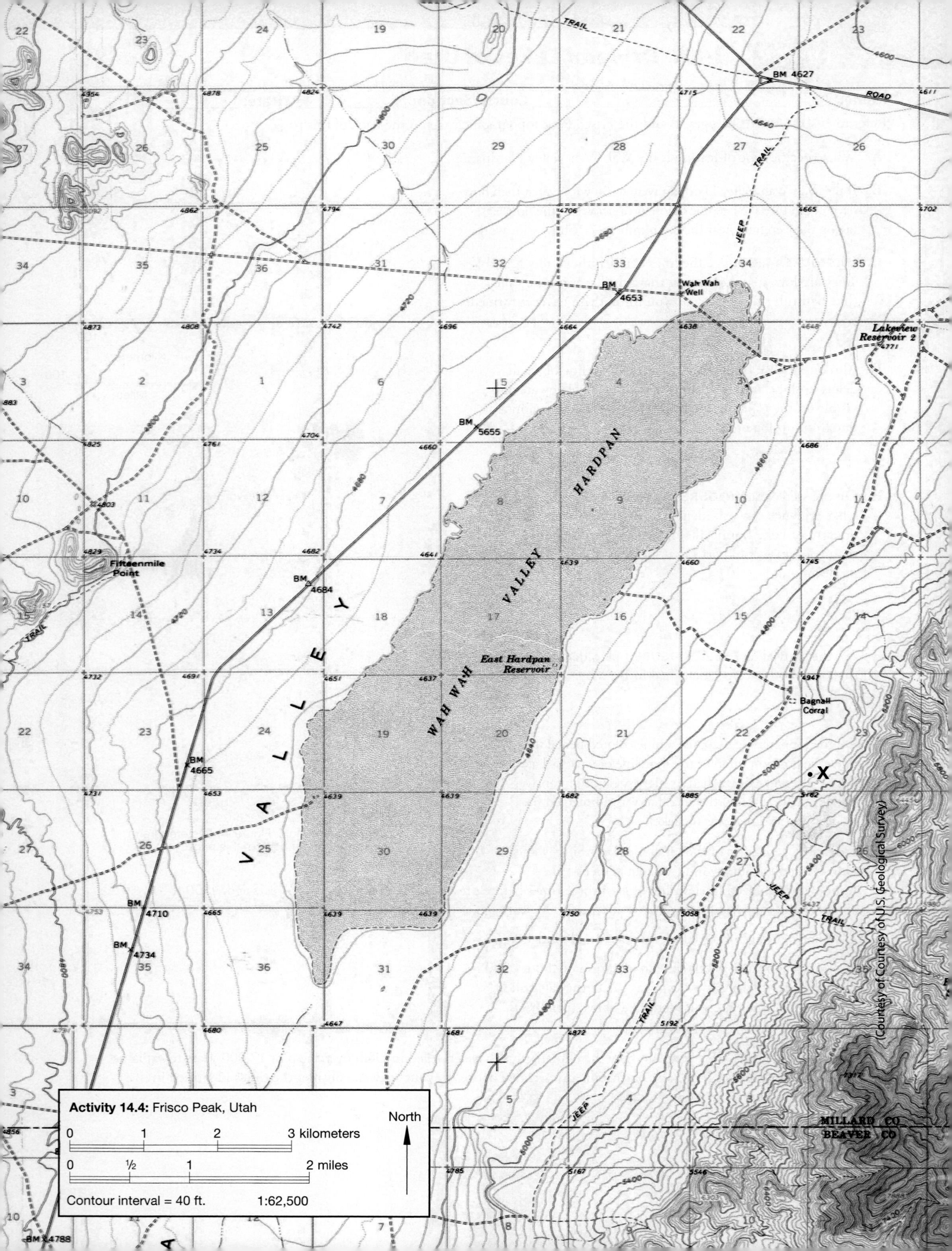
Activity 14.4: Frisco Peak, Utah
0 1 2 3 kilometers
0 ½ 1 2 miles
Contour interval = 40 ft.
1:62,500
North
WAH WAH VALLEY HARDPAN
VALLEY
East Hardpan Reservoir
Fifteenmile Point
Wah Wah Well
Lakeview Reservoir 2
Bagnall Corral
BM 4627
BM 4653
BM 5655
BM 4684
BM 4665
BM 4710
BM 4734
BM 4788
TRAIL
ROAD
JEEP TRAIL
JEEP
MILLARD CO
BEAVER CO
X
(Courtesy of Courtesy of U.S. Geological Survey)

LABORATORY 15

Coastal Processes, Landforms, Hazards, and Risks

CONTRIBUTING AUTHORS
James G. Titus • *U.S. Environmental Protection Agency*
Donald W. Watson • *Slippery Rock University*

Storm waves are eroding sand from this Hatteras, North Carolina beach. Homeowners risk the loss of their properties when severe storms occur. (Photo by Michael Collier)

BIG IDEAS

A coastline is the boundary between land (geosphere) and the ocean or a lake (hydrosphere), but it is also affected by the atmosphere, organisms (biosphere, including humans), and sometimes glaciers (cryosphere). The shapes of coastal landforms depends on how the land is affected by the other spheres. Specific factors like waves, erosion, sediment supply, storms, and sea-level changes, are particularly effective in shaping coastal landforms and may pose hazards to humans or their property. Therefore, artificial structures are used to manage shorelines and protect coastal properties.

FOCUS YOUR INQUIRY

THINK About It What factors affect the shape and position of shorelines?

ACTIVITY 15.1 Coastline Inquiry *(p. 376)*

ACTIVITY 15.2 Introduction to Shorelines *(p. 376)*

THINK About It How successful are efforts to protect shorelines from erosion by building artificial structures?

ACTIVITY 15.3 Shoreline Modification at Ocean City, Maryland *(p. 381)*

THINK About It How will rising sea levels affect communities along shorelines?

ACTIVITY 15.4 The Threat of Rising Seas *(p. 381)*

Introduction

When viewed from an airplane or satellite, coastlines appear to be very simple—the obvious linear boundaries between land and the ocean or a lake. But at ground level, one cannot help but notice that coastlines are dynamic systems characterized by constant change. There is constant interaction among not only land (geosphere) and water (hydrosphere), but also the atmosphere, organisms (biosphere), and sometimes ice (cryosphere). Wind is blowing, water is flowing, rocks are being eroded, sediment is moving about, and landscapes are being shaped. Organisms, including humans, are abundant. The United Nations Environment Programme (UNEP) has found that more than half of the world's population lives within 60 km of the ocean, and three-quarters of all large cities are located on the coast. There, humans find many resources along copastlines, but they also face hazards associated with living at the dynamic interface of many spheres.

ACTIVITY

15.1 Coastline Inquiry

THINK About It | **What factors affect the shape and position of shorelines?**

OBJECTIVE Compare and contrast photographs of coastlines and determine what factors primarily affect them.

PROCEDURES

1. **Before you begin**, read the Introduction and Dynamic Natural Coastlines below. Also, this is **what you will need:**

    ____ Activity 15.1 Worksheet (p. 385) and pencil

2. **Complete the worksheet in a way that makes sense to you.**
3. **After you complete the worksheet**, be prepared to discuss your observations and classification with other geologists.

Dynamic Natural Shorelines

Some examples of shorelines are pictured in **FIGURE 15.1**. In each case, the land is acted upon by water, wind, organisms, and sometimes ice, in ways that vary in both intensity and time. For example, there is constant water and air motion, but their intensities vary throughout the day in relation to tides and the weather. At one part of the day or year, the **erosional processes** (those that remove sediment and cut into rock, reefs, and marshes) may be dominant over the **depositional processes** (those that cause sediment to accumulate and marshes or reefs to grow). At another part of the day or year, the depositional processes may be dominant over the erosional ones. Over longer periods of time, one or the other process (erosion or deposition) is generally dominant, so most coastlines are either receding (moving landward, eroding back) or advancing (building seaward).

Factors Affecting Coastlines

There are many specific factors that affect the shapes of coastal landforms and the overall positions of coastlines, but here are some of the most important factors:

- **What the land is made of** determines how much the agents of change must work on the land to shape it. The land may be hard rock, clay, sand, large loose rocks, or a combination of these. The land may also be "armored" with large boulders (called "rip rap") or rigid concrete structures added by humans.
- **Supply of sediment** carried to a specific location along a coastline by rivers, coastal currents, or people often determines whether the coastline is sandy or of bare rock and whether the position of the coastline is advancing or receding. A coastline cannot advance seaward if it lacks a supply of sediment to do so. Some sediment may be eroded from the land itself, as when waves undercut a cliff and rocks collapse into the water.
- **Waves** carry sediment onto beaches when they are gentle (*low energy waves*), but they remove sediment from beaches and erode the land when they are large and forceful (*high energy waves*). Particles moved by waves and blasted against rocky surfaces will cause abrasion (smoothing and wearing down of the rocky surfaces). The direction of the waves is a factor in what direction sediment is moved and what parts of a coastline are eroded the most.
- **Wind** interacts with the surface of the water to generate the waves and blows beach sand into dune forms on the adjacent land.
- **Currents** running along the coastline (*longshore currents*), in streams reaching the coastline (*stream currents on deltas*), and back-and-forth through coastal environments (*tidal currents*) move sediment about and redeposit it on beaches, sand bars and spits, and tidal flats.
- **Storms** are highly energized systems, so they are one of the main factors that determines the shape of coastal landforms. A single storm, like a hurricane, can significantly erode one part of a coastline and deposit a large volume of sediment on another part of the coastline.
- **Organisms (including humans)** modify their environment. Corals construct reefs that armor the coastline against erosion. Marsh plants and mangroves trap and bind sediment with their roots and absorb the energy of storms. Humans use a variety of methods to preserve and build up coastlines, but they also destroy marshes and reefs and otherwise degrade elements of the coastline.

ACTIVITY

15.2 Introduction to Shorelines

THINK About It | **What factors affect the shape and position of shorelines?**

OBJECTIVE Identify and interpret natural shoreline landforms and distinguish between emergent and submergent shorelines.

PROCEDURES

1. **Before you begin**, read Submergent vs. Emergent Coastlines below. Also, this is **what you will need:**

    ____ ruler, calculator

    ____ Activity 15.2 Worksheets (pp. 386–387) and pencil

2. **Then follow your instructor's directions** for completing the worksheets.

a. Maryland coastline with saltmarsh (NOAA)

b. San Francisco, California coastline (NOAA)

c. Oregon coastline (NOAA)

d. North Carolina coastline (NOAA)

e. Destin, Florida urbanized coastline

f. Florida Keys coastline with mangrove plants

g. Maine coastline (NOAA photo by Albert E. Theberge)

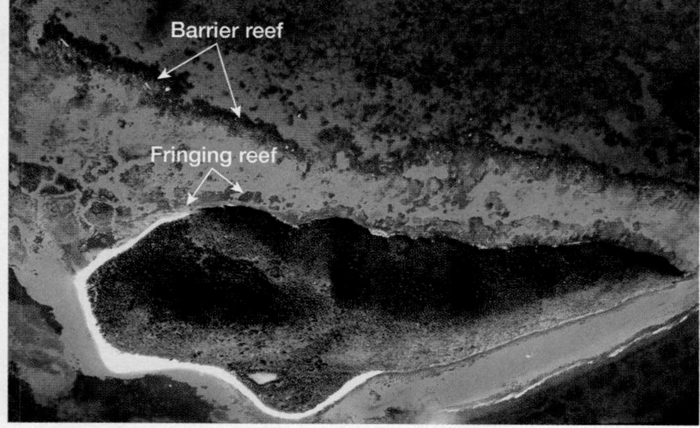

h. Caribbean island coastline with reefs (NOAA)

FIGURE 15.1 Photographs of eight different coastlines.

Submergent vs. Emergent Coastlines

Over decades of time, geologists characterize coastlines as submergent (retrogradational) or emergent (progradational): in one of two ways:

- A **submergent coastline**—is one that is being flooded, eroded back, or is otherwise receding (moving landward, retrograding). This can occur on short timescales due to erosion by waves, but it also occurs over longer periods of time due to sea-level rise. The sea level may rise may be caused by the water level actually rising (global sea-level rise, called *transgression*) or by the land getting lower (called *subsidence*).
- An **emergent coastline**—is one that is advancing (moving out into the water, prograding). This can occur when sediment and reefs build up to sea level, and then build seaward. It can also occur when sea level actually falls globally (called *regression*) or when the seafloor rises (called *uplift*). Uplift can occur because the region is tectonically active. It can also occur where the crust and mantle are rebounding upward after an ice sheet melts from atop them.

Submergent coastlines may display some emergent features, and vice versa. For example, the Louisiana coastline is submergent, enough so that dikes and levees have been built in an attempt (that failed in Hurricane Katrina) to keep the ocean from flooding New Orleans. However, the leading edge of the Mississippi Delta (at the mouth of the Mississippi River) is emergent (progradational)—building out into the water. This is because of the vast supply of sediment being carried there and deposited from the Mississippi River.

FIGURES 15.2 and **15.3** illustrate some features of *emergent* and *submergent* shorelines that you will need to identify in **FIGURE 15.4**, **15.5**, and **15.6** Study these features and their definitions below.

- **Barrier island**—a long, narrow island that parallels the mainland coastline and is separated from the mainland by a lagoon, tidal flat, or salt marsh (submergent, **FIGURE 15.3**).
- **Beach**—a gently sloping deposit of sand or gravel along the edge of a shoreline. Wide beaches are associated with emergent coastlines (**FIGURE 15.2**) and narrow beaches are associated with submergent coastlines (**FIGURE 15.3**).
- **Washover fan**—a fan-shaped deposit of sand or gravel transported and deposited landward of the beach during a "washover" of the land or island during a storm or very high tide.
- **Berm crest**—the highest part of a beach; it separates the *foreshore* (seaward part of the shoreline) from the *backshore* (landward part of the shoreline). This can occur on either type of coastline but is best developed on emergent coastlines that do not experience washover events.
- **Estuary**—a river valley flooded by a rise in the level of an ocean or lake (submergent, **FIGURE 15.3**). A flooded glacial valley is called a *fjord*.

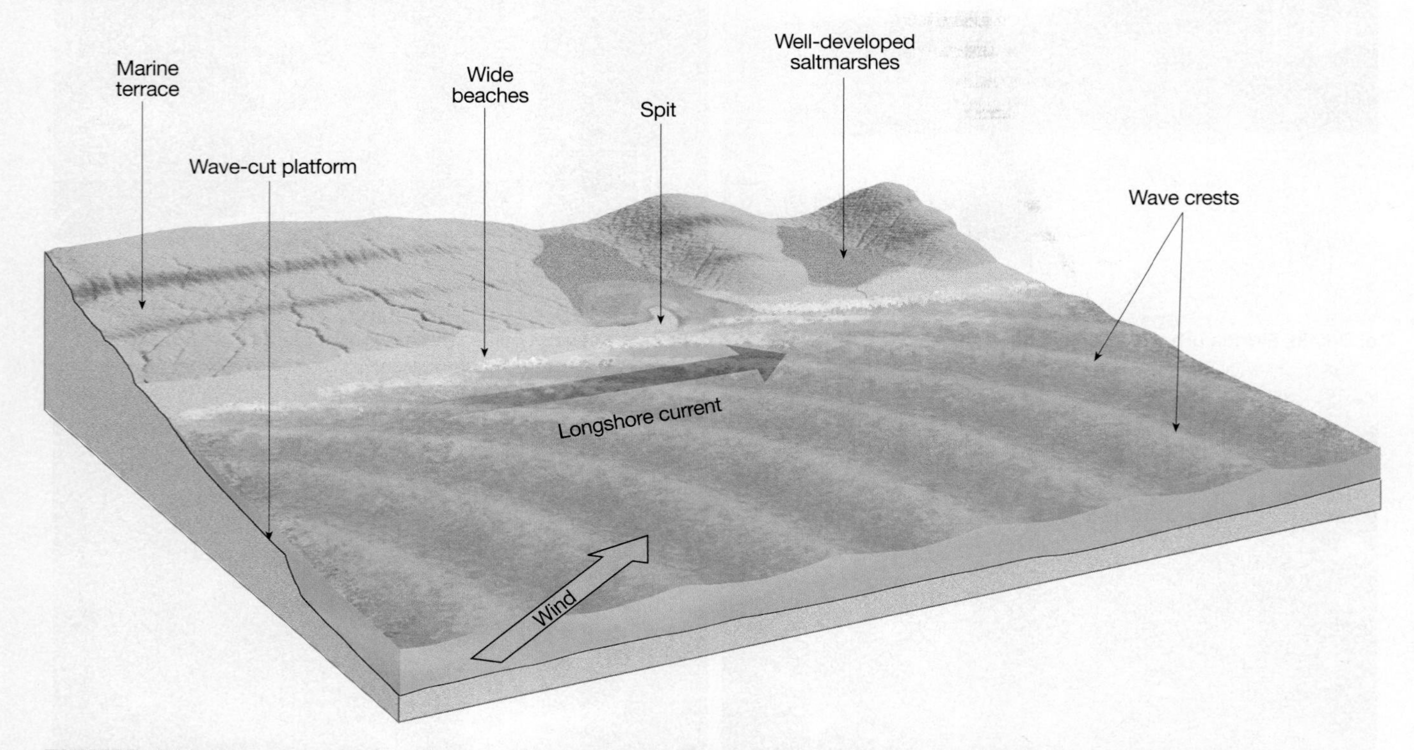

FIGURE 15.2 Emergent coastline features. An emergent coastline is caused by sea level lowering, the land rising, or both. Emergence causes tidal flats and coastal wetlands to expand, wave-cut terraces are exposed to view, deltas prograde at faster rates, and wide stable beaches develop.

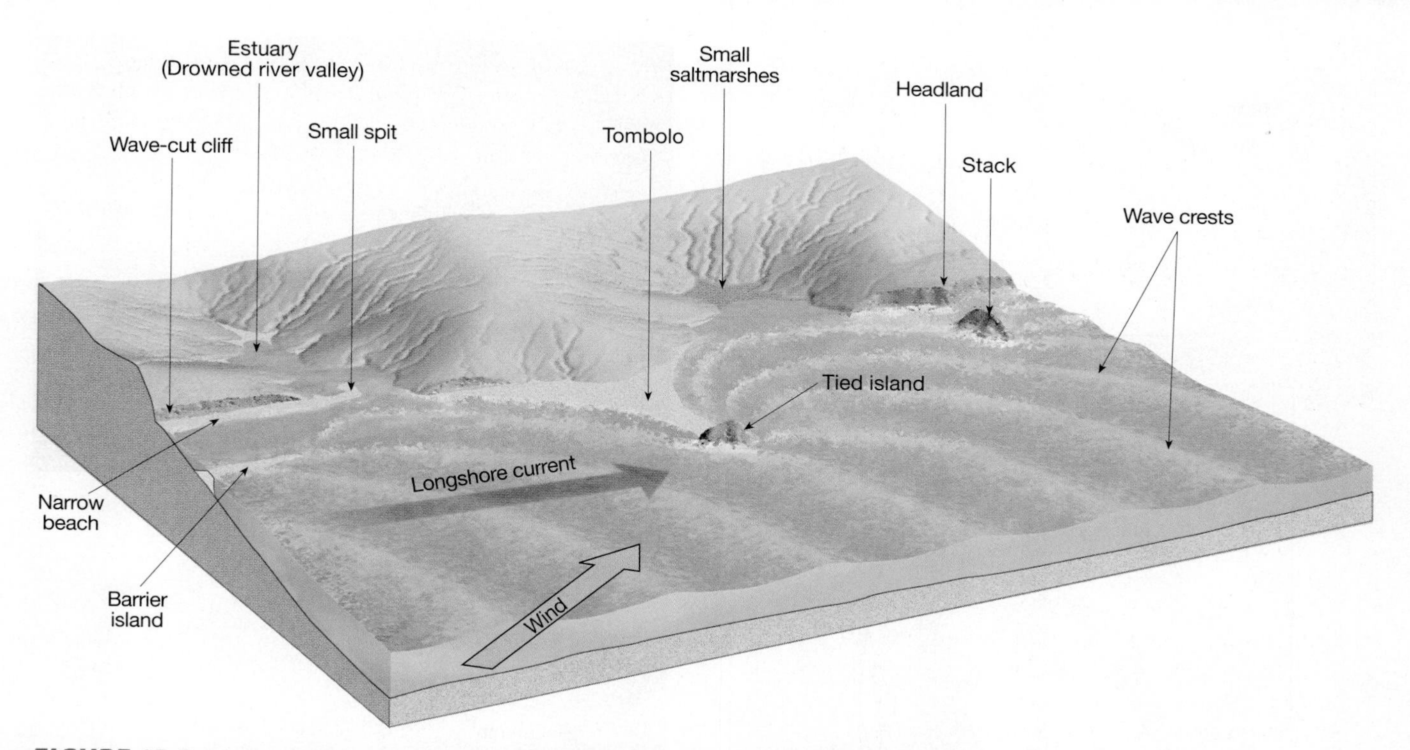

FIGURE 15.3 Submergent (drowning) coastline features. A submergent coastline is caused by sea level rising (transgression), sinking of the land, or both. As the land is flooded, the waves cut cliffs, valleys are flooded to form estuaries, wetlands are submerged, deep bays develop, beaches narrow, and islands are created.

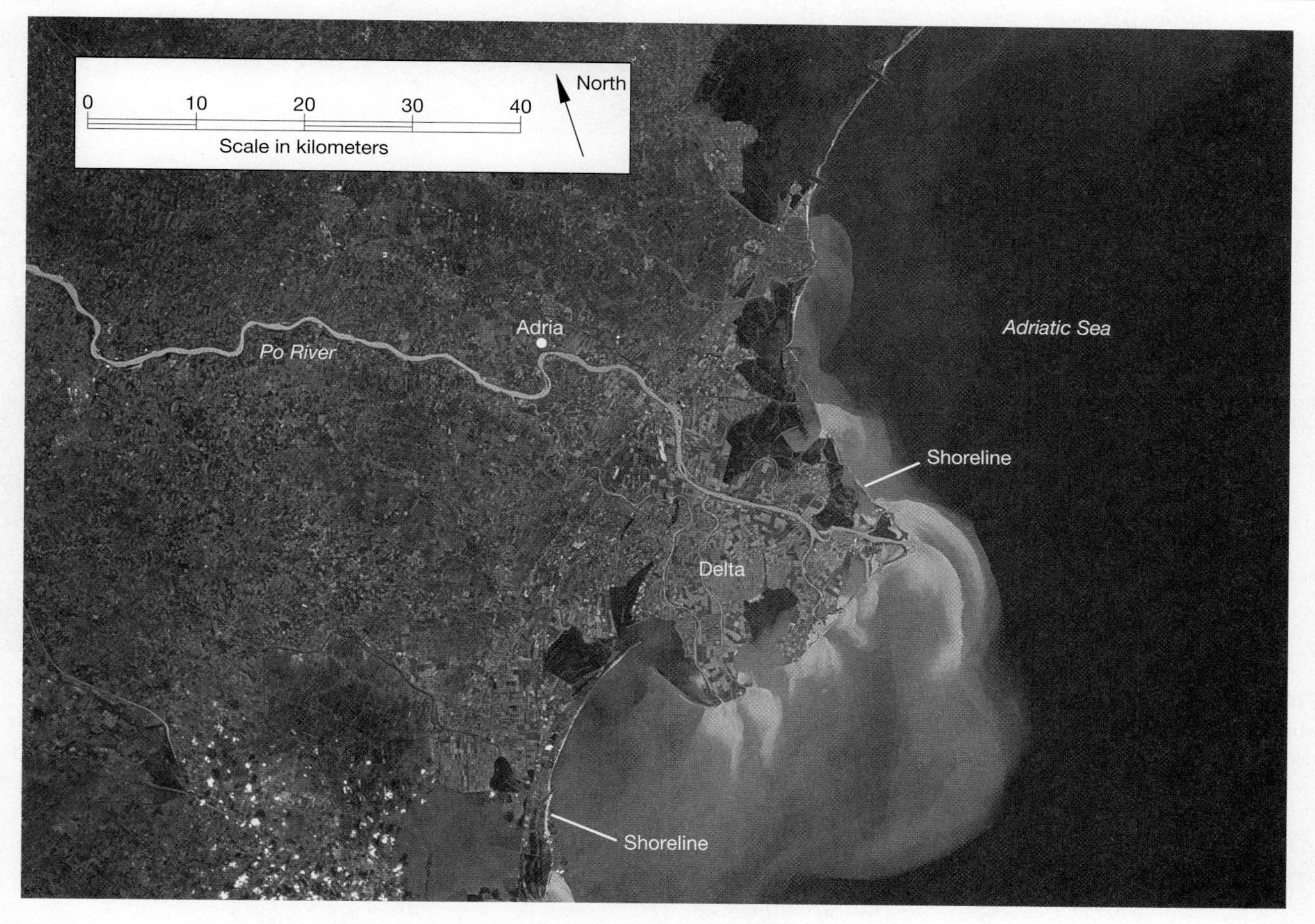

FIGURE 15.4 Space Shuttle photograph of the Po Delta region, northern Italy.
(Courtesy of NASA)

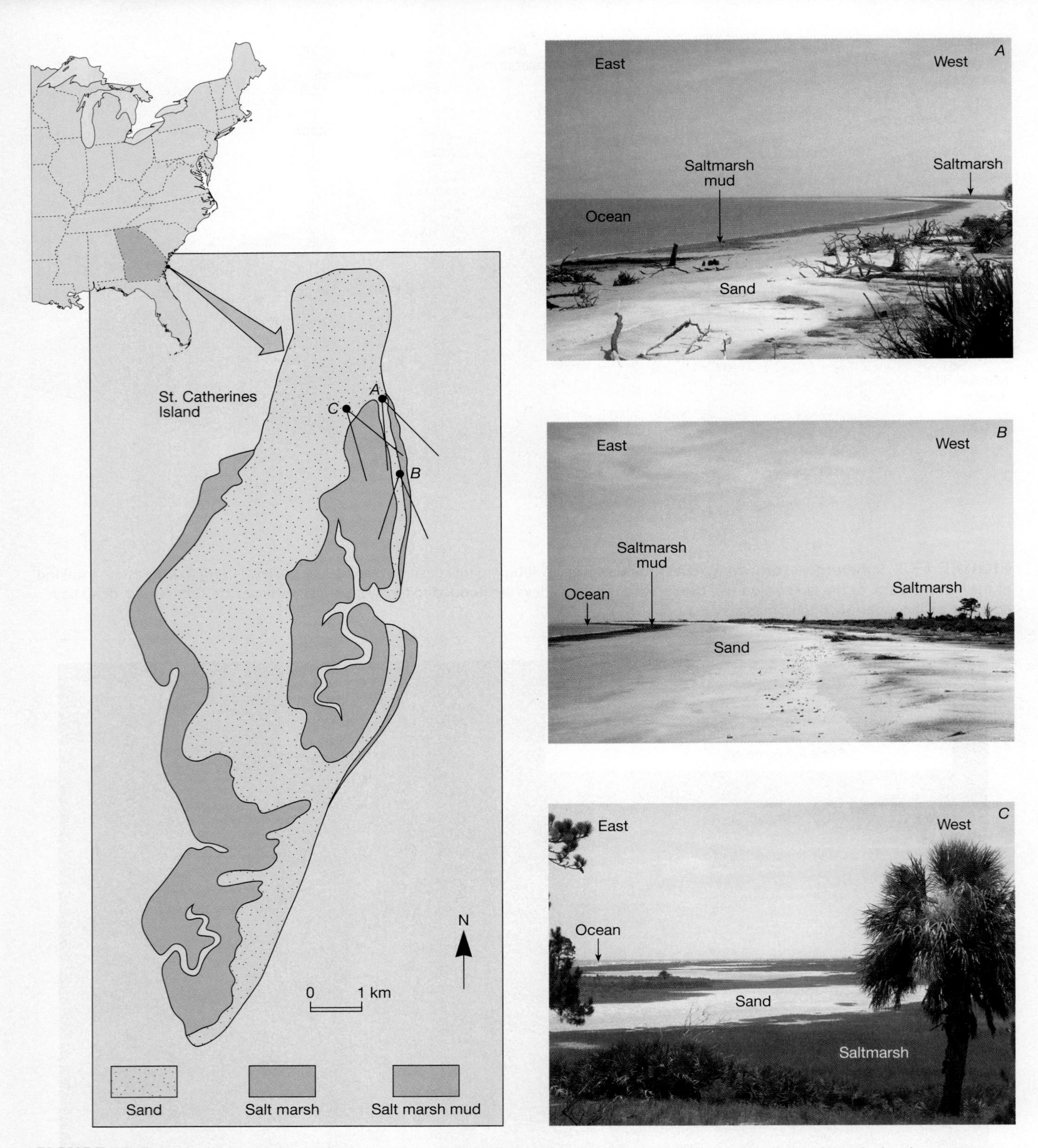

FIGURE 15.5 St. Catherines Island, Georgia. Notice the distribution of sand, salt marsh (living), and salt marsh mud, plus the points where the three photographs were taken. **A.** View south–southeast from point **A** on map, at low tide. Dark-brown "ribbon" adjacent to ocean is salt marsh mud. Light-colored area is sand. **B.** View south from point **B** on map at low tide. **C.** View southeast from point **C** (Aaron's Hill) on map.

- **Delta**—a sediment deposit at the mouth of a river where it enters an ocean or lake. (emergent, FIGURE 15.2).
- **Headland with cliffs**—projection of land that extends into an ocean or lake and generally has cliffs along its water boundary (submergent, FIGURE 15.3).
- **Spit**—a sand bar extending from the end of a beach into the mouth of an adjacent bay (emergent, FIGURE 15.2).
- **Tidal flat**—muddy or sandy area that is covered with water at high tide and exposed at low tide. Tidal flats are best developed along emergent coastlines.
- **Saltmarsh**—a marsh that is flooded by ocean water at high tide. Saltmarshes are best developed along emergent coastlines (FIGURE 15.2).
- **Wave-cut cliff** (or *sea cliff*)—a seaward-facing cliff along a steep shoreline, produced by wave erosion. Wave-cut cliffs are best developed along submergent coastlines (FIGURE 15.3).
- **Wave-cut platform**—a bench or shelf at sea level (or lake level) along a steep shore, and formed by wave erosion. Wave-cut platforms are best developed along emergent coastlines (FIGURE 15.2).
- **Marine terrace**—an elevated wave-cut platform that is bounded on its seaward side by a cliff or steep slope (and formed when a wave-cut platform is elevated by uplift or regression; emergent coastline, FIGURE 15.2).
- **Stack (sea stack)**—a small rocky island with steep or vertical sides, carved from a bedrock headland by wave erosion (FIGURE 15.3).
- **Tombolo**—a sand bar that connects an island with the mainland or another island. Tombolos are best developed along submergent coastlines (FIGURE 15.3).
- **Tied island**—an island connected to the mainland or another island by a tombolo (usually submergent, FIGURE 15.3).

ACTIVITY 15.3 Shoreline Modification at Ocean City, Maryland

THINK About It **How successful are efforts to protect shorelines from erosion by building artificial structures?**

OBJECTIVE Identify the common types of artificial structures used to modify shorelines and explain their effects on coastal environments.

PROCEDURES

1. **Before you begin**, read Human Modification of Shorelines below. Also, this is **what you will need**:
   ___ ruler, calculator
   ___ Activity 15.3 Worksheets (pp. 388–389) and pencil
2. **Then follow your instructor's directions** for completing the worksheets.

Human Modification of Shorelines

Humans build several common types of coastal structures in order to protect harbors, build up sandy beaches, or extend the shoreline. Study the following four kinds of structures and their effects in FIGURE 15.6.

- **Sea wall**—an embankment of boulders, reinforced concrete, or other material constructed against a shoreline to prevent erosion by waves and currents.
- **Breakwater**—an offshore wall constructed parallel to a shoreline to break waves. The longshore current is halted behind such walls, so the sand accumulates there and the beach widens. Where the breakwater is used to protect a harbor from currents and waves, sand often collects behind the breakwater and may have to be dredged.
- **Groin** (or *groyne*)—a short wall constructed perpendicular to shoreline in order to trap sand and make or build up a beach. Sand accumulates on the up-current side of the groin in relation to the longshore current.
- **Jetties**—long walls extending from shore at the mouths of harbors and used to protect the harbor entrance from filling with sand or being eroded by waves and currents. Jetties are usually constructed of boulders and in pairs (one on each side of a harbor or inlet).

ACTIVITY 15.4 The Threat of Rising Seas

THINK About It **How will rising sea levels affect communities along shorelines?**

OBJECTIVE Describe the probability of global sea-level rise and analyze the coastal hazards and increased risks it may cause.

PROCEDURES

1. **Before you begin**, read The Threat of Rising Seas below. Also, this is **what you will need**:
   ___ calculator
   ___ Activity 15.4 Worksheet (p. 390) and pencil
2. **Then follow your instructor's directions** for completing the worksheets.

The Threat of Rising Seas

The National Oceanic and Atmospheric Administration (NOAA), using measurements from satellite radar altimeters, estimates that global sea level has been rising at a rate of 2.9 mm/yr since 1992. Meanwhile, a 2013 report by Working Group 1 of the Intergovernmental

Panel on Climate Change (IPCC) of the United Nations Environment Programme (UNEP) has suggested that sea level will continue to rise and is now expected to achieve a mean global rise of at least 0.17–0.32 m (6.7–13.6 inches) by 2046–2065. Note that these figures are for mean (average) changes of global sea level. Specific locations may experience more or less of a rise in sea level. For example, although TOPEX/Poseidon satellite altimetry indicates that sea level is rising at a global rate of 2.9 mm/yr, NOAA tide gauge records indicate that sea level at Ocean City, Maryland (**FIGURE 15.7**), has been rising nearly twice as fast (5.48 mm/yr since 1975).

Sea level also fluctuates both above and below mean sea level during daily tidal cycles and storm surges. A **storm surge** is an abnormal rise of water pushed landward by high winds and/or low atmospheric pressure associated with storms. The storm surge is over and above the normal tide, and NOAA expresses it as the height above the expected tide level. NOAA also measures **storm tide**—which it defines as the water level height caused by a combination of the normal tide level and the storm surge. Storm surges can cause the ocean to rise by about 1–10 feet above the normal astronomical tide, depending on the magnitude of the storm and other factors. However, except for hurricanes, most storm surges are in the range of 2–3 feet.

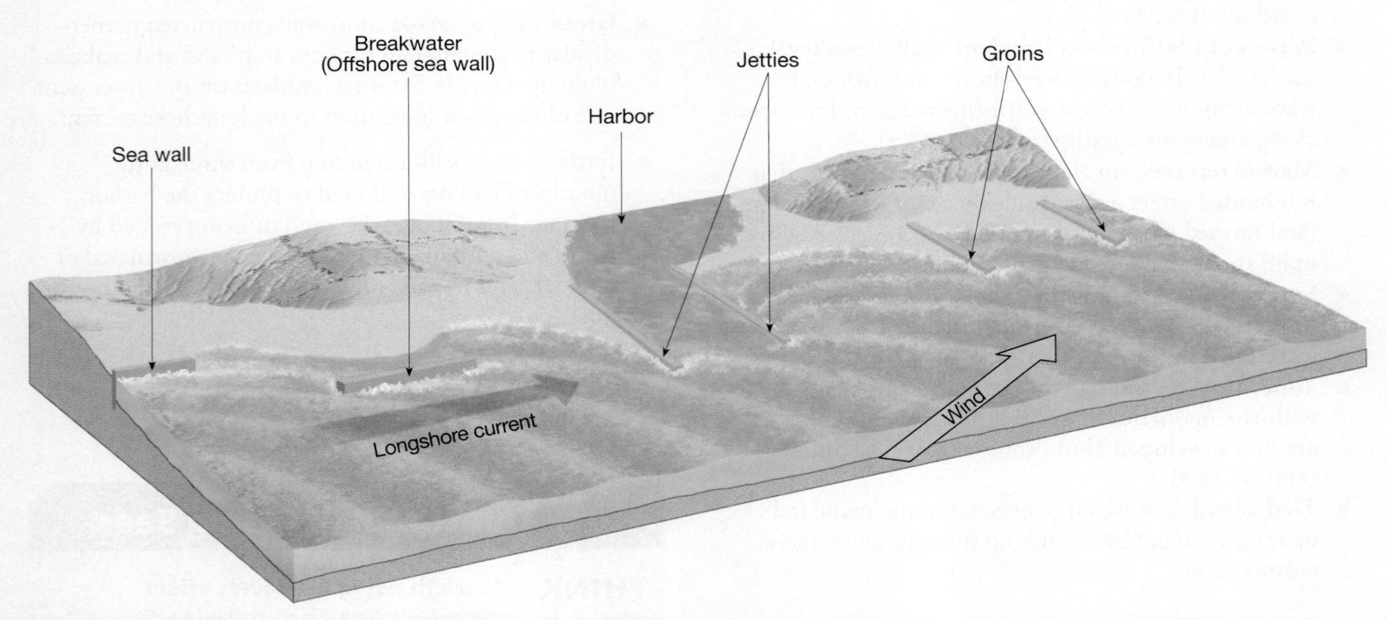

FIGURE 15.6 Coastal structures. Sea walls are constructed along the shore to stop erosion of the shore or extend the shoreline (as sediment is used to fill in behind them). **Breakwaters** are a type of offshore sea wall constructed parallel to shoreline. The breakwaters stop waves from reaching the beach, so the longshore drift is broken and sand accumulates behind them (instead of being carried down shore with the longshore current). **Groins** are short walls constructed perpendicular to shore. They trap sand on the side from which the longshore current is carrying sand against them. **Jetties** are long walls constructed at entrances to harbors to keep waves from entering the harbors. However, they also trap sand just like groins.

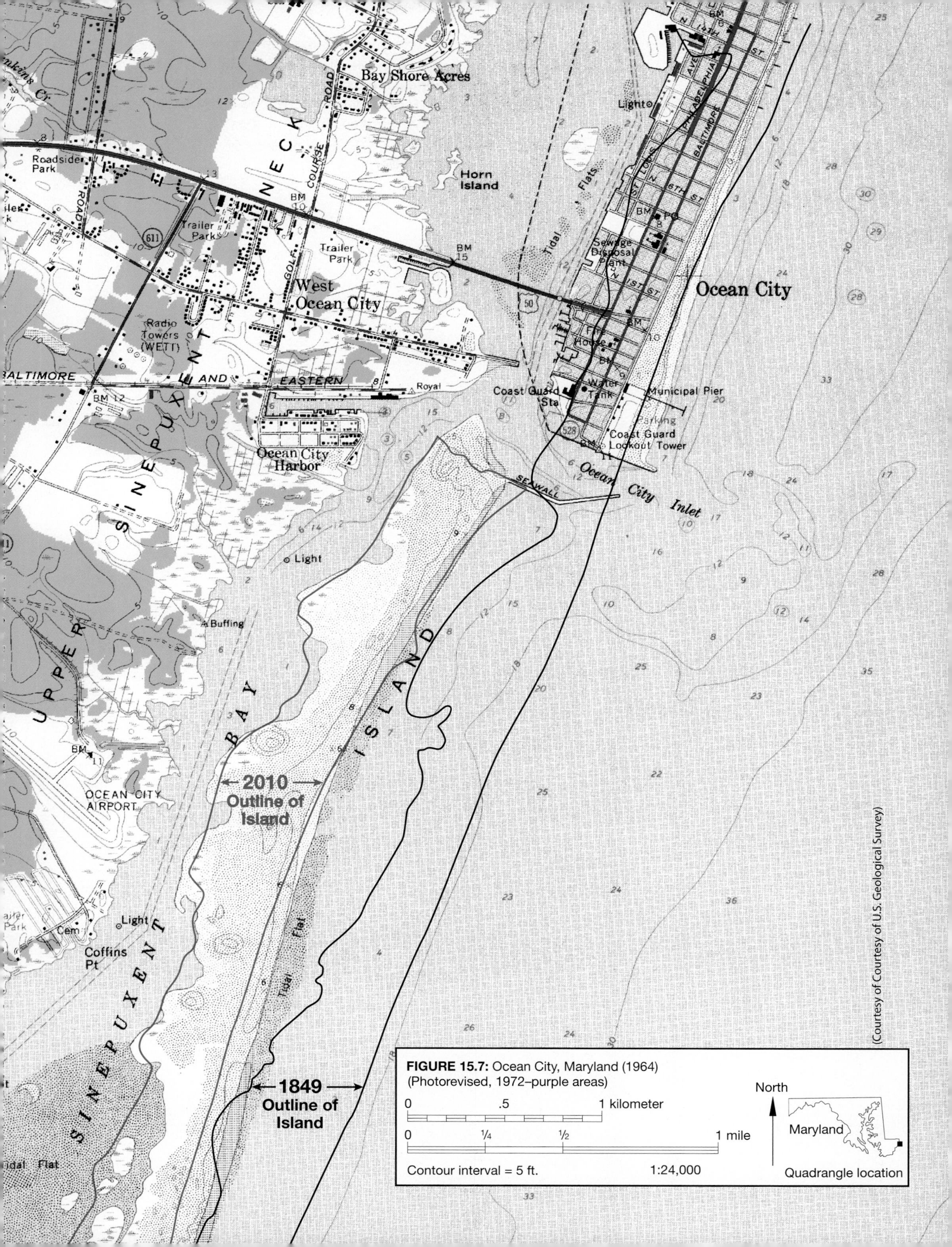

FIGURE 15.7: Ocean City, Maryland (1964) (Photorevised, 1972–purple areas)

ACTIVITY 15.1 Coastline Inquiry

Name: ______________________ **Course/Section:** ______________________ **Date:** ____________

A. Refer to the photographs of coastlines in **FIGURE 15.1** and the list of Factors Affecting Coastlines on page 376.

1. Describe what each shoreline is made of (if visible). Then name the two or three main factors that are primarily affecting the coastline and describe how they combine to shape the coastline.

a. Maryland coastline with saltmarsh grasses rooted in clay.

b. San Francisco, California, coastline.

c. Oregon coastline.

d. North Carolina coastline.

e. Destin, Florida, urbanized coastline.

f. Florida Keys coastline with mangrove plants.

g. Maine coastline (note person for scale).

h. Caribbean island coastline with fringing reefs (i.e., reefs attached to the island) and a barrier reef.

B. REFLECT & DISCUSS **FIGURES 15.1C** and **15.1D** are both sandy coastlines. Which one is building out into the water and what is causing that to happen? Which one seems to be receding landward, and what seems to be causing that to happen?

ACTIVITY 15.2 Introduction to Shorelines

Name: ______________________ **Course/Section:** ______________________ **Date:** ____________

A. Refer to the Space Shuttle photograph of the Po Delta, Italy (**FIGURE 15.4**). The city of Adria, on the Po River in Northern Italy, was a thriving seaport during Etruscan times (600 B.C.). Adria had such fame as to give its name to the Adriatic Sea, the gulf into which the Po River flows. Over the years, the Po River has deposited sediment at its mouth in the Po Delta. Because of the Po Delta's progradation, Adria is no longer located on the shoreline of the Adriatic Sea. The modern shoreline is far downstream from Adria.

1. What has been the average annual rate of Po Delta progradation in centimeters per year (cm/yr) since Adria was a thriving seaport on the coastline of the Adriatic Sea? (Show your work.)

2. Based on the average annual rate calculated above (A1), how many centimeters would the Po Delta prograde during the lifetime of someone who lived to be 60 years old? (Show your work.)

3. **REFLECT & DISCUSS** Sea level is rising and submerging coastlines adjacent to the Po Delta. Why do you think the delta is prograding out into the Adriatic Sea?

B. Refer to the map and photographs of Saint Catherines Island, Georgia (**FIGURE 15.5**). Note that on the southwestern and east-central parts of the island there are large areas of salt marsh. Living salt marsh plants are present there, as shown on the right (west) in **FIGURES 15.5A** and **B**. Also, note the linear sandy beach in **FIGURES 15.5A** and **B**, bounded on its seaward side (left) by another strip of salt marsh mud. However, all of the living, surficial saltmarsh plants and animals have been stripped from this area. This is called **relict** salt marsh mud (mud remaining from an ancient salt marsh).

1. What type of sediment is probably present beneath the beach sands in **FIGURES 15.5A** and **B**?

2. Explain how you think the beach sands became located landward of the relict saltmarsh mud.

3. Portions of the living saltmarsh (wetland) in **FIGURE 15.5C** recently have been buried by bodies of white sand that was deposited from storm waves that crashed over the beach and sand dunes. What is the name given to such sand bodies?

4. Photograph 15.6C was taken from a landform called Aaron's Hill. It is the headland of this part of the island. What will eventually happen to Aaron's Hill? Why?

5. Based upon your answer in part 4, would Aaron's Hill be a good location for a resort hotel? Explain your answer.

6. **REFLECT & DISCUSS** Based upon your inferences, observations, and explanations above, what will eventually happen to the living salt marsh in **FIGURES 15.5B** and **C**?

ACTIVITY 15.3 Shoreline Modification at Ocean City, Maryland

Name: ______________________ **Course/Section:** ______________________ **Date:** ____________

Ocean City is located on a long, narrow barrier island called Fenwick Island. During a severe hurricane in 1933, the island was breached by tidal currents that formed Ocean City Inlet and split the barrier island in two. Ocean City is still located on what remains of Fenwick Island. The city is a popular vacation resort that has undergone much property development over the past 50 years. The island south of Ocean City Inlet is called Assateague Island. It has remained undeveloped, as a state and national seashore.

Rising sea level at Ocean City has increased the risk of beach erosion there. Therefore, Ocean City constructed barriers to trap sand. Examine the portion of the Ocean City, Maryland, topographic quadrangle map provided in **FIGURE 15.7**. Purple features show changes made in 1972 to a 1964 map, so you can see how the coastline changed from 1964–1972. Also note the black and red outlines of the barrier island as it appeared in 1849 and 2010 according to the U.S. Geological Survey.

A. After the 1933 hurricane carved out Ocean City Inlet, the Army Corps of Engineers constructed a pair of jetties on each side of Ocean City Inlet to keep it open. The southern jetty is labeled "seawall" on the map. Sand filled in behind the northern jetty, so it is now a seawall forming the straight southern edge of Ocean City on Fenwick Island (a straight black line on the map). Based on this information, would you say that the longshore current along this coastline is traveling north to south, or south to north? Explain your reasoning.

B. Notice that Assateague Island has migrated landward (west), relative to its 1849 position (**FIGURE 15.7**). This migration began in 1933.

1. Why did Assateague Island migrate landward?

2. Field inspection of the west side of Assateague Island reveals that muds of the lagoon (Sinepuxent Bay) are being covered up by the westward-advancing island. What was the rate of Assateague Island's westward migration from 1933–1972 in feet/year and meters/year? (Show your work.)

3. Based on your last answer above (B2), and extrapolating from 1972, in what approximate year should the west side of Assateague Island have merged with saltmarshes around Ocean City Harbor? (Show your work.)

4. Notice from the 2010 position of Assateague Island (red outline on **FIGURE 15.7**) that it has not merged with saltmarshes of the mainland. What natural processes and human activities may have prevented this?

C. Notice the groins (short black lines) that have been constructed on the east side of Fenwick Island (Ocean City) in the northeast corner of **FIGURE 15.7** (above 2 km north of the inlet).

1. Why do you think these groins have been constructed there?

2. What effect could these groins have on the beaches around Ocean City's Municipal Pier (southern end of Fenwick Island)? Why?

D. Hurricanes normally approach Ocean City from the south–southeast. In 1995, one of the largest hurricanes ever recorded (Hurricane Felix) approached Ocean City but miraculously turned back out into the Atlantic Ocean. How does the westward migration of Assateague Island increase the risk of hurricane damage to Ocean City?

E. Compare the position of Assateague Island in **FIGURE 15.7**, from 1972 (purple position of the island) to 2010 (red outline of the island). The northern two kilometers of the island remained in a relatively stable position from 1972–2010, but the rest of Assateague Island did not. Explain what happened to the southern three kilometers of Assateague Island on **FIGURE 15.7** from 1972–2010 and infer why it may have happened.

F. REFLECT & DISCUSS The westward migration of Assateague Island could be halted and probably reversed if all of the groins, jetties, and sea walls around Ocean City were removed. How would removal of all of these structures place properties in Ocean City at greater risk to environmental damage than they now face?

ACTIVITY 15.4 The Threat of Rising Seas

Name: ____________________ **Course/Section:** ____________________ **Date:** ____________

In planning for coastal management and safe and economical coastal development, responsible planning commissions and real estate developers should "play it safe" and assume that sea level will continue to rise. There are many predictions of future rises in global mean sea level, but regional trends should also be considered as in these examples.

A. Imagine that you are planning to buy a shorefront property in Ocean City, Maryland, this year. You plan to use the property for family vacation getaways over the next 50 years and then sell the property. The front door of the property was four feet above mean sea level in 2010.

1. According to the U.S. National Oceanic and Atmospheric Administration, the historic rate of sea level rise here since 1975 has been 5.48 +/– 1.67 mm/yr. Using the "plus or minus" error, what has been the minimum rate and the maximum rate of mean sea level rise here in mm/yr?

a. ____________________ mm/yr minimum rate **b.** ____________________ mm/yr maximum rate

2. Using the minimum and maximum rates above, calculate how much sea level will rise in mm and inches at Ocean City over the next 50 years.

a. ____________________ mm minimum **b.** ____________________ inches minimum

c. ____________________ mm maximum **d.** ____________________ inches maximum

3. Mean sea level is the average position of sea level between low and high tides. High tides occasionally reach 2.9 feet (0.88 m) above mean sea level here, and storm surges often raise sea level an additional foot (0.3 m). When Hurricane Sandy passed offshore of Ocean City in 2012, the storm surge caused a total storm tide of 3.59 feet. Given these natural day-to-day variations in sea level, and the prospect of sea level rise calculated above, would it be a wise decision to purchase the shorefront property that you planned to buy? Explain your reasoning.

4. The City of Ocean City expects the following temporary increases in sea level due to storm surges in hurricanes. How would this affect your purchasing decision? Why?

Category 1 hurricane: 74–95 mph winds, Storm Surge: 4–5 feet

Category 2 hurricane: 96–110 mph winds, Storm Surge: 6–8 feet

Category 3 hurricane: 111–130 mph winds, Storm Surge: 9–12 feet

Category 4 hurricane: 131–155 mph winds, Storm Surge: 13–18 feet

Category 5 hurricane: 156 mph + winds, Storm Surge: more than 18 feet

5. REFLECT & DISCUSS Given the fact that most existing topographic maps of coastal areas have contour intervals of 5 feet, what would you suggest as the contour line below which construction of the living/working floor of homes should not occur along the Ocean City coast? Explain.

BIG IDEAS

Earthquakes are vibrations of Earth that occur naturally when magma moves beneath volcanoes, volcanoes erupt, or rocks suddenly rupture along faults. The vibrations radiate outward as seismic waves, which are detected at seismic stations and recorded as seismograms. The seismograms can be used to calculate distance from the earthquake, and seismograms from three or more seismic stations can be used to determine where the earthquake occurred. Geologists use models to study the effects of earthquakes on natural materials and human structures and mitigate earthquake hazards.

FOCUS YOUR INQUIRY

THINK About It How do bedrock and sediment behave during earthquakes and how does this affect human-made structures?

ACTIVITY 16.1 Earthquake Hazards Inquiry *(p. 392)*

THINK About It How can seismic wave data be used to locate the epicenter of an earthquake?

ACTIVITY 16.2 How Seismic Waves Travel through Earth *(p. 392)*

ACTIVITY 16.3 Locate the Epicenter of an Earthquake *(p. 393)*

THINK About It How do geologists use remote sensing, geologic maps, seismograms, and first motion studies to analyze fault motions?

ACTIVITY 16.4 San Andreas Fault Analysis at Wallace Creek *(p. 394)*

ACTIVITY 16.5 New Madrid Blind Fault Zone *(p. 394)*

16

LABORATORY

Earthquake Hazards and Human Risks

CONTRIBUTING AUTHORS

Thomas H. Anderson • *University of Pittsburgh*
David N. Lumsden • *University of Memphis*
Pamela J.W. Gore • *Georgia Perimeter College*

An earthquake in Sichuan, China pulled this road apart, damaged this home, and caused boulders to fall from adjacent hillsides. (USGS image by Sarah C. Behan)

Introduction

Earthquakes are shaking motions and vibrations of Earth caused by large releases of energy that accompany volcanic eruptions, explosions, and movements of Earth's bedrock along fault lines. News reports usually describe an earthquake's **epicenter,** which is the point on Earth's surface (location on a map) directly above the **focus** (underground origin of the earthquake, in bedrock) (**FIGURE 16.1**).The episodic releases of energy that occur along fault lines strain the bedrock like a person jumping on a diving board. This strain produces waves of vibration and shaking called **seismic waves** (earthquake waves). Seismic waves originate at the earthquake's focus (**FIGURES 16.1**) and travel in all directions through the rock body of Earth and along Earth's surface. They are detected at seismic stations (**FIGURES 16.1** and **16.2**) and recorded as seismograms (**FIGURE 16.3**). The up and down motions on the seismograms in **FIGURE 16.3** record the passing of seismic waves.

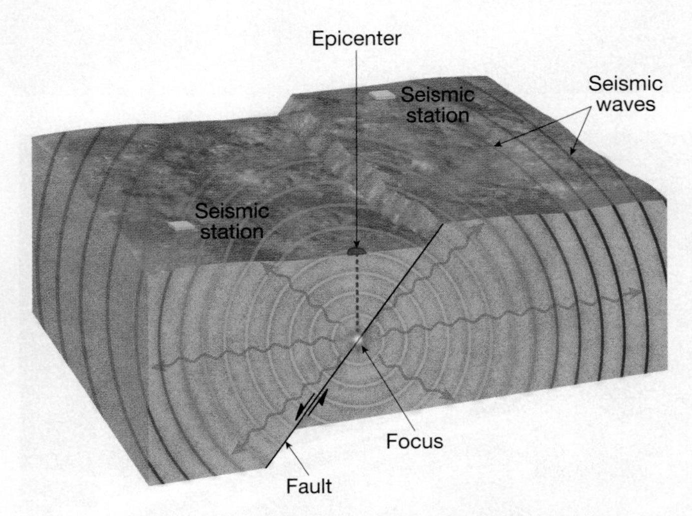

FIGURE 16.1 The focus and epicenter of an earthquake.

ACTIVITY 16.1 Earthquake Hazards Inquiry

THINK About It How do bedrock and sediment behave during earthquakes and how does this affect human-made structures?

OBJECTIVE Experiment with models to determine how earthquake damage to buildings is related to the Earth materials on which they are constructed. Apply your experimental results to evaluate earthquake hazards and human risks in San Francisco.

PROCEDURES

1. **Before you begin**, read the Introduction and Earthquake Hazards and Risks below. Also, this is **what you will need:**
   ____ Activity 16.1 Worksheet (p. 397) and pencil
   ____ Materials provided in the lab: cups, sand, coins, dropper bottles with water
2. **Complete the worksheet in a way that makes sense to you.**
3. **After you complete the worksheet**, be prepared to discuss your observations and classification with other geologists.

ACTIVITY 16.2 How Seismic Waves Travel through Earth

THINK About It How can seismic wave data be used to locate the epicenter of an earthquake?

OBJECTIVE Graph seismic data to construct and evaluate travel time curves for P-waves, S-waves, and L-waves, then use seismograms and travel time curves to locate the epicenter of an earthquake.

PROCEDURES

1. **Before you begin**, read Graphing Seismic Data and Locating the Epicenter of an Earthquake below. Also, this is **what you will need:**
   ___ ruler, calculator
   ___ Activity 16.2 Worksheet (p. 399) and pencil
2. **Then follow your instructor's directions** for completing the worksheets.

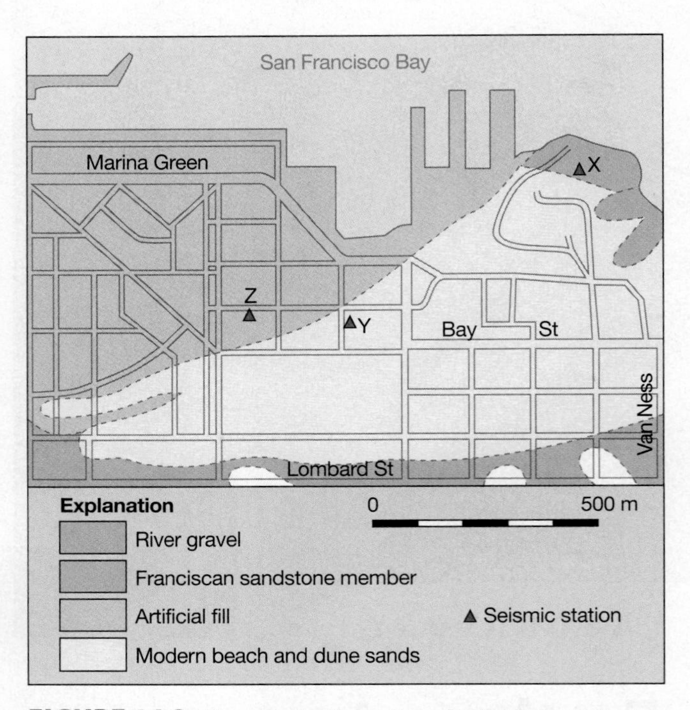

FIGURE 16.2 Map of seismic stations for Activity 16.1. The map also shows the nature and distribution of Earth materials on which buildings and roads have been constructed for a portion of San Francisco, California. (Courtesy of U.S. Geological Survey)

Earthquake Hazards and Risks

Earthquakes can have very different effects on buildings in different parts of the same city. To understand why, geologists compare earthquake damage to maps of bedrock. They also simulate earthquakes in the laboratory to observe their effects on models of construction sites, buildings, bridges, and so on. All of this information is used to construct earthquake hazard maps and revise building codes for earthquake prone regions.

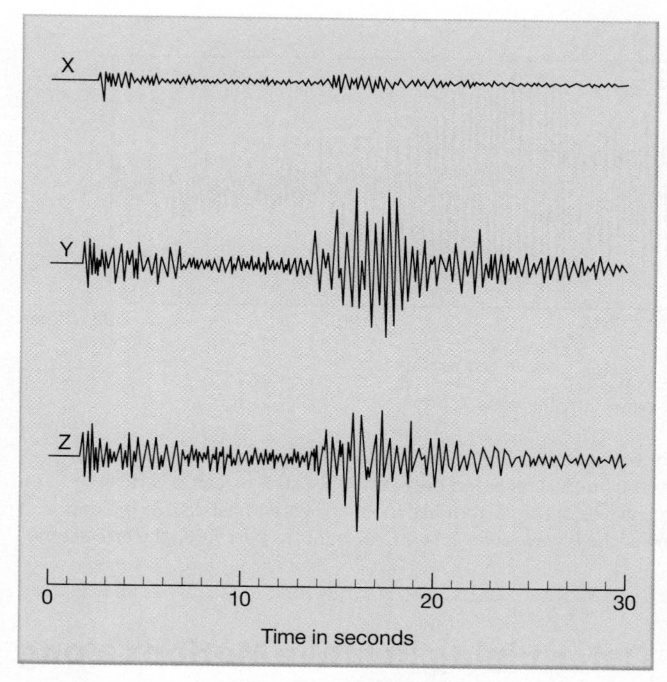

FIGURE 16.3 Seismograms recorded at stations X, Y, and Z on Figure 16.3. The seismograms record a strong (Richter Magnitude 4.6) aftershock of the Loma Prieta, California, earthquake. During the earthquake, little damage occurred at **X**, but significant damage to houses occurred at **Y** and **Z**. (Courtesy of U.S. Geological Survey)

ACTIVITY 16.3 Locate the Epicenter of an Earthquake

THINK About It | **How can seismic wave data be used to locate the epicenter of an earthquake?**

OBJECTIVE Graph seismic data to construct and evaluate travel time curves for P-waves, S-waves, and L-waves, then use seismograms and travel time curves to locate the epicenter of an earthquake.

PROCEDURES

1. **Before you begin**, read Graphing Seismic Data and Locating the Epicenter of an Earthquake below. Also, this is **what you will need:**
   ___ ruler
   ___ Activity 16.3 Worksheet (p. 401) and pencil
   ___ Drafting compass will be provided in the laboratory.
2. **Then follow your instructor's directions** for completing the worksheets.

Graphing Seismic Data and Locating the Epicenter of an Earthquake

An earthquake produces three main types of seismic waves that radiate from its focus/epicenter at different rates. Seismographs are instruments used to detect these seismic waves and produce a **seismogram**—a record of seismic wave motions obtained at a specific recording station (**FIGURE 16.4**). The seismograms are analyzed using graphs of travel time versus distance (**FIGURE 16.5**) to determine the seismic station's distance from the earthquake epicenter.

3 Main Kinds of Seismic Waves

Seismographs can detect and record (as seismograms) several types of seismic waves.

- **P-waves:** *P* for primary, because they travel fastest and arrive at seismographs first. (They are compressional, or "push–pull" waves.) P-waves are *body waves*, meaning that they travel through Earth's interior (rather than along its surface) and radiate in all directions from the focus.
- **S-waves:** *S* for secondary, because they travel more slowly and arrive at seismographs after the P-waves. (They are perpendicular, shear, or "side-to-side" waves.) S-waves are body waves, like the P-waves.
- **L-waves** or *Love waves* (named for A. E. H. Love, who discovered them): L-waves are not body waves like those above. L-waves travel along Earth's surface (a longer route than the body waves) and thus are recorded after the S-waves and P-waves arrive at the seismograph.

Interpreting Seismograms

FIGURE 16.4 is a seismogram recorded at a station located in Australia. Seismic waves arrived there from an earthquake epicenter located 1800 kilometers (1125 miles) away in New Guinea. Notice that the seismic waves were recorded as deviations (vertical zigzags) from the nearly horizontal line of normal background vibrations. Thus, the first pulse of seismic waves was P-waves, which had an **arrival time** of 7:14.2 (i.e., 14.2 minutes after 7:00). The second pulse of seismic waves was the slower S-waves, which had an arrival time of 7:17.4. The final pulse of seismic waves was the L-waves that traveled along Earth's surface, so they did not begin to arrive until 7:18.3. The earthquake actually occurred at the New Guinea epicenter at 7:10:23 (or 10.4 minutes after 7:00, written 7:10.4). Therefore the **travel time of the main seismic waves** (to go 1800 km) was 3.8 minutes for P-waves (7:14.2 minus 7:10.4), 7.0 minutes for S-waves (7:17.4 minus 7:10.4), and 7.9 minutes for L-waves (7:18.3 minus 7:10.4).

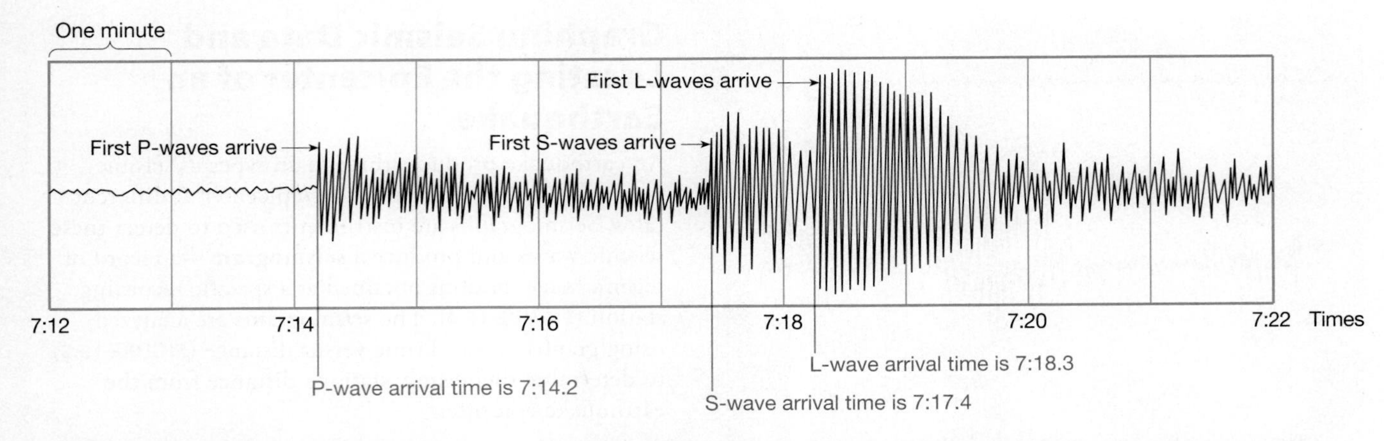

FIGURE 16.4 Seismogram of a New Guinea earthquake. Most of the seismogram, recorded at a location in Australia, shows only minor background deviations (short zigzags) from a horizontal line, such as the interval recorded between 7:12 and 7:14. Large vertical deviations indicate motions caused by the arrival of P-waves, S-waves, and L-waves of the earthquake (note arrows with labels). By making detailed measurements with a ruler, you can determine that the arrival time of the P-waves was 7:14.2 (14.2 minutes past 7:00), the arrival time of the S-waves was 7:17.4, and the arrival time of the L-waves was 7:18.3.

ACTIVITY 16.4 San Andreas Fault Analysis at Wallace Creek

THINK About It **How do geologists use remote sensing, geologic maps, seismograms, and first motion studies to analyze fault motions?**

OBJECTIVE Analyze and evaluate active faults using remote sensing and geologic maps.

PROCEDURES

1. **Before you begin**, this is **what you will need:**
   ___ ruler
   ___ Activity 16.4 Worksheet (p. 403) and pencil with eraser

ACTIVITY 16.5 New Madrid Blind Fault Zone

THINK About It **How do geologists use remote sensing, geologic maps, seismograms, and first motion studies to analyze fault motions?**

OBJECTIVE Interpret seismograms and first motion studies to infer fault motion in the blind New Madrid Fault within the North American Plate.

PROCEDURES

1. **Before you begin**, read Determining Relative Motions Along the New Madrid Fault Zone. Also, this is **what you will need:**
   ___ ruler
   ___ Activity 16.5 Worksheet (p. 404) and pencil

Determining Relative Motions along the New Madrid Fault Zone

The relative motions of blocks of rock on either side of a fault zone can be determined by mapping the way the pen on a seismograph moved (up or down on the seismogram) when P-waves first arrived at various seismic stations adjacent to the fault. This pen motion is called **first motion** and represents the reaction of the P-wave to dilation (pulling rocks apart) or compression (squeezing rocks together) as observed on seismograms (see **FIGURE 16.6**, left).

If the first movement of the P-wave was up on a seismogram, then that recording station (where the seismogram was obtained) experienced compression during the earthquake. If the first movement of the P-wave was down on a seismogram, then that recording station was dilational during the earthquake. What was the first motion at all of the seismic stations in **FIGURE 16.3**? (Answer: The first movement of the pen was up for each P-wave, so the first motion at all three sites was compressional.)

By plotting the first motions observed at recording stations on both sides of a fault that has experienced an earthquake, a picture of the relative motions of the fault emerges. For example, notice that the first motions observed at seismic stations on either side of a hypothetical fault are plotted in relation to the fault in **FIGURE 16.6** (right side). The half-arrows indicate how motion proceeded away from seismic stations where dilation was recorded and toward seismic stations where compression was recorded (for each side of the fault).

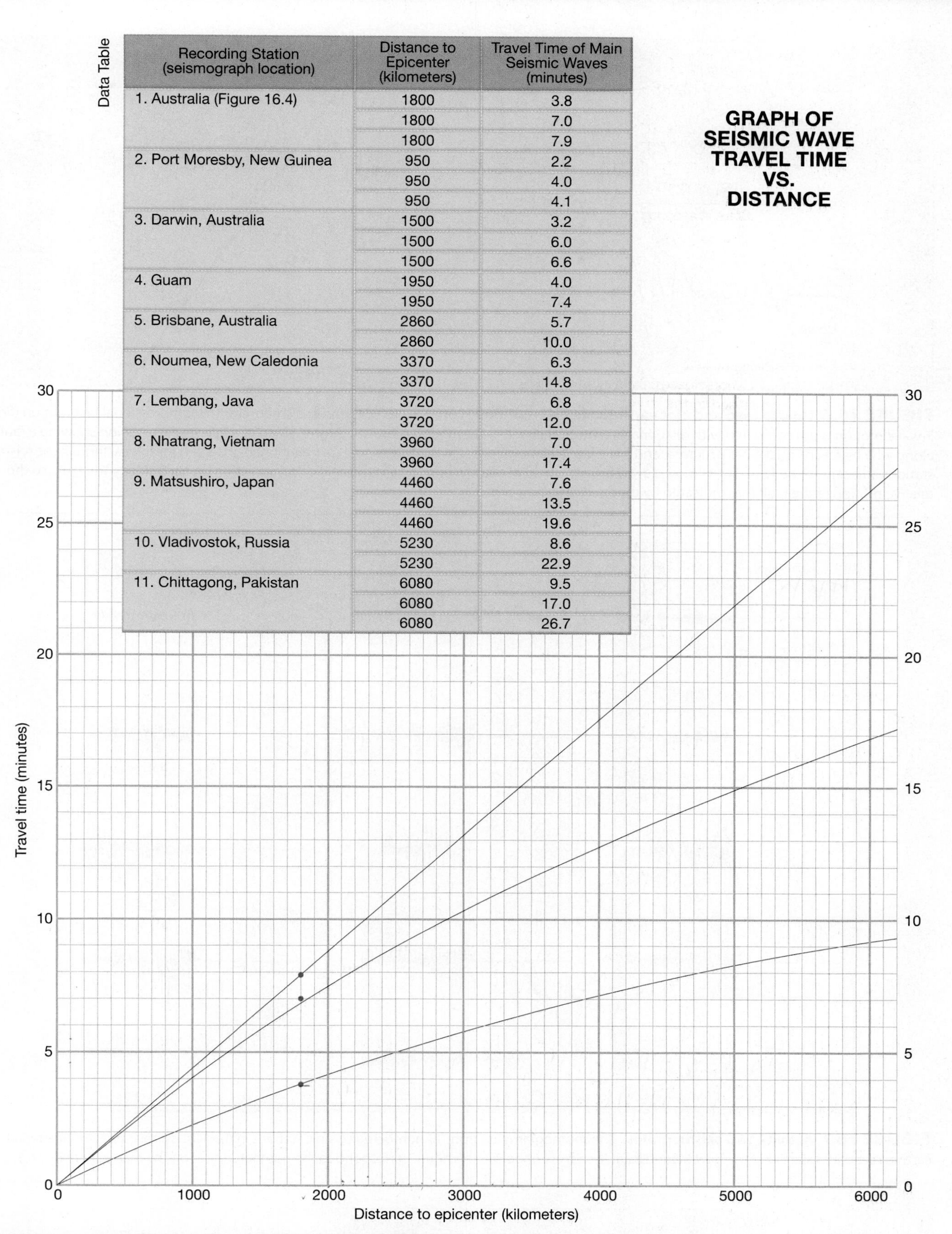

Data Table

Recording Station (seismograph location)	Distance to Epicenter (kilometers)	Travel Time of Main Seismic Waves (minutes)
1. Australia (Figure 16.4)	1800	3.8
	1800	7.0
	1800	7.9
2. Port Moresby, New Guinea	950	2.2
	950	4.0
	950	4.1
3. Darwin, Australia	1500	3.2
	1500	6.0
	1500	6.6
4. Guam	1950	4.0
	1950	7.4
5. Brisbane, Australia	2860	5.7
	2860	10.0
6. Noumea, New Caledonia	3370	6.3
	3370	14.8
7. Lembang, Java	3720	6.8
	3720	12.0
8. Nhatrang, Vietnam	3960	7.0
	3960	17.4
9. Matsushiro, Japan	4460	7.6
	4460	13.5
	4460	19.6
10. Vladivostok, Russia	5230	8.6
	5230	22.9
11. Chittagong, Pakistan	6080	9.5
	6080	17.0
	6080	26.7

FIGURE 16.5 Seismic wave data and time-distance graph paper. The data are for an eathquake that occurred in New Guinea (at 3° North latitude and 140° East longitude) at a Greenwich Mean Time of 10.4 minutes past 7:00 (7:10.4). The travel time of a main seismic wave is the time interval between when the earthquake occurred in New Guinea and when that wave first arrived at a recording location. The surface distance is the distance between the recording location and the earthquake epicenter. Graph is for plotting points that represent the travel time of each main seismic wave at each location versus the surface distance that it traveled.

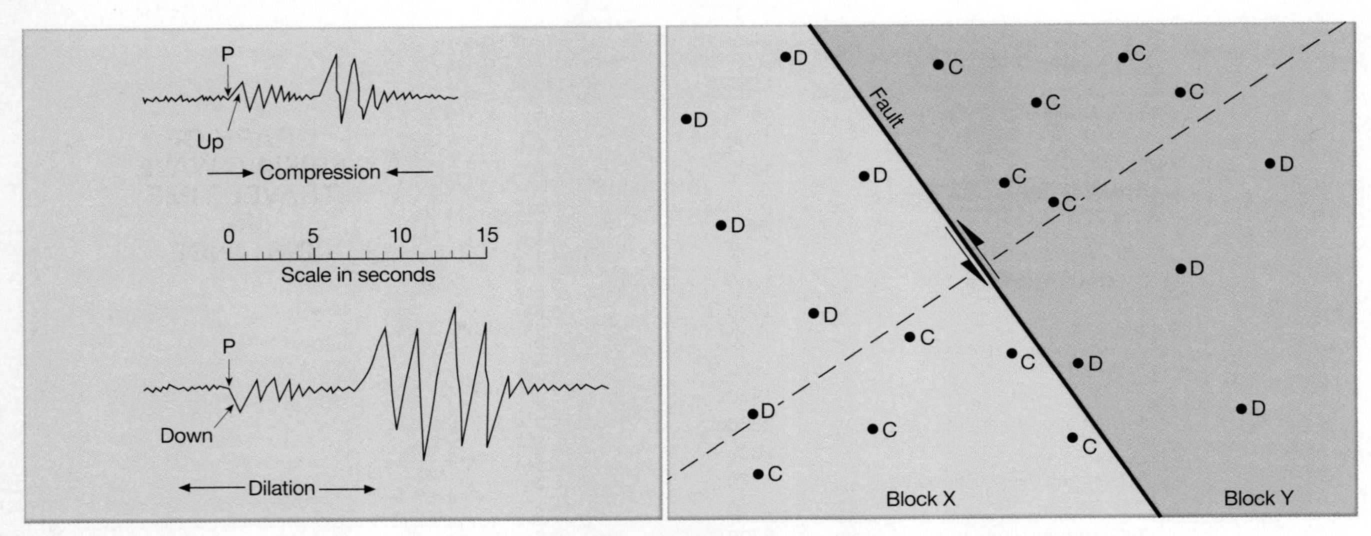

FIGURE 16.6 How to read and plot earthquake first motions. Left—Sketch of typical seismograms for compressional first motion (first P-wave motion is up) compared with dilational first motion (first P-wave motion is down). **Right**—Map of a hypothetical region showing a fault along which an earthquake has occurred, and the P-wave first motions (C = compressional, D = dilational) observed for the earthquake at seismic stations adjacent to the fault. Stress moves away from the field of dilation and toward the field of compression on each side of the fault, so the relative motion of the fault is as indicated by the smaller half-arrows.

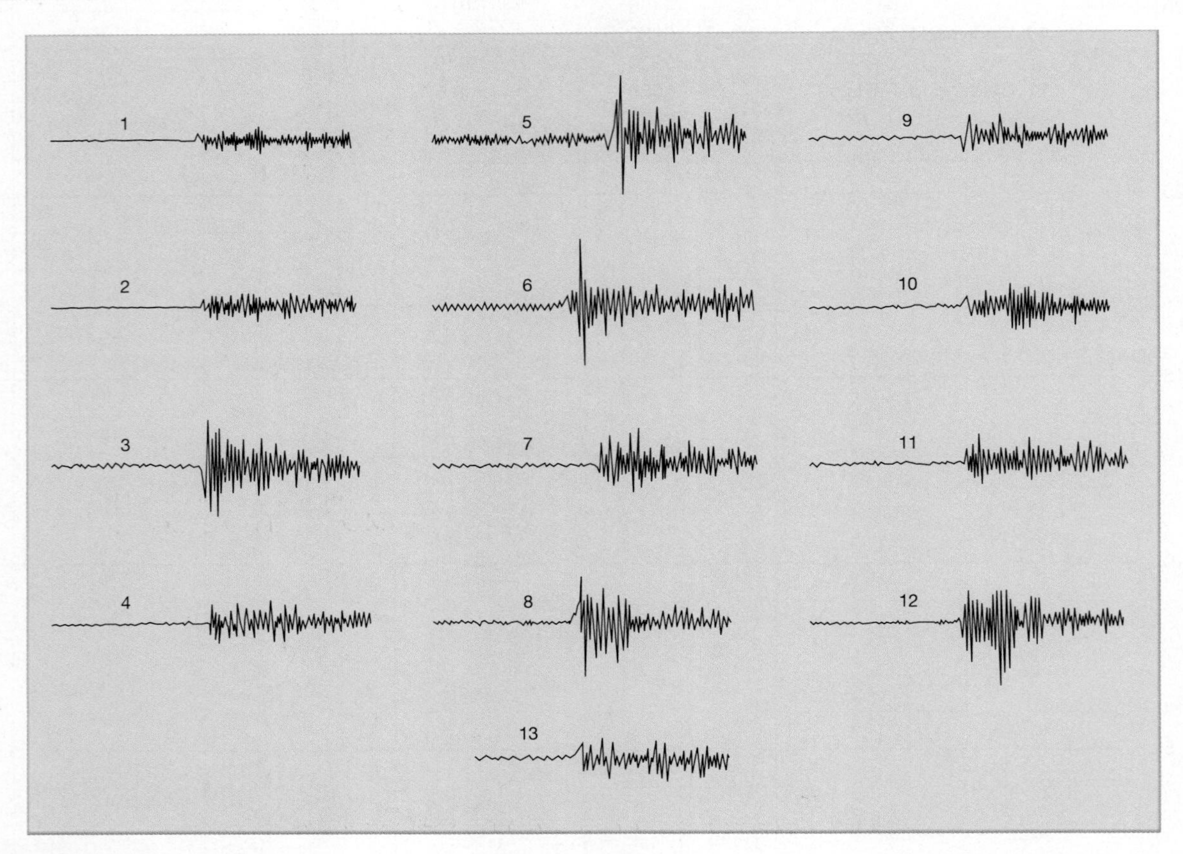

FIGURE 16.7 Activity 16.5 seismograms. The seismograms are from 13 numbered seismic stations in the Mississippi Embayment after an earthquake that occurred in the New Madrid Fault System. Numbers in this figure correspond to the numbered sites on the map in Activity 16.5.

ACTIVITY 16.1 Earthquake Hazards Inquiry

Name: ______________________ **Course/Section:** ______________________ **Date:** ____________

A. Obtain a small plastic or paper cup. Fill it three-quarters full with a dry sand. Place several coins upright in the sediment so they resemble vertical walls of buildings constructed on a substrate of uncompacted sediment. This is Model 1. Observe what happens to Model 1 when you *simulate an earthquake* by tapping the cup on a table top while you also rotate it counterclockwise.

1. What happened to the vertically positioned coins in the uncompacted sediment of Model 1 when you simulated an earthquake?

2. Now make Model 2. Remove the coins from Model 1, and add a small bit of water to the sediment in the cup so that it is moist (but not soupy). Press down on the sediment in the cup so that it is well compacted, and then place the coins into this compacted sediment just as you placed them in Model 1 earlier. *Simulate an earthquake* as you did for Model 1. What happened to the vertically positioned coins in the compacted sediment of Model 2 when you simulated an earthquake?

3. Based on your experimental Models 1 and 2 above, which kind of Earth material is more hazardous to build on in earthquake-prone regions: compacted sediment or uncompacted sediment? (Justify your answer by citing evidence from your experimental models.)

4. Consider the moist, compacted sediment in Model 2. Do you think this material would become *more* hazardous to build on, or *less* hazardous to build on, if it became totally saturated with water during a rainy season? To find out, design and conduct another experimental model of your own. Call it Model 3, describe what you did, and tell what you learned.

5. REFLECT & DISCUSS Write a statement that summarizes how water in a sandy substrate beneath a home can be beneficial or hazardous. Justify your reasoning with reference to your experimental models.

B. San Francisco, California is located in a tectonically active region, so it occasionally experiences strong earthquakes. **FIGURE 16.2** is a map showing the kinds of Earth materials upon which buildings have been constructed in a portion of San Francisco. These materials include hard compact Franciscan Sandstone, uncompacted beach and dune sands, river gravel, and artificial fill. The artificial fill is mostly debris from buildings destroyed in the great 1906 earthquake that reduced large portions of the city to blocks of rubble. Imagine that you have been hired by an insurance company to assess what risk there may be in buying newly constructed apartment buildings located at **X, Y,** and **Z** on **FIGURE 16.2**. Your job is to infer whether the risk of property damage during strong earthquakes is **low** (little or no damage expected) or **high** (damage can be expected). All that you have as a basis for reasoning is **FIGURE 16.2** and knowledge of your experiments with models in Part A of this activity.

1. Is the risk at location **X** low or high? Why?

low because of bedrock

2. Is the risk at location **Y** low or high? Why?

Moderate because

3. Is the risk at location **Z** low or high? Why?

High because it is artificial

C. On October 17, 1989, just as Game 3 of the World Series was about to start in San Francisco, a strong earthquake occurred at Loma Prieta, California, and shook the entire San Francisco Bay area. Seismographs at locations **X, Y,** and **Z** (see **FIGURE 16.2**) recorded the shaking, and the resulting seismograms are shown in **FIGURE 16.3**. Earthquakes are recorded on the seismograms as deviations (vertical zigzags) from a flat, horizontal line. Thus, notice that much more shaking occurred at locations **Y** and **Z** than at location **X.**

1. The Loma Prieta earthquake caused no significant damage at location **X,** but there was moderate damage to buildings at location **Y** and severe damage at location **Z.** Explain how this damage report compares to your above predictions of risk (Part B).

accurate

2. The Loma Prieta earthquake shook all of the San Francisco Bay region. Yet **FIGURE 16.3** is evidence that the earthquake had very different effects on properties located only 600 m apart. Explain how the kind of substrate (uncompacted vs. firm and compacted) on which buildings are constructed influences how much the buildings are shaken and damaged in an earthquake.

The softer the material slows down the waves & amplifies the altitude.

D. REFLECT & DISCUSS Imagine that you are a member of the San Francisco City Council. Name two actions that you could propose to **mitigate** (decrease the probability of) future earthquake hazards such as the damage that occurred at locations **Y** and **Z** in the Loma Prieta earthquake.

ACTIVITY 16.2 How Seismic Waves Travel through Earth

Name: ______________________ **Course/Section:** ______________________ **Date:** ____________

Notice the seismic data provided with the graph in **FIGURE 16.5**. There are data for 11 recording stations where seismograms were recorded after the same New Guinea earthquake (at 3° North latitude and 140° East longitude). The **distance from epicenter** (surface distance between the recording station and the epicenter) and travel time of main seismic waves are provided for each recording station. Notice that there are 3 lines of data from most of the recording stations. They show the travel times for all three main kinds of seismic waves (P-waves, S-waves, and L-waves). However, instruments at some locations recorded only one or two kinds of waves (P-waves, or P- and S-waves). Location 1 is the Australian recording station where the seismogram in **FIGURE 16.4** was obtained.

A. On the graph in **FIGURE 16.5**, plot points from the data table in pencil to show the travel time of each main seismic wave in relation to its distance from the epicenter (when recorded on the seismogram at the recording station). For example, the data for location 1 have already been plotted as red points on the graph. Recording station 1 was located 1800 km from the earthquake epicenter and the main waves had travel times of 3.8 minutes, 7.0 minutes, and 7.9 minutes. Plot points in pencil for data from all of the remaining recording stations in the data table, and then examine the graph.

Notice that your points do not produce a *random pattern*. They fall in *discrete paths* close to the three narrow black lines (or curves) already drawn on the graph. These black lines (or curves) were formed by plotting many thousands of points from hundreds of earthquakes, exactly as you just plotted your points. Explain why you think that your points, and all of the points from other earthquakes, occur along three discrete lines (or curves).

There is a traveling distance time.

B. Study the three discrete, narrow black lines (or curves) of points in **FIGURE 16.5**. Label the line (curve) of points that represents travel times of the P-waves. Label the line or curve that connects the points representing travel times of the S-waves. Label the line or curve that connects the points representing travel times of the L-waves. Why is the S-wave curve steeper than the P-wave curve?

S wave takes longer to travel.

C. Why do the L-wave data points that you plotted on **FIGURE 16.5** form a straight line whereas data points for P-waves and S-waves form curves? (*Hint:* The curved lines are evidence of how the physical environments and rocks deep inside Earth are different from the physical environments and rocks just beneath Earth's surface.)

D. Notice that the origin on your graph (travel time of zero and distance of zero) represents the location of the earthquake epicenter and the start of the seismic waves. The time interval between first arrival of P-waves and first arrival of S-waves at the same recording station is called the **S-minus-P time interval.** How does the S-minus-P time interval change with distance from the epicenter?

The farther from the epicenter the greater the interval

E. Imagine that an earthquake occurred this morning. The first P-waves of the earthquake were recorded at a recording station in Houston at 6:12.6 a.m. and the first S-waves arrived at the same Houston station at 6:17.1 a.m. Use the travel time graph (**FIGURE 16.5**) to determine an answer for each question below. 6:12.6

1. What is the S-minus-P time interval of the earthquake?

4.5

2. How far from the earthquake's epicenter is the Houston recording station located?

3,000 km

3. **REFLECT & DISCUSS** You have determined the distance (radius of a circle on a map) between Houston and the earthquake epicenter. What additional data would you require to determine the location of the earthquake's epicenter (point on a map), and how would you use the data to locate the epicenter?

ACTIVITY 16.3 Locate the Epicenter of an Earthquake

Name: ______________________ **Course/Section:** ______________________ **Date:** ____________

A single earthquake produced the seismograms below at three different locations (Alaska, North Carolina, and Hawaii). Times have been standardized to Charlotte, North Carolina to simplify comparison. See if you can use these seismograms and a seismic-wave travel time curve (**FIGURE 16.5**) to locate the epicenter of the earthquake that produced the seismograms.

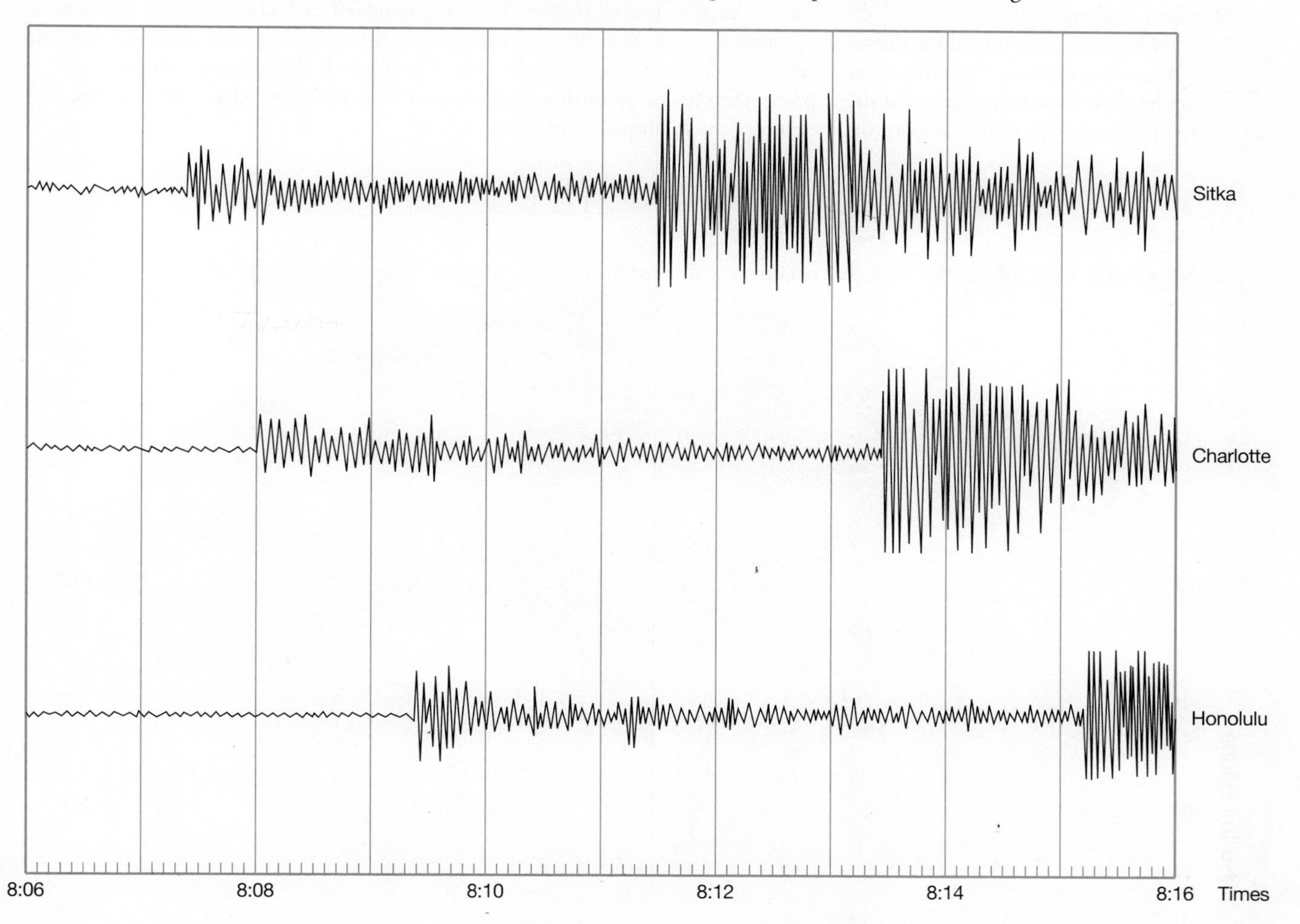

A. Estimate to the nearest tenth of a minute in the seismographs above, the times that P-waves and S-waves above first arrived at each seismic station (Sitka, Charlotte, Honolulu). Then, subtract P from S to get the S-minus-P time interval:

	First P arrival	First S arrival	S-minus-P
Sitka, AK	8:07.3	8:11.4	4.1
Charlotte, NC	8:08.0	8:13:4	5.4
Honolulu, HI	8:09.4	8:15.2	5.8

11.4 − 7.3 = 4.1   13.4 − 8 = 5.4   15.2 − 9.4 = 5.8

B. Using the S-minus-P time intervals above, and the travel time curves in **FIGURE 16.5**, determine the distance from the epicenter (in kilometers) for each seismic station.

Sitka, AK: 2700 km   Charlotte, NC 4000 km   Honolulu, HI 4200 km

C. Next, find the earthquake's epicenter on the map below using the distances just obtained.

1. First use the geographic coordinates below to locate and mark the three recording stations on the world map below. Plot these points as exactly as you can.

Sitka, AK: 57°N latitude, 135°W longitude
Charlotte, NC: 35°N latitude, 81°W longitude
Honolulu, HI: 21°N latitude, 158°W longitude

2. Use a drafting compass to draw a circle around each recording station. Make the radius of each circle equal to the *distance from epicenter* determined for the station in Part B. (Use the scale on the map to set this radius on your drafting compass.) The circles you draw should intersect approximately at one point on the map. This point is the epicenter. (If the three circles do not quite intersect at a single point, then find a point that is equidistant from the three edges of the circles, and use this as the epicenter.) Record the location of the earthquake epicenter:

N Latitude 35° W Longitude 121°

3. What is the name of a major fault that occurs near this epicenter? San Andreas Fault

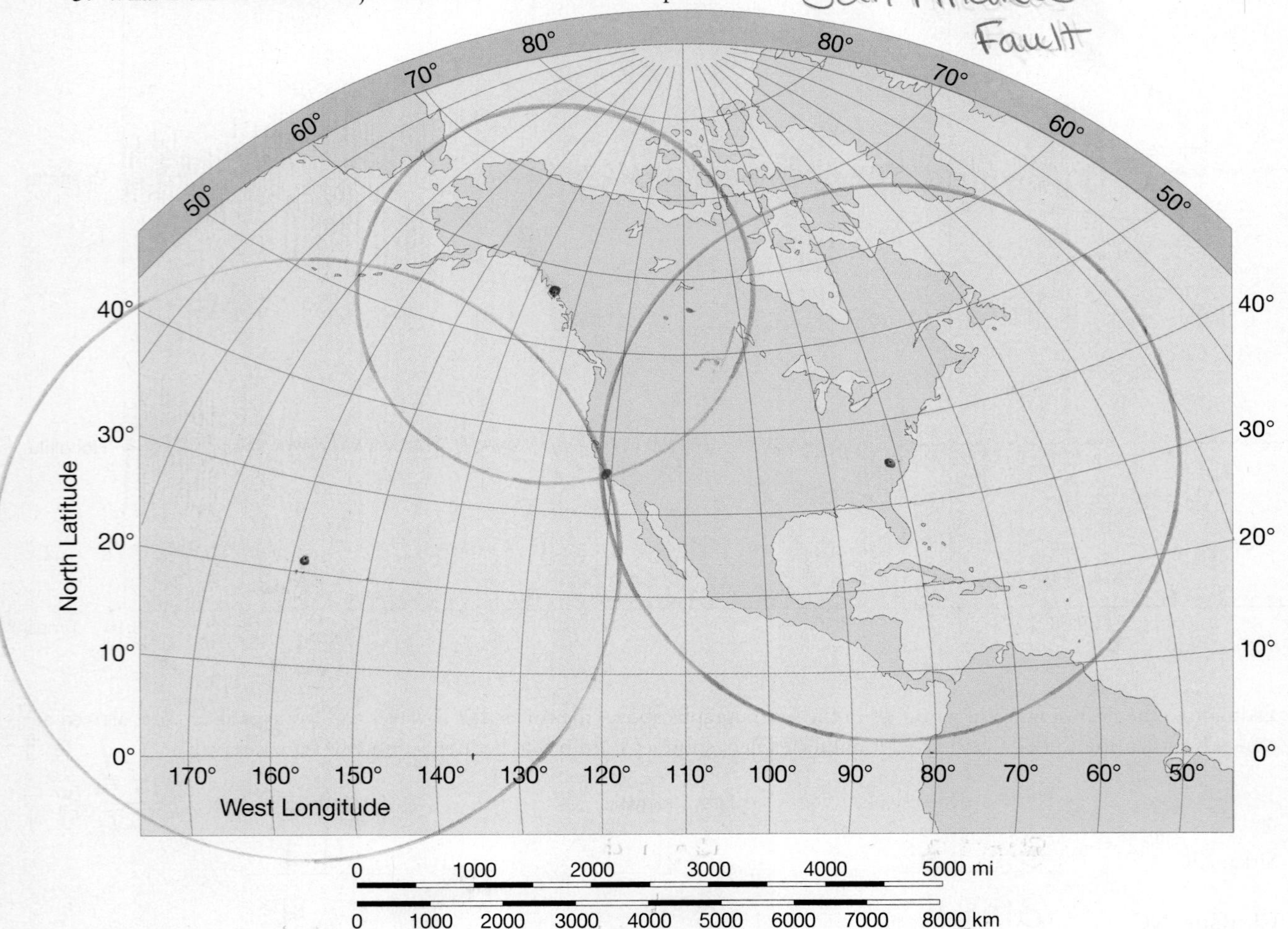

D. REFLECT & DISCUSS Your three circles may not have intersected exactly at a single point. How could you increase the precision of your results, so a point is generated?

better percise tools & scales